Pest and Disease Management Handbook

Edited by

David V Alford
BSc PhD

Published for the British Crop Protection Council
by Blackwell Science

b

Blackwell
Science

Blackwell Science Ltd
Editorial Offices:
Osney Mead, Oxford OX2 0EL
25 John Street, London WC1N 2BS
23 Ainslie Place, Edinburgh EH3 6AJ
350 Main Street, Malden
MA 02148 5018, USA
54 University Street, Carlton
Victoria 3053, Australia
10, rue Casimir Delavigne
75006 Paris, France

Other Editorial Offices:

Blackwell Wissenschafts-Verlag GmbH
Kurfürstendamm 57
10707 Berlin, Germany

Blackwell Science KK
MG Kodenmacho Building
7–10 Kodenmacho Nihombashi
Chuo-ko, Tokyo 104, Japan

First published 2000

Set in 10/12.5pt Times
by DP Photosetting, Aylesbury, Bucks

DISTRIBUTORS

Marston Book Services Ltd
PO Box 269
Abingdon
Oxon OX14 4YN
(*Orders:* Tel: 01235 465500
Fax: 01235 465555)

USA
Blackwell Science, Inc.
Commerce Place
350 Main Street
Malden, MA 02148 5018
(*Orders:* Tel: 800 759 6102
781 388 8250
Fax: 781 388 8255)

Canada
Login Brothers Book Company
324 Saulteaux Crescent
Winnipeg, Manitoba R3J 3T2
(*Orders:* Tel: 204 837-2987
Fax: 204 837-3116)

Australia
Blackwell Science Pty Ltd
54 University Street
Carlton, Victoria 3053
(*Orders:* Tel: 03 9347 0300
Fax: 03 9347 5001)

A catalogue record for this title is available from the British Library

ISBN 0-632-05503-0

Library of Congress
Cataloging-in-Publication Data is available

For further information on
Blackwell Science, visit our website:
www.blackwell-science.com

Contents

Foreword

The British Crop Protection Council (BCPC) is a registered charity (formed in 1967) now having the principal objective of promoting the development, use and understanding of effective and sustainable crop protection practice. It brings together a wide range of organisations interested in the improvement of crop protection. The members of the Council represent the interests of government departments, the agrochemical industry, farmers' organizations, the advisory services and independent consultants, distributors, the research councils, agricultural engineers, environment interests, consumer groups, training and overseas development.

For over 30 years, the Council has published independently or with collaborators a range of literature: conference proceedings, information manuals, guides and indices covering a great many aspects of crop protection. Among these have been the highly successful series of handbooks, *Pest and Disease Control*, and *Weed Control*. Each has run to several editions, evidence of their value to many sectors of UK agriculture. This has been achieved for each edition by careful choice of topics and contributors, to ensure that the contents are totally relevant to current issues and practices in the ever-changing agricultural scene.

This freshness is evident in the new edition of the *Pest and Disease Management Handbook*. Indeed, the small but significant alteration in the title from the previous 1989 edition (namely, the substitution of 'control' by 'management') is indicative of the changed perceptions of and attitudes towards crop protection over the past decade.

The BCPC has been fortunate in obtaining the services of Dr D V Alford, with his distinguished career in applied entomology, as editor, and of a group of eminent colleagues, each bringing up-to-date knowledge of field practice to their respective chapters.

I strongly recommend this new edition as a worthy successor in the series, and especially its use alongside the revised titles in the extensive BCPC book catalogue, in particular *Boom and Fruit Sprayers Handbook*, *Hand-held and Amenity Sprayers Handbook*, *The UK Pesticide Guide*, *Using Pesticides* and *The BioPesticide Manual*.

Trevor Lewis CBE
Lawes Trust Senior Fellow
IACR-Rothamsted

Preface

This handbook updates the third edition of the *Pest and Disease Control Handbook*, a series that began life as the *Insecticide and Fungicide Handbook for Crop Protection*, first published in 1963. The original title ran to five editions: 1963, 1965, 1969, 1972 and 1976; the second ran to three: 1979, 1983 and 1989.

This handbook differs from its immediate predecessor in excluding a range of introductory chapters that covered general topics such as the future of crop protection, the safe and efficient use of pesticides, the application of pesticides, and the principles of insecticide and fungicide evaluation. These have been replaced by a new introductory chapter on the principles of pest and disease management. This chapter (and the title of the handbook) acknowledges the advances being made in integrated crop management and the trends towards the more rational use of pesticides on UK crops. Although the main thrust of the handbook is pest and disease management, in a few cases (e.g. potato tubers) mention is made of physiological disorders. In recognition of the polarization of protected crops, the original chapter on protected crops has been subdivided into one on protected vegetable crops and mushrooms, and another on protected ornamentals. Also, 'turf grass', originally included in the chapter on hardy ornamentals, has been moved to that dealing with grassland (Chapter 4); similarly, the topic of 'bedding plants' is now included under protected flowering ornamentals (Chapter 10). Finally, to maintain emphasis on field, plantation and protected crops, chapters on 'forestry pests and diseases' and 'pests of stored cereals and oilseed rape' have been excluded.

Chemical recommendations within the various crop-based chapters relate primarily to on-label approvals, in each case mention being made of the common name of the active ingredient. Occasionally (but more extensively in the case of horticultural crops), mention is also made of specific off-label approvals (SOLAs); these are distinguished from on-label approvals by the addition of '(off-label)', followed by the SOLA reference number, after each such entry. In some instances, authors also refer to uses under the provisions of the off-label extension of use arrangements: *Revised Long Term Arrangements for Extension of Use (2000)*. Although approved, off-label uses are not endorsed by manufacturers and such treatments are made entirely at the risk of the user.

Unlike previous handbooks, dose rates for pesticides have been excluded. This is in line with the mandatory requirement for users to consult manufacturers' product labels before applying pesticides. Where a chemical pesticide is mentioned in the text of the handbook, this does not necessarily imply that all products containing the active ingredient have approval (on-label or off-label) for the use stated. Pesticide recommendations, and regulations governing their use, are under constant review, and for further information readers should consult an up-to-date copy of *The UK Pesticide Guide*, published annually by CAB International and BCPC. Readers are also reminded that, under the Control of Pesticides Regulations 1986, it is illegal to use any pesticide except as officially approved, and approvals are constantly changing. Some pesticide manufacturers, for example, are not supporting data calls made by MAFF PSD as part of the current review of anticholinesterase compounds (mainly carbamate and organophosphorus pesticides). As a consequence, approvals for non-supported compounds have been revoked and the permitted usage (the use-up period) of some formulations will expire at the end of 2000 or some time in 2001. It is essential, therefore, to keep up to date with current recommendations and to consult the current manufacturer's label before applying any pesticide.

Although inclusion of a classification scheme for pests (on a chapter by chapter basis) proved reasonably straightforward, that for pathogens introduced numerous difficulties as there appears to be no universally accepted system. The system finally adopted follows that recommended by Dr P. Kirk (CAB International), who kindly checked through the various lists. Guidance on nomenclature was also provided by Dr R.T.A. Cook, Mr R.P. Hammon and Dr D.E. Stead (CSL).

David V. Alford
Editor

Disclaimer

While every effort has been made to ensure that the information in this handbook is accurate, no liability can be accepted for any error or omission in the content or for any loss, damage or other accident arising from the use of the pesticides (chemical or otherwise) cited. The omission of the name of a pesticide from the text or from a table does not necessarily mean that it is not approved and available for use within the UK.

Abbreviations

agg.	aggregation (botanical)
bv.	biovar.
c.	*circa* (= approximately)
CDA	controlled droplet application
cm	centimetre(s)
CSL	Central Science Laboratory
cv.	cultivar
cvs	cultivars
DM	dry matter
DMI	demethylation inhibitor
DNA	deoxyribonucleic acid
DSS	decision support system
DTC	dithiocarbamate (fungicide)
EBDC	ethylene bis-dithiocarbamate (fungicide)
EU	European Union
FAO	Food and Agriculture Organization
f. sp.	forma specialis (*see* glossary)
GMT	Greenwich Mean Time
GS	growth stage (of a crop)
h	hour(s)
ha	hectare(s)
HDC	Horticultural Development Council
HGCA	Home Grown Cereals Authority
HSE	Health and Safety Executive
HV	high volume
HWT	hot water treatment
IACR	Institute of Arable Crops Research
ICM	Integrated Crop Management
IGER	Institute of Grassland and Environmental Research
IPM	integrated pest management
kg	kilogram(s)
km	kilometre(s)
LV	low volume
LVM	low-volume mister
m	metre(s)
m^2	square metre(s)
m^3	cubic metre(s)
MAFF	Ministry of Agriculture, Fisheries and Food
MBC	benzimidazole (fungicide)

mm	millimetre(s)
MRL	maximum residue level
NFT	nutrient film technique
NFU	National Farmers' Union
NIAB	National Institute of Agricultural Botany
nm	nanometre(s)
OC	organochlorine (insecticide)
OP	organophosphate (insecticide)
p.	page
PC	personal computer
PCN	potato cyst nematode
pH	a quantititive expression of acidity/alkalinity (*see* glossary)
PHSI	Plant Health and Seeds Inspectorate
pp.	pages
ppm	parts per million
PSD	Pesticide Safety Directorate
pv.	pathovar (*see* glossary)
PVC	polyvinyl chloride
SAC	Scottish Agricultural College
SOLA	specific off-label approval
sp.	species (singular)
spp.	species (plural)
ssp.	subspecies
PGRO	Processors and Growers Research Organisation
RNA	ribonucleic acid
SBI	sterol biosynthesis inhibitor
STRI	Sports Turf Research Institute
syn.	synonym
t	tonne(s)
UHT	ultra heat treated
UK	United Kingdom
ULV	ultra low volume
UN	United Nations
US	United States
UV	ultraviolet
var.	variety
WTO	World Trade Organization
WWW	World Wide Web
µm	micrometre(s), micron(s)
$<$	less than
$>$	greater than

Chapter 1
Principles of Pest and Disease Management in Crop Protection

K.F.A. Walters and N.V. Hardwick
Central Science Laboratory, York

Introduction

In recent years, the move towards global trading and pricing, coupled with a range of other factors, has resulted in reduced farm incomes and placed farmers under intense financial pressures. Arable farmers therefore need to make efficient use of variable inputs such as insecticides, fungicides, fertilizers, seeds and energy. This trend has coincided with increased public concern about environmental protection issues, which has led to a demand for significant reductions in the amount of pesticides applied to crops and the development of more environmentally acceptable alternatives. To achieve the objective of minimizing the environmental impact of the key inputs required to maximize income, many farmers now adopt the concept of integrated crop management (ICM) to combine efficient production with greater environmental sustainability.

For many years, the use of pesticides has offered a reliable and cost-effective approach to the control of damaging pests and diseases affecting arable crops. Although several organisms have become resistant to some products, this has usually been overcome by the development of new classes of pesticides and the establishment of improved management techniques. The latter has often been based on the careful selection and integration of the products used, or on a knowledge of the biology of the subject organism. For example, optimal timing can be used to maximize the effect of a pesticide application and therefore reduce the need for repeat treatments. In addition, advances in other fields (such as the development of improved farm application machinery or computer models which enable improved integration of the wide range of information upon which decision making on pest and disease management relies) offer the prospect of more cost-effective and, therefore, competitive approaches to farming. Despite these advances, relatively inexpensive prophylactic applications are still widely adopted, resulting in a significant level of unnecessary pesticide use which reduces farm profit margins and can be environmentally damaging.

This handbook provides a summary of modern approaches to pest and disease management that will assist the farmer or advisor by providing a compendium of information that is essential for robust, cost-effective decision making in this important area of crop production.

Decision making

Appropriate management of farm resources, including the use of variable inputs such as pesticides, will ultimately affect the production capacity of the farm. The optimal use of such inputs depends on the objectives of the farmer, which vary depending on farmer perceptions and circumstances. For some farmers profit maximization is an appropriate objective, whereas other farmers may be more interested in achieving a safer return, sacrificing some income to lessen the risk of a crop not being profitable.

The crop grown or the nature of the pest or disease that attacks them also affects decision making. Some crops have to reach minimum quality standards, and failure to achieve them will result in the produce commanding a much lower price when sold. Further, the introduction of novel crops may promote a pest or disease that has hitherto been considered unimportant. For example, the energy crop *Miscanthus* supports barley yellow dwarf virus (BYDV) and one of its vectors: cereal-leaf aphid (*Rhopalosiphum maidis*). This aphid species is not considered to be a major pest in the UK (see Chapter 2), but if *Miscanthus* is grown more widely then the insect may develop into a more serious threat. Thus, pest management decisions must take into account a wide range of factors, including some relating to other crops on the farm.

Approaches to the management of indigenous and non-indigenous pests and diseases frequently have different objectives. Management of indigenous pests often aims at reducing the infestation to levels at which the cost of further control measures will be greater than the economic advantage gained by applying them. The objective when managing outbreaks of non-indigenous pests is to prevent establishment and, thus, widespread crop damage in the UK. Hence, eradication or at least containment in the outbreak area is required. These criteria often govern whether conventional (i.e. chemical) control, biological control, or a combination of both (integrated pest management) are appropriate.

Pest management

Weather and pest populations

In order to optimize the use of the various control measures available to the farmer, it is important to understand why pest outbreaks occur. If we know what causes an upsurge in pest numbers, then it may be possible to avoid, or at least reduce, their intensity in the future. Frequently, fairly predictable cycles can occur (caused in the main by delayed density-dependent processes) but, on occasions, stochastic events (such as those due to weather conditions) may increase pest numbers unpredictably.

Climatic variables such as temperature, wind and rain can have significant consequences on pest numbers. They tend to act in a density-independent fashion, and instead of resulting in pest populations reaching equilibrium can lead to more radical increases or decreases in pest numbers. The body temperatures of all invertebrate pests vary with their surroundings and, as a result, many processes that affect the development of outbreaks are regulated to some degree by environmental temperatures. Processes such as growth, development, age-specific mortality, length of adult life and reproductive rates are significantly affected by temperature. For example, within the range of temperatures experienced in agricultural fields during UK summers, aphids tend to grow and develop faster under warmer conditions, potentially leading to more generations feeding on the crop. In addition, each adult will produce more offspring under similar warm conditions. A combination of all these responses often results in larger outbreaks of aphids feeding on the crop and depressing yields in warm summers. Temperature can also affect the rate of movement of invertebrate pests, with low temperatures precluding flight or even walking between crop plants. As many insects reduce crop yields owing to their ability to act as vectors of plant viruses, higher temperatures during critical periods of crop growth can result in an increased need to apply a control measure. For example, a warm autumn may increase both the numbers and rate of movement of the aphid vectors of barley yellow dwarf virus (BYDV) in winter wheat, resulting in an increased rate of infection. Further, the number of generations of pests in a year can vary according to temperature. Well known examples include that of codling moth (*Cydia pomonella*), which usually has just one generation per year in the UK but in favourable years may have a partial but significant second.

The low temperatures experienced during northern European winters can have a detrimental effect on pest populations, even when the pest overwinters in non-crop habitats. Various lethal and sub-lethal effects of low temperature have been recorded, which can reduce survival and, thus, population levels in both spring and early summer. Although most arable pests are well adapted to survive all but the most extreme winter conditions, work on the overwintering success of some insects has enabled forecasting systems for pest pressure in spring and/or the time of arrival in crops of migratory pests to be based partly on winter temperatures.

As well as the pests themselves, temperature conditions experienced during both the summer and winter also affect their natural enemies (notably, parasitoids and predators). The consequences for natural enemy numbers, synchrony with pest populations, searching efficiency and other factors can be difficult to predict but may be critical if they are to be incorporated into an IPM approach. In such cases, an understanding of relevant aspects of both the natural enemy and pest biology is essential if reliable systems are to be implemented.

Rainfall can have both direct and indirect effects on invertebrate pests. Heavy rain can dislodge insects from their host plants and may cause rapid changes in the numbers of, for example, pea aphids (*Acyrthosiphon pisum*) feeding on a crop. Some species of both Coleoptera and Hemiptera have even been recorded as being killed by violent thunderstorms. However, low rainfall can result in desiccation and death. Seasonal variation in rainfall can influence plant flushing and growth, whereas some drought-stressed plants are rendered more susceptible to insect attack. Rainfall also affects humidity and soil moisture, which combine with local temperature and wind to determine microclimate conditions that can, for example, influence both the degree of damage caused by slugs and the need for control.

It has long been recognized that rainfall can influence the efficacy of control measures applied against pests. The growing interest in the use of pathogenic organisms as biological control agents offers an interesting example of the importance of rainfall effects in decision making. Viruses are largely species-specific pathogens, which (after application) can remain attached to foliage and infective for considerable lengths of time. It has been shown that rain can wash the pathogen from the leaves of a plant, thus reducing the effectiveness of the treatment. However, under certain circumstances, some of the virus particles will fall on to lower leaves that may have remained relatively unprotected by the initial application, thus providing a degree of redistribution of the protectant activity. Thus, effects of rainfall on invertebrates are complex and may not exert a noticeable influence on pest populations at the time of the rainfall event. Instead, as with leatherjackets (Tipulidae), rain may affect insect numbers some months later. However, it remains a component of decision making on pest management that cannot be ignored.

Long-distance dispersal of insects by wind has been shown to be important in many farming systems around the world, but (in most cases) pest dispersal to UK arable crops tends to be over relatively short distances. Long-distance aerial dispersal does occur, and has been well documented, but the most damaging pest groups (such as aphids) tend to be a more localized problem. In these cases, low-velocity wind can assist movement and host finding but higher wind speeds can also delay the flight of insects to crops, as many species will attempt to take off only into winds of a limited range of velocities. Wind can also affect invertebrate pests by influencing the relative humidity or moisture levels in a microhabitat and by causing physical disturbance of surface-active pests when plants brush together.

An important property of wind for pests is its ability to convey chemical messages to insects from point sources. Most plants emit volatiles that are carried on the wind and give information to the pest on the location or state of a plant. Several insect pests have also evolved the use of semiochemical signals, pheromones, to locate mates. In still air, these volatile compounds remain in close proximity to the emitter and diffuse over only short distances. However, in a light wind they form a chemical plume or concentration gradient that can be followed to the source. Recently, some parasitoids have been found to use the same chemical signals to locate their hosts.

Pest management methods

Although chemical insecticides remain the most commonly used method of pest management in many UK crops, a range of alternative options are available to the farmer and advisor. Currently, a major objective of the farmer and advisor is to optimize the response to chemical treatments whilst minimizing the number of applications made.

Chemical control

In ICM systems the above-mentioned aim is addressed, in part, by basing decisions on insecticide use on crop monitoring coupled with action thresholds. The action threshold is the pest density that warrants initiation of the control strategy. This is not necessarily the application of a control treatment but could, for example, represent the point at which computerized decision support models should start to be run. Three other related thresholds are also recognised: (a) the economic damage threshold, i.e. the amount of damage that justifies the cost of artificial control; (b) the economic injury level (EIL), i.e. the lowest population that will cause economic damage; and (c) the economic threshold, i.e. the population level at which control measures should be implemented to prevent populations reaching the EIL. The economic threshold can differ from the EIL under certain circumstances, for example where cabbage seed weevil (*Ceutorhynchus assimilis*) adults are controlled to prevent the damaging larvae of this pest reaching significant numbers. Thresholds and cost-effective assessment techniques are available for many of the major arable pests and provide an important background for practical decision making.

An important component of sustainable use of chemicals is the careful choice of the product applied. All products are carefully screened for efficacy and environmental and other effects before they are registered in the UK, but other factors can determine product choice. Insecticides vary in their mode of action, with some acting through direct contact with the pest's body, those with fumigant activity penetrating the body through spiracles and other orifices, while others can be ingested as the insect eats contaminated plant material or imbibes droplets from the leaf surfaces. Pests that feed on internal tissues of plants (especially those, such as aphids, that imbibe sap) can be killed by a systemic pesticide. The length of time that the pesticide remains active after application also needs to be considered when selecting a product. In addition to product choice, application rates can sometimes be reduced without there being a detrimental effect on subsequent control of the pest. For example, application of sub-label recommended rates of some aphicides to cereals will result in more aphids surviving than after full-rate treatments. However, enhanced activity of natural enemies following such reduced-rate treatments can ensure that effective control is maintained. Combining careful product choice with optimal timing of applications can also result in increased natural control of pests, further reducing the need to apply chemical pesticides. For example, over 50% mortality of the damaging cabbage seed weevil (*Ceutorhynchus assimilis*) larvae in commercial oilseed rape fields can be achieved by promoting the naturally occurring parasitoid wasp *Trichomalus perfectus* through strict adherence to pest thresholds, selection of compatible chemicals and optimal spray timing. Such systems rely on a clear understanding of the interactions between different species in the agro-ecosystem, and can be difficult and time-consuming to develop.

Timing of application can also enhance the efficacy of a treatment and reduce the need for repeat applications. Young insects are often more susceptible to chemical sprays, but changing complexity of the crop canopy or the behaviour of pests can result in their spending at least part of their lives feeding in a area where chemicals sprays will not penetrate. Optimal timing of pesticide applications often relies on accurate monitoring or prediction systems which identify when the vulnerable or damaging life stage of the pest is becoming prevalent. For example, efficient control of pea moth (*Cydia nigricana*) is achieved by the use of pheromone traps to determine the timing of spraying.

Further reductions in the amount of pesticide applied can be achieved by careful product placement on the crop, and in recent years advances in application technology have allowed dramatic improvement in this area. Hydraulic nozzles that form a mist of droplets are still widely used, and there is a strong relationship between droplet size in the spray cloud, the volume of spray used per unit target area and the efficiency of the operation. Greater efficiency can be achieved by reducing the range of droplet sizes sprayed, and techniques for controlled droplet application (CDA) and ultra low volume (ULV) application have been developed. In most situations there will be little choice in the equipment available for pesticide application on a farm. However, where there is scope, thought should be given to improving product placement.

In recent years, insect growth regulators (synthetic analogues of hormones which control the growth and development of insects) have become available. These tend to be active against a narrower range of insects than many conventional pesticides, and are therefore compatible with the objective of minimizing environmental effects of pest management. Despite their mode of action, there have been cases where prolonged exposure to insect growth regulators has resulted in some species becoming tolerant to the product, illustrating the need to remain vigilant when using these chemicals.

The use of semiochemicals in pest management strategies can be a useful method of reducing the number of applications of conventional chemical pesticides. Pheromones are known for a large number of insects, including pest species, and many traps have been designed and are commercially available to take advantage of the opportunities they offer. The chemicals are usually contained in slow-release formulations that are placed within a capture device. Mass trapping has been attempted, where pests are lured into traps (using either a sex or aggregation pheromone) and killed, but the approach has been successful in only a few cases and under well defined biological and physical conditions. However, semiochemicals have been widely and successfully used for monitoring pests. Estimates of pest numbers in a crop can be combined with thresholds to determine the need to apply a control measure and to optimize timing (e.g. in the case of pea moth, cited above). Another technique that has been applied successfully is mating disruption. Male insects often locate females by following a plume of sex-pheromone. By releasing a synthetic sex-pheromone, host-finding behaviour can be disrupted, preventing mating and thus reducing the number of insects in the next generation. Once again, although some examples of very successful mating disruption are available, the technique must be used with care. Recently, research into push–pull techniques (stimulo-deterrent diversionary strategies) has demonstrated a potentially useful alternative role in ICM systems. The approach being developed involves the use in late autumn of semiochemicals to attract parasitoids of pest species from open fields to more protected overwintering habitats. This enhances the winter survival of the parasitoids, which in the following year are then able to move back into fields in early spring and depress pest populations. Initial results of research investigating this technique are promising, and larger-scale development work is underway.

Biological pest control

Classical biological control techniques and their integration into crop production systems were originally developed for high-value crops. There is now widespread use of bio-control in many horticultural crops, and a range of bio-control agents are available commercially. Inundative releases of biological control agents and techniques using banker plants are commonplace in protected crops and have been the subject of several authoritative reviews. Currently, there is great interest in applying the principles of biological control in arable crops, but it is uneconomic to rear and release bio-control agents in large field crops. Instead, attention is focused on enhancing the numbers and activity of naturally occurring bio-control agents.

In crops such as oilseed rape natural enemies have been shown to have considerable potential for limiting or reducing pest populations, and ICM strategies which take account of these insects provide an ideal basis for developing sustainable pest management approaches. However, not all natural enemies will play a role in such strategies. Experiments on polyphagous beetles and spiders in UK cereal fields during summer have investigated the effect of reducing the activity of these groups by up to 85% (compared with control areas), but no difference was found in the numbers of grain aphids (*Sitobion avenae*). Thus, careful selection of the target pests and control agents is essential if successful systems are to be developed. If a biological control agent is to be successful, then it must be able to respond to differing pest densities by increasing the number killed as pest numbers increase, and careful manipulation of the environment to promote the natural enemy can be used to improve their impact. This may be achieved by the provision of suitable habitats or alternative hosts or prey, or by identifying and avoiding farming activities that depress the natural enemy populations. The control of cabbage seed weevil (*Ceutorhynchus assimilis*) by manipulation of the naturally occurring parasitoid wasp *Trichomalus perfectus*, as described above, is an example of the latter approach.

The influence of farm practices on pests

Several aspects of farm practice will affect the abundance of both pests and their natural enemies. Paramount among these is crop rotation, which has conventionally been adopted to reduce the carry-over of pest and disease problems from one crop to another. However, rotations are also likely to promote certain natural enemies. For example, various species of ground-dwelling carabid beetle that are commonly found in cereal fields have been shown to recover more slowly after each successive insecticide application made to first and second wheats. However, if these crops are followed by oilseed rape (a common break crop in the cereal rotation), the structure of the crop protects the carabids from insecticides and their recovery time quickly returns to normal.

Crop isolation can also play a role in reducing the impact of pests and diseases. The growing of seed potato crops in Scotland is a good example of crop isolation in practice. Virus-free seed potatoes are more difficult to produce in areas where the aphid vectors of these viruses are prevalent, but the climate in Scotland results in fewer aphids flying into seed crops during the critical periods, making the achievement of effective aphid (and thus virus) control easier. Minimum distances between potato crops also limit the availability of viruses in the environment, and further reduce contamination.

Primary cultivations depend largely on soil type and weather conditions but methods such as ploughing, discing, rotary cultivation, harrowing, direct drilling and broadcasting of seed into stubble are available. Minimal cultivation tends to be favoured in ICM systems because of factors such as reduced mineralization and leaching of nitrogen, the introduction of alternative strategies for weed control and improved physical properties of top soil. There is some evidence that adopting non-inversion tillage throughout a rotation can contribute to the conservation and enhancement of both soil-dwelling natural enemies (such as carabid beetles) and some hymenopterous parasitoids of pests such as pollen beetle (*Meligethes aeneus*). However, some pests can be suppressed by ploughing, to some extent countering the beneficial effects of enhanced natural enemy populations. Thus, selection of an appropriate cultivation method needs to take into account consideration of both pests and natural enemies, as well as other factors.

Choice of cultivar can have a significant effect on pest problems encountered during the growing cycle. Some cultivars display increased tolerance of pest damage or even confer a degree of resistance (often through antibiosis), or non-preference, to pests but there are also indirect effects that are cultivar specific. The architecture of the crop canopy of oilseed rape is determined in part by the growth habit of the cultivar sown, as well as by other factors such as plant spacing and nitrogen applications. In future it may be possible to manipulate crop canopy to reduce the impact of pests on the crop or to increase the effect of natural enemies. Provided the current public concerns surrounding the techniques of genetic modification can be adequately addressed, new cultivars may be developed more rapidly in future, resulting in a greater scope to tackle pest and disease problems through cultivar choice.

Effective crop establishment can play a substantial role in reducing early-season pest problems. For example, in winter-sown oilseed rape, damage from pigeon, slug and flea beetle activity can result in complete crop failure. Sowing during August or early September can reduce such losses, but drilling too early can make the crop susceptible to damage from cabbage root fly (*Delia radicum*). Once again, adverse weather conditions during critical periods can dictate sub-optimal sowing dates, and result in little opportunity to consider pest problems. Poor crop establishment can also result in modified pest management thresholds after the initial establishment period. For example, action thresholds for cabbage stem flea beetle (*Psylliodes chrysocephala*) are lower on a backward or poorly growing crop than on a healthy vigorous one.

Several studies have highlighted the potential for enhancing natural enemy numbers using uncultivated field margins. Flowering strips can encourage the build-up of syrphid fly and parasitoid populations, but age and composition appear to be important. Increased control of pests currently appears to be limited to areas of crops close to the field margin, and further work is required before the technique can be used in commercial fields. Trap crops, which divert pests away from commercial crops and concentrate them in a small area where they can be treated with pesticides or other agents, have also been considered as a possible component of ICM systems. These could be developed in conjunction with the push–pull strategies described above to enhance their efficiency.

Non-indigenous pests

In addition to many of the methods available for indigenous pests, some management options are unique to non-indigenous species. Alien pests are frequently introduced via the international trade in plants and plant products. To reduce the movement of pests around Europe a system of establishing that such commodities are substantially free of damaging pests before being transported has been established. In addition, products entering the EU or individual countries are inspected and legislative measures taken if they are found to be carrying a restricted pest. For example, such measures may include re-export, destruction or fumigation. If outbreaks of non-indigenous pests occur on the nurseries receiving the plants or plant products, then measures for containment and eradication of the outbreak are adopted. Historically these have often included the use of chemicals, but recently biological control agents have also been deployed successfully.

Pest forecasting

Where pest outbreaks can be forecast accurately, advanced planning can allow more effective measures to be taken. For example, optimal timing of pesticide applications can be made. Several forecasting methods have been developed to support decision-making on pest management, but many of those that have been introduced successfully have been based on extensive knowledge of the biology of the pest and on large databases of information that has been gathered in and around the crops attacked. Effective methods of forecasting the size and timing of migrations of aphids that result in crop colonization or virus spread have been based on the long-established suction trap network of the Rothamsted Insect Survey, illustrating the value of such long-term datasets.

The use of computerized models, together with user-friendly interfaces to make them more accessible to practitioners, is now well developed. An early example was the EPIPRE model, which demonstrated the potential of such systems and has been followed by many others. In the UK, several models have been introduced and taken up by the industry. The PESTMAN system provides decision support for top-fruit growers; recently, in the arable sector, a pea aphid model (PAM) has been released. The latter model illustrates the importance of working with end-users throughout the development of such systems. The growers defined the problem (the need to predict optimal timing of insecticide applications on vining peas), and commented on and tested each version of the model during its development. The result was enthusiastic uptake when the system was released, with farmers controlling a very large proportion of the vining pea area in the UK obtaining a copy.

The next generation of models are currently undergoing validation, and will integrate several pest or disease problems into a single decision-making process. For example, an oilseed rape model which integrates decisions on all the major pests and takes some account of disease control decisions has been developed. This contains features that enable dual-purpose pesticide applications controlling more than one pest, and minimization of non-target effects of pesticides on beneficial invertebrates, to be incorporated routinely into farming practice. The use of such systems is rapidly developing and offers a wide range of advantages to farmers and to advisors.

Crop diseases

Disease is a symptom and not a cause. Causes of diseases can be biotic or abiotic. The causes of biotic diseases are the mycoplasmas, viruses, bacteria and fungi (plant pathogens). Abiotic causes include extremes of climate (temperature, precipitation and wind), atmospheric pollution, chemical injury as a consequence of pesticide application or as a recipient of drift, nutritional imbalance and structural problems with the growing medium. Some of the abiotic causes, however, do pave the way for invasion by the biotic.

In order to be successful in disease management it is essential to identify the cause of the problem. Identification of the cause, or diagnosis, is the means to an end – the end being the control, or at least amelioration, of the disease.

Abiotic diseases can be split into those that can be dealt with by crop husbandry techniques, e.g. waterlogging by improved drainage, low pH by liming, soil structural problems by sub-soiling. Others are unpredictable and will occur by accident or owing to lack of foresight, e.g. discharge from a power plant or incinerator, spray drift from a neighbouring field, contaminated spray tanks because of inadequate washing, and even the bizarre, such as lightning strikes. The prime key to managing these diseases is, as with the biotic, the correct identification of the cause so that remedial action can be instigated (generally in the following season).

Control or management of the biotic causes of disease involves a number of processes, each of which needs to be understood. The disease tetrahedron (Fig. 1.1) is the classic way of understanding the interaction between the main elements which combine to produce disease.

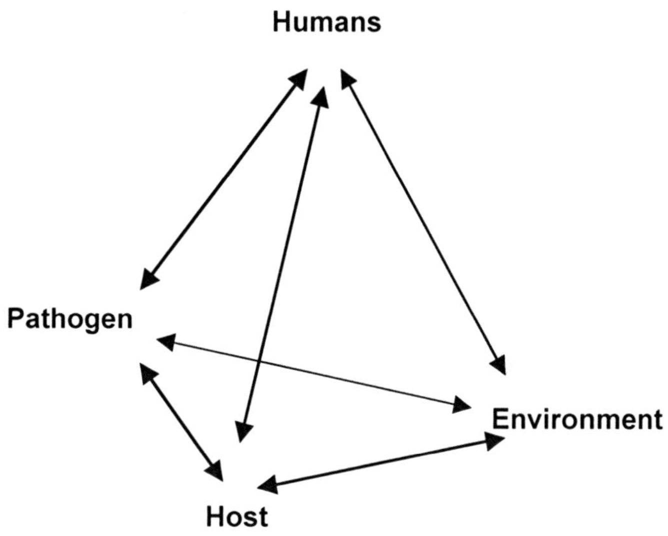

Fig. 1.1 The disease tetrahedron describing the interaction between people, the environment, host and pathogen. *(Redrawn from Zadoks and Schein, 1979.)*

Disease will not occur where there is no pathogen, no receptive host and no environment favourable to infection by pathogens. Humans can influence each to a varying degree, and that is where management affects the severity of disease. All plants carry their own burden of diseases. However, it is rare in nature to see heavily diseased plants. It is only when plants of similar genotypes are grown together for the economic production of food or ornament that disease becomes a major problem, leading to the 'boom and bust cycles' of disease increase and cultivar susceptibility. A gene-for-gene hypothesis has been proposed as an attempt to understand the nature of disease resistance and pathogenicity.

Pathogens have to reach a suceptible host, and there are a number of ways in which this can be prevented, the main method being exclusion. On the global scale this means restricting entry of commodities that may carry alien pathogens. This is achieved by a system of international plant health legislation and phyto-sanitary certificates. The WTO states that plant health regulations must not be used to provide an artificial barrier to international trade. While recognizing the rights of individual countries to protect themselves from potentially damaging alien pests or diseases, any exclusions must be based on science and the measures must be appropriate to the risk. On the local scale, pathogen contact can be prevented by isolation or protection. Methods of isolation include crops grown in areas or on land which has had no previous history of the crop, and protection by being grown in glasshouses or polyethylene tunnels. There is also the concept of 'disease escape', in the sense that the crop is grown where the pathogen exits but at a time when conditions may not be favourable for infection. This is particularly appropriate for the aphid-vectored viruses, when crops emerge or are harvested before the aphid vectors become active.

The host can resist attack in a number of ways, both passive and active. The active include production of phytoalexins and specific acquired resistance. The passive involves the cuticle on the leaf surface and the habit of the plant, e.g. whether the plant has an open or a closed canopy architecture, so affecting the microclimate (which may be prove conducive for infection by a particular pathogen). These factors are all under genetic influence and can be affected by plant breeding.

The environment plays a major role in pathogen infection and disease development: wind and rain to disperse the pathogen; rain to provide leaf wetness; and sun to provide optimum air and soil temperatures and optimum humidities.

The above are the natural events which impinge on disease severity and its development. Over-arching these events are the various processes of human intervention. To produce crops economically generally calls for uniform stands of plants of the same genotype to ease sowing, harvesting, storage and other intermediate operations. This uniformity provides the ideal situation for the development of disease epidemics. This has been met by the increasing use of fungicides, particularly on arable crops. In the UK, fungicides were not cleared for use on winter wheat until 1975. Since then their use has risen to over 98% of crops and with an average of 2.5 applications per crop. In 1986, the first year for which national data were available, 50% of oilseed rape crops were sprayed at least once with a fungicide. By 1998, this had risen to 95% crops, with a mean number of spray applications per crop reaching 2.2. With potatoes, fungicides applied to the growing crop were for blight control and 98% of crops are now sprayed, the number of spray applications ranging from two to 14, with an average of eight.

The successful production of crops has become increasingly dependent on the application of fungicides to control diseases. In spite of this, the first line of defence against pathogens should always be the selection of a cultivar most resistant to the pathogen capable of causing the most economic damage in a particular location. However, for many of the pathogens, disease resistance frequently has to be compromised, and that it is not always deployed is due mainly to customers increasingly demanding cultivars for a specified end use. This often leaves the grower with little choice in being able to select cultivars with appropriate levels of disease resistance, e.g. the extremely blight-susceptible cv. Russet Burbank demanded for French fry production. This problem is compounded further by increasing consumer demand for pesticide-free produce. The increasing interest in organic produce and more sustainable systems (the former driven by adverse reaction to genetically modified plants and general health concerns) brings with it the need to consider other methods of production. Producing disease-free crops without recourse to fungicides offers major challenges for plant breeders and plant pathologists.

Disease management

Disease management is about reducing the impact of inoculum on the plant. This can be achieved in one of two ways. The first (and the most effective method) is by isolating the host from the pathogen, and the second (and less effective method) is by introducing control measures once the pathogen is present. This process has been described thus:

$$\Delta_t = \frac{230}{r} \log_{10} \frac{x_0}{x_{0s}}$$

where Δ_t is the delay, r is the infection rate, x_0 is the proportion infected without sanitation, x_{0s} with sanitation (e.g. removal of infected debris at the soil surface by ploughing or diseased plants by roguing), and it follows that reducing r will be assisted by a reduction in x_{0s}.

However, it has been suggested that, because of the logarithmic basis of sanitation from the practical point of view of disease control, there are diminishing returns from the application of too high a degree of sanitation and effort should then switch to reducing the rate of infection.

Preventing contact between pathogen and host at its extreme is the subject of plant health and phytosanitary regulations aimed at preventing entry of alien diseases. At the national level there are limits to restricting contact. Isolation is difficult where wind-blown spores are concerned. The production of high-grade potato seed tubers in upland areas and in the north of the UK (where aphid numbers are generally low or arrive too late in the season to transmit the severe viruses) has been effective in most seasons in reducing disease. The consequences of the earlier drilling of cereals has had the opposite effect, with crops in the north of England now at risk from barley yellow dwarf virus (BYDV); this contrasts with the 1970s when BYDV was generally confined to the southern counties. Reducing inoculum by the various sanitation methods will always leave some affected material behind. It has been shown, for example, that as little as the equivalent of 0.1 eyespot-affected barley straws/m^2 is sufficient to produce the inoculum required for primary infection and disease development in winter barley. Pathogen adaptation to the current agro-ecosystem in the UK means that, in most seasons, both inoculum and weather are rarely limiting and so the current trend is to resort to fungicides as the prime disease management tool. This has unfortunate consequences when weather conditions do not permit or delay spray application, and the epidemic spreads either unchecked or the optimum timing is missed.

Disease forecasting

Disease forecasting, which has a long and chequered history, has as its prime aim the precise timing of control measures to prevent disease reaching epidemic proportions. Ideally, these should prevent disease reaching the exponential phase of growth. Sometimes, the rapid increase in disease can only be postponed, even by the application of the most potent of control measures, but the delay can be sufficient to enable the achievement of a satisfactory harvest. Forecasting requires an ability to respond to an assessment of disease risk. This has been achieved mainly by the developments in chemistry since the observation by Millardet in the late 1880s that chemical application to vines could do more than discourage pilfering of the grapes! The advent at that time of Bordeaux mixture as the first fungicide increased the ability of plant pathologists to use observations on the epidemiology of plant pathogens and specific combinations of meteorological variables to establish disease/yield loss relationships and predict disease epidemics. Forecasting was first used in the UK in the mid-1940s when (in the south-west of England) Beaumont evaluated the 'Dutch rules' for forecasting potato blight. However, Beaumont found that the Dutch scheme gave warnings too far ahead of the subsequent outbreak of blight. He therefore modified it by replacing dew occurrence with relative humidity as the former was more difficult to determine.

The main fungicides to control blight were copper compounds, which happened to be phytotoxic to the developing potato foliage and reduced yield in the absence of blight. In areas outside the south-west, where blight epidemics did not occur with the same regularity, it became important to delay the application of fungicides when the risk from blight was not present. The introduction of less phytotoxic fungicides (e.g. dithiocarbamates, chlorothalonil, fluazinam and the phenylamides) encouraged routine application. The Beaumont Period and its subsequent replacement (the Smith Period) have never been fully adopted, because data have to be collected from the synoptic network, processed and the resulting computations interpreted and the information disseminated – leading, inevitably, to delays. Despite various methods of trying to communicate risk to growers via press notices and direct mailing, inevitably not all farmers were aware of the warning. Also, the risk from blight was considered to be too great to rely on forecasts, on the erroneous perception that freedom from blight can be achieved by blanket treatment of the crop. This may be possible under low disease pressure but not when conditions are particularly favourable to the pathogen.

As weather has the major impact on the process of infection and is the driving force behind the development of epidemics, a prerequisite for a successful forecasting scheme is the readily availability of accurate meteorological data. The development of inexpensive microprocessors with large storage capacities has facilitated the construction of portable weather stations suitable for field use. This has enabled researchers to develop forecast models for use individually or as part of local networks. However, the development of increasingly sophisticated models for forecasting potato blight has not necessarily improved accuracy over the simple systems. In fact, systems such as the Smith Period based on thresholds of temperature and humidity have proved to be very robust.

The introduction of fungicides for use on cereals in the 1970s required some guidelines on their use. Disease/yield-loss relationships had not been determined and, therefore, the economic value of control by fungicides not established. Forecast models for individual disease were attempted. However, a more pragmatic approach was developed, centred around key fungicide timing in relation to yield responses. Key application timings were identified as the first-node development stage (GS 31), flag-leaf emerged (GS 39) and ear emerged (GS 59). Various combinations were employed to identify which gave the best disease control and the highest yields, results indicating that the standard three-spray programme, applied at these development stages, was likely to produce the maximum yield increase. In practice, higher gross margins could probably be obtained from a two-spray programme by more careful selection of active ingredients and timing of application.

The empirical approach taken in the timing of spray applications in various combinations was amenable in providing an integrated approach to disease management, and this led to the concept of managed disease control. Other schemes, or decision support systems (DSSs) have been tried, notably EPIPRE and PRO_PLANT. However, such systems are probably short-lived, as farmers and consultants get used to the output and begin to make value judgements for themselves. Also, some of the reasons suggested for the non-adoption of such schemes have been that they (a) paid insufficient attention to user needs, (b) were difficult to use, and (c) were not perceived to be of any major benefit to the end user. An additional reason was that, as with specific forecasting schemes, they were prone to error. The value judgements use by farmers and consultants have been described as 'determinants of spray decisions'. The process of crop inspection during the growing season, to assess disease levels as a basis for estimating future disease development (and, therefore, to serve as an indicator of the need for intervention), is not valid. Neither is the final level of disease reached at the end of the season useful as an assessment of likely yield loss. This has always been accepted and has been recognized as a crude but practical approach to decision making. However, these parameters do suggest ways in which this process could be improved by using an index of green leaf area over time to calculate healthy leaf area duration, as green leaf is the 'factory' that ultimately produces the yield. This would require more detailed examination of crops than has hitherto been practised, and also more convenient methods of measuring leaf area than are currently available.

Other factors can contribute to risk analysis, such as the various schemes for indicating risk from sclerotinia stem rot (*Sclerotinia sclerotiorum*) of oilseed rape. Schemes have ranged from identifying the agronomic criteria which predispose the crop to infection, through monitoring the germination of apothecia, to assessing the number of petals infected with ascospores.

PC-based information

The memory capacity of PCs has increased such that it has enabled vast quantities of detailed meteorological recordings to be taken and stored at intervals of 10 minutes or less. Together with analysis of epidemiological data, this has allowed for the development of sophisticated forecasting schemes. However, and importantly, software development has enabled the grower to input his own data and to provide a personal visualization of the output.

Elements of the agronomy and pathology of crops can be combined with weather data into complex DSSs. A UK-wide DSS, the Decision Support System for Arable Crops (DESSAC), is a powerful computer program which holds all the basic cropping information into which specifically designed modules, such as disease control in wheat, can be plugged. In Denmark, disease forecast models have been incorporated into a computer-based program, PC-Plant Protection. Dissemination of the output has been in the form of newsletters and leaflets, and recently via the Internet on a system called Pl@nteinfo (www.pl@nteinfo.dk). The Internet also provides the opportunity for bureau services to be developed where data, e.g. meteorological information from weather stations, can be gathered automatically and processed (and responses delivered via the Internet to farmers' PCs) via schemes such as the Dutch Plant-Plus system.

Individual crop disease reports have also been placed on corporate web sites, such as an oilseed rape light leaf spot forecast scheme at IACR-Rothamsted (www.iacr.bbsrc.ac.uk/lightleafspot).

The Internet also provides access sites that offer compendia of diseases (e.g. Co-operative Extension, Institute of Agriculture and Natural Resources, University of Nebraska Lincoln at www.ianr.unl.edu/pubs/plantdisease/) as an aid to disease diagnosis. The UN's Food and Agriculture Organization (FAO) created an electronic, interactive, multimedia compendium of plant protection information in 1987. Their current Global Plant & Pest Information System (GPPIS) is a web-based version of the earlier work and can be found at www.pppis.fao.org. CAB International have developed a Crop Protection Compendium, Global Module, which is also an encyclopaedic multimedia tool. It is published on CD-ROM, linked to the World Wide Web (www.cabi.org/CATALOG/CDROM/cropcom.htm). Further, CSL has developed an on-line interactive pesticide database (LIAISON). This subscription service contains information on currently registered agrochemical products for all crops grown in Great Britain, including specific off-label approval (SOLA) recommendations. Web-based systems have the advantage over books, leaflets and CD-ROMs in that they can be updated immediately new information is available. This gives increased confidence to the user that the information is accurate and up to date.

Sustainability

It has been suggested that the increasing trend towards the global industrialization of agriculture will lead to a non-sustainable system, not only of agriculture and its interaction with the environment but also of economics and society. A number of diversification strategies can be employed at species, cultivar and genetic level. These can best be managed on a system of polyculture which can be based on a rotational or intercropping regime.

Reliance on fungicides as the main means to control diseases is imprudent. By way of example, septoria tritici leaf spot (*Mycosphaerella graminicola*) can be controlled and there should be little excuse for high levels of disease occurring in commercial crops. The years 1998 and 1999 saw the worst *M. graminicola* epidemics since 1985, with mean national levels reaching over 7% of the second-leaf area affected. Currently, over 70% of crops are sown with a National Institute of Agricultural Botany (NIAB) rating of 5 or less for *M. graminicola* (in a range of 1–9, where 1 is highly susceptible and 9 highly resistant) despite, as indicated above, widespread fungicide use. In fact, current foliar disease levels in cereals are little changed from the mid-1970s when fungicides were first introduced, which begs the question of whether we are making the correct management decisions.

Fungicides alone will not control disease when conditions are most suitable for the pathogen. Most of the major pathogens which affect UK crops are favoured by wet weather, the very conditions that may preclude timely spray application. Therefore, there must be an integrated approach to disease management, making full use of resistant cultivars, rotation, nutrition and sowing dates to reduce disease pressure. In addition, the concept of good farm hygiene and the use of tested disease-free seed have been advocated.

Cultivar diversification and appropriate mixtures are a means of exploiting disease resistance that is available and a means to reduce risk. That they are not widely practised is partly due to reasons of convenience in the case of the former and of buyers in the latter. These factors may change with increasing pressure to reduce the cost of production as commodity prices fall, and with consumer and/or government pressure to restrict pesticide inputs.

The increasing interest by consumers in the perceived health benefits of organically produced food will present challenges to plant pathologists, as the area of production increases and the possible buffering effects of surrounding treated crops diminish. The strategies for disease control outlined above will assume increasing importance, and their value will need to be communicated effectively.

Conclusions

The principles of good pest and disease management lie in a thorough understanding of disease ætiology, epidemiology and pest biology. Both pest and disease control decisions should be based on sound scientific principles, observation and the socio-economic circumstances of the grower. A mechanistic approach, dictated by scientific experiment, will not always result in the correct action being taken. Results are retrospective and are specific to site and season; while they inform, the identical circumstances from which the data are derived are not repeatable. The 'determinants of spray decisions' and general pest and disease management decisions, by their very nature, have to be intuitive but should be guided by the results of research. Care should be taken when using imported pest and disease forecasting schemes, as they do not tend to 'travel' and are prone to error when used in areas and on data for which they have not been developed. More reliable results can be obtained from systems tested in the country or area in which they are developed.

There is an unresolved conflict between optimum disease management and crop appearance. How is it possible to determine whether management decisions have been the correct ones to produce a crop with the best gross margin? Neither the relative freedom of the crop from pest or disease at the end of the season nor yield provides a good indication. The achievement of a yield above expectation, while satisfying in its own right, does not mean it could not have been bettered, or that a lower yield would not have resulted in a larger gross margin.

The philosophy of 'if in doubt do' is a difficult one to overcome, even with sound scientific evidence. Similarly, the desire for insurance because the prophylactic approach will generally cover costs, and there is always the problem that the cost of inspection and assessment will outweigh the cost of treatment. The concept of sustainability is a difficult argument to win when faced with farmers who require high yields or high quality and consultants who require repeat business in order to survive.

The information now available to aid pest and disease management decisions is legion, not only via the printed word but also electronically. The WWW means that access to information is instant. Most major agrochemical companies have web sites and some provide news bulletins, weather information, discussion topics and links to other useful sites (e.g. Cyanamid at www.agricentre.co.uk./site/agricentre.taf). The web is also an excellent vehicle for providing bureau services and interactive DSSs. The prospects for improving sustainable pest and disease management of crops involving a combination of chemical and non-chemical methods have never been more promising.

Chapter 2
Pests and Diseases of Cereals

J.N. Oakley
ADAS Rosemaund, Herefordshire

W.S. Clark
ADAS Boxworth, Cambridgeshire

Introduction

Increasing economic pressures have tightened financial margins in cereal growing and added further to the need to use crop protection chemicals rationally. Farmers can no longer afford to withstand losses from pests or diseases or to spend money on unnecessary pesticide applications. Many now employ consultants to check on pests, diseases and weeds within their crops and to advise on the need for, and timing of, sprays. This more managed approach to control requires an appreciation of the various factors which place crops at risk of attack, and a planned programme of crop-walking at critical times. There is often a considerable time-lag between damage occurring to a crop and obvious symptoms showing in the field, so that treatments applied in response to perceived damage are frequently too late to be effective. Successful control depends on anticipation, assessment and timeliness of appropriate action. The alternative approach of scheduled preventive treatments regardless of risk cannot be justified economically. They could also have unwanted side-effects on beneficial organisms (e.g. parasitoids and predators) and so prove counterproductive.

As yet, no cereal pests have developed resistance to pesticides in Britain but resistance is now widespread in several important cereal diseases. In planning their control strategy, farmers should first consider cultural methods and varietal (cultivar) resistance, so as to minimize dependence on chemical control. Rotations and the position of cultivars in them can be arranged to reduce risk. The HGCA UK Recommended List gives lists of cultivars and indicates their resistance to the main cereal diseases. Farmers intending to use low-input or organic systems should select cultivars with the least difference between fungicide-treated and untreated yields in these tables, and should avoid those susceptible to diseases prevalent in their area.

No later corrective action can undo mistakes made in establishing the crop. Losses from many pests and diseases are greatly reduced in vigorous, well manured crops with unrestricted rooting. A high priority, therefore, should be given to preparing a firm, even seedbed and drilling the crop correctly and in the optimum time period. Good quality seed, treated against seed-borne diseases, should be used. Where weather conditions prevent ideal crop establishment a careful watch should be kept for seed and seedling pests (such as slugs) whose depredations can be particularly severe in poorly established crops.

Some pests and diseases attack only one cereal species, whereas others are more wide-ranging. Where several cereal species are attacked, cross-references are given in this chapter to the main host under which the pest or disease is described. Specific control recommendations are given for pests. For diseases, the spectrum of activity of each fungicide is shown in tables, to allow the selection of chemicals to cover whatever range is required. Tank mixes of different chemicals to cover several problems should be used only where specific recommendations for this are made on product labels. Reduced application rates are less appropriate for pests than for diseases, for which a good coverage of the plant is required; manufacturers' recommendations on spray volume and quality should be followed.

Polyphagous predators, such as ground beetles (Carabidae), feed on a range of prey and are thought to provide useful early-season control of aphid infestations; they also reduce numbers of other pests. As pest control agents, they have the advantage over prey-specific predators, as they are not dependent on pest abundance for their own success. Other groups of polyphagous predators of importance in cereal crops include rove beetles (Staphylinidae), soldier beetles (Cantharidae) and spiders (Araneae). Money spiders (Linyphiidae) are the most common group of spiders in cereal fields, being able to recolonize fields after cultivation by 'ballooning' on threads of silk from other non-crop habitats that act as important refuges for many predators. Ground beetles are very vulnerable to some molluscicides, and money spiders are extremely vulnerable to pyrethroid insecticides.

In contrast to polyphagous predators, specific predators and parasitoids (e.g. numerous parasitoid wasps and some rove beetles) attack a more restricted range of pests and they generally migrate greater distances to seek their specific target prey. Some parasitoid wasps can locate their aphid hosts by detecting plants suffering attack by these insects. The population dynamics of specific predatory and parasitoid species is often determined by that of the species attacked. Generally, predator or parasitoid populations fluctuate greatly according to their prey populations, often following a cycle closely reflecting that of their prey. As a general rule, it is thought that specific predators provide the greater control of pest outbreaks, as their numbers rise and fall in relation to the pest population. As with polyphagous predators, experimental work has concentrated on cereal aphids, for which the main groups of specific natural enemies are ladybirds (Coccinellidae), hover flies (Syrphidae) and parasitoid wasps (e.g. Aphidiidae, Braconidae and Chalcidae). Many minor cereal pests suffer greatly from parasitoids, with 50% or more of individuals parasitized in most seasons. Outbreaks of such pests may occur if this natural balance is disturbed by broad-spectrum pesticides applied to control other pests. Specific predators and parasitoids need to search through crops very actively to locate hosts, making them more vulnerable to pesticides than the less active pests upon which they prey.

The impact of pesticides on predators and parasitoids in cereal crops may be reduced if insecticide use is restricted to cases where the economic thresholds detailed in this chapter are exceeded. A prophylactic approach to insecticide use may be unlikely to target pests accurately, many of which are vulnerable to control measures for only a relatively short period in their life-cycle – perhaps between their colonizing the crop and starting to feed upon it. The more damaging insecticides carry label restrictions as to the number and timing of applications, and application to a buffer strip at the field margin to restrict the impact on hedgerows and other non-crop habitats. In the later stages of crop growth, careful choice of spray volume and quality (to restrict the coverage to the upper parts of the crop) can give good control of the target pests, whilst minimizing the impact on ground-dwelling (epigaeic) predators and other non-target invertebrates. Predators and parasitoids tend to be most active in crops some days or weeks after the pests have invaded. Where the application of a pesticide is delayed by bad weather, the appropriateness of the revised timing of application should be reviewed before the pesticide is applied, as the pest may no longer be vulnerable but its natural enemies may be at greater risk.

Wheat

Pests

Aphids

Several species of aphid occur on cereal crops, including bird-cherry aphid (*Rhopalosiphum padi*), cereal-leaf aphid (*Rhopalosiphum maidis*), fescue aphid (*Metopolophium festucae*), grain aphid (*Sitobion avenae*) and rose/grain aphid (*Metopolophium dirhodum*). Bird-cherry aphid and grain aphid are both important vectors of barley yellow dwarf virus (BYDV) – see Diseases section. Cereal-leaf aphid, which is restricted to grasses and cereals, is rarely found on wheat but is sometimes numerous on barley; fescue aphid occurs mainly on grasses (see Table 2.1).

On wheat, summer infestations by grain aphid or rose/grain aphid can cause substantial yield loss. When present on the crop before flowering, the aphids damage the crop by reducing the numbers of grains in the ear; when present from flowering to the end of the grain-filling period, the aphids reduce the size of the grain. Improvements in disease control and cereal cultivars have prolonged the grain-filling period, allowing aphid infestations to develop later than previously and so causing greater yield loss. Grain aphids infest both the ear and the upper leaves, whereas rose/grain aphids remain on the underside of the leaves. Threshold levels for aphicide application have been set at (a) half or more of the tillers infested by aphids up to the flowering stage, and (b) two-thirds of tillers infested between flowering and the end of the grain-filling period. Treatments are applied on the presumption that aphid numbers will continue to increase, causing

Table 2.1 The relative importance of aphids on cereal crops

	Pest species				
Crop	bird-cherry aphid†	cereal-leaf aphid	fescue aphid	grain aphid†	rose/grain aphid
barley	(*)	–	(*)	*	(*)
oats	(–)	–	(*)	*	(*)
rye	(–)	–	–	*	–
triticale	(–)	–	–	*	(*)
wheat	(–)	–	–	*	(*)

† Also an important vector of barley yellow dwarf virus (see text).
* Often damaging.
(*) Occasionally or locally damaging.
(–) Often present but of little importance as a cause of direct damage.
– Rarely or never present.

further yield loss. A rider is applied to the threshold levels in that they apply only if aphid numbers are increasing rapidly when the threshold is reached. Heavy rainfall can drastically reduce numbers and if natural enemies, such as ladybirds (Coccinellidae), hover fly larvae (Syrphidae) or hymenopterous parasitoids (e.g. Aphidiidae), are numerous in the crop, it may be worth reassessing the situation after a few days to check whether the aphid problem is declining naturally.

Chemical control options include (a) pirimicarb, which is less harmful to many of the natural enemies, (b) organophosphorus insecticides (such as chlorpyrifos and dimethoate), which are more broad-spectrum in nature, and (c) various pyrethroids (such as alpha-cypermethrin, cypermethrin, deltamethrin and lambda-cyhalothrin), which are intermediate in their spectrum of toxicity. All the aphicides are effective against aphids present on the ear, but the contact-acting pyrethroids are relatively ineffective against aphids on the undersides of leaves where spray coverage is minimal.

Cereal ground beetle (*Zabrus tenebrioides*)
See under Barley, p. 36.

Cereal leaf beetle (*Oulema melanopa*)
Larvae may be found feeding on the upper leaves of cereal crops in June. The larvae remove strips of the epidermis from the upper sides of leaves, causing very obvious physical damage. Whilst the damage has a strong visual impact it has little effect on the crop unless 25% or more of the leaf area is destroyed. This degree of damage has yet to be recorded in the UK and control measures are not advocated.

Common rustic moth (*Mesapamea secalis*)
The larvae (caterpillars) of this pest may attack cereal crops following grass. They typically transfer from the ploughed-down sward to the new crop in the autumn, at the same time as frit fly (see under Oats, p. 44) and, usually, cases of damage involve a complex of frit fly, common rustic moth and other ley pests. Whereas each frit fly larva completes its feeding in one tiller before pupating in the winter, common rustic moth caterpillars continue feeding to the end of May, each destroying 12–15 tillers before becoming fully fed. Feeding causes the death of the central leaf and a 'deadheart' symptom, as seen with other stem-boring pests. A common rustic moth caterpillar enters the shoot through a ragged hole in the base and leaves a mass of green frass behind when moving on to invade other shoots – these features aid diagnosis of the cause of damage. Initially, the caterpillars are mainly green in colour but in their final instar they develop violet lines along their backs.

Control may be obtained by a spray of chlorpyrifos, applied as recommended for use against frit fly and other ley pests. Where (in a case of ley pest damage) common rustic moth caterpillars have damaged a significant proportion of the shoots, an action threshold of 5% of damaged shoots (rather than the 10% used for attacks predominantly involving frit fly larvae) may be appropriate.

Frit fly (*Oscinella frit*)
See under Oats, p. 44.

Gout fly (*Chlorops pumilionis*)
This pest has two generations per year. The autumn generation of adult flies lay their eggs on the leaves of newly emerged cereal crops. Crops are at greatest risk if sown early, especially in sheltered locations. The larvae hatch from the eggs after about a week and enter the shoot at the top. Larval feeding causes the shoot to swell and take on a 'gouted' appearance. As well as the affected shoot, the other tillers on the plant also die. However, because neighbouring plants produce extra tillers, crops can compensate for damage of up to 25% of gouted plants. Adults of the spring generation are on the wing in May and early June, and they also lay their eggs on the leaves of cereal plants. The larvae feed within the leaf sheath on the straw, and damage can cause ears to fail to grow above the flag leaf with a loss of the lower grain sites.

Damage may be reduced by sprays of chlorpyrifos, as for control of frit fly (see p. 44), if applied before egg hatch.

Grain thrips (*Limothrips cerealium*)
This abundant pest (see also under Barley, p. 38) often causes concern and is present on most wheat crops. The thrips feed mainly on the underside of the glumes, and later they attack the ripening grain, causing a dark brown spotting and flecking of the flour. No threshold level has been established for wheat and, because serious damage is very unusual, control is not generally advocated.

Leatherjackets

See under Barley, p. 37.

Slugs

Several species attack cereals, the most common being field slug (*Deroceras reticulatum*). The most important damage occurs at or below the soil surface where the grain may be hollowed out after drilling or the shoot severed above the seed. Grazing on the leaves above ground may be of importance if the crop is growing slowly but, generally, this has little effect after the crop has reached the four-leaf growth stage.

The risk of slug damage is greatest on loose, cloddy seedbeds on heavy soils, following leafy crops such as oilseed rape, peas or cereals. The best form of control is to prepare a firm, even seedbed and to sow the crop at the best time to ensure rapid germination. The seed should be sown at a depth of at least 25 mm and, if the soil is loose, it should be consolidated with a ring roller to restrict access to the seed by slugs. Where adverse weather conditions prevent the formation of a good seedbed, chemical control may be necessary.

The timing of treatment needs to be flexible, to take account of surface activity by slugs (these are at their most active on humid, still nights). The most effective time to apply pellets is immediately after drilling, and this should be considered in fields prone to damage where seedbeds are poor. Should wet weather delay drilling after primary cultivation, slug pellets can be applied to the worked surface. Pellets can be admixed with the seed in situations where seed hollowing is likely to occur. Pellet application made after crop emergence is likely to be less cost effective, but may be needed if damage is severe and if the crop is slow to reach the four-leaf growth stage.

Prepared pelleted baits are available containing metaldehyde, methiocarb or thiodicarb. Recent research has improved the palatability of baits and their weather hardiness, but as acceptability to slugs will be reduced once soil has splashed up on to the pellets, they are likely to work best when applied after rain rather than before it.

Swift moths

Caterpillars of garden swift moth (*Hepialus lupulinus*) and, less frequently, ghost swift moth (*H. humuli*) sometimes attack cereal crops. Damage may be experienced in crops following grass – see Chapter 4, p. 92, for further details. The pests are most frequently observed where patches of couch grass (*Elytrigia repens*) have been controlled, forcing the swift moth caterpillars feeding there to seek other food sources. They often damage several plants along a row, severing the stems of seedlings at or about ground level. The caterpillars rapidly disappear backwards down their silk-lined burrows when disturbed and are therefore often overlooked. No chemical control measures are advocated. Damage is generally restricted to discrete patches and is of insufficient economic impact to justify the use of an insecticide.

Wheat blossom midges

Two species occur on cereals and these are considered separately below.

Orange wheat blossom midge (Sitodiplosis mosellana)

This pest rose to prominence in the UK in 1993, when serious outbreaks affected the majority of wheat crops. Numbers have remained high in 'hot spots' where conditions tend to be warmer and damper in May, favouring the midges. Problems tend to be most persistent on farms growing a large proportion of wheat in sequences of two or more crops between breaks. The adult midges are slender and orange-coloured, and about 3 mm long. In years of abundance they may be seen at dusk laying their eggs on wheat ears during the period of ear emergence. Midge larvae can remain in the soil for several years, being induced to pupate (prior to adult emergence) by warm, wet soil conditions at the end of May or early June. Thundery weather at this time can provide ideal conditions for development, and may be followed a week to 10 days later by a mass flight of newly emerged midges.

Eggs hatch in about a week and the larvae immediately move down to feed on the swelling grain. They feed by exuding an α-amylase enzyme that loosens the pericarp enclosing the grain, which then becomes more susceptible to sprouting before harvest. High levels of infestation, affecting 10% or more of the grain, can cause direct yield loss; if damp weather occurs before harvest, midge-induced sprouting can reduce Hagberg falling number to below acceptable levels for milling.

Wheat blossom midges need warm mild evenings to allow them to fly and lay eggs on wheat. Under cooler conditions midges tend to accumulate in crops waiting for better weather. If this does not arrive before the ear-emergence stage is over, then few eggs are laid and the crop may escape damage. It is important, therefore, to base decisions on the need for treatment on direct observations of midges laying eggs in the evening rather than on day-time counts of midge numbers in crops, which may be misleading. Treatment thresholds have been established as follows:

- for feed crops – one or more midges laying eggs on an average of one in three ears;
- for milling and seed crops – one or more midges laying eggs on an average of one in six ears.

Spray treatment may kill both adult midges and their eggs laid on the ear, but will not kill larvae established on the grain; therefore, treatments should be applied within a week of observing numbers of egg-laying midges above the appropriate threshold. Approved chemical treatments include chlorpyrifos; some control may also be given by pyrethroid insecticides applied to control aphids. All are broad-spectrum in nature and should not be applied where midge numbers are below threshold as they also kill parasitoids that attack the pest and can provide longer-term suppression of infestations.

Yellow wheat blossom midge (*Contarinia tritici*)
Larvae of this midge feed on the anthers of wheat ears. They prevent fertilization of the grain, which does not develop in damaged florets. The larvae are a bright lemon-yellow in colour, as are the adult midges that lay their eggs on the emerging ears in the early stages of ear emergence (GS 51–53). Serious attacks are restricted to fields growing long runs of cereal crops, and none have been recorded in the UK since the 1970s. For control measures see under Orange wheat blossom midge, p. 25.

Wheat bulb fly (*Delia coarctata*)
Adult flies are on the wing from June onwards. Eggs are laid on bare soil from late July to late August. The eggs are laid mainly on full or partial fallows or on land worked by early August following earlier-harvested crops or set-aside. Eggs are also laid under crops such as potatoes or sugar beet with bare soil exposed beneath the canopy.

The larvae are white and noticeably blunt at the hind end. They usually invade crops from early February to mid-March, although egg hatch may be advanced or retarded by mild or cold weather. They enter at the base of the shoot and begin to feed immediately. Following attack the centre leaf withers within a few weeks to show the classic 'deadheart' symptom. Soon after deadhearts begin to show, the larvae move on to destroy further shoots.

In addition to wheat, other cereal crops (barley, rye and triticale) may be attacked; oats, however, are immune. Populations vary widely from year to year and from area to area. Wheat bulb fly is mainly a pest in certain areas of eastern England, where a large proportion of wheat is grown in susceptible situations. Pockets of infestation are also to be found further west, in areas growing potatoes and field vegetables.

Crops suffer most damage if they have not tillered before an attack begins, because the attacked plants are likely to die. Sowings from November onwards should be protected in affected areas if they follow crops such as potatoes or sugar beet, which put them at particular risk. Seed treatment with tefluthrin is particularly effective for these late sowings, and should give sufficient protection against most attacks. Treatment is recommended wherever egg numbers are likely to exceed 1.25 million/ha. Treated seed should not be sown deeper than 40 mm or control may be reduced.

Earlier-sown crops suffer less from attack because, although some tillers may be lost, fresh tillers will be produced and so compensate for damage; plant populations on such crops, therefore, are reduced only by very heavy attacks. Seed treatments are less effective on earlier sowings. Where egg populations exceed 5 million/ha a chlorpyrifos spray treatment, applied at the start of larval invasion of the plants, may be beneficial. Sprays should be applied following warnings of egg hatch in January or February.

As a supplement to earlier treatment in fields suffering a severe attack, or as a single treatment for moderate infestations, a dimethoate spray treatment may be applied when the first deadhearts are observed. The treatment threshold depends on the growth stage, being 10% of tillers attacked before the onset of tillering (GS 20), 15% of tillers when the first tiller is present (GS 21) and 20% of tillers for more advanced crops.

Spring cereals may suffer heavy damage if sown during the period when larvae are invading plants. Consequently, seed should be treated with tefluthrin if spring crops are to be sown in susceptible situations.

Wireworms

Several species of click beetle, mainly *Agriotes* spp., live in old grassland. Their larvae, known as wireworms, are tough, yellow in colour, and develop over four years. Damage to cereals can occur for the first three years after ploughing-up of old grassland. Numbers were greatly reduced after the introduction of organochlorine insecticides in the 1950s but are now gradually recovering to former levels. As wireworm populations continue to recover, the incidence of damage in arable rotations is becoming more frequent, aided by the use of set-aside and the enhancement of the agri-environment to increase numbers of predatory beetles that are of value as antagonists of pests such as aphids.

Shoots are damaged below soil level. Often, there is a distinct hole at the side of the plant at the base; alternatively, the shoots may be chewed and frayed just above the seed. Attacked plants become yellow and die. Damage is usually less severe on headlands, owing to better consolidation of the soil. Wireworms feed throughout the year but damage tends to be more severe in late spring and, to some extent, in the autumn. Crop loss can be reduced by sowing into a firm seedbed and rolling when damage is seen (so long as the crop is not too far advanced).

Imidacloprid or tefluthrin may be used as seed treatments to control moderate wireworm infestation levels of 1.25 million/ha. A soil treatment with gamma-HCH may be used to reduce populations above this level, provided that potatoes or carrots are not to be grown in the field, owing to the risk of taint.

Yellow cereal fly (*Opomyza florum*)

The adult flies occur in the autumn and lay their eggs on the soil in early-sown wheat crops. Eggs hatch in February and the larvae then enter the shoots at the top. After a resting period they begin feeding in late March, destroying the growing point and causing a 'deadheart' symptom. The larvae complete their development in a single shoot, so that by the time symptoms of damage are seen it is too late to apply control measures. Crop loss is rare in the UK and specific control measures are not generally advocated. However, pyrethroid insecticides, applied in November to control BYDV vectors (see p. 28), can give good control of yellow cereal fly.

Diseases

Barley yellow dwarf virus (BYDV)

Symptoms of this important virus disease are similar to those described under barley (see p. 39), but wheat is generally less severely affected. The tips of affected leaves may develop a red/purple discoloration.

Bird-cherry aphid (*Rhopalosiphum padi*) and grain aphid (*Sitobion avenae*) are the main vectors of BYDV in the UK. The two species of aphid differ significantly in their biology and they carry different strains of the virus, so epidemics can be produced under very distinct scenarios. To control infection efficiently it is important to consider the risk of infection from three sources: green-bridge infection, introduction by bird-cherry aphid and introduction by grain aphid.

Green-bridge infection

Green-bridge infection comes about through wingless aphids surviving on plants from previous grass crops, or from cereal volunteers and grass weeds in stubbles, and then transferring to the newly sown crop. Such infections are usually very patchy in nature and the distribution of damage often reflects patterns of cultivation in the field. Aphid transfer from the previous crop is facilitated if the previous crop is cultivated shallowly, allowing plants to recover and grow through into the new crop. Control is dependent on cultivating or killing the previous crop with a desiccant herbicide early enough for the plants to be completely dead before the new crop starts to grow. Products containing either glyphosate or paraquat have been shown to provide effective stubble cleaning for this purpose.

Aphids as vectors

BYDV vectored by bird-cherry aphid is of greatest important in warm autumns and on earlier-sown crops. Infection arises from winged migrant aphids carrying the virus from sources in grassland and infecting plants on to which they deposit their wingless nymphs. The progeny of this first, wingless generation then spread through the crop, infecting fresh plants. The aphid exists in two forms. One form overwinters as eggs laid on bird-cherry (*Prunus padus*) and predominates in northern areas – this form is not important as a vector of BYDV. A second form overwinters as live aphids on cereals and grasses but, being vulnerable to frosts, is resident mainly in southern Europe, populations spreading up annually to infest crops in the UK. In milder districts and winters this form may survive on UK crops throughout the winter to form a greater proportion of the population in the following year. Bird-cherry aphid reproduces quickly in warmer autumn conditions, and epidemics caused by it are characterized by a trend to early sowing, with warm, dry conditions in September and October allowing a rapid build-up of populations. Aphids continue to increase and spread virus for as long as mean daily temperatures remain above 3°C; the aphids are vulnerable to frost, populations on average being halved if minimum temperatures reach −0.5°C. Epidemics can be considered at an end once three frosts of below this value have been recorded. Aphicide sprays are best applied 4 weeks after crop emergence, when the first generation of wingless aphids will have reached maturity and started to produce young nymphs that may spread to, and infect, other plants. An earlier treatment may be required if migration pressure is high and more than 5% of plants become infested with aphids before the 4-week period is up. There are no threshold levels of aphid infestation below which treatment may be safely avoided, as the aphids are difficult to detect in the autumn and the proportion carrying BYDV will not be known. As a general rule, crops sown up to 10 October may be vulnerable under most autumn conditions; additionally, crops sown later in October may be considered vulnerable in unusually warm autumns.

BYDV vectored by grain aphid is of most importance in mild winters. Grain aphid is slower to reproduce than bird-cherry aphid but is more frost hardy – temperatures of −8°C being required to halve populations. In mild winters, populations that are initially low can gradually increase, slowly spreading BYDV through the crop. Infection with BYDV can cause significant yield loss up to the start of stem extension (GS 31), and in mild winters it may be worth treating unsprayed crops up to this stage if an aphid infestation is noticed.

Chemical control of aphid vectors

Control of BYDV vectors may be obtained by the use of a wide range of pyrethroid insecticides, including alpha-cypermethrin, cypermethrin, deltamethrin, esfenvalerate, lambda-cyhalothrin, tau-fluvalinate and zeta-cypermethrin. Sprays should be applied 4 weeks after crop emergence to crops at risk, unless 5% or more of plants are found to be infested before this. A second spray may be required if the first spray is applied before 11 October. Alternatively, the seed may be treated with imidacloprid; however, owing to the cost of this treatment, its use is likely to be restricted to crops thought to be at particular risk.

Following mild winters, when large numbers of aphids may have overwintered on grasses, spring-sown cereal crops may be at risk of BYDV infection. The epidemiology of the virus on such crops varies from the usual autumn situation in that most infection results from transfer by winged aphids, rather than from wingless ones that have a slower rate of spread within a crop. Chemical control has proved of little value in spring, as the crop soon outgrows any protection afforded. Yield loss is greatest on crops infected at an early growth stage, so that early sowing is the best method of reducing losses. Cultivars of spring barley differ in their susceptibility to BYDV and this characteristic is included in the NIAB recommended list. Where late sowing is unavoidable, cultivars with a high degree of tolerance may suffer less yield loss than others.

Black point (*Alternaria* spp.)

The embryo end of the grain shows a brown/black discoloration and, although grain size and germination are unaffected, this can render grain unmarketable for milling purposes. Although the disease is frequently associated with *Alternaria* spp., several other fungi have also been isolated from affected grain. The disease is sporadic in occurrence. Most cultivars have reasonably good disease resistance but some, such as Hereward, frequently show high levels of the disease. There are no fungicides recommended for the control of this disease.

Brown foot rot and ear blight (*Fusarium* spp. and *Monographella nivalis* – anamorph: *Microdochium nivale*)
The most common species found associated with the stem base is *Monographella nivalis* (anamorph: *Microdochium nivale*). Emerging young plants may be killed outright (fusarium seedling blight), causing poor establishment. On established plants infection is often seen during winter and late summer as a dark-brown discoloration of the lower nodes or internodes. In very dry summers a foot root can be caused by *Fusarium culmorum* or, less commonly, by *Fusarium graminearum*. In this case a pink spore-mass can sometimes be seen on the lower internodes. If wet weather occurs during flowering, ears can also become infected, causing complete or partial bleaching of the ear with little or no grain development ('whiteheads'). As with the foot-rot phase, pink spore masses can sometimes be seen within the glumes of affected ears. When black fruiting bodies are present amongst the pink fungal growth on the ear, the 'perfect' stage of *F. graminearum* is frequently implicated and the condition is known as 'scab'. Sprays of conazole fungicides applied during or soon after flowering can give some control of the ear-blight phase. The seed-borne phase of the disease may be controlled with seed treatments, see Table 2.2.

Table 2.2 Fungicide seed treatments for winter wheat

	Target disease				
Active ingredient	bunt	fusarium seedling blight	loose smut	septoria seedling blight	take-all
bitertanol + fuberidazole	*	*	–	–	–
carboxin + thiram	*	*	–	*	–
fludioxonil	*	*	–	–	–
fluquinconazole + prochloraz	–	*	–	*	*
guazatine	*	(*)	–	*	–
silthiofam	–	–	–	–	*
triadimenol + fuberidazole	*	(*)	*	*	(*)

* Will give control.
(*) Will give partial control.
– Not recommended.

Brown root rot (*Pythium* spp.)
This disease is of some importance in parts of America; although widespread in the UK, it is not thought to cause significant losses. Identification without the aid of a microscope is very difficult and, since the fungus is only weakly pathogenic, it may be considered an opportunist, attacking only plants already weakened by some other cause. Root tips are brown with generally poor growth and showing a yellowing of the foliage. The disease occurs on particular soil types, especially where the phosphate index is low, and may be aggravated by cool, wet soil conditions. In the UK the disease rarely causes significant yield loss and, although seed treatments can reduce the symptoms, the effects on yield are small. There are no foliar sprays that give control of the disease.

Brown rust (*Puccinia recondita*)
This disease is frequently seen in the autumn and winter months, particularly in mild winters. The disease usually remains at a low incidence until May/June. Symptoms are small, round, orange-brown pustules randomly scattered on the leaves, occasionally with a pale yellow halo. The disease is more common in very warm summers. The disease is well controlled by many foliar fungicides, particularly by the conazoles and morpholines and by azoxystrobin. Fungicides applied at the flag-leaf emergence timing are particularly important in preventing the disease becoming established on the upper leaves. There is a considerable range of disease resistance in modern wheat cultivars, some showing very high levels of resistance to brown rust.

Bunt (*Tilletia caries*)
Plants affected by bunt (also called stinking smut) are very difficult to see in the field, as visual symptoms are very subtle. Affected plants tend to be slightly shorter than healthy ones, and have slightly 'bluish' ears. Grains are replaced by spore balls containing a mass of black, greasy spores which, when fresh, smell of rotting fish. Usually, all grain sites in the ear are affected although, occasionally, partial ears are affected. During harvesting, spore balls are broken open, releasing many millions of spores per ear, which stick readily to healthy grains. The spores remain on the outside until the grain germinates; they then infect the emerging coleoptile. The disease can survive in dry soils, particularly if intact ears are buried. However, in the UK there are very few cases of survival of bunt spores in soil for more than a few weeks. The disease is well controlled by seed treatments; see Table 2.3.

Ergot (*Claviceps purpurea*)
See under Rye, p. 49.

Eyespot (*Tapesia acuformis* – anamorph: *Ramulispora acuformis*, and *T. yallundae* – anamorph: *R. herpotrichoides*)
In young plants, symptoms are seen as indistinct, light-brown smudges on the leaf sheath. Later symptoms are the presence of brown, oval-shaped lesions. Occa-

Table 2.3 Fungicides for use on winter wheat

Active ingredient	Target disease					
	brown rust	eyespot	glume blotch	leaf spot	powdery mildew	yellow rust
azoxystrobin	*	–	*	*	(*)	*
benomyl	–	–	(*)	(*)	(*)	–
carbendazim	–	–	(*)	(*)	(*)	–
chlorothalonil	–	–	*	*	–	–
cyprodinil	–	*	–	–	*	–
epoxiconazole	*	(*)	*	*	(*)	*
fenpropidin	*	–	–	–	*	*
fenpropimorph	*	–	–	–	*	*
fluquinconazole	*	–	*	*	(*)	*
flusilazole	*	*	*	*	(*)	*
flutriafol	*	–	*	*	(*)	*
iprodione	–	–	(*)	–	–	–
kresoxim-methyl	–	–	(*)	(*)	*	–
mancozeb	(*)	–	(*)	(*)	–	(*)
maneb	(*)	–	(*)	(*)	–	(*)
metconazole	*	–	*	*	(*)	*
nuarimol	–	–	–	–	(*)	–
prochloraz	–	*	*	*	(*)	–
propiconazole	*	–	*	*	(*)	*
sulfur	–	–	–	–	(*)	–
tebuconazole	*	–	*	*	(*)	*
triadimefon	*	–	–	–	(*)	*
triadimenol	*	–	(*)	(*)	(*)	*
tridemorph	–	–	–	–	*	(*)
trifloxystrobin	*	–	*	*	*	(*)

* Will give control.
(*) Will give partial control.
– Not recommended.

sionally, a grey/black mycelial mass can be seen in the centre of the lesion. Severe attacks can kill young tillers but, more commonly, the fungus slowly penetrates successive leaf sheaths during the season, eventually reaching the stem. As the plant ripens, the oval lesion on the stem takes on a pale straw colour and, if the attack is severe, the weakened straw can bend at that point, causing lodging. 'Whiteheads' may be produced but, generally, these occur on individual tillers rather than in patches – unlike take-all (see p. 35). The disease is common, but severe attacks are relatively infrequent.

Fungicides available for control are listed in Table 2.3. Early drilling of winter crops increases the risk of infection more than any other factor, although successive cereal cropping will also increase the risk of severe disease. Some cultivars are now being introduced with better resistance to the disease.

Leaf stripe (*Cephalosporium gramineum*)
Diseased plants are usually seen after flag-leaf emergence, when the upper leaves exhibit a single yellow stripe which runs longitudinally and often extends on to the leaf sheath. Frequently, all leaves on an infected tiller show striping. If the stem is cut transversely at the nodes, brown staining of the vascular tissue may be seen. Occasionally, affected plants are stunted and grain fill is generally poor. The disease is soil-borne, infecting plants through damaged roots. For this reason the symptoms are often seen around gateways and other areas of compaction, and are common in fields where grass has been recently ploughed out and remaining insect pests, particularly wireworms (*Agriotes* spp.), have damaged the roots of the following wheat crop. The disease is very sporadic in occurrence and when present rarely, if ever, causes significant yield loss. There are no fungicides active against this disease. However, treatments applied for the control of wireworms will often result in reductions in the incidence of the disease.

Loose smut (*Ustilago nuda*)
Symptoms are similar to those described under barley, see p. 42, where the disease is more common. However, cross-infection between wheat and barley does not occur. Control is obtained with seed treatments, see Table 2.2, p. 30.

Powdery mildew (*Erysiphe graminis* f. sp. *tritici*)
Symptoms of infection are usually seen as fluffy, white pustules on the leaf and stem. In high temperatures, which are unfavourable for the disease, lesions become less fluffy and take on a light-brown coloration. Soon after ear emergence the disease is often seen to infect the glumes. On the glumes the fungus develops as a white, fluffy mass and then, finally, as a fine, mid-brown mat of mycelium; eventually, small, black resting bodies (cleistothecia) develop within the mycelial mat. At this time these resting bodies are also common on stems and older leaf lesions. Specific races of the fungus infect wheat, barley, oats, rye and grasses, therefore, cross-infection does not occur. For chemical control options, see Table 2.3, p. 32.

Septoria nodorum leaf spot and glume blotch (*Leptosphaeria nodorum* – anamorph: *Stagonospora nodorum*)
This disease is commonly referred to as *'Septoria nodorum'*. Although seed infection can lead to seeding losses (septoria seedling blight), the disease is rarely seen to any extent before mid-season. Lesions on the leaves begin as oval, yellow spots with brown margins, but these commonly coalesce to form large areas of dead tissue. Under high disease pressure, early symptoms may show as discrete and dark, brown–black spots, each 1–2 mm in diameter. The fruiting bodies (pycnidia), when present, are difficult to see. They are small and flesh-coloured, are embedded in the leaf tissue, and are best observed in the field by viewing lesions with a hand lens against a good source of light. Ears can become infected, resulting in a purple/brown discoloration of the glumes (= glume blotch). Severe

infection results in shrivelled grain. The disease can be very damaging in warm wet seasons and, consequently, is most common and damaging in the south-west of England. Many of the foliar-applied fungicides, particularly the conazoles, can give good control of the disease. Although the disease is very damaging on the ear, it is very important to control the disease on the upper leaves to reduce inoculum which could otherwise spread to the ears. In high-risk areas it is important to apply fungicides to the flag leaf and to the ear.

Septoria tritici leaf spot (*Mycosphaerella graminicola* – anamorph: *Septoria tritici*)
This disease has now become the main foliar disease of wheat crops in the UK. Symptoms are green/grey, oval lesions on the leaf, often with the fruit bodies (pycnidia) present as pinhead-sized black spots within the lesion. Occasionally, early lesions are delimited by the veins in the leaf, leading to short, angular, striping symptoms. The disease is spread by water splash and by physical contact between lower infected leaves and upper leaves. Symptoms develop 4–6 weeks after infection. Ear infection is very rare. Various fungicides are effective, see Table 2.3, p. 32.

Sharp eyespot (*Ceratobasidium cereale* – anamorph: *Rhizoctonia cerealis*)
This fungus can attack cereals at the seedling stage, causing general browning on younger roots, and distinct lesions on older roots. The most common symptoms are seen as lesions on the leaf sheath and stem. The fungus often invades the lower part of the stem, producing lesions that are easily confused with those of eyespot (see p. 31). On younger plants, lesions of sharp eyespot tend to have a darker and more sharply defined margin than those of eyespot. In addition, with sharp eyespot, the leaf sheath is often shredded in the lesion centre. If the fungal mycelium can be seen in the centre of the lesion, that of sharp eyespot is pale pink to purple-brown, whereas that of eyespot is a dark grey/black. Lesions often develop higher on the stem than do those of eyespot, and this may also help with identification.

Although the disease can be severe in some crops, causing weakening of the straw, the disease is generally not as damaging as eyespot. The disease is very sporadic and it is not possible to predict when it is likely to be damaging. Consequently, it is rare for specific treatments to be applied. The disease tends to be most damaging when infection occurs early in the spring. The fungicides azoxystrobin and prochloraz, applied in the spring against other diseases, can give incidental control of sharp eyespot, but the level of control is very variable.

Sooty moulds (*Alternaria* and *Cladosporium* spp.)
The general term 'sooty moulds' is given to saprophytic fungi that colonize senescing crops at the end of the season. They are commonly found on plants that have ripened prematurely owing to stem-base or root disease, but can also be

found on healthy plants when harvest is delayed by wet conditions and the plants are no longer protected by fungicides.

The most common fungi are *Alternaria* spp. and *Cladosporium herbarum*. Sooty moulds do not affect yield, but may discolour grain which can be important if the grain is to be used for bread making. Sooty moulds can also develop on honeydew excreted by aphids.

Take-all (*Gaeumannomyces graminis*)
This soil-borne fungus attacks the roots. In severe outbreaks this may result in yellowing and stunting of young plants. The disease is more commonly seen after ear emergence, during late grain filling, when affected plants are seen in patches. Plants show severe stunting, and senesce prematurely with poor grain fill. Affected plants often have little or no grain in the ears, which are frequently bleached – often referred to as 'whiteheads'. These ears are often subsequently colonized by sooty moulds, resulting in blackening of the ears late in the season. In severe attacks, the fungus invades not only the roots but also the base of the stem, causing a superficial blackening.

The fungus builds up on successive cereal crops, reaching a peak in anything from the second to the sixth successive crop. Once the peak is reached, infection levels may decline in certain soils; this is thought to be caused by the build-up of antagonistic micro-organisms. The disease is favoured by early drilling, and any factor limiting root growth (e.g. high or low soil pH, nutrient imbalance and poor drainage). In addition, a 'puffy' seedbed can allow the fungus to grow rapidly through the soil, leading to severe infection. The fungus survives on crop debris; also, couch grass (*Elytrigia repens*) is susceptible to infection and can act as an important source of the disease.

Cereal species vary in their susceptibility to the disease. Wheat is the most susceptible cereal. Barley is less affected, and rye is practically resistant. Oats are affected by a different strain of the fungus (*G. graminis* var. *avenae*), which is relatively rare, and thus oats are often considered resistant to the disease. Some difference in varietal (cultivar) susceptibility, although not well documented, is thought to exist, but best control is by good soil management and crop rotation. Where third and successive cereal crops are grown, it is advisable to grow barley in the high-risk years, as this crop will be less affected than wheat.

Seed treatments based on fluquinconazole and silthiofam are available which can reduce the impact of the disease in crops that are in high-risk positions in the rotation (see Table 2.2, p. 30).

Yellow rust (*Puccinia striiformis* f. sp. *tritici*)
In the last decade, this disease has caused very severe losses, particularly in eastern England. Serious outbreaks are invariably associated with the widespread growing of susceptible cultivars, coupled with mild winters.

The fungus exists as numerous strains, with no cross-infection between cereal species and grasses. Some strains are capable of infecting only certain cultivars of wheat, but new strains can arise very rapidly and so varietal (cultivar) resistance is often short lived.

Very early symptoms in the autumn appear as scattered pustules that are frequently confused with brown rust. Later, in the spring and summer, symptoms are more typically seen as lines of bright yellow pustules on the leaves, running parallel with the veins. Stems and ears may also become affected. Late in the season these pustules blacken, as teliospores are formed. Mild winters and cool, moist weather in the spring and early summer favour the disease. Very hot, dry conditions often slow down the disease considerably.

Growing cultivars with a range of resistance genes is a sensible approach, especially where winter and spring wheat are grown in close proximity. In areas where epidemics are common, early application of fungicides is a wise precaution. For available materials, see Table 2.3, p. 32.

Barley

Pests

Aphids
See under Wheat, p. 21.

Cereal ground beetle (*Zabrus tenebrioides*)
This once rare pest is becoming increasingly common in cereal crops in southern England. The larvae cause damage superficially similar to that caused by leatherjackets although, on close inspection (and typical of cereal ground beetle), leaves can be found pulled down into the larval burrows. The larvae are active from November to May. They feed at night and spend the day at the bottom of burrows in the soil, each about 150 mm deep. They look similar to larvae of predatory ground beetles, with which they may be confused. The adult beetles lay their eggs in the autumn, in the stubbles of cereal fields and in grassland. The beetles are slow to disperse, and damage is restricted to fields following several cereal crops or grass.

Cultural control can be obtained by the use of a broad-leaved break crop or by ploughing early to bury stubble volunteers. Chemical control may be obtained by a spray of chlorpyrifos, applied as recommended for leatherjackets (p. 37).

Frit fly (*Oscinella frit*)
See under Oats, p. 44.

Gout fly (*Chlorops pumilionis*)
See under Wheat, p. 23.

Leatherjackets

Leatherjackets are the larvae of crane flies, notably *Tipula* spp. and *Nephrotoma* spp. The biology of the various species varies in detail but in the case of *Tipula paludosa* (one of the most abundant species), eggs are laid in grassland in the autumn and soon hatch. When the grass is ploughed for cereals, the leatherjackets feed at first on the ploughed-up turf but then attack the new crop. One species (*Tipula oleracea*) may also damage crops sown after oilseed rape, as the low-flying adults can be trapped within the canopy and are then forced to lay eggs in a situation they would otherwise avoid. This source of damaging populations has been observed most frequently in Scotland.

Cereal plants are damaged at, or just below, ground level; injured tissue appears torn rather than cut. Leatherjackets are usually easy to find near the soil surface by damaged plants (cf. swift moth caterpillars, p. 24). Spring cereals are usually damaged in March and April, when leatherjackets are most active, but winter cereals can be damaged in mild periods from November onwards.

Large numbers of crane flies, damp weather during egg-laying and a mild winter all tend to result in greater numbers of leatherjackets in the following year. Since crane flies have an annual life cycle, damage is restricted to the first year after ploughing grass or oilseed rape. Ploughing of grassland in July or August, before the main egg-laying period, reduces the risk of leatherjackets, and also of frit fly attack, but increases the danger of wheat bulb fly in areas where this is a problem.

Chemical control of leatherjackets may be obtained with a spray of chlorpyrifos, applied when the first damage is seen. Control may be worth while if five leatherjackets or more are found per metre of row. Successful control depends on the leatherjackets actively feeding at the surface on the nights following application. Activity is restricted on nights when the minimum temperature falls below 5°C, and treatment should be withheld until the next mild period if minima below this value are forecast.

Saddle gall midge (*Haplodiplosis marginata*)

Larvae feed on the stems of cereal plants, under the leaf sheath. The damage may be felt as a bump on the stem; peel back the leaf sheath to confirm the cause. The blood-red larvae cause a saddle-shaped depression on the stem, with a raised bump at each end. All species of cereals may be attacked. The damage reduces yield by limiting the flow of sap to the ear and, in barley, may cause the stem to buckle and break, further increasing yield loss.

In May, the adult midges lay their blood-red eggs in raft-shaped groups on the upper side of the leaves of cereal plants and grasses. The eggs hatch after about a week and the larvae then migrate to the stem to commence feeding. They feed for about a month before leaving the plant to overwinter in cells in the soil, where they may remain dormant for several years. The adult female midges do not fly very far (50 m maximum) and damage is restricted to fields growing cereal crops for several years without a break. Heavy clay soils are favoured by the pest, as a high clay content is necessary for the formation of an overwintering cell.

No specific chemical control measures are available in the UK. However, sprays of pyrethroid insecticides are recommended elsewhere and, if cleared for use on cereals at the appropriate time, may be used in the UK. For control to be effective the sprays would need to be applied before the eggs have hatched and larvae have migrated to feed on the stem. Control is thought to be worth while if an average of five or more eggs are found per tiller. A non-cereal break crop grown in the rotation every 4 years should greatly reduce numbers and prevent damage.

Slugs

See under Wheat, p. 24.

Thrips

Thrips (notably barley thrips, *Limothrips denticornis*, and grain thrips, *L. cerealium*) are probably the most numerous of cereal pests, and several black adults or yellow nymphs can be found crawling within the florets and leaf sheaths of virtually any ear examined. Small numbers cause little damage to crops, but large numbers can cause significant damage to the stems and ears. Thrips damage may also induce infection with secondary fungi. In barley, feeding on the ear and stem within the boot is the most important form of damage, and control measures need to be applied when the awns first appear from the flag-leath sheaf to obtain effective control. Treatment may be worth while if two or more thrips can be found per ear at this stage. Control may be obtained by applying a spray of chlorpyrifos.

Wheat bulb fly (*Delia coarctata*)

See under Wheat, p. 26.

Wireworms

See under Wheat, p. 27.

Diseases

Barley mosaic viruses

Barley mosaic was not identified in the UK until 1980, but since then has been seen in most areas where winter barley is grown intensively. The disease was once thought to be caused by a single virus but is now known to be caused by two distinct viruses: barley yellow mosaic virus (BaYMV) and barley mild mosaic virus (BaMMV). Both viruses are spread by the common soil-borne fungus *Polymyxa graminis*. Symptom expression requires a period of cold weather, and usually occurs from January onwards. Patches of pale growth appear in the crop, which may be mistaken for nutrient deficiency, waterlogging or acidity. Occasionally, large areas of fields or entire fields may show symptoms. On close examination, small, pale-green flecks can be seen on the younger leaves, which soon turn yellow and, eventually, brown. Symptoms are dependent on cold weather, and may disappear completely during a mild spell. Severely affected patches of a crop may well remain stunted.

Infection is confined to barley, and symptoms are seen only in autumn-drilled crops. Many resistant cultivars of winter barley are now available. The virus has been shown to be capable of survival within the fungal vector for several years, even in the absence of a susceptible crop. Where the disease is present in patches in a field the spread of infected soil should be kept to a minimum. There is no method of chemical control.

Barley yellow dwarf virus (BYDV)

This virus is transmitted by certain cereal aphids: bird-cherry aphid (*Rhopalosiphum padi*) and grain aphid (*Sitobion avenae*). Symptoms vary in intensity depending on the strain of the virus, the age of the plant at infection, and the cultivar of cereal being grown. Sometimes, the virus can be found in plants exhibiting no obvious symptoms of infection.

Infected young plants show a golden yellowing of the leaf tips, which gradually extends down the leaf blade. Occasionally, dark-brown flecking is seen on yellowed leaves. Plants are frequently stunted, can show increased tillering and, occasionally, can be killed.

Surviving plants, or those showing later infection, show a golden-yellow coloration of the leaves, stunting and poor ear development. Infection generally shows as affected circular patches of plants, resulting from localized movement of aphid vectors, but infection can also be fairly generalized, depending on the aphid species concerned and the time of infection.

There are varietal (cultivar) differences in susceptibility to the virus but the most practical control measures at present are the use of insecticide sprays, linked with crop monitoring or forecasting of viruliferous aphids in the area. Insecticidal seed treatments are also available for use in consistently high-risk areas. For details of control measures, see under Wheat, p. 28.

Black point (*Alternaria* spp.)

See under Wheat, p 29.

Brown foot rot and ear blight (*Fusarium* spp.)

Symptoms of brown foot rot can be common in winter barley, although the disease rarely causes losses. Symptoms of ear blight are much less common than in wheat. For further details, see under Wheat, p. 30.

Brown rust (*Puccinia hordei*)

This disease appears as scattered orange/brown pustules, occasionally surrounded by a small, yellow halo. The disease is commonly seen in mild winters but rarely develops significantly until the summer period. It develops in hot, dry conditions and is rarely serious, except in very susceptible cultivars and then especially in southern and eastern England. As with yellow rust, the disease can spread from autumn- to spring-sown crops if the latter are grown in the vicinity. Fungicides available for control are given in Table 2.4.

Table 2.4 Fungicides for use on barley

	Target disease					
Active ingredient	brown rust	eyespot	leaf blotch	net blotch	powdery mildew	yellow rust
azoxystrobin	*	–	*	*	*	(*)
benomyl	–	–	*	–	(*)	–
carbendazim	–	–	*	–	(*)	–
chlorothalonil	–	–	*	–	–	–
epoxiconazole	*	(*)	*	*	(*)	*
fenpropidin	*	–	*	–	*	*
fenpropimorph	*	–	*	–	*	*
flusilazole	*	*	*	*	(*)	*
flutriafol	*	–	*	*	(*)	*
iprodione	–	–	–	*	–	–
kresoxim-methyl	–	–	*	–	*	–
mancozeb	(*)	–	–	–	–	(*)
maneb	(*)	–	–	–	–	(*)
metconazole	*	–	*	*	(*)	*
nuarimol	–	–	–	–	(*)	–
prochloraz	–	*	*	*	(*)	–
propiconazole	*	–	*	*	(*)	*
quinoxyfen	–	–	–	–	*	–
sulfur	–	–	–	–	(*)	–
tebuconazole	*	–	*	*	(*)	*
thiophanate-methyl	–	–	*	–	(*)	–
triadimefon	*	–	*	–	(*)	*
triadimenol	*	–	*	–	(*)	*
tridemorph	(*)	–	–	–	*	–
trifloxystrobin	*	–	*	*	*	(*)
triforine	–	–	–	–	(*)	–

*   Will give control.
(*) Will give partial control.
–   Not recommended.

Covered smut (*Ustilago hordei*)
This disease is rare, but can occur if infected, untreated seed is sown. Symptoms are similar to those seen in wheat. For available seed treatments, see Table 2.5.

Ergot (*Claviceps purpurea*)
Ergot is very rare in barley – see under Rye, p. 49.

Table 2.5 Fungicide seed treatments for barley

Active ingredient	Target disease			
	brown foot rot and seedling blight	leaf stripe	loose smut	seedling net blotch
carboxin + thiram	*	(*)	(*)	–
fludioxonil	*	(*)	–	–
flutriafol + ethirimol + thiabendazole	–	(*)	*	*
guazatine	(*)	–	–	–
guazatine + imazalil	(*)	*	–	*
imazalil	–	*	–	*
tebuconazole + triazoxide	–	*	*	*
triadimenol + fuberidazole	*	(*)	*	–

* Will give control.
(*) Will give partial control.
– Not recommended.

Eyespot (*Tapesia acuformis* – anamorph: *Ramulispora acuformis*, and *T. yallundae* – anamorph: *R. herpotrichoides*)
This disease is common in autumn-sown crops, owing to the frequent place of this crop in a cereal rotation, and early drilling. The disease can be particularly difficult to identify in the early spring, with general discoloration of the stem base common in barley. For further details, see under Wheat, p. 31.

Halo spot (*Selenophoma donacis*)
Halo spot is found mainly in western coastal areas, where outbreaks occur in wet summers after flag-leaf emergence. The disease appears as small leaf spots (1–3 mm long) that are often square or rectangular in shape, and pale brown in the centre with dark purple/brown, well-defined margins. Pycnidia occur in lines along the veins, within the central area of a lesion. Spots generally occur towards the tips and along the edges of leaves. The pycnidia also affect the leaf sheath and ear (especially the awns). Halo spot often occurs with *Rhynchosporium* (see below) but can be distinguished from the latter by the smaller size of its spots and the presence of pycnidia within the lesions. The disease affects barley and various species of grass, but the forms on grasses do not cross-infect or spread to barley. The form on barley is specific to that crop and does not affect other cereals.

There is little available information on varietal (cultivar) resistance. Good stubble hygiene, such as ploughing in affected straw, would help reduce inoculum but infestation levels are rarely high enough to warrant specific control measures.

Broad-spectrum foliar fungicides such as conazole/MBC mixtures often give incidental control of the disease, so that no specific chemical treatment is required. Where the disease occurs, sprays applied soon after ear emergence would be most effective in preventing the disease becoming established on the upper leaves and awns.

Leaf stripe (*Pyrenophora graminea* – anamorph: *Drechslera graminea*)
This disease is seed-borne. Each of the emerging leaves of infected seedlings shows pale striping, and seedlings can occasionally be killed. The leaves of plants that survive show a pale striping that turns yellow and, eventually, dark brown. Affected leaves may split along the lesions. If an infected plant produces ears, these are poorly developed.

Control is by selection of good-quality seed, and the use of effective seed treatments (see Table 2.5, p. 41).

Loose smut (*Ustilago nuda*)
This disease is present at low infestation levels in many crops of winter barley. Only as the ear emerges does the disease become obvious, when black spore masses replace the grain sites on infected ears. All tillers of an infected plant are affected. The spores are readily dispersed by the wind, leaving a bare spike that is not so conspicuous in the growing crop.

Crops should be grown from smut-free or certified seed. A seed test is available from Official Seed Testing Stations at Cambridge and Edinburgh.

Net blotch (*Pyrenophora teres* – anamorph: *Drechslera teres*)
Although the disease is seed-borne, the most important source of inoculum is straw debris from previous or adjacent crops. Seed-borne and early airborne infection is first seen as very small, dark-brown flecks on the leaves. As these lesions mature, they form one of two distinct symptoms. The traditional 'netting' symptoms, which gave rise to the common name, have a dark-brown to black criss-crossing network against a yellow background. The other symptom is now much more common, appearing as short, dark-brown stripes, often delineated by the veins. The fungus, in fact, exists in two forms, a 'net' form and a so-called 'spot' form. The 'net' form produces both netting and striping symptoms. The less common 'spot' form produces small, dark-brown, elliptical spots, each with a chlorotic halo. Wet weather can lead to high levels of disease, and substantial yield loss, particularly if the disease affects the awns. Available fungicides are listed in Table 2.4, p. 40, and seed treatments in Table 2.5, p. 41.

Powdery mildew (*Erysiphe graminis* f. sp. *hordei*)
For details, see under Wheat, p. 33. This disease can be particularly severe in late-sown spring crops, especially those grown near mildew-affected winter barley crops. Early-sown winter crops on light land often carry high levels of infection in the autumn when weather is mild, and the restriction to root development which the disease causes can lead to winter kill. Spraying to control the disease at this stage, however, is rarely necessary, except in situations where winter kill is common. Fungicides available for control are listed in Table 2.4, p. 40.

Rhynchosporium leaf blotch (*Rhynchosporium secalis*)
This disease begins with symptoms of grey, water-soaked lesions on leaves. As the lesions age, they develop well defined, dark-brown margins. Lesions often occur in the leaf axil. This can be particularly damaging, as a single lesion can lead to death of the whole leaf. Lesions may also occur on the lower leaf sheaths, producing symptoms that may be confused with those of eyespot. This disease is particularly common in the wetter parts of the UK. There is a wide range of disease resistance in current cultivars, although in high-risk areas disease resistance may not keep disease levels low. In high-risk areas with susceptible cultivars it is frequently necessary to apply a two-spray fungicide programme to control the disease. Sprays applied in the early spring prevent the disease becoming established on newly emerging leaves. A second spray, applied soon after ear emergence, protects the upper leaves. Under high disease pressure a mixture of conazole and morpholine fungicides is needed to give good control of the disease.

Sharp eyespot (*Ceratobasidium cereale* – anamorph: *Rhizoctonia cerealis*)
See under Wheat, p. 34.

Snow rot (*Typhula incarnata*)
This disease is quite common in parts of the country where winter barley is intensively grown and snow cover is common. Symptoms are often first seen as patches of dead plants after the snow covering has melted. The disease usually produces dense, white fungal growth on the lower, dead, leaf tissue and clusters of resting bodies (sclerotia), which are pink to brown in colour and 2–3 mm in diameter; these sclerotia occur on the stem base and on the lower leaf sheaths. The disease can be mistaken for snow mould (caused by *Monographella nivalis* – anamorph: *Microdochium nivale*) (see under Forage grasses, pink snow mould, Chapter 4, p. 93) which also causes plant death (especially after snow cover) and a dense, white fungal growth on the affected plants, but no sclerotia. The disease is widespread on light soils, and growing successive crops of winter barley on the same land will increase the risk of an attack. Cultivars differ in their susceptibility to snow rot, although information on modern ones is scarce.

There is some evidence that conazole fungicides applied during the autumn or early winter can reduce infection. This is likely to be worth while only where infection is common each year.

Take-all (*Gaeumannomyces graminis*)
See under Wheat, p. 35.

Yellow rust (*Puccinia striiformis* f. sp. *hordei*)
This disease is relatively rare on barley, with distinct races existing on barley, wheat, oats and grasses. Symptoms are similar to those seen on wheat (see p. 35). For fungicides available for control, see Table 2.4, p. 40.

Oats

Pests

Cereal cyst nematode (*Heterodera avenae*)
This formerly important pest has declined greatly in the UK, following the use of resistant cultivars and the influence of naturally occurring fungal pathogens. Populations of cereal cyst nematode will also decline where non-host crops are grown for at least two years in the rotation. Although associated mainly with oats, infestations also occur on other cereal crops, including (in order of susceptibility) wheat, barley and rye. Nowadays, however, damage is rarely seen.

Frit fly (*Oscinella frit*)
This pest is primarily a grassland species, which infests mainly ryegrasses, on which it can successfully pass all three generations per year. The flies may also be attracted to spring-sown cereal crops, on which they will also lay their eggs (in the axils of unfolded leaves).

The first generation is on the wing in late May and may be attracted to spring oats. The young larvae feed at the base of the centre shoot, causing it to turn yellow and die. Small plants are killed, whereas larger ones are induced to produce many short tillers which gives the crop a 'grassy' appearance. Damage to seedling spring oats can be avoided by sowing before mid-April, as plants with four leaves still unfolded are more resistant to attack.

The second-generation flies emerge in time to lay their eggs on the newly developing panicles. Larvae feed on the swelling grain, producing the 'fritted grain' of spring oats, and can cause grain loss in other cereals. The adult frit flies may emerge from the grain after harvest, and contamination may cause rejection by grain merchants. No further damage is caused in store and the adult flies soon disperse.

Third-generation adults, on the wing from August to mid-October, may lay their eggs on newly sown cereals, although most of the dipterous eggs found on these crops are those of the gout fly (see p. 23). The larvae may be very numerous in ryegrass leys and, in years of high abundance, in cereal volunteers within stubbles. Largest numbers are found in two-year-old Italian ryegrass crops. Where these are followed by a cereal crop, the larvae can transfer from the previous host. The larvae enter the shoot at the top and feeding causes the typical 'deadheart' symptom. Unlike those of wheat bulb fly (p. 26), frit fly larvae complete their development in a single cereal shoot. Various other pests attack cereal crops following leys, and as some of the other pests continue to feed through the winter and spring it is important to identify the causal agent(s) before deciding on the need for control.

Chemical control may be obtained by a spray of chlorpyrifos applied (a) to the previous crop or stubble before cultivation, or (b) to the soil surface before emergence, or (c) to the newly emerged crop. Control may be cost effective if 10% or more of plants are damaged at the one- to two-leaf growth stage.

Stem nematode (*Ditylenchus dipsaci*)
This pest feeds within the shoots of oats and rye, causing swelling and distortion. Many other plants are hosts to the oat race, including beans, peas, vetches, sugar beet, rhubarb, strawberry and onion and several weeds. Control is by crop rotation, weed control and by the use of resistant cultivars, such as Gerald, Image and Lexicon.

Diseases

Barley yellow dwarf virus (BYDV)
Oats are more severely affected than either wheat or barley. Plants can be severely stunted and leaves of affected plants become purple/red. For further details, see under Wheat, p. 28, and under Barley, p 39.

Brown foot rot and ear blight (*Fusarium* spp.)
This disease can be quite serious in some years. See under Wheat, p. 30. For available seed treatments, see Table 2.6.

Table 2.6 Fungicide seed treatments for oats

Active ingredient	Target disease			
	brown foot rot and seedling blight	covered smut	leaf spot	loose smut
bitertanol + fuberidazole	*	*	–	–
carboxin + thiram	*	*	–	–
guazatine	(*)	–	–	–
guazatine + imazalil	(*)	–	*	–
tebuconazole + triazoxide	–	*	–	–
triadimenol + fuberidazole	*	*	*	*

* Will give control.
(*) Will give partial control.
– Not recommended.

Covered smut (*Ustilago hordei* f. sp. *avenae*)
Oat plants grown from seed infected by this fungus appear to have blackened ears at the time of ear emergence. In fact, the grain sites are replaced by masses of spores that usually remain surrounded by a membrane until this is broken during harvest. Occasionally, the spore masses are exposed, making differentiation

between covered smut and loose smut (see p. 47) very difficult without microscopical examination. For available seed treatments, see Table 2.6.

Crown rust (*Puccinia coronata*)
This is the common rust of oats, but it does not affect other cereals. Distinct strains are associated with several grasses, but these strains do not cross-infect to oats. The fungus has two alternate hosts: alder buckthorn (*Frangula alnus*) and common buckthorn (*Rhamnus catharticus*), on which the aecial stage occurs. Symptoms on oats are seen as bright orange pustules, mainly on the leaves but also on the stem and panicle. Later in the season, black spores (teliospores) form on the stem and leaves. These teliospores germinate in the spring to produce basidiospores which then infect the alternate hosts. Aeciospores, in turn, are produced on the alternate host and these then infect oats. The disease is favoured by warm, humid weather, and so rarely develops to any extent until mid-summer. Many cultivars are susceptible to crown rust. Although destruction of the alternate hosts is recommended in parts of continental Europe, it is not a practical proposition in the UK and the value of it is not known. Controlling volunteers and deep ploughing of stubble from infected crops would reduce the risk of carryover. Growing susceptible spring oats next to winter oats also should be avoided.

Conazole and morpholine fungicides give good control of crown rust. The disease usually develops late in the season, and sprays earlier than the start of flag leaf emergence are rarely required.

Ergot (*Claviceps purpurea*)
This disease is very rare in oats – see under Rye, p. 49.

Eyespot (*Tapesia acuformis* – anamorph: *Ramulispora acuformis*, and *T. yallundae* – anamorph: *R. herpotrichoides*)
This disease is fairly common but rarely serious in oats. For further details, see under Wheat, p. 31.

Halo blight (*Pseudomonas syringae* pv. *coronafaciens*)
This disease is caused by a bacterium, and is common at low disease levels in northern and western parts of the UK. The disease is seed-borne, initially causing spotting on seedling leaves. The disease is then spread by wind and water splash to the upper leaves and panicles. Small, dark-green to brown, water-soaked spots with a yellow halo appear on the leaves. The disease is rarely serious and no control measures exist.

Leaf spot (*Pyrenophora avenae* – anamorph: *Drechslera graminis*)
Seedlings affected by this disease show narrow, brown stripes with purple margins on the first three or four seedling leaves; also, the first leaf may be distorted and twisted. Plants can die at this stage or, even earlier, before they emerge from

the soil. Plants that are affected less severely survive, showing a brown striping on the lower leaves. These stripes produce spores that splash up to infect the upper leaves, producing the secondary spotting symptoms. Leaf spots are oval, and red-brown with purple margins. Spores from leaf spots on the upper leaves can be splashed on to the developing grain, producing the seed-borne phase of the disease. For available seed treatments, see Table 2.6, p. 45.

Loose smut (*Ustilago avenae*)
This disease is very rare. When the ears emerge, black spore masses are seen to replace the grain. These spore masses can be partially or completely covered by a thin membrane, thus resembling covered smut (see above). There is a risk of developing loose smut if crops are grown repeatedly from untreated seed. Seed testing is available. If seed is tested and found to be free of the disease or to have only very low levels, then the seed could be sown without seed treatment. There are no specific cultural measures that can be undertaken to reduce the disease.

Oat mosaic virus (OMV)
This disease is similar in symptoms and epidemiology to barley yellow mosaic virus – see under Barley, p. 38. Occasionally, oat golden stripe virus (OGSV) is also found. This shows symptoms when oat mosaic virus is present and, as the name suggests, manifests itself as a bright, golden-yellow stripe on the leaves (particularly the flag and second leaf) of infected plants. Changes in cultural practice have a limited effect on the disease. Deep ploughing spreads the disease less rapidly than tine cultivation but otherwise has no beneficial effect. Early sowings are more prone to infection. Spring-sown crops, although they become infected by the fungus, do not show symptoms of the disease. Cultivars differ in their susceptibility to the disease but there are no chemical control measures.

Powdery mildew (*Erysiphe graminis* f. sp. *avenae*)
This is a common and often severe disease. For symptoms see under Wheat, p. 33. Powdery mildew can be very damaging to oats, particularly with mild winters where the disease epidemic starts early in the spring. Many cultivars of winter and spring oat are susceptible to the disease. Early sowing tends to favour the disease, so delaying sowing can help prevent early epidemics. There are few cultivars with good disease resistance. Many mildew fungicides give good control of the disease (see Table 2.3, p. 32, and Table 2.4, p. 40). Early treatment in the spring, during stem extension, is important if the disease has overwintered and is well established. A second spray at ear emergence may be needed in susceptible cultivars.

Speckled blotch (*Leptosphaeria avenaria* – anamorph: *Septoria avenae*)
This disease is widespread and can be severe, especially in wet seasons. It is characterised by round, or oval-shaped, dark-brown to purple spots with orange borders that occur on the leaves and leaf sheaths. Inside these are the brown to

black fruiting bodies (pycnidia), which give rise to the common name of the disease. Under wet conditions spotting of the panicle occurs and the stalk may be attacked. This results in the straw rotting and breaking. Under these conditions the glumes can also become infected, and the grain may become discoloured. The disease can then be carried over on the seed. Many seed treatments will control the seed-borne phase of the disease. Conazole fungicides applied in the spring and after panicle emergence will give control of the disease.

Take-all (*Gaeumannomyces graminis* var. *avenae*)
Although oats are not affected by the strain of the take-all fungus normally found in cereal growing areas, the crop can be affected by this specialized strain. This strain of the fungus is also capable of infecting wheat and barley. Because the disease is uncommon, specific control measures are not necessary.

Rye and triticale

Pests

Aphids
See under Wheat, p. 21.

Frit fly (*Oscinella frit*)
See under Oats, p. 44.

Leatherjackets (*Nephrotoma* spp. and *Tipula* spp.)
See under Barley, p. 37.

Stem nematode (*Ditylenchus dipsaci*)
See under Oats, p. 45.

Wheat bulb fly (*Delia coarctata*)
See under Wheat, p. 26.

Wireworms (*Agriotes* spp.)
See under Wheat, p. 27.

Diseases

Brown foot rot and ear blight (*Fusarium* spp.)
See under Wheat, p. 30.

Bunt (*Tilletia caries*)
See under Wheat, p. 31.

Ergot (*Claviceps purpurea*)
Rye is the most susceptible of the cereal species and can be severely affected by this disease. The ergot is the resting stage of the fungus, and is seen as a large, dark, purple–black structure protruding from a diseased spikelet, where it completely replaces one or more grain sites. Each ergot can be as much as 20 mm long. The same fungus is often found in the spikelets of many grasses, and these strains of the fungus are sometimes capable of attacking cereals. Grasses growing in headlands, and grass weeds, particularly black-grass (*Alopecurus myosuroides*), are a common source of infection.

Ergots contain toxic alkaloids derived from ergotine, many of which (although used in medicine) are capable of causing acute illness in animals and humans.

Because ergots are short lived (rarely surviving for more than 12 months), some control is achieved by crop rotation and by deep ploughing. Spraying with benzimidazole-related fungicides at, or just prior to, anthesis may be beneficial.

Eyespot (*Tapesia acuformis* – anamorph: *Ramulispora acuformis*, and *T. yallundae* – anamorph: *R. herpotrichoides*)
See under Wheat, p. 31.

Rhynchosporium leaf blotch (*Rhynchosporium secalis*)
This disease is common on rye and triticale, but rarely severe. For further details, see under Barley, p. 43.

Sharp eyespot (*Ceratobasidium cereale*)
See under Wheat, p. 34.

Take-all (*Gaeumannomyces graminis*)
Rye appears to be the most resistant of the cereals to take-all. For further details, see under Wheat, p. 35.

List of pests cited in the text*

Agriotes spp (Coleoptera: Elateridae)	click beetles
Chlorops pumilionis (Diptera: Chloropidae)	gout fly
Contarinia tritici (Diptera: Cecidomyiidae)	yellow wheat blossom midge
Delia coarctata (Diptera: Anthomyiidae)	wheat bulb fly
Deroceras reticulatum (Stylommatophora: Limacidae)	field slug
Ditylenchus dipsaci (Tylenchida: Tylenchidae)	stem nematode
Haplodiplosis marginata (Diptera: Cecidomyiidae)	saddle gall midge
Hepialis humuli (Lepidoptera: Hepialidae)	ghost swift moth
Hepialus lupulinus (Lepidoptera: Hepialidae)	garden swift moth
Heterodera avenae (Tylenchida: Heteroderidae)	cereal cyst nematode
Limothrips cerealium (Thysanoptera: Thripidae)	grain thrips
Limothrips denticornis (Thysanoptera: Thripidae)	barley thrips
Mesapamea secalis (Lepidoptera: Noctuidae)	common rustic moth
Metopolophium dirhodum (Hemiptera: Aphididae)	rose/grain aphid

Metopolophium festucae (Hemiptera: Aphididae)	fescue aphid
Nephrotoma spp. (Diptera: Tipulidae)	spotted crane flies
Opomyza florum (Diptera: Opomyzidae)	yellow cereal fly
Oscinella frit (Diptera: Chloropidae)	frit fly
Oulema melanopa (Coleoptera: Chrysomelidae)	cereal leaf beetle
Rhopalosiphum maidis (Hemiptera: Aphididae)	cereal leaf aphid
Rhopalosiphum padi (Hemiptera: Aphididae)	bird-cherry aphid
Sitobion avenae (Hemiptera: Aphididae)	grain aphid
Sitodiplosis mosellana (Diptera: Cecidomyiidae)	orange wheat blossom midge
Tipula oleracea (Diptera: Tipulidae)	a common crane fly
Tipula paludosa (Diptera: Tipulidae)	a common crane fly
Zabrus tenebrioides (Coleoptera: Carabidae)	cereal ground beetle

* The classification in parentheses refers to order and family.

List of pathogens/diseases (other than viruses) cited in the text*

Alternaria spp. (Hyphomycetes)	black point
Alternaria spp. (Hyphomycetes)	sooty moulds
Cephalosporium gramineum (Ascomycota)	leaf stripe of wheat
Ceratobasidium cereale (Basidiomycetes)	sharp eyespot
Cladosporium herbarum (Hyphomycetes)	sooty moulds
Cladosporium spp. (Hyphomycetes)	sooty moulds
Claviceps purpurea (Ascomycota)	ergot
Drechslera graminea (Hyphomycetes)	– anamorph of *Pyrenophora graminea*
Drechslera graminis (Hyphomycetes)	– anamorph of *Pyrenophora avenae*
Drechslera teres (Hyphomycetes)	– anamorph of *Pyrenophora teres*
Erysiphe graminis f. sp. *avenae* (Ascomycota)	powdery mildew of oats
Erysiphe graminis f. sp. *hordei* (Ascomycota)	powdery mildew of barley
Erysiphe graminis f. sp. *tritici* (Ascomycota)	powdery mildew of wheat
Fusarium culmorum (Hyphomycetes)	foot rot of wheat
Fusarium graminearum (Hyphomycetes)	foot rot of wheat
Fusarium spp. (Hyphomycetes)	brown foot rot and ear blight
Gaeumannomyces graminis (Ascomycota)	take-all of barley, rye and wheat
Gaeumannomyces graminis var. *avenae* (Asomycota)	take-all of oats
Leptosphaeria avenaria (Ascomycota)	speckled blotch of oats
Leptosphaeria nodorum (Ascomycota)	glume blotch of wheat
Microdochium nivale (Hyphomycetes)	– anamorph of *Monographella nivalis*
Monographella nivalis (Ascomycota)	seedling blight
Mycosphaerella graminicola (Ascomycota)	septoria tritici leaf spot
Pseudomonas syringe pv. *coronafaciens* (Gracilicutes: Proteobacteria)†	halo blight of oats
Puccinia coronata (Teliomycetes)	crown rust of oats
Puccinia hordei (Teliomycetes)	brown rust of barley
Puccinia recondita (Teliomycetes)	brown rust of wheat
Puccinia striiformis f. sp. *hordei* (Teliomycetes)	yellow rust of barley
Puccinia striiformis f. sp. *tritici* (Teliomycetes)	yellow rust of wheat
Pyrenophora avenae (Ascomycota)	leaf spot of oats
Pyrenophora graminea (Ascomycota)	leaf stripe of barley
Pyrenophora teres (Ascomycota)	net blotch of barley
Pythium spp. (Oomycetes)	brown root rot of wheat

Ramulispora acuformis (Hyphomycetes)	– anamorph of *Tapesia acuformis*
Ramulispora herpotrichoides (Hyphomycetes)	– anamorph of *Tapesia yallundae*
Rhizoctonia cerealis (Hyphomycetes)	– anamorph of *Ceratobasidium cereale*
Rhynchosporium secalis (Hyphomycetes)	leaf blotch of barley
Selenophoma donacis (Coelomycetes)	halo spot of barley
Septoria avenae (Coelomycetes)	– anamorph of *Leptosphaeria avenaria*
Septoria tritici (Coelomycetes)	– anamorph of *Mycosphaerella graminicola*
Stagonospora nodorum (Coelomycetes)	– anamorph of *Leptosphaeria nodorum*
Tapesia acuformis (Ascomycota)	'rye' type eyespot
Tapesia yallundae (Ascomycota)	'wheat' type eyespot
Tilletia caries (Ustomycetes)	bunt or stinking smut of wheat
Typhula incarnata (Basidiomycetes)	snow rot of barley
Ustilago avenae (Ustomycetes)	loose smut of oats
Ustilago hordei (Ustomycetes)	covered smut of barley
Ustilago hordei f. sp. *avenae* (Ustomycetes)	covered smut of oats
Ustilago nuda (Ustomycetes)	loose smut of barley and wheat

* For fungi, the classification in parentheses refers to class, although this is not possible within the phylum Ascomycota where classes have yet to be satisfactorily defined (see *Mycological Research*, February 2000). Some fungi have an asexual (anamorph) and a sexual (teleomorph) state, and the convention is to refer to them by their teleomorph name. However, where anamorph names are still in common use, these are listed and cross-referenced to the teleomorph name. Strictly, fungi classified as Coelomycetes and Hyphomycetes should be known as 'hyphomycetous anamorphs' and 'coelomycetous anamorphs' of the relevant teleomorph taxon (e.g. hyphomycetous anamorphic Sclerotiniaceae, for *Botrytis fabae*), respectively. These problems highlight the continual changes in the classification of the fungi.

† Bacteria – the classification in parentheses refers to division and class.

Chapter 3
Pests and Diseases of Oilseeds, Brassica Seed Crops and Field Beans

A. Lane
Independent Consultant, Church Aston, Shropshire

P. Gladders
ADAS Boxworth, Cambridgeshire

Introduction

Arable break crops are an important component of UK agriculture. Whilst they must be profitable in their own right, they also bring advantages for subsequent cereal crops through increased fertility, decreased disease pressure from take-all and trash-borne pathogens, and opportunities to control problem weeds in the rotation. Oilseed rape and field beans are well suited to the heavier soils. The performance of spring-sown break crops can be variable if dry conditions prevail in the spring or summer. Linseed shows drought tolerance and with EU support has become popular on a range of soils. In the UK it was grown on 99 500 ha in 1998. However, reducing financial support for linseed under Agenda 2000 is likely to see this crop decline over the next few years. Winter linseed has been grown on up to 20 000 ha per year since its introduction in 1996, but its full yield potential has yet to be realized, mainly because of pasmo disease (*Mycosphaerella linicola*) which became a problem soon after the crop was introduced.

Oilseed rape, linseed and field beans are the major break crops in the UK and were grown on 750 000 ha in 1998. Oilseed rape, at 530 000 ha, was the third most extensive crop after wheat (2 045 000 ha) and winter barley (771 800 ha). Oilseed rape is mainly winter-sown, although the area of spring-sown oilseed rape does vary and increases in years when autumn weather prevents drilling of the winter crop. Specialist cultivars of oilseed rape are also grown for industrial purposes on set-aside land. A range of minor oilseed crops are grown on small areas each year, including: borage, echium, evening primrose, linola, lupin, poppy, soya bean and sunflower. In the UK, pesticides recommended for use on oilseed rape may also be used on certain minor crops, including mustard, linseed, evening primrose, honesty, linola and flax. In addition, specific off-label approvals (SOLAs) may be available for some pesticides for use on minor oilseed crops but copies of the approval must be obtained before using the pesticide.

There are increasing costs to maintain or develop recommendations for pesticides on break crops, and in future there are likely to be fewer approved products available. In 1999, for example, approval was withdrawn for all seed treatments containing gamma-HCH and for some benzimidazole fungicides. The loss of seed treatments based on gamma-HCH reduces the options available for the control of flea beetles, especially flax flea beetles. As approvals are constantly changing, it is important, therefore, for users to remain up to date with current recommendations.

Control of pests and diseases in break crops is likely to become more important in future as these crops are traded at world prices. The protection of yield and quality, and hence profitability, will require improved targeting of inputs. Research over the last decade has provided new understanding about control requirements, forecasting and risk assessments. Economic thresholds and monitoring techniques are available, although the complexity of decision making would be overcome by the availability of decision support systems.

The expansion of oilseed rape has been accompanied by many changes in husbandry practices and cultivars. Winter oilseed rape is now the second most widely grown crop after winter wheat and most crops are double-low types (seed with low erucic acid and low glucosinolate content), the oil being suitable for human consumption and the meal (a valuable protein source) used in animal feeds. Other types are grown for industrial purposes on set-aside or for specialist niche markets. New cultivars of oilseed rape continue to be developed, including hybrids with higher yield potential. Improved resistance to diseases such as canker and light leaf spot is desirable to reduce the need for fungicides. Consistently higher yields will be required in future to ensure profitability at world prices. This can be achieved by plant breeding, combined with improved management of the crop and optimisation of pest and disease control.

Efforts to improve profitability by reducing establishment costs have seen renewed interest in sowing oilseed rape into standing wheat. This early sowing of winter oilseed rape has implications for pests such as aphids and cabbage root fly (*Delia radicum*), and the diseases dark leaf spot (*Alternaria brassicae*), light leaf spot (*Pyrenopeziza brassicae*) and powdery mildew (*Erysiphe cruciferarum*), which all appear to be more common in August sowings. Early sown crops can also be vulnerable to frost damage if they produce early stem-extension growth. However, the economic significance of these interactions remains to be established.

Good husbandry is still essential for the successful cultivation of rape-seed. A rotation with at least 4 years between successive crops is desirable, to reduce the risks of a build-up of soil- and debris-borne diseases. In recent years, much shorter rotations have been adopted on some farms, which has increased the risk of yield loss from diseases. Where attacks of sclerotinia stem rot (*Sclerotinia sclerotiorum*) have occurred in a field, the interval between susceptible crops (e.g. oilseed rape, peas, potatoes, linseed, spring beans and various vegetable crops) should be increased as far as is practicable. Ploughing or incorporation of oilseed rape stubble after harvest is also important, to reduce spread of diseases (particularly canker) to newly sown rape crops. To minimize the risk of spread of pests and diseases, crops should be grown as far as possible from the previous year's crop (although this might not be the most appropriate strategy for maximizing the effects of naturally occurring beneficial organisms – parasitoids, pathogens and predators).

The incidence of pests and diseases shows considerable variation from field to field and from season to season. Regular monitoring of crops is strongly advised during the autumn (for cabbage stem flea beetle and slugs, and for phoma leaf spot and light leaf spot), and during the flowering period (for pollen beetle and seed weevil, and for sclerotinia and alternaria). Information on the current pest and disease situation is available in ADAS Crop Action reports. These reports are also published in the agricultural press and are available via the Internet.

Although pests and diseases can be found in most crops, routine treatment with pesticides is not justified for economic and environmental reasons. Treatments should be applied according to manufacturers' recommendations when the appropriate action threshold for treatment has been reached. Self-propelled high clearance sprayers, or sprayers mounted on tractors with narrow wheels and belly-shields, should be used to minimize crop damage.

Oilseed rape is susceptible to a wide range of pests and diseases that also affect other brassica crops. Crops of ware or fodder brassicas grown for feed, for example, have many similar pest and disease problems. Surveys of Brussels sprouts during 1983–1985 provided circumstantial evidence for the spread of light leaf spot from oilseed rape to vegetable brassicas. The health and quality of seed produced by seed crops is of paramount importance and justifies greater expenditure on pesticides than would be needed on oilseed rape grown for crushing.

Field beans are currently grown on about 110 000 ha in the UK and are most popular in eastern England and in the Midlands. A major problem with field beans is the marked fluctuation in yield from year to year. Consistently high yields from both winter and spring field beans will be essential if these crops are to maintain their current status.

Oilseed crops – oilseed rape

Pests

Many invertebrates (insects, nematodes, slugs) can be found in or associated with oilseed rape crops. It is important to identify the potentially damaging ones and to decide whether they are likely to cause economic damage. Pests most likely to be important on oilseed rape in the UK are shown in Table 3.1. Most of these pests of oilseed rape can now be found wherever the crop is grown.

Crops should not be sprayed as a routine but only when careful examination of pest numbers on or in plants has shown that infestations have reached threshold levels at which it is considered economic to spray. Treatment thresholds have

Table 3.1 Pests of oilseed rape

	Winter rape	Spring rape
Stem borers		
cabbage stem flea beetle	+	□
cabbage stem weevil	–	(+)
rape winter stem weevil	(+)	□
Inflorescence/pod pests		
brassica pod midge	(+)	(+)
cabbage seed weevil	+	(+)
pollen beetle	(+)	+
Other pests		
aphid virus vectors	(+)	–
cabbage aphid	(+)	+
cabbage root fly	(+)	–
leaf miners	–	–
nematodes	(+)	–
slugs	+	(+)

\+ Often damaging.
(+) Occasionally or locally damaging.
– Present, but of little importance.
□ Not present.

been determined for several pests, based on damage assessment work. Where thresholds are not yet available, guidelines for treatment are given based on experience. Synthetic pyrethroid insecticides are commonly used to control a number of pests of oilseed rape. Even though these products are relatively inexpensive, they should be used only when pest thresholds are exceeded; **routine spraying is not advised**. Many of the pyrethroid insecticides have recommendations for the control of inflorescence pests, to be applied up to and during the flowering period. To avoid harming bees and other beneficial insects, sprays must be applied according to the product label. Crops in flower should never be sprayed with an insecticide, unless there is a specific recommendation on the product label to do so. In addition, insecticides applied during flowering must not be mixed with other pesticides which may render the insecticide toxic to bees.

Target plant populations for oilseed rape crops going into the winter are about $80/m^2$ for conventional cultivars and about $50/m^2$ for hybrids. Crops of hybrid types, therefore, with their lower plant populations, require extra monitoring for establishment pests, to ensure that plant losses are not excessive. Research is in progress to establish pest treatment thresholds for hybrid cultivars of oilseed rape.

Aphids

Cabbage aphid (*Brevicoryne brassicae*) and peach/potato aphid (*Myzus persicae*) are the only aphid species of any importance to be found in oilseed rape. They can affect crop growth when present in large numbers and can also transmit virus diseases. *M. persicae* usually infests crops in the autumn and early winter, rather than in spring and summer. Also, direct feeding damage is uncommon as this aphid tends to be present only in small colonies, distributed throughout the crop. It is the most important vector of beet western yellows virus (BWYV), which it introduces into oilseed rape in the autumn and, depending on temperatures, spreads throughout the winter months. BWYV infection, which is often symptomless, can be found in many crops, sometimes affecting large numbers of plants. Whilst yield losses from BWYV have been demonstrated, the relationship between aphid numbers, virus infection and crop loss is uncertain; benefits are most likely to follow prevention of early virus infection. Recent surveys have shown infection levels of BWYV to be generally low.

Autumn infestations of *B. brassicae* can affect crop establishment (especially when aphids invade crops soon after emergence), attacked plants being discoloured, stunted and distorted. This aphid will also transmit BWYV, but is more important as a vector of cauliflower mosaic virus (CaMV) and turnip mosaic virus (TuMV). These two viruses, although less common than BWYV, produce characteristic symptoms of infection (including leaf distortion and stunted growth), usually found in small patches throughout the crop. Yield losses from infected plants are likely to be high.

To reduce the risk of direct feeding damage and aphid-borne virus infection, crops should be sprayed in September or early October with a pyrethroid insecticide if aphids are easily found. Deltamethrin and lambda-cyhalothrin have specific recommendations for the control of aphid virus vectors. Pyrethroid insecticides applied in the autumn to control infestations of cabbage stem flea beetle (see p. 58) will give incidental control of aphids and so reduce virus infection. Autumn aphid infestations tend to be more of a problem on very-early-sown crops that are emerging as aphids are migrating to their winter hosts. Avoiding such early sowings will reduce the risk of aphid damage.

Infestations of *B. brassicae* in oilseed rape are more common during the summer months after a mild winter, especially on spring rape, the aphids often forming dense colonies at the top of the plant racemes and predominantly near the edge of the crop. Infested stems are very obvious, but usually only a small number of plants are affected. Control measures are rarely justified unless large colonies develop before pods are set; this may occur in hot, dry summers. Spring rape crops should be monitored from the early bud stage until early pod set (GS 3.5–5.5) and sprayed if more than 10% of plants are infested with obvious aphid colonies. Chemical treatment of winter rape is unlikely to be necessary in most years.

Pyrethroid insecticides applied during the bud/flowering stages to control pollen beetle and cabbage seed weevil will also suppress aphid numbers. However, pirimicarb (which has a specific recommendation for cabbage aphid control) is preferred to minimize impact on parasitoids of pollen beetle and seed weevil which are likely to be active in crops during this period.

Brassica pod midge (*Dasineura brassicae*)
A widespread pest, mainly of winter rape, and currently at low levels of infestation. Eggs are laid by the small delicate midge in pods which have already been damaged by the feeding and ovipositing of cabbage seed weevils and, occasionally, in pods weakened by pollen beetle larvae and fungal infection. Feeding inside, the midge larvae cause the pod to become yellow, swollen and distorted and eventually to split with premature shedding of the seed. Damage tends to be concentrated on the headlands of fields and in sheltered areas; serious crop losses in the UK are rare. As it is currently not possible to monitor for the adult midges, there is no recommended threshold; instead, previous history of the pest on the farm should be used as a guide to the need for control measures. Sprays applied to control cabbage seed weevil (see below) will generally reduce infestations of pod midge.

There are at least two generations of pod midge per year; when spring rape is grown in the same area as the winter crop, there is the possibility of a third generation and this may serve to increase the overall midge population on the farm. As pod midges are weak fliers, growing oilseed rape in a wide rotation, and avoiding the close cropping of spring and winter cultivars, may limit the build-up of midge populations.

Cabbage root fly (*Delia radicum*)
This is a widespread and serious pest of vegetable brassicas (see p. 190). Early sowings of winter rape, in which plants emerge in late August and early September, may also be vulnerable to attack. Spring-sown rape is rarely affected.

Eggs are laid by the flies near the emerging seedlings, and the maggots feed on the developing roots; attacked plants may be stunted and they wilt in dry conditions. Control is not usually necessary unless there is a history of damage; this is more likely to occur in traditional areas of vegetable-brassica production. As it is not possible to treat the crop once an attack is underway, a granular insecticide should be applied at or before sowing; both carbofuran and chlorpyrifos are recommended. Carbofuran is also likely to give some control of early aphid infestations, and reduce damage caused by both cabbage stem flea beetle and rape winter stem weevil.

Cabbage seed weevil (*Ceutorhynchus assimilis*)
This weevil is a widespread and important pest of winter oilseed rape and other brassica seed crops; damage to spring rape is uncommon. Adult weevils invade crops during flowering in May, and lay eggs in young pods. Each larva, typically one per pod, eats about a quarter of the seeds before leaving the pod and pupating in the soil. Adult weevils of the new generation emerge in early August; they sometimes damage vegetable brassicas by making punctures in cauliflower curds, cabbage leaves and Brussels sprout buttons. Egg-laying and feeding punctures made by seed weevils in oilseed rape pods provide points of entry for infestations of brassica pod midge (see above).

During the last few years, seed weevil damage to oilseed rape has been low and control measures rarely justified. However, populations of this pest fluctuate considerably between years, and between crops, even on the same farm. Therefore, crops should be monitored regularly (on at least two occasions) for adult weevils during the flowering period but sprayed only if treatment thresholds are exceeded. Crop monitoring is best done when conditions are optimal for maximum weevil activity, i.e. crop dry, with little wind and temperatures above 15°C. On farms where the often-associated brassica pod midge is not a problem, treatment should be applied at the threshold of two or more weevils per plant. Where pod midge causes regular damage, the threshold for seed weevil control is adjusted downwards, to one weevil per plant. The same thresholds apply to spring rape; however, as seed weevil infests these crops at an earlier growth stage, monitoring should be done from the green-bud stage onwards.

Pyrethroid insecticides are commonly used for seed weevil control in rape. They are applied to kill the adult weevils but are ineffective against larvae feeding inside the pods. For optimum control, sprays of alpha-cypermethrin, lambda-cyhalothrin or zeta-cypermethrin are recommended for application during the flowering period, when thresholds are exceeded. Pyrethroid sprays applied later are unlikely to be effective and may harm important parasitoids. Phosalone sprays are recommended during the later stages of flowering. Treatments applied to control seed weevil will also control pod midge. On spring rape, treatments applied to control pollen beetle (see below) will generally give control of seed weevil (when present). Recent surveys have shown that populations of the seed weevil parasitoid *Trichomalus perfectus*, which can exert considerable natural control, have begun to increase since UK crops are no longer sprayed with post-flowering organophosphorus insecticides. In some crops, levels of parasitism can exceed 70%. It is important, therefore, to encourage such parasitoids by ensuring that insecticides for seed weevil/pod midge control are applied only when absolutely necessary and never after flowering.

Cabbage stem flea beetle (*Psylliodes chrysocephala*)

This important pest affects the establishment of winter rape and other over-wintering brassicas. It is widely distributed but infestations tend to be most severe in East Anglia and in south-east England. Adult flea beetles infest winter rape soon after crop emergence (in late summer/early autumn) and feed on the young seedlings producing a characteristic shot-holing effect. Damage can be severe if the weather is dry and crop growth slow, but it is rarely necessary to apply insecticides to control adult feeding damage. It is the larvae, hatching from eggs laid in the soil and which invade plants from October to March (depending on soil temperatures), which cause most damage. They tunnel at first into the leaf stalk and then down into the plant stem, large numbers severely affecting plant growth and sometimes causing collapse of plants.

There are no thresholds to determine the need for control of adult flea beetles, although water traps can be used to monitor crop invasion and activity within the crop. With the recent revocation of the approval for seed treatments containing gamma-HCH, which formerly minimized adult flea beetle damage, control of the beetles relies totally on foliar sprays. Where flea beetle damage is severe and crop growth is slow, sprays of the pyrethroids alpha-cypermethrin, bifenthrin, cypermethrin, deltamethrin, lambda-cyhalothrin or zeta-cypermethrin are advised. These treatments will also give some control of aphid infestations and may reduce virus infection.

Treatment to control the flea beetle larvae, using the same insecticides, is recommended when an average of five or more larvae per plant is found in the autumn/early winter period in a well established crop. This threshold equates to 60% of leaf petioles showing feeding scars. Where the crop is backward and thin, a lower treatment threshold of three larvae per plant (equivalent to 30% leaf scarring) is advised. To reduce the risk of flea beetle attack, very early sowings (before mid-August) which can attract large infestations, should be avoided. Crops sown into a well-prepared seedbed will emerge and establish quickly and will be less susceptible to damage.

It is important to distinguish between larvae of cabbage stem flea beetle (these have a distinct head and legs) and larvae of leaf miners (see below), which have no obvious head and are apodous and of little importance.

Cabbage stem weevil (*Ceutorhynchus pallidactylus*)

Eggs of this pest are laid in leaf stalks and stems of oilseed rape and other brassicas. Larvae can be found in May and June, tunnelling within the stem, where they destroy the pith and facilitate stem colonization by canker (see p. 62). Infestations in spring rape are common, and can reduce plant vigour and yield. Larvae are frequently found in winter rape but infestations are not thought to be damaging under UK conditions. There are no established thresholds for this pest, but where it is a regular problem, spray spring rape with deltamethrin before flowering, usually at the green- to yellow-bud stage, to control the adult weevils. Pyrethroid insecticides applied at this time to control pollen beetle (see below) will give some incidental control of cabbage stem weevil.

Flea beetles (*Phyllotreta* spp.)

These small, black beetles are very common and may damage late-sown spring rape seedlings in hot, dry conditions. Since the recent revocation of the approvals for seed treatments containing gamma-HCH, a previously successful safeguard against this pest, the only control option is to spray when damage is seen. Sprays of lambda-cyhalothrin are recommended.

Leaf miners

Larvae of the cabbage leaf miner (*Phytomyza rufipes*) are commonly found in the leaf veins and stalks of oilseed rape plants in the autumn. Affected leaves turn yellow and fall off the plant. The loss of these leaves, usually the lower ones, is not important and control measures are not advised.

Another leaf miner (*Scaptomyza flava*) causes a conspicuous whitish blotch on the leaf, resembling nitrogen scorch. Damage is quite common but of no economic importance and control measures are not required.

N.B. It is important, however, to distinguish between larvae of life miners and those of cabbage stem flea beetle, as the latter can be damaging.

Nematodes

Several species can damage the roots of oilseed rape, causing patchy growth. Juveniles of cyst-forming nematodes, brassica cyst nematode (*Heterodera cruciferae*) and beet cyst nematode (*Heterodera schachtii*), invade and damage oilseed rape roots from early autumn. Infestations are likely to be most frequent in vegetable brassica growing areas and where beet cyst nematode predominates, i.e. East Anglia. Other than crop rotation, there are no specific recommendations for the control of cyst nematodes in oilseed rape. However, soil fumigants applied elsewhere in the rotation for the control of potato cyst nematode (PCN) (see p. 131) will reduce infestations of other cyst nematode species.

Patches of poor growth have also been associated with migratory nematode species, i.e. needle nematodes (*Longidorus* spp.) and stubby-root nematodes (*Trichodorus* spp. and *Paratrichodorus* spp.), which are known to cause similar damage (Docking disorder) to sugar beet grown on light sandy soils. No pesticide treatment is available.

Pollen beetle (*Meligethes aeneus*)

The very common shiny, greenish-black beetles are active from April to June. Adults biting into the flower buds of rape plants to feed and to oviposit, together with the feeding larvae, damage the flower buds; this results in blind stalks, in place of set pods, on the racemes. Slightly damaged buds may produce distorted and weakened pods, which may become more susceptible to attack by brassica pod midge (see above). Once crops are in flower, they are unlikely to be damaged by pollen beetle, even though large number of beetles may be found feeding in the open flowers. There is only one generation per year.

Winter rape crops are usually in flower and beyond the susceptible green- to yellow-bud stages (GS 3.3–3.7) before pollen beetles become active. Extensive bud damage, therefore, is unlikely unless the crop is very backward and poorly growing. In addition, it has been demonstrated that current winter rape cultivars have considerable ability to compensate for pollen beetle damage by setting more pods. The guideline for control in winter rape is therefore set high at 15 or more beetles per plant at the susceptible bud stages. Surveys have shown that infestations rarely approach this figure. Treatment of backward and poorly growing crops, which are less able to compensate for damage, is recommended at five or more beetles per plant.

Pollen beetle is a much more serious pest of spring rape (which is often at the rosette stage when first colonized), because the beetles are most numerous during the susceptible bud stages and spring-rape plants are less able to compensate for damage. Sprays are recommended at three or more beetles per plant between GS 3.3 and 3.7. A second spray may be necessary if infestations again reach this threshold before flowering commences. In Scotland, where infestations develop very early and where the growing season is shorter, treatment is recommended at a lower threshold of one beetle per plant during the early bud stages.

There is increasing evidence that where spring-sown varietal (cultivar) associations (composite hybrids) are grown, pollen beetles concentrate on the small percentage of pollinizer plants within the crop, so reducing pollen supplies to the sterile hybrid plants and thereby affecting seed set and yield. In this situation, it is suggested that crops be sprayed when pollen beetle infestations on the pollinizers average one per plant during the bud stages. Insecticides recommended for the control of pollen beetle are alpha-cypermethrin, cypermethrin, deltamethrin, lambda-cyhalothrin, phosalone, tau-fluvalinate and zeta-cypermethrin.

Rape winter stem weevil (*Ceutorhynchus picitarsis*)

Since the early 1980s, when this pest caused severe damage to crops of winter rape in northern and eastern counties, infestations have remained at low levels and have been restricted to only a few areas. Its life history is similar to that of cabbage stem flea beetle. The adult weevils invade crops from late September onwards but cause no obvious damage. Eggs are laid in the leaf stalks throughout the autumn and winter and the larvae, which have a distinct head but no legs, tunnel into the leaf stalks and eventually into the plant crown. The terminal bud may be destroyed and the attacked plant stunted or killed when large numbers of larvae are present.

Adults and small larvae are difficult to find in the crop and there is no threshold for treatment. As a guideline, treatment is advised when 10% of plants become infested with larvae during the late autumn/early winter period. Sprays of alpha-cypermethrin, bifenthrin, cypermethrin or deltamethrin are advised. Where the pest is a persistent problem, granules of carbofuran, applied at sowing, are recommended but such treatment may not be cost effective. It is likely that treatments applied specifically to control cabbage stem flea beetle larvae will give incidental control of rape winter stem weevil.

Slugs

On heavier soils, slugs, usually field slug (*Deroceras reticulatum*), may attack oilseed rape seedlings in the autumn, causing leaf shredding and loss of plants. In many situations, however, the crop will outgrow slug damage. Damage tends to be worst on crops emerging late, especially in a seedbed with a cloddy tilth. Severe damage to spring rape is uncommon, as crops usually establish much quicker than autumn-sown crops.

Slug damage to oilseed rape has increased in recent years, owing largely to a combination of factors – including the now widespread cropping of double-low cultivars which are inherently more susceptible to attack, more straw incorporation and frequent cropping of rape after set-aside, which encourages build-up of slug populations. Preparation of a firm and fine seedbed, together with early sowing to encourage rapid crop establishment, is recommended on slug-prone soils. Pellets of metaldehyde, methiocarb or thiodicarb should be applied before or at sowing if slugs are active on the soil surface. Treatment post-emergence may be necessary if conditions favour slug activity and severe leaf grazing is occurring to seedlings. Once crops are established and plants have at least two true leaves, slug attacks are rarely severe enough to justify treatment.

Diseases

Oilseed rape plants are often affected by several diseases, and fungicides may be required to control them at critical stages. In recent years, diseases have caused considerable loss of yield in many winter oilseed crops, despite the widespread use of fungicides. It is clear that many treatments were not being used effectively, and improved guidelines are being developed. One of the major problems to overcome is the marked regional and seasonal variation in the occurrence of diseases. Regular monitoring of crops is essential, particularly in the autumn. The use of cultivars with good disease resistance is beneficial and can be more cost-effective than using pesticides. Avoid planting crops adjacent to stubbles of the previous year's oilseed rape crops or large areas of volunteer plants. Early sowings are more prone to a range of diseases, including powdery mildew, alternaria leaf spot and light leaf spot. Sowing before 20 August, therefore, is not generally recommended.

Accurate identification of diseases, for which colour photographs are very useful, is essential. In making decisions to apply fungicides, a cost–benefit analysis should be made, to ensure that the estimated yield loss justifies not only the cost of the fungicide but also the cost of application. It is essential to minimize damage by spray application – this is often equivalent to 3% of yield for a spray application through a 12-m boom at the late flowering stage. Specialist high-clearance sprayers, or tractors fitted with belly-shields and dividers for the wheels, should be used for ground applications to minimize wheeling damage. A minimum application volume of 200–220 litres of water/ha is usually recommended for fungicide sprays.

Canker, light leaf spot and sclerotinia stem rot are the most important diseases in winter oilseed rape. Recent research has identified improved strategies to control these diseases, and fungicides can be cost-effective.

Canker (*Leptosphaeria maculans* – anamorph: *Phoma lingam*)

Canker remains a common disease of oilseed rape and has been an important cause of yield loss since 1977 and 1978, when high levels were found in susceptible cultivars. Most commercial cultivars now have moderate resistance to canker but severe infections still occur in southern, eastern and central England. There are usually fewer problems in northern England and very little canker is seen in Scotland. Stubbles of the previous year's oilseed rape crop are the main source of new canker infections. Airborne spores (ascospores) produced on infected stem and root debris are discharged in large numbers from September to April during rainfall and for a short period afterwards. The ascospores infect the cotyledons and leaves of young rape plants, producing leaf spots (5–15 mm diameter) which are beige to white and bear small, black, fruiting bodies (pycnidia). Leaf spots appear within a week at the optimum temperature of 20°C, but it may take over 4 weeks for leaf spots to appear at 3°C. Although the pycnidia produce numerous small spores, which are dispersed by rain splash, these appear to be of little importance under UK conditions. Leaf spots can be found from October onwards in winter crops and persist until extension growth and flowering (April/May). In recent years, the time of appearance of phoma leaf spot has varied between late September and December or January. Above-average rainfall in August and September is associated with early epidemics, and dry weather during this period delays the onset of leaf spotting. Early epidemics pose the main threat to yield, as they result in early cankers (at or before flowering) which are capable of causing premature ripening.

The canker fungus spreads through the leaf from the infection site and grows down the leaf stalk to the main stem, where canker lesions develop about 6 months after the autumn leaf spotting. Under optimum conditions, the canker fungus is able to grow down the petiole at a rate of 5 mm per day. This is a crucial factor, as fungicides appear to be able to control the fungus whilst it is within the leaf, but cannot eradicate it once it has reached the stem. The relationship between leaf spotting and canker is rather variable between crops and will be influenced by factors such as temperature, leaf size, duration of leaf retention and cultivar. Cankers are dark brown, slightly sunken areas at the base of the stems. They can be found from spring onwards but usually appear most commonly during June and July. Yield reductions of up to 1 t/ha occur when cankers girdle the stem, causing lodging and premature ripening. It is recognized that there are two types of the canker pathogen in the UK and these may be distinct species. Both types are capable of producing leaf spots, with the A group producing distinct canker lesions and the B group causing milder symptoms, such as blackening of the stem pith. The A group is dominant in most areas.

The fungus can affect all parts of the plant, producing beige to brown areas with pycnidia on lateral shoots, buds, flowers and pods. The fungus can also invade the seed, and its seed-borne phase is an important part of the disease cycle because it enables the pathogen to be introduced into new cropping areas and facilitates the spread of new strains of the fungus. Seed should be treated with iprodione where seed-borne infection is present. Fungicides applied to control seed-borne phoma will not protect seedlings after emergence.

In many areas, infected stubbles are the most important source of canker, which is best controlled by cultural methods. All rape debris should be chopped, and then buried by ploughing or cultivation as soon as possible after harvest. This will not give complete control but will reduce the risk of severe attacks. Some control can also be achieved by isolating newly sown crops, as far as possible, from the previous year's crop. Loss of yield is most likely to occur when oilseed rape is sown adjacent to unploughed rape stubbles. As canker can survive for at least seven years in root debris, a long rotation between brassica crops is also important for the control of this disease. Resistant cultivars should be grown whenever possible and, though these will not provide complete control of canker, they will enable the number of fungicide sprays to be reduced.

A wide range of fungicides will give some control of phoma leaf spot and canker, but good control requires treatment as soon as phoma leaf spots start to appear (10–20% plants affected) with a further application when new leaf spots appear. Sprays of difenoconazole or flusilazole + carbendazim are considered to have useful eradicant activity, whereas other products (such as prochloraz ± carbendazim or tebuconazole ± carbendazim) are most effective when used as protectant sprays. Treatments are generally recommended as a split-dose programme, applied in the autumn and spring, though this emphasis may change as current research suggests that spring infection has little effect on yield.

Clubroot (*Plasmodiophora brassicae*)
Oilseed rape is very susceptible to this serious soil-borne disease of brassicas. The continued expansion of oilseed rape production has brought the crop on to land heavily cropped with other brassicas, and a few severe attacks of clubroot have occurred in rape crops in both England and Scotland. The incidence of this disease can be minimized by maintaining a soil pH of 7.0–7.3 and by not growing brassicas more frequently than one year in five. Soil tests for clubroot are available, and may be appropriate before selecting new land for cropping if the detailed cropping history is not known. Where clubroot has occurred, brassica crops should not be grown for at least 8 years and land should be limed to reduce the risks of future problems.

Damping-off (*Pythium* spp. and *Thanatephorus cucumeris* – anamorph: *Rhizoctonia solani*) ***and seed decay***
The failure of seeds to germinate, and the death of seedlings, can seriously reduce emergence, particularly where there are poor seedbeds and soil conditions are cold and wet. Seed can be protected by thiram seed treatment.

Dark leaf and pod spot (*Alternaria brassicae* and *A. brassicicola*)
This was the most damaging disease of winter oilseed rape in the early 1980s, and some damaging attacks have occurred again in the south and east in recent years. Disease surveys have shown that the incidence of alternaria (mainly *A. brassicae*) increased with intensification of oilseed rape cultivation from 1976 to 1981. Alternaria causes black spots on leaves, which slowly enlarge to form circular brown spots with concentric light and dark zones (target spots). Loss of yield is most likely when similar symptoms develop on the pods. Infection of young pods causes loss of yield by reducing pod size but, more typically, the disease causes premature ripening of pods, which shatter prior to or during harvest.

Severe yield losses are most likely to occur with a combination of wet weather during flowering (May) and severe crop lodging. Given the dependence on weather, the incidence and severity of attacks is likely to show considerable seasonal and regional variation, and the most damaging attacks have been in southern England.

Fungicides are often justified when alternaria is present on the upper leaves during flowering. Many of the fungicides applied to control canker and light leaf spot also give useful control of alternaria during the winter and spring. Good control of pod infection and increases in yield have been obtained with iprodione applied at mid- to late-flowering. Many growers now prefer to use a broad-spectrum treatment at early flowering to mid-flowering, for the control of both sclerotinia and pod diseases.

Sprays of difenoconazole, iprodione, prochloraz, propiconazole or tebuconazole, applied alone or as mixtures with carbendazim (or with thiophanate methyliprodione) between mid-flowering (i.e. 20 pods on the main raceme at least 2.5 cm long) and the end of flowering, should give good control of pod spot. There can also be some reduction in alternaria following a spray of vinclozolin ± carbendazim. Treatments should be applied when alternaria can be found on the upper leaves or is just starting to affect the pods. Harvest intervals vary from 3 to 6 weeks, depending on the product, though many farmers now apply these treatments at early flowering to mid-flowering (primarily for control of sclerotinia) as this timing minimizes losses from wheeling damage. If there is a 'high' disease risk, apply a second spray at least 3 weeks after the first treatment (observing the 21-day harvest interval).

Affected debris and other brassicas, including strips of game cover, are important sources of alternaria in oilseed rape. The disease is commonly seed-borne, and effective control can be obtained with iprodione seed treatments as used for the control of canker (see p. 62).

Downy mildew (*Peronospora parasitica*)

This is the commonest disease of oilseed rape, especially in the autumn (on cotyledons and the first true leaves), and in the spring (when downy mildew builds up on lower leaves during the period of extension growth to early flowering – March to early May). It causes yellowing of the upper leaf surface, which bleaches with age and may become translucent after frost. The white spore-bearing structures of the fungus are produced mainly on the lower leaf surface. Such effects are often transient, and the disease declines when affected leaves drop off. It is difficult to predict where economic responses to fungicidal control will occur, and control measures are aimed at protecting vulnerable seedlings which might succumb to frost kill. For control, use protectant sprays of carbendazim + mancozeb, carbendazim + maneb, carbendazim + mancozeb + sulfur, chlorothalonil, chlorothalonil + metalaxyl and mancozeb (not more than two applications) applied in the autumn.

Grey mould (*Botryotinia fuckeliana* – anamorph: *Botrytis cinerea*)
This common fungus affects the leaves, stems, flowers and pods of oilseed rape plants. In the spring, infections are often secondary to damage caused by nitrogen fertilizer or frost. Crops drilled early in August may start to flower during the winter, when their shoots are vulnerable to frost damage and secondary fungal rots such as *Botrytis*. During flowering, particularly in wet weather, *Botrytis* commonly colonizes petals and pollen, adhering to leaves and bracts, and then invades the foliage to cause grey lesions. Fortunately, only a small proportion of these leaf infections spreads to the main stem and cause premature ripening. Grey mould can be found on bleached pods as the crop reaches maturity. This symptom is often, though not always, associated with damage caused by brassica pod midge (*Dasineura brassicae*) (see p. 57) or cabbage seed weevil (*Ceutorhynchus assimilis*) (see p. 57).

Specific control measures for *Botrytis* are considered worthwhile in some crops in Scotland but are rarely justified in England. Benomyl, chlorothalonil (pre-flowering only), iprodione (alone or with carbendazim or thiophanate methyl), prochloraz ± carbendazim and vinclozolin ± carbendazim carry label recommendations for control of *Botrytis*. In most cases, *Botrytis* will be a secondary target for these fungicides. In oilseed rape, strains of *Botrytis cinerea* which are resistant to benzimidazole fungicides may be present, and these would not be controlled by benomyl, carbendazim or thiophanate-methyl.

Light leaf spot (*Pyrenopeziza brassicae*)
In the mid 1970s, severe outbreaks of this disease were associated with the use of dalapon herbicide on susceptible cultivars. Typical symptoms are pale green or bleached areas on leaves, which slowly extend and coalesce, causing leaf death. Small, white spore droplets of the fungus, which resemble a spray deposit, can often be found around the edges of affected leaf tissue. Light leaf spot is the most important disease in winter oilseed rape in Scotland and northern England, and can cause significant loss of yield in other parts of England and in Wales. A regional forecasting scheme for England and Wales has been developed which estimates the proportion of crops at risk from light leaf spot in spring. The forecast takes account of the carryover of inoculum from one crop to the next and weather factors. The risk of light leaf spot is usually highest in northern England and lowest in eastern England, with intermediate severity in the west, south-west and south. The risk of yield loss from light leaf spot can be reduced considerably by growing resistant cultivars. In England, this may obviate the need for spraying altogether, whereas in Scotland some spraying may still be required.

Ascospores of light leaf spot produced on dead leaves and other infected residues, particularly stems, are thought to enable the pathogen to spread into new crops in autumn and winter. The duration of this dispersal phase may be limited, as fungicides applied during the late autumn or winter can provide long-lasting control; spring rape is rarely affected. Once established in the crop, further cycles of infection can take place via splash dispersal of the asexual spores (conidia). Symptoms are often first seen in January when small groups of plants ('foci') can be found in crops, though they are easily confused with frost damage or nitrogen fertilizer scorch. Further spread is favoured by wet weather, as the spores are dispersed by rain splash. All aerial parts of the plant (including stems, bracts, buds, flowers and pods) can be affected. Stem symptoms are superficial pink streaks, with fine black speckling at the edge of the lesions that become conspicuous prior to harvest when the leaves have abscised. Light leaf spot can also be seed-borne.

Severe attacks of light leaf spot can significantly reduce plant populations during the winter and reduce seed yields by 50%. Early treatment, preferably in the autumn, is needed to achieve effective control of severe infection. The fungus is well adapted to low temperatures and appears to cause more damage in cold winters.

A range of fungicides carry recommendations for control of light leaf spot. Resistance to benzimidazoles has been found in Scotland, and MBC products alone cannot be relied upon for control in Scotland. Azole fungicides show particularly good activity against light leaf spot; these include cyproconazole, difenoconazole*, flusilazole*, prochloraz*, propiconazole and tebuconazole* (*may be used with carbendazim). Benomyl, carbendazim, carbendazim + vinclozolin, carbendazim + maneb, carbendazim + maneb + sulfur, and iprodione (with carbendazim or thiophanate-methyl) are also approved. Sprays should be applied at the first signs of light leaf spot in the autumn or early winter, with a further spray if active disease is found in late winter or spring. A split-dose approach usually gives the most consistent results. At early stem extension in the spring, a scattering of infection throughout the crop is generally required to achieve a worthwhile yield response. If active disease is still present at flowering, a broad-spectrum treatment (to protect the pods) should be considered.

Powdery mildew (*Erysiphe cruciferarum*)

In the autumn, the characteristic fluffy-white colonies of powdery mildew can often be found on the undersurface of leaves of August drillings. Stem and pod symptoms can be very common in hot, dry summers. In winter oilseed rape the disease is not thought to be important but in spring oilseed rape, where the whole plant can be affected from flowering onwards, there could be some yield loss. The importance of this disease in oilseed rape has not been established. The risk of autumn infection can be reduced by drilling after mid-August and by destroying volunteer oilseed rape plants in the stubble of the previous rape crop.

There are no specific recommendations for control of powdery mildew, but several fungicides (e.g. azoles) applied for other diseases on oilseed rape are likely to give partial control.

Root rot (*Phytophthora megasperma*)

This soil-borne disease occasionally causes rotting of roots during winter and premature ripening of small patches of plants where there is soil compaction and impeded drainage. Attention to soil conditions and drainage, particularly on headlands, is the most effective way of avoiding this disease.

Stem rot (*Sclerotinia sclerotiorum*)
In the UK, sclerotinia has become more important and some severe outbreaks occur in England in most years. Particularly widespread problems occurred in 1991 and fungicides have been more widely used at flowering ever since. In parts of France and Germany, stem rot is a major problem in oilseed rape, especially where short rotations are practised. It is a soil-borne disease that is able to survive for at least eight years in soil by means of small, black, resting bodies called sclerotia (these measure 1–2 mm × 3–8 mm). Sclerotia near the soil surface produce small yellowish-brown fruiting bodies (apothecia) from March onwards, which produce airborne spores that spread locally within a crop and to adjacent fields. Infection is largely dependent on fallen petals that are carrying spores of sclerotinia sticking to leaves. The incidence of stem rot shows considerable seasonal variation and is favoured by above-average temperatures in April and May, and by showery weather during flowering.

The first symptoms of infection are pale brown blotches on the leaves. The fungus may spread from leaves to the stem, where it produces pure white, elongated areas from mid-May onwards. Stem symptoms can be distinguished from those of *Botrytis* (see p. 66) by the absence of grey mould on the lesion and by the presence of black sclerotia within the central cavity of the affected stem. Yield losses occur when stem lesions girdle the stem causing lodging, stem break and premature ripening. Yield from affected plants is about half that of healthy plants and 10–20% infection is required for fungicide sprays to be cost effective.

Oilseed rape should not be grown on land with a recent history of sclerotinia. The risks of damaging attacks are increased by short rotations of susceptible crops, including beans, carrots, celery, peas, potatoes, linseed and sunflowers. There should be at least 4 years between such crops. It is difficult to predict the level of sclerotinia attack, and risk assessment should consider individual farm and field histories, the presence of germinating sclerotia of *S. sclerotiorum* nearby and weather conditions. Farms with previous attacks on > 20% plants in oilseed rape are considered to be at high risk in most years. If germinated sclerotia have produced fruiting bodies (apothecia) during flowering of oilseed rape, and the weather is showery or unsettled, this will favour infection. Tests to determine the incidence of sclerotinia on petals have been used by ADAS to improve risk assessments. There is now interest in developing a rapid diagnostic test that would provide results for individual fields and guide decision making. Iprodione, prochloraz, tebuconazole and vinclozolin alone or as formulated mixtures with carbendazim (and iprodione + thiophanate-methyl) are approved for sclerotinia control. Rates of application for some products are varied according to risk. MBC fungicides have useful activity against sclerotinia and there are no reports of MBC-resistant strains in the UK. Treatments have little eradicant activity;

timing is very critical, and is usually optimal at early to mid-flower and coincides with early petal fall. Very occasionally, a second application may be needed.

Following an attack of sclerotinia, disease carryover can be reduced by deep ploughing (to bury the crop debris) and by using a succession of non-susceptible crops such as cereals or grass.

***Verticillium* wilt** (*Verticillium longisporum*)

This has not been recorded in oilseed rape in the UK, but it is present in various European countries, including France, Germany and Sweden. This pathogen has only recently been distinguished from *Verticillium dahliae* and it is primarily a pathogen of brassicaceous (cruciferous) plants. It is believed to be present in the UK (in vegetable brassica-growing areas) and vigilance is required in oilseed rape where typical symptoms are yellowing of one side of the leaf and the presence of grey microsclerotia in stem-base tissues. Crop rotation and general hygienic measures will be required to restrict spread of this disease once field outbreaks occur.

Virus diseases

Beet western yellows virus (BWYV) is the most common virus disease of oilseed rape and is often virtually symptomless. Beet western yellows virus is spread in the early autumn by aphids, particularly peach/potato aphid (*Myzus persicae*) (see p. 56). Its effect on yield in commercial crops has been assessed in experimental work, and early autumn infection has potential to reduce yield by 10%. Cauliflower mosaic virus (CaMV) is generally present at low levels, but occasionally it results in extensive patches of stunted growth and may occur together with the more severe turnip mosaic virus (TuMV). Both viruses are transmitted predominantly by cabbage aphid (*Brevicoryne brassicae*) (see p. 56). TuMV and the related broccoli necrotic yellows virus (BNYV) appear to be less widespread than CaMV. Typically, CaMV causes symptoms ranging from severe stunting to mosaic of the foliage, necrotic spots and streaks on the stems and pods, and distortion of pods and stems. These spots can resemble those caused by alternaria. TuMV causes similar symptoms or produces a lethal reaction in some cultivars.

White leaf spot (*Mycosphaerella capsellae* – anamorph: *Pseudocercosporella capsellae*)

Attacks occur regularly from autumn onwards in parts of southern England and, occasionally, in the Midlands and in the north. The leaf symptoms are initially small, irregular, brown and black spots which become paler as the lesions enlarge to form white circular spots, 10–20 mm diameter. On pods and stems, the symptoms are black spots which develop brownish centres as the lesions enlarge (similar to alternaria but distinguished by a dark reticulation within the brown pod-spot and a less well defined margin). This disease is of local importance and the crop may escape serious infection if the disease is not splash dispersed up the canopy during stem extension growth. Prochloraz ± carbendazim are approved for control of white leaf spot.

Oilseed crops – linseed and flax

Spring linseed has been widely grown for many years but winter linseed was introduced in 1996 and has been grown on up to 20 000 ha per year. In addition to linseed, there is a small area of flax (grown for its fibre) and also a small area of linola (grown for edible oil).

Pests

During the rapid expansion of the linseed crop area in the early 1990s, several pest problems became apparent, their importance increasing with the crop area. Since then the area grown has fluctuated according to economic trends and weather patterns, and now includes spring- and winter-sown cultivars, together with flax for fibre production and a small area grown for edible oil (linola). Treatment thresholds for most pests of linseed have yet to be established, so guidelines for treatment are given based on experience.

The current off-label arrangements for the use of pesticides in minor crops, including linseed, permit products approved for use on oilseed rape to be used on linseed and other minor oilseed crops at the grower's risk. However, insecticides classified as harmful or dangerous to **bees must not be used on linseed or any other crop during flowering**, including those products that may already have approval for use on rape in flower.

Capsids (e.g. common green capsid, *Lygocoris pabulinus*)
Capsids are an occasional problem in linseed, but are usually restricted to the headland where they are associated with hedgerow vegetation. The bugs feed on the leaves and flower buds of linseed, causing leaf distortion, delayed or reduced flowering, poor pod set and some loss in yield. They sometimes feed on the seed capsules, damaging the seeds within. Control is rarely necessary but where there is a history of damage on the farm, crops should be monitored before flowering and the **headlands only** sprayed if capsids are readily found. A pyrethroid insecticide, approved for summer use on oilseed rape, is recommended.

Flax flea beetles
The large flax flea beetle (*Aphthona euphorbiae*), together with the small flax flea beetle (*Longitarsus parvulus*), are the most damaging pests of spring-sown linseed. Both flea beetle species occur in varying proportions in different crops and are widely distributed. Winter crops are not affected.

After overwintering, the adult beetles migrate to linseed crops in the spring, feeding on the seedlings as they emerge to cause shot-holing on the cotyledons and young leaves. Damage is most severe when crops are attacked just below soil level as the seeds begin to germinate. Severe losses can result from such early attacks if infestations are heavy. Flea beetle larvae, arising from eggs laid around the plants, feed on the roots but are not thought to cause significant damage unless crops are under severe drought stress. The new generation of adult beetles emerges at the end of summer and may infest establishing winter linseed crops, but little damage, if any, is caused.

Since the revocation of approvals for seed treatments containing gamma-HCH, there are no cost-effective measures available to limit damage caused by very early flea beetle attacks. In areas of high risk, the only option is to incorporate gamma-HCH into the soil surface before sowing. Elsewhere, crops should be monitored during the early stages of establishment, and treatment applied where flea beetles are causing damage. A second spray may be necessary where attacks are sustained. A pyrethroid insecticide approved for summer use on oilseed rape is recommended.

Leatherjackets (e.g. larvae of *Tipula* spp.)
(See under Cereal pests, p. 37). Linseed sown in the spring after ploughed-out grassland or after a very weedy stubble may be attacked by leatherjackets. Most damage occurs in late spring, when emerging seedlings are eaten at or below ground level. Winter linseed is likely to be well established by this time, so serious crop damage is unlikely. In areas where surveys have indicated a high risk of damage, a spray of gamma-HCH, incorporated into the soil immediately before sowing, is recommended. For attacks in progress, methiocarb pellets, principally used for slug control (see below), will give some control of leatherjackets.

Slugs
On heavier soils, slugs (usually field slug, *Deroceras reticulatum*) can damage linseed crops, especially winter linseed in the autumn, affecting crop establishment. Damage to linseed, however, is usually not as serious as that to oilseed rape. Pellets of metaldehyde, methiocarb or thiodicarb are recommended for application before sowing when slugs are active on the soil surface and when damage is expected, or after crop emergence when slugs are feeding on establishing seedlings.

Thrips
The overwintering, wingless generation of field thrips (*Thrips angusticeps*) is common on light, stony soils, and can infest spring-sown crops such as linseed soon after germination; attacked plants becoming stunted and distorted. Sprays of alpha-cypermethrin, cypermethrin, deltamethrin or lambda-cyhalothrin can be applied at the first signs of damage in areas of high risk, although a high level of control is rarely achieved. Preparation of a good seedbed, together with early

sowing, will aid rapid crop establishment so that plants will outgrow thrips damage.

Members of the summer generation of *T. angusticeps* infests linseed crops before and during flowering, their feeding causing distortion of the leaves, flower buds and seed capsules. Crops should be monitored carefully before flowering for early signs of thrips feeding and, if necessary, sprayed with a pyrethroid insecticide recommended for pollen beetle control in oilseed rape (see p. 60). **Linseed crops must not be sprayed with an insecticide during flowering.**

Diseases

Alternaria blight (*Alternaria linicola*)

Alternaria linicola is a common seed-borne pathogen, which can reduce germination and establishment by killing seedlings. This problem is most serious when crops are sown in cold, wet soils. Two other *Alternaria* species (*A. alternata* and *A. infectoria*) also occur on linseed; they are less serious pathogens than *A. linicola*, and possibly merely saprophytes. *Alternaria linicola* also causes black leaf spots from emergence onwards and may be difficult to recognize where pasmo is also present. Warm, wet conditions favour rapid development of *A. linicola* and, whilst symptoms can be found on the lower leaves of seedlings, further activity may not be apparent until the capsules are formed. The incidence of alternaria, together with *Botryotinia*, is determined on seeds prior to certification and seed lots with > 5% infection are rejected. Disease control requirements, therefore, are based not only on yield but also on the potential to obtain a seed premium.

Control of seed-borne infection is particularly important and seed treatment with prochloraz is effective. Resistance to the fungicide iprodione has been confirmed in *A. linicola* in recent years and the effectiveness of an iprodione seed treatment may therefore be variable. Foliar sprays applied at mid-flowering have given economic yield responses in wet seasons and alternaria is part of the disease complex being controlled. Products approved for use on oilseed rape may be used on linseed under extrapolation arrangements for minor crops, and those with alternaria activity would be appropriate.

Grey mould (*Botryotinia fuckeliana* – anamorph: *Botrytis cinerea*)

Grey mould is ubiquitous but problems are particularly associated with wet conditions during flowering. Foliage, stems and capsules are attacked, infection being stimulated by the presence of petals. The association of large yield responses to fungicide sprays with wet seasons is attributable, in part, to control of *Botrytis*. Some reduction in *Botrytis* can be achieved with tebuconazole applied at flowering.

Pasmo (*Mycosphaerella linicola* – anamorph: *Septoria linicola*)

Although, historically, this disease has been regarded as a threat to linseed crops, it remained a minor disease until winter linseed was introduced into the UK. In June 1997, pasmo caused extensive leaf and stem infection after the end of flowering, which resulted in the appearance of large, brown patches in winter linseed crops. Typical symptoms of pasmo are grey or black, circular leaf spots containing small dark fruiting bodies (pycnidia) on the foliage and superficial, brown stem lesions. The first leaf symptoms can be found in autumn on cotyledons and it is thought that airborne ascospores spread infection from residues of the previous crops. The fungus can also be seed-borne. Fungicides used on oilseed rape have given large yield responses where pasmo was controlled (>1 t/ha) and treatments are advised at mid-flower, with an earlier spring treatment if the disease is well established on the lower leaves. There are no specific recommendations for pasmo, although a range of treatments have been investigated recently. The inclusion of carbendazim with tebuconazole (see powdery mildew) has given good results and other fungicides approved at flowering in oilseed rape (see under Oilseed rape – Stem rot, p. 68, and Dark leaf and pod spot, p. 64) can also be used.

Powdery mildew (*Sphaerotheca lini*)

Severe powdery mildew infection produces dense covering of the foliage but sparse colonies in the initial stages of the epidemic are difficult to identify. Typically, symptoms are first seen during flowering and the disease develops very rapidly to cover the entire plant. The spring linseed cv. Antares is particularly susceptible and winter linseed is also affected. Control of powdery mildew has been achieved with fungicides but this has not always given a worthwhile yield response. Tebuconazole has a specific recommendation for powdery mildew control. In many situations crops will be at risk from a range of diseases and selection of a broad-spectrum fungicide at mid-flowering will give some control of powdery mildew (see under Pasmo above).

Sclerotinia stem rot (*Sclerotinia sclerotiorum*)

This occurs occasionally in linseed but the crop appears to be much less susceptible than oilseed rape. Flowers naturally lose their petals after only one day and this may reduce risk of infection by petal-borne inoculum compared with oilseed rape.

Soil-borne diseases

Linseed and flax can be affected by a range of soil-borne pathogens (*Fusarium oxysporum* f. sp. *lini*, *Fusarium* spp., *Phoma exigua* var. *linicola*, *Pythium* spp. and *Thanatephorus cucumeris*) and these are associated with previous frequent cropping with linseed or flax. Badly affected crops show patches of poor growth, and laboratory diagnosis will often be required to identify the problem. Verticillium wilt (*Verticillium dahliae*) affects a range of crops, with a limited number of reports in linseed. Typical symptoms of early senescence, brown streaks and grey microsclerotia on the stem may be overlooked. This persistent soil-borne pathogen has a wide host range, including potatoes, and long rotations between susceptible crops may be the only practical means of control.

Brassica seed crops (excluding oilseed rape)

Crops included in this category are fodder and horticultural brassicas for seed production, and mustard crops grown for human consumption.

Pests

A succession of pests may attack brassica seed crops, from plant emergence to pod set, and most of these have been described in detail in the sections on pests of oilseed rape (see p. 54) and horticultural brassicas (see Chapter 7). Precise threshold levels for control have not been established for most pests of brassica seed crops so only broad guidelines can be given, based on experience. Control measures for many of the pests may be similar to those given for oilseed rape, but the product label should be checked to confirm that these recommendations are not specific to oilseed rape. Under the current off-label arrangements, pesticides approved for use on oilseed rape may be used on mustard, at the grower's risk. Brassica seed crops must not be sprayed with an insecticide once flowering has started.

Brassica pod midge (*Dasineura brassicae*)
(See p. 57). All brassica seed crops are susceptible to attack, except white mustard. Damage to brown mustard and other seed crops is quite common, especially around field headlands, but occurs at generally low levels. Effective control of cabbage seed weevil will also reduce pod midge damage and headland treatments alone may be adequate. Deltamethrin and phosalone have label recommendations for use on seed brassicas, including mustard.

Cabbage seed weevil (*Ceutorhynchus assimilis*)
(See p. 57). All spring-sown and overwintered brassica seed crops may be attacked, except white mustard (which is immune to attack by seed weevil larvae, although adults may be found on the crop). Chemical treatment of overwintered crops will not be necessary in most seasons. Brown mustard, and other susceptible spring-sown crops, should be sprayed if there are one or more weevils per plant by the yellow-bud stage. Phosalone has a label recommendation for use on seed brassicas.

Cabbage stem flea beetle (*Psylliodes chrysocephala*)
(See p. 58). This pest can be found on overwintered crops in many areas, but infestation levels are currently low. Treat as for oilseed rape.

Cabbage stem weevil (*Ceutorhynchus pallidactylus*)
(See p. 59). Although cabbage stem weevil attacks overwintered crops, it is usually a more serious pest on spring-sown brassicas, particularly those sown between late April and late May. Of the spring-sown crops, white mustard is usually less affected than brown mustard. One spray at the yellow-bud stage (but before the flowers open), with an insecticide recommended for cabbage seed weevil control in oilseed rape, is advised. Sprays applied at this time to control pollen beetle will also reduce infestations of stem weevil.

Flea beetles (*Phyllotreta* spp.)
(See p. 59).

Mustard beetle (*Phaedon cochleariae*)
Infestations by this shiny metallic-blue beetle have recently been at very low levels, as the area grown to mustard in the UK has declined. After hatching from eggs laid in late May, the larvae feed on the leaves, flower buds and developing pods, white mustard being more prone to damage than brown. There may be two generations per season. Sprays applied to control pollen beetle will also control mustard beetle, but must be applied before flowering.

Pollen beetle (*Meligethes aeneus*)
All brassica seed crops are attacked. On overwintered brassicas, the pest is usually more serious on crops with a long flowering period and less serious on crops that produce buds, flowers and pods quickly. Treatment is considered worth while if there are 15 or more pollen beetles per plant before flowering. On spring-sown crops (especially mustard), which are more vulnerable to pollen beetle attack, treatment is advised when there are three beetles per plant at early green-bud stage and five per plant at yellow bud. A two-spray programme may be necessary where attacks are prolonged. On both overwintered and spring-sown crops, deltamethrin and phosalone have label recommendations for pollen beetle control, to be applied before flowering. These treatments will also give control of cabbage seed weevil and brassica pod midge.

Rape winter stem weevil (*Ceutorhynchus picitarsis*)
This pest caused severe damage to swede and turnip seed crops at the time when infestations were first found in oilseed rape. Currently, infestations are localized and at very low levels. Treat as for oilseed rape (see p. 61).

Diseases

The common diseases of brassica seed crops include those described for oilseed rape and, in addition, diseases of horticultural brassicas – such as ringspot (*Mycosphaerella brassicicola*) (also occasionally found in oilseed rape) and white blister (*Albugo candida*) (see Chapter 7). There are few recommendations for the use of fungicides on brassica seed crops, and treatments used on oilseed rape are not necessarily applicable. There are no recommendations for applications of fungicide to mustard crops for human consumption.

Canker (*Leptosphaeria maculans*)
Severe stem infections are often recorded in seed crops of kale and other brassicas, causing lodging, premature ripening and loss of yield. As with oilseed rape, the prompt burial of debris after harvest, crop isolation, fungicide seed treatments with iprodione and a sound crop rotation are essential to reduce the risks of severe attacks (see p. 62).

Damping-off (*Pythium* spp. and *Thanatephorus cucumeris* – anamorph: *Rhizoctonia solani*) ***and seed decay***
These diseases could be particularly important where valuable seed stocks or seed of cultivars in short supply are sown at low seed rates and encounter difficult conditions at emergence. Seed treatment with iprodione, used for oilseed rape (see p. 64), is not available for other crops. However, there is a specific recommendation for thiram seed treatment on mustard.

Dark leaf and pod spot (*Alternaria brassicae* and *A. brassicicola*)
Both species of alternaria have caused serious reductions in yield in many seed crops and can cause complete crop loss. In the early 1980s, problems in brassica seed crops were linked to the build-up of *Alternaria* spp. in oilseed rape. The application of iprodione is recommended on brassica seed crops between mid-flowering and the end of flowering or when first pod symptoms are seen, with a further one or two treatments if there is a high disease risk. A maximum of three sprays may be applied to any one crop and the harvest interval is 21 days. These sprays should also give control of *Botrytis* (see p. 66) and sclerotinia (see p. 68) and also reduce the incidence of seed-borne alternaria in the harvested seed. The isolation of seed crops from other brassicas should also reduce the risk of severe attacks.

Field beans

A number of important pests and diseases of field beans can be transmitted by seed. Both bought-in and home-saved seed should be free from stem nematode and preferably also free from *Ascochyta fabae*. The priorities for pest and disease control differ between winter and spring field beans, and both crops are subject to marked seasonal variation in the severity of pest and disease attacks.

Pests

Bean beetle (*Bruchus rufimanus*)
Over recent years, a large number of autumn-sown (winter) and spring-sown field bean crops have been damaged by the larvae of the bean seed beetle. The damage

is characterized by a circular hole in the seed where the adult beetles have emerged; this may occur in the field or after harvest when the beans are in store. The most significant effect of the damage is to reduce the value of the crop for the human consumption export trade or for seed. However, studies have shown that the germination of attacked beans when used as seed is largely unaffected. Nevertheless, as bruchid damage is unsightly, infested stocks are often rejected for seed.

The adult beetles, black or dark brown in colour and with a characteristic hump-backed appearance, emerge from overwintering sites during late May/ early June and lay their eggs on the surface of developing pods. The emerging larvae bore through the pod wall into the developing seed. The larvae remain inside the seed throughout their development and pupate *in situ*.

Recent experimental work has shown that bruchid damage can be reduced by well-timed insecticide sprays applied in the field to kill the adult beetles before significant egg laying occurs. Crops should be carefully monitored during flowering and a spray of deltamethrin applied as the first pods begin to set, which is when beetles are active in the crop. Monitoring is best done when beetles are likely to be most active, usually when daytime temperatures reach 18–20°C. A second application is advised 7–10 days later. Good spray penetration into the crop is important for the insecticide to reach the lower pods. Sprays should be applied in the early evening to avoid direct contact with bees. Experimental work continues to evaluate monitoring systems and develop forecast spray dates.

Bean stem midge (*Resseliella* sp.)
Crops of winter beans in eastern England have occasionally suffered yield losses, following crop lodging that has occurred in the presence of stem infestations of the orange-red larvae of this pest and an associated fungus (*Fusarium*) (see p. 81). The midge larvae are the primary cause of damage, with the fungus infecting the midge feeding sites in wet weather, but there are no recommendations for chemical control.

Black bean aphid (*Aphis fabae*)
Populations fluctuate considerably from year to year. The aphid overwinters as eggs on spindle (*Euonymus europaeus*), and winged migrant aphids fly into bean crops during late May and early June. Annual forecasts of the probability of attack, based on the number of winter eggs on spindle bushes, are co-ordinated by Imperial College at Silwood Park. Further information is available during the growing season from suction traps operated by the Rothamsted Insect Survey. Data from trapping fine-tunes the treatment advice. In years of heavy infestation, spring-sown field beans can be damaged severely, with considerable loss of yield, but winter-sown crops are unlikely to be seriously affected.

In years when the forecast indicates that damaging attacks are *'probable'*, it may be cost-effective to apply a preventive treatment of disulfoton granules before flowering begins. Foliar-applied granular insecticides are more persistent than sprays and more effective as preventive treatments. When the forecasts indicate that attacks are *'possible'* or *'unlikely'*, a preventive treatment is advised only to those crops with more than 5% plants on the south-west headland infested at the beginning of flowering. If the infestation is detected early enough, a headland-only treatment may be sufficient.

Where preventive treatments have not been applied, eradicant treatments may be necessary when more than 10% of plants are infested with obvious aphid colonies that extend to the developing pods. As eradicant treatments are likely to be used when the crop is in flower, there is a risk of killing bees and losing their pollinating benefits. When flowering has commenced, only a pirimicarb-based insecticide should be considered, and this applied by high-clearance equipment with narrow wheels to minimize crop damage.

Green aphids

Vetch aphid (*Megoura viciae*) and pea aphid (*Acyrthosiphon pisum*) both colonize field beans. They do little direct damage but are important vectors of bean viruses (see p. 81). Treatments used to control black bean aphid will also control green aphids.

Pea & bean weevil (*Sitona lineatus*)

This very common, light-brown weevil attacks newly emerged peas and beans, and is particularly troublesome during early, dry, warm springs. Feeding damage by the adults is characterized by the appearance of U-shaped notches around the leaf margins. Economic damage is unlikely to result from adult feeding, except when the terminal shoots of late-sown or backward crops are attacked. Damage to root nodules by the weevil larvae may be important in spring beans and can result in yield loss. Spring-sown crops are more vulnerable than winter crops, as the latter are usually well established by the spring when weevils invade crops.

In areas where severe leaf notching is seen in most years, treatment may be justified. However, to be effective, insecticide sprays must be aimed at the adults as soon as leaf damage is seen and before eggs are laid. Early treatment is therefore essential, and to determine the timing and need for treatment the use of a recently developed monitoring system is recommended. The system is based on traps baited with a weevil chemical attractant (pheromone), which detects adult weevils emerging from hibernating sites. Research has shown that insecticide treatment is justified only when peak catches in the traps occur during the first emergence of the spring-sown crop. Sprays of cypermethrin, deltamethrin, lambda-cyhalothrin or zeta-cypermethrin are recommended.

Stem nematode (*Ditylenchus dipsaci*)

This widely distributed pest is abundant where host crops have been grown frequently. Many races of the nematode are known and those which affect field beans can also damage peas, oats, onions and strawberry. The nematode also breeds in many common weeds. Infested bean plants show reddish or blackened patches on the stem base, and attacked plants may be stunted and prone to lodging. Severe infestations can result in yield loss. Seed-borne infestations usually have little effect on the vigour of the first bean crop but they are important in introducing the nematode on to uninfested land and so affecting subsequent crops.

Seed for sowing should be saved only from uninfested crops. Samples are now routinely checked in the laboratory for nematode infestation. There are no recommendations for chemical control in the field.

Diseases

Chocolate spot (*Botryotinia fuckeliana* – anamorph: *Botrytis cinerea* and *Botrytis fabae*)

This is a very serious disease of field beans in wet seasons and can cause serious yield loss of autumn-sown crops. Spring beans are usually less severely affected than winter beans. The leaf symptoms are small, brown spots that coalesce and produce large, black blotches (the aggressive phase of the disease) if high humidity prevails.

Severe epidemics are favoured by prolonged wet weather, particularly in the spring and during flowering. Dense, forward crops can be affected severely during the autumn and winter, so crops are often drilled after the middle of October to avoid early infection.

The disease is carried over on debris of previous crops in adjoining fields, on volunteer plants and on seed. An inadequate supply of potash may aggravate the incidence of the disease.

Frequently, *Botrytis fabae* is the main cause of chocolate spot, but *Botrytis cinerea* also appears to be involved in many crops and can be isolated from leaves, flowers, pods and stems. Strains of both species resistant to the benzimidazole fungicides have been found in commercial crops. Therefore, sprays of these fungicides should be kept to a minimum, as multiple treatments lead to a build-up of resistant strains of the pathogens that cannot be controlled. Since the late 1970s, a mixture of chlorothalonil + benzimidazole fungicide, applied during flowering, has given good results. Recommendations for many benzimidazole products were withdrawn during 1999 and only benomyl is currently available. Chlorothalonil, cyproconazole + chlorothalonil, iprodione, tebuconazole and vinclozolin are also approved for chocolate spot control, and will often be used in mixtures at reduced rates. Fungicide protection is required during flowering and for up to 3 weeks after flowering. Spraying is advised when chocolate spot activity is detected on the lower leaves, to be repeated 2–3 weeks later if active disease is still present.

Damping-off (*Pythium* spp.) ***and seed decay***

Poor or uneven emergence can occur in crops sown into poor seedbeds that are cold and wet. Seed treatments are not required routinely and can be selected from thiram (primarily for damping-off), thiabendazole + thiram (for ascochyta control) or metalaxyl + thiabendazole + thiram (for ascochyta and downy mildew control), according to the perceived disease risks.

Downy mildew (*Peronospora viciae*)
Severe attacks have been more common in spring field beans than in autumn-sown crops. Typically, the leaves show large blotches that are initially pale green on the upper surface and have dense purplish-grey fungal growth on the underside. These lesions eventually turn necrotic. Where primary infection is soil-borne, severely affected plants can be found soon after emergence, sometimes in distinct patches. Spread of this disease is favoured by cool, showery or humid weather and the disease cycles rapidly at intervals of 10–14 days. The entire shoot tip and the pods may be affected. Control is difficult to achieve with fungicides as they give only a limited period of protection, especially when plant growth is rapid. Chlorothalonil + metalaxyl should be applied as a spray when the disease is first seen and a second application made 14 days later if necessary.

Leaf spot, stem rot and pod rot (*Didymella fabae* – anamorph: *Ascochyta fabae*)
This fungus produces black leaf spots that often have a grey centre with concentric zones of black pinhead-sized structures (pycnidia) within the leaf spot. Similar spots also occur on seed pods and stems, where they become sunken as the fungus penetrates the tissues. Serious yield losses are likely to occur when stem rotting is prevalent. It has now been recognized that there is a sexual stage of *A. fabae* that produces airborne spores on crop residues during autumn and winter. There is potential for spread of the disease between fields and this appears to explain the widespread infection seen in 1997/98. Previously seed-borne infection, and spread from volunteer plants, had been considered the main sources of the disease. It is still important not to plant seed infected with ascochyta, as seed treatments give limited control. Thresholds of 1% seed infection (2% with a seed treatment) have been standard for many years. Farmers can have their seed tested for the presence of ascochyta by the Official Seed Testing Station, Cambridge. Cultivars of winter beans with useful resistance to *Ascochyta* are now available. Seed treatment with benomyl, metalaxyl + thiabendazole + thiram or thiabendazole + thiram may be used to control seed-borne infection. In the growing crop there are no specific approvals, but sprays for control of chocolate spot may have some activity against leaf spot and pod rot.

Other leaf diseases
There are occasional reports of leaf spot (*Cercospora zonata*) and net blotch (*Pleospora herbarum*) in field beans, and these are often overlooked because symptoms are rather similar to chocolate spot. There are no chemical control measures for these diseases.

Root and stem rots (*Fusarium* spp., *Phytophthora megasperma* and *Thielaviopsis basicola*)
Severe root rotting can occur when beans are sown into compacted seedbeds which restrict root development. *Fusarium* spp. can be damaging in both winter and spring beans, especially when crops come under moisture stress. In eastern England, *Fusarium* spp. have been common on roots and stems of winter bean crops in some years and developed as secondary colonists of stem tissue damaged by larvae of the bean stem midge (*Resseliella* sp.), see p. 77.

A number of other soil-borne pathogens occasionally cause damage to field beans, including *Phytophthora megasperma* and *Thielaviopsis basicola*. There are no recommendations for chemical control of these diseases.

Rust (*Uromyces viciae-fabae*)
Infections usually occur late in the season in winter beans and generally have a limited effect on yield. On spring beans, rust appears at a much earlier stage of crop development and can cause significant (> 50%) loss of yield. The disease spreads by means of brown, powdery spores that are produced mainly in small pustules on leaves. Rust increases rapidly when temperatures are high and is the most important disease of spring beans. Destruction of haulm and volunteers (as recommended for *Ascockyta fabae*, p. 80) will reduce disease spread. Rust is probably worst where soil is low in potash; a minimum level of Index 2 for potash is recommended. Fungicides can give very good control of rust and reduced dose rates have been used successfully where rust is just starting to build up. Cyproconazole, fenpropimorph and tebuconazole are all very effective against rust, whilst chlorothalonil shows useful protectant activity.

Sclerotinia rot (*Sclerotinia sclerotiorum* and *S. trifoliorum*)
On spring beans the main pathogen appears to be *S. sclerotiorum*, which has a wider host range, whilst a specialized form of the clover rot fungus (*S. trifoliorum* var. *fabae*) is found mainly in winter beans. The first sign of the disease is the death of plants in spring, and this can affect large patches in some years. Typically, one or two plants collapse and adjacent plants die as the fungus spreads slowly from plant to plant down the row. Sclerotinia rots the stems and produces black resting bodies (sclerotia) on or in the plant tissue. The sclerotia can persist in soil for at least 8 years and severe attacks are usually associated with too frequent cropping with legumes or other susceptible crops. In addition to the soil-borne phase, there may also be some spread by airborne spores in the autumn (on winter beans) and in the spring (on spring beans). There are no recommendations for chemical control of this disease.

Virus diseases
Field beans can be affected by several viruses but broad bean true mosaic virus (BBTMV) and broad bean stain virus (BBSV) are usually the most important. They both cause leaf puckering and some leaf mottling. Early infection causes

stunting and failure to set pods. Infection before flowering has been shown to reduce the yield of individual plants by up to 80%. BBSV infection can also produce small, brown patches or bands on the seeds. Both viruses can be seed-borne and are transmitted within the crop by weevils such as *Apion* spp., especially bean flower weevil (*A. vorax*) and pea & bean weevil (*Sitona lineatus*). Whilst winter beans may be affected by virus they are less likely to suffer yield losses than spring beans.

Although chemicals are available that will control the weevil vectors of BBSV and BBTMV, there is little information on their efficiency and timing to prevent virus spread.

There are no specific tolerances for seed-borne viruses in field bean seed but growers should not use seed from crops known to be virus-infected.

List of pests cited in the text*

Acyrthosiphon pisum (Hemiptera: Aphididae)	pea aphid
Aphis fabae (Hemiptera: Aphididae)	black bean aphid
Aphthona euphorbiae (Coleoptera: Chrysomelidae)	large flax flea beetle
Apion spp. (Coleoptera: Apionidae)	flower weevils
Apion vorax (Coleoptera: Apionidae)	bean flower weevil
Brevicoryne brassicae (Hemiptera: Aphididae)	cabbage aphid
Bruchus rufimanus (Coleoptera: Bruchidae)	bean beetle
Ceutorhynchus assimilis (Coleoptera: Curculionidae)	cabbage seed weevil
Ceutorhynchus pallidactylus (Coleoptera: Curculionidae)	cabbage stem weevil
Ceutorhynchus picitarsis (Coleoptera: Curculionidae)	rape winter stem weevil
Dasineura brassicae (Diptera: Cecidomyiidae)	brassica pod midge
Delia radicum (Diptera: Anthomyiidae)	cabbage root fly
Deroceras reticulatum (Stylommatophora: Limacidae)	field slug
Ditylenchus dipsaci (Tylenchida: Tylenchidae)	stem nematode
Heterodera cruciferae (Tylenchida: Heteroderidae)	brassica cyst nematode
Heterodera schachtii (Tylenchida: Heteroderidae)	beet cyst nematode
Longidorus spp. (Dorylaimida: Longidoridae)	needle nematodes
Longitarsus parvulus (Coleoptera: Chrysomelidae)	small flax flea beetle
Lygocoris pabulinus (Hemiptera: Miridae)	common green capsid
Megoura viciae (Hemiptera: Aphididae)	vetch aphid
Meligethes aeneus (Coleoptera: Nitidulidae)	pollen beetle
Myzus persicae (Hemiptera: Aphididae)	peach/potato aphid
Paratrichodorus spp. (Dorylaimida: Trichodoridae)	stubby-root nematodes
Phaedon cochleariae (Coleoptera: Chrysomelidae)	mustard beetle
Phyllotreta spp. (Coleoptera: Chrysomelidae)	flea beetles
Phytomyza rufipes (Diptera: Agromyzidae)	cabbage leaf miner
Psylliodes chrysocephala (Coleoptera: Chrysomelidae)	cabbage stem flea beetle
Resseliella sp. (Diptera: Cecidomyiidae)	bean stem midge
Scaptomyza flava (Diptera: Drosophilidae)	a brassica leaf miner
Sitona lineatus (Coleoptera: Curculionidae)	pea & bean weevil
Thrips angusticeps (Thysanoptera: Thripidae)	field thrips
Tipula spp. (Diptera: Tipulidae)	crane flies
Trichodorus spp. (Dorylaimida: Trichodoridae)	stubby-root nematodes

* The classification in parentheses represents order and family.

List of pathogens/diseases (other than viruses) cited in the text*

Albugo candida (Oomycetes)	white blister of brassicas
Alternaria brassicae (Hyphomycetes)	dark leaf and pod spot of brassicas
Alternaria brassicicola (Hyphomycetes)	dark leaf and pod spot of brassicas
Alternaria linicola (Hyphomycetes)	alternaria blight of linseed
Ascochyta fabae (Coelomycetes)	– anamorph of *Didymella fabae*
Botryotinia fuckeliana (Ascomycota)	chocolate spot of beans, (common) grey mould
Botrytis cinerea (Hyphomycetes)	– anamorph of *Botryotinia fuckeliana*
Botrytis fabae (Hyphomycetes)	chocolate spot of beans
Cercospora zonata (Hyphomycetes)	leaf spot of beans
Didymella fabae (Ascomycota)	leaf and pod spot of beans
Erysiphe cruciferarum (Ascomycota)	powdery mildew of brassicas
Fusarium oxysporum f.sp. *lini* (Hyphomycetes)	fusarium wilt of linseed
Fusarium spp. (Hyphomycetes)	root and stem rot of beans and linseed
Leptosphaeria maculans (Ascomycota)	canker of brassicas
Mycosphaerella brassicicola (Ascomycota)	ringspot of brassicas
Mycosphaerella capsellae (Ascomycota)	white leaf spot of brassicas
Mycosphaerella linicola (Ascomycota)	pasmo disease of linseed
Peronospora parasitica (Oomycetes)	downy mildew of brassicas
Peronospora viciae (Oomycetes)	downy mildew of beans
Phoma exigua var. *linicola* (Coelomycetes)	canker of linseed
Phoma lingam (Coelomycetes)	– anamorph of *Leptosphaeria maculans*
Phytophthora megasperma (Oomycetes)	root rot of brassicas and beans
Plasmodiophora brassicae (Plasmodiophoromycetes)	clubroot of brassicas
Pleospora herbarum (Ascomycota)	net blotch of beans
Pseudocercosporella capsellae (Hyphomycetes)	– anamorph of *Mycosphaerella capsellae*
Pyrenopeziza brassicae (Ascomycota)	light leaf spot of brassicas
Pythium spp. (Oomycetes)	damping-off of seedlings
Rhizoctonia solani (Hyphomycetes)	– anamorph of *Thanatephorus cucumeris*
Sclerotinia sclerotiorum (Ascomycota)	stem rot of beans and oilseed rape
Sclerotinia trifoliorum (Ascomycota)	stem rot of beans
Sclerotinia trifoliorum var. *fabae* (Ascomycota)	stem rot of beans
Septoria linicola (Coelomycetes)	– anamorph of *Mycosphaerella linicola*
Sphaerotheca lini (Ascomycota)	powdery mildew of linseed
Thanatephorus cucumeris (Basidiomycetes)	wirestem of brassicas
Thielaviopsis basicola (Hyphomycetes)	black root rot of beans
Uromyces viciae-fabae (Teliomycetes)	rust of beans
Verticillium dahliae (Hyphomycetes)	wilt of linseed
Verticillium longisporum (Hyphomycetes)	wilt of oilseed rape

* For fungi, the classification in parentheses refers to class, although this is not possible within the phylum Ascomycota where classes have yet to be satisfactorily defined (see *Mycological Research*, February 2000). Oomycetes are now classified in Chromista with the brown algae, rather than as true fungi. Plasmodiophoromycetes are now classified as Protozoa rather than as true fungi. Some fungi have an asexual (anamorph) and a sexual (teleomorph) state, and the convention is to refer to them by their teleomorph name. However, where anamorph names are still in common use these are listed and cross-referenced to the teleomorph name. Strictly, fungi classified as Coelomycetes and Hyphomycetes should be known as 'hyphomycetous anamorphs' and 'coelomycetous anamorphs' of the relevant teleomorph taxon (e.g. hyphomycetous anamorphic Sclerotiniaceae, for *Botrytis fabae*), respectively. These problems highlight the continual changes in the classification of the fungi.

Chapter 4
Pests and Diseases of Forage and Amenity Grass and Fodder Crops

G.C. Lewis and R.O. Clements
Institute of Grassland and Environmental Research, North Wyke, Devon

Introduction

Grassland for agricultural use (excluding rough grazing) accounts for nearly 40% of the total agricultural area of 18.5 million ha in the UK. Fodder crops, including maize, whole-crop cereals, brassicas and beet, account for a further 3%. Despite the abundance of forage and fodder crops in the UK, their relatively low value compared with human food crops has resulted in little research on pest and disease control. For many of the pests and diseases included in this chapter, the damage they cause has not been quantified and the need for control measures has not been established adequately. Although chemical treatments are available for many of the pests and diseases cited below, in general such treatments have been developed for human food crops. The relatively low value of forage and fodder crops means that chemical control must be employed efficiently in order to be economic. Essential to the efficient use of chemical control is a correct diagnosis of the pest or disease present, or the damage that they cause, at as early a stage as possible. Once damage becomes easily visible, it is likely that the response to chemical control will not justify the cost. Therefore, an assessment of the risk of damage by particular pests or diseases is of great benefit. For such an assessment it is necessary to know the life-cycle and epidemiology of the pest or disease concerned, the factors that predispose a crop to damage, the pests and diseases prevalent in the local area, and past occurrences of pest and disease damage on individual farm fields. This information will allow the prediction of which pests or diseases are likely to be a problem and when – it is a case of 'forewarned is forearmed'. Another factor to be considered is that the status of some pests varies between the different forage or fodder crops, largely as a reflection of differences in plant population. For example, forage maize sown at 11 plants/m^2 is much more vulnerable to pests such as leatherjackets and wireworms than forage grass sown at 1000 or more plants/m^2.

Risk assessment charts have been compiled for the two major pests of grassland, frit fly (including grass & cereal flies) and leatherjackets (see Tables 4.1 and 4.2). These charts are valuable for employing the strategy of integrated pest and disease management, which combines cultural, biological and chemical methods. The risk of attack by pests or diseases can be reduced significantly by cultural

Table 4.1 Risk assessment chart for damage to forage grass by frit fly and by grass & cereal flies

Risk category	Rating*
Grass species	
Italian ryegrass	4
Perennial ryegrass	3
Others	0 (no significant risk)
Previous ground cover	
Grass (predominantly ryegrasses)	4
Grassy stubble	3
Cereals	2
Other	1
Locality	
Predominantly grassland	4
Mixed arable/grass	3
Predominantly arable	2
Entirely arable	1
Past history	
Damage noted previously on farm	3
Damage noted previously on neighbouring farms	2
No history of problem in area	1
Date of sowing	
Early August	2
Mid- to late August	4
Early September	3
Mid-September	2
Other times	1

* Cumulative score of 13 or more – chemical control advisable. Cumulative score of 8–12 – chemical control at low dose rate advisable. Cumulative score of <8 – treatment may not be necessary, unless damage is noted at the early stages of crop growth.

methods that take into account the life-cycle and epidemiology of the organisms concerned. Included in these cultural methods is the use of resistant cultivars. Unfortunately, research on resistant cultivars of forage and fodder crops has declined, and the information currently available to farmers and growers is limited. The biological methods used in integrated pest and disease management include predators or other antagonists of pests or pathogens. Therefore, it is important to consider the impact of control methods on non-target organisms. In some instances there may be benefits from chemical control of pests and diseases, but in others predators or other antagonists may be eliminated so that the application of a pesticide becomes counterproductive. Cultural and biological control methods alone may not be sufficient to eliminate a pest or disease but damage may be reduced to a level at which chemical treatment may become unnecessary.

The need for integrated pest and disease management is greatest when crops

Table 4.2 Risk assessment chart for damage to forage grass by leatherjackets

Risk category	Rating*
Crop	
Winter/spring cereals	3
Established grassland	3
Newly sown grassland	3
Fodder brassicas	1
Fodder root crops	1
Previous ground cover	
Established grass	3
Grassy stubble	3
Cereals	2
Others	1
Locality	
Predominantly grassland	3
Mixed grass/arable	2
Predominantly arable	1
Other	1
Past history	
Damage noted previously on farm	3
Damage noted previously on neighbouring farms	3
No history of problem in area	1
Weather in late summer/autumn	
Warm and damp	4
Cold and damp	2
Warm and dry	1
Cool and dry	1

* Cumulative score of more than 13 – routine chemical control advisable. Cumulative score of 8–12 – chemical control advisable if damage if noted in the early stages of crop growth. Cumulative score of < 8 – chemical control may not be required.

are grown 'organically', that is without inputs of inorganic agrochemicals. In recent years there has been a sharp increase in organic farming within the UK. The three main strategies employed are to provide optimum conditions for crop growth, to utilize crop rotations and to provide a diverse environment within and around crops. Plants that are under stress from incorrect or inappropriate management are more at risk from pests and diseases. This risk could be reduced by attention to soil structure, pH and fertilization, selection (if available) of high-quality seed of cultivars that show resistance to those pests and diseases that are likely to be prevalent, and use of appropriate sowing techniques and timing to encourage robust plant growth. Rotation of arable crops reduces the build-up of pests and diseases specific to particular crops. One feature of forage and fodder crops is that they are often kept for more than one year before ploughing, and sometimes for many years. This situation allows the build-up of pests and diseases; established grassland, for example, probably carries a wider range of pests and diseases than any other arable crop. To reduce the risk to new sowings, these crops should not be sited in close proximity to old ones and newly established crops should be harvested before old ones. Finally, diversification of habitat on the farm will promote the establishment of populations of predators and parasitoids of pests. All of these strategies depend upon the knowledge of how a particular pest or disease is affected by biotic and abiotic factors.

The strategies described above for control of pests and diseases can also be applied to amenity grass. The botanical composition of amenity grass is little different from agricultural grassland; it is the management that sets them apart. Frequent mowing (a feature of amenity grass) eliminates some foliar pests and diseases that are a problem in agricultural grass but creates a favourable environment for other diseases.

Forage grasses

About 80% of grassland (excluding rough grazing) in the UK is five or more years old. This indicates the long-term nature of grassland. Of the 20% of grassland less than five years old, about one quarter is newly sown. Newly sown grassland is more vulnerable to pests and diseases than established grassland, because of the relatively small amount of plant tissue present at the seedling stage; also, sowings following previous grassland are particularly prone to attack because of the carryover of pests and diseases. During the establishment phase, pests often cause large reductions in plant stand and, consequently, large reductions in subsequent herbage yield. Bare patches caused by pests allow the ingress of weeds and accelerate deleterious changes in sward botanical composition. In the case of established grassland, the impact of pests and diseases is likely to be greater, and control measures economic, under intensive rather than extensive management. Therefore, although established grassland occupies a substantial area of agricultural land in the UK, only a small proportion is likely to be subject to pest and disease control.

In forage grasses, losses from pests and diseases often go unnoticed because, in farming practice, it is difficult to compare the yields of damaged and undamaged crops. In addition, reductions in yield, persistency or forage quality have been quantified for only a very few pests and diseases. The output of forage grasses is difficult to assess because it is mostly measured in terms of animal production rather than in association with the crop itself. Damage by pests and diseases to grassland is often not obvious. However, significant losses in yield were demonstrated in extensive work during the 1970s and 1980s, involving researchers from IGER, IACR-Rothamsted and ADAS. The application of pesticides consistently increased annual dry matter yield of lowland grassland by 1.0–1.2 t/ha. Responses to pesticide in upland grassland were slight and inconsistent, which was attributed to low populations of pests and/or susceptible grasses. However, severe, localized pest damage does occur sporadically in upland grassland.

Most chemicals available for use on forage grasses have restrictions on the type of grassland that can be treated and/or the number and timing of treatments. It is essential, therefore, before application, for potential users to consult product labels. Management has a large impact on pests and diseases in grassland, particularly defoliation by cutting or grazing; this often removes much of the pest or disease presence on the herbage, obviating the need for specific control measures.

Grass and fodder crops grown for seed production may be affected by pests and diseases that attack the inflorescence or developing seed. These crops have a higher cash value than forage crops. Also, foliar diseases of grasses can be more severe in seed crops because cutting of such crops is infrequent.

Pests

Antler moth (*Cerapteryx graminis*)

The larvae (caterpillars) of antler moth are brown with a pale stripe and up to 40 mm long. The adult moth lays eggs in the autumn, most of which hatch in the following spring. The caterpillars then feed at or below soil level until midsummer. Occasionally the caterpillars reach epidemic proportions in upland regions, when large areas of grassland can be destroyed. During such epidemics, the caterpillars sometimes migrate in large groups. Damage is more common in the northern regions of Britain.

No chemical control is available in the UK. Digging a ditch in front of the advancing caterpillars has been suggested as a means of 'control'.

Aphids

Several species occur in forage grasses, including bird-cherry aphid (*Rhopalosiphum padi*), fescue aphid (*Metopolophium festucae*) and grain aphid (*Sitobion avenae*). These aphids vary in colour from shades of green to brown. There are winged and wingless forms, and it is the latter that disperse within and between crops. The aphids feed on the sap of all the important forage grasses, causing patches of stunted growth and transferring viruses from infected to healthy plants – *R. padi* is the main vector for barley yellow dwarf virus (see p. 96). Aphids multiply rapidly in warm, dry weather and are more likely to be important in seed crops in periods of prolonged dry weather following mild winters.

Chemical control can be obtained by the application of sprays of pirimicarb or, for seed crops only, deltamethrin (off-label) (SOLA 1693/96) or dimethoate. Dimethoate should be applied when plants are growing well, to get the best systemic activity. Cultural control can be obtained by cutting and removing aphid-infested herbage.

Bibionid flies

The brown larvae of bibionid flies (especially those of fever fly, *Dilophus febrilis*, and of St. Mark's fly, *Bibio marci*) are sometimes mistaken for small leatherjackets, but the difference is that bibionids have a distinct dark head. They can occur in very large numbers in grassland, but only occasionally and then in localized areas. The larvae feed on dead organic matter and living plants. Although bibionids may feed on the roots of grasses, the main damage appears to be loosening of the roots, particularly in poorly compacted soils, leading to poor growth and increased susceptibility to 'winter kill'. Heavy applications of farmyard manure or slurry will promote large populations of bibionids.

No chemical control is available in the UK. However, when soil conditions allow, the use of a heavy roller will squash the larvae and resettle loosened grass. The risk of damage can be reduced by improving rooting depth and avoiding the presence of excess organic matter.

Chafer grubs

These pests (larvae of chafers) often occur in grassland. Cockchafer (*Melolontha melolontha*) and garden chafer (*Phyllopertha horticola*) are the most commonly found species. The larvae are up to 45 mm long and have a characteristic curved shape. They are pale in colour with brown heads, and there are three pairs of legs. Adults emerge from the soil in May/June and lay eggs; the resulting larvae feed on the roots of grasses until late autumn. Populations in grassland can reach 70 per m^2, and damage can be so severe that large areas of grass can be easily lifted out of the soil. Damage is often accentuated by large numbers of birds searching for the larvae. Damage is more likely on light land and in sheltered upland areas. In dry weather, affected patches quickly turn brown because of the root damage. Fine-leaved grasses are often preferentially attacked, whereas cocksfoot and perennial ryegrass show some degree of resistance.

No chemical control is available in the UK. The only recourse when damage is severe is to cultivate the grass and re-seed.

Common leaf weevil (*Phyllobius pyri*)

The white, apodous larvae occasionally occur in large numbers and damage the roots of perennial ryegrass and *Festuca* spp. Grassland sited near woodland is particularly at risk. Feeding by adult weevils on the foliage is of little consequence.

No chemical control is available in the UK. Repeated rolling of the sward when damage is first recognised can give some control and will reduce further damage.

Frit fly (*Oscinella frit*)

Larvae of this species (see Chapter 2, p. 44), along with those of the close relative *Oscinella vastator* and of various grass & cereal flies (genera *Geomyza*, *Meromyza* and *Opomyza*), occur in forage grasses, although the eponymous species *O. frit* (an important cereal pest) is relatively rare in grasses. *O. vastator* is the most common in ryegrasses. Adult oscinellid flies are small and insignificant and the females lay their eggs on or near seedlings or mature grass plants. The resulting larvae, which are up to 3 mm long, bore into the base of seedlings and tillers, causing death or greatly reduced vigour. Populations of larvae in new sowings can reach several thousand per m^2 and exceed the number of seedlings present. The risk of damage to established grass is increased where swards are grazed rather than cut for silage. Seedlings of Italian ryegrass are more susceptible to damage than those of perennial ryegrass and, although established crops are less affected, the relatively low tiller population of the former can result in significant damage. If grass is sown after grass, established seedlings will be attacked by larvae migrating out of the old sward, in addition to those hatching from eggs laid by incoming adult flies. The shorter the interval between ploughing and sowing, the greater the survival rate of larvae. Direct drilling poses the greatest risk, because a larger proportion of larvae can migrate from the old sward killed by herbicide than can surface from plough depth in a conventional seedbed. A gap of 4 weeks or more between ploughing or sward destruction and sowing will minimize larval migration to seedlings. The populations of adult oscinellid flies, and other grass & cereal flies, are likely to be greater if the sowing is in a grassland rather than arable area. Finally, oscinellid flies, and other grass & cereal flies, lay eggs at certain times of the year and seedlings emerging at these times are at particular risk of attack. Damage to new sowings is particularly prevalent in autumn sowings.

Chemical control can be obtained by the application of sprays of chlorpyrifos before, at or soon after seedling emergence, or cypermethrin soon after emergence. Chlorpyrifos can reduce populations of carabid beetles, which are predators of many pest species. A risk-assessment chart has been devised (see Table 4.1, p. 85), incorporating the various factors that influence the risk of serious damage. This chart enables an integrated pest management programme to be operated.

Garden grass veneer moth (*Chrysoteuchia culmella*)

Larvae of this generally abundant pest often cause damage in permanent grassland, severing the plants at or below ground level. In serious cases, areas of dead or dying grass may extend over several hectares. This insect often occurs in company with various other grassland pests, including chafer grubs and leatherjackets.

No specific chemical treatment is available but spraying against leatherjackets (see below) may have some beneficial effect on this pest. Repeated rolling of the sward when damage is first recognized can also give some control.

Leatherjackets

Leatherjackets (larvae of crane flies, e.g. *Tipula paludosa*) are brownish-grey larvae up to 40 mm in length, with no legs and an indistinct head. The larvae are soil dwelling, and in grassland populations can reach several hundred per m^2. Adult crane flies lay eggs in autumn and the resulting larvae feed through the winter. Damage is greatest in spring, when the larvae are at maximum size and are feeding actively. Plants are severed at or just below soil level, causing patches of yellowing plants that later die. Severe damage can be caused to both newly sown and established grassland. Lighter attacks, without obvious bare patches, may delay the first flush of vegetative growth in spring, which is important where 'early bite' is required for livestock. One indication of a severe infestation is the presence of flocks of starlings or other birds probing the soil for the larvae.

Chemical control can be obtained by the application of sprays of chlorpyrifos or gamma-HCH, the latter being applied to the seedbed prior to sowing. In established grassland, a linear relationship has been reported between response to chlorpyrifos (applied in early winter or spring) and leatherjacket numbers, with a maximum response in excess of 1 t DM/ha. Cultural control can be obtained by sowing in July or early August to avoid the main egg-laying period. In addition, leatherjacket activity can be restricted by rolling after sowing, to ensure good consolidation of the soil. The threshold level for economic damage to established grassland has been estimated at 1 million leatherjackets/ha. ADAS provide a service for assessing populations in individual fields. Alternatively, an estimate of the populations of leatherjackets present can be made by inserting sections of plastic drainpipe into the soil and filling the pipe with a brine solution. The leatherjackets in the soil float to the surface of the brine, where they can be counted. By consulting a chart, the number of leatherjackets found is converted to the population per hectare and if this exceeds the above-mentioned threshold chemical control is recommended. A risk assessment chart has been devised (see Table 4.2, p. 86), incorporating the various factors that influence the risk of serious damage.

Red-legged earth mite (*Penthaleus major*)

This minute pest has a purple body and eight bright-red legs. Attacks occur only infrequently and then usually in eastern England. Damage appears as extensive silvering and senescence of foliage, with patches of thin growth in late autumn, particularly on older grassland.

No chemical control is available in the UK.

Slugs

Slugs, especially field slug (*Deroceras reticulatum*), damage forage grasses. The field slug is up to 50 mm long, feeds above ground, and is active at low temperatures. Damage is most severe on heavy soils that are high in organic matter, and in wet seasons. Direct-drilled grass is particularly at risk because the minimal cultivation has little effect on the resident slug population and the slits provide an ideal protective habitat. Damage to established grass is inconsequential.

Chemical control can be obtained by the application of granules of aluminium sulfate or, on ryegrass leys only, methiocarb. Methiocarb also controls leatherjackets. The need for chemical control of slugs can be assessed by placing small quantities of granules, covered by a tile, at intervals over the field. If dead slugs are detected at most sites, chemical control may be necessary. An alternative method is to place squares of insulated plastic sheets at intervals over the field. Slugs are attracted to the warm, moist conditions under the sheets; if they are found under most sheets, chemical control may be necessary. A fine, firm seedbed, free of residues from the previous crop, will reduce crevices in which slugs can hide.

Swift moths (*Hepialus* spp.)
The white caterpillars of garden swift moth (*Hepialus lupulinus*) and ghost swift moth (*Hepialus humuli*) (up to 35 mm and 50 mm long, respectively) feed on the roots of grasses, occasionally damaging newly sown and established grass. The adult moths lay eggs during the summer and the resulting caterpillars feed through the winter and the following spring. In the case of ghost swift moth, the caterpillars usually continue to feed for a second year before pupating.

No chemical control is available in the UK. However, cultivation in preparation for sowing will reduce populations of the larvae.

Wireworms
Wireworms are the larvae of click beetles, of which many species (e.g. *Agriotes lineatus*, *A. obscurus*, *A. sputator*, *Athous haemorrhoidalis* and *Ctenicera* spp.) occur in grassland. The larvae are a shiny, golden-brown colour and up to 25 mm in length. They have a life-cycle of 3–5 years and therefore are most numerous in long-established grassland where the lack of soil disturbance allows their numbers to increase. Wireworms feed on grass plants mainly in spring and autumn. Generally, they cause little damage, but young plants may be destroyed in reseeded grassland.

Chemical control of wireworms can be obtained by spraying with gamma-HCH. The chemical should be applied to the seedbed and then incorporated into the soil. In contrast to frit fly and grass & cereal fly larvae (i.e. stem-boring dipterous larvae) (see p. 89), wireworms cause less damage if grass is sown soon after ploughing, so insecticide application may be of more benefit when there is a delay between ploughing and sowing.

Diseases

Breeding for resistance to diseases in ryegrass in the UK has concentrated on the foliar fungal diseases and indications of resistance in cultivars are available. In the case of crown rust, a continued effort in breeding is required to counter the appearance of new strains of the fungus able to break down plant resistance. Management of the crop can have a large influence on disease levels and can be manipulated to reduce the effects of disease. Defoliation removes much of the disease inoculum and reduces the moist microclimate within a tall standing crop that favours disease spread. Manipulation of sward defoliation and fertilization can reduce the risk and extent of damage caused by foliar diseases.

Cultivars with a susceptibility to a particular disease should be avoided in areas where that disease is prevalent. Growing susceptible cultivars in mixture with resistant ones is another approach.

Drechslera leaf spot (*Drechslera andersenii, D. festucae, D. phlei, Pyrenophora dictyoides* – anamorph: *Drechslera dictyoides*, and *P. lolii* – anamorph: *Drechslera siccans*)
This disease is widespread and common throughout the UK, and can occur at all times of year. Five species of *Drechslera* have been reported on ryegrass in England and Wales, of which *D. andersenii* and *D. siccans* are the commonest. *D. dictyoides* causes a net blotch on the leaves of meadow fescue, *D. festucae* produces large, chocolate brown spots on tall fescue, and *D. phlei* causes extensive leaf streaking on cocksfoot and on timothy. Drechslera leaf spot is typically a disease of high incidence but low severity, although the spots vary in size and frequency depending on the species of host and fungus, and on environmental conditions. Infection can reduce forage quality. In ryegrass, losses of yield and quality can be significant, even with low levels of infection. Infection can be seed-borne.

Chemical control can be obtained by the application of sprays of propiconazole, but only on crops grown for silage or for seed. For silage crops, one application is permitted each year, before any cut. For seed crops, two applications are permitted each year, one in spring and one in autumn.

Ergot (*Claviceps purpurea*)
This is the most widespread and important of the diseases that attack the inflorescence of grasses, and is particularly important in ryegrasses. The fungus has a very wide host range within the Poaceae, although several strains exist. The importance of infection is not in its effect on the plant, but in the production of black fungal sclerotia (ergots) that contain alkaloids toxic to mammals. When ingested by cattle or sheep, the ergots can cause gangrene, lameness and abortion. The disease is most common in old pastures. Ergots overwinter on the soil surface and germinate in the early summer of the following year. The spores that are produced are carried by wind to infect the inflorescence. The infection prevents the formation of seed, and ergots are produced instead.

No chemical control is available in the UK. Cultural control can be obtained by defoliation to prevent the formation of significant numbers of grass seed heads, thus minimizing the risk of ergot poisoning. When crops known to have been contaminated with ergots are re-seeded, deep ploughing will bury the ergots and prevent them from producing fruiting bodies above the ground. In addition, a rotation of 2–3 years between susceptible crops will reduce infection from buried ergots.

Pink snow mould (*Monographella nivalis* – anamorph: *Microdochium nivale*)
This disease is of importance in regions where there is prolonged snow cover in winter. The disease may be present before snow cover develops, but severe

damage occurs under the snow. After snow melt, patches of bleached, water-soaked leaves are visible, often covered with a pinkish mycelium. The disease affects cocksfoot, ryegrasses and timothy, and is favoured by cool, humid conditions. Grey snow mould (see p. 101) may also be present.

No chemical control is available in the UK. Cultural control can be obtained by limiting the development of disease prior to snow cover. This is achieved by avoiding late applications of nitrogen fertilizer and defoliating the crop before winter, thus preventing the humid microclimate that exists within a substantial leaf canopy.

Powdery mildew (*Blumeria graminis*)

Most grass species are affected by this disease, which appears as a white, superficial, powdery covering on the leaves, particularly on the upper surface. Grass yield and quality can be reduced and infection is most severe in dense crops, and following dry periods. The disease exists in many strains, and cultivars vary greatly in their susceptibility. The NIAB recommended list of perennial, Italian and hybrid ryegrasses gives indications of varietal (cultivar) resistance to mildew. In Scotland, mildew is the most common disease of established ryegrass, particularly Italian ryegrass, and ratings for mildew resistance are given in the SAC list of grass varieties (cultivars) for Scotland.

Chemical control can be obtained by the use of sprays of propiconazole, triadimefon or sulfur. The use of propiconazole is restricted to crops grown for silage or seed. For silage crops, one application is permitted each year, before any cut. For seed crops, two applications are permitted each year, one in spring and one in autumn. Cultural control can be obtained by eliminating the shade and humid microclimate within dense crops through reduced nitrogen inputs.

Pre- and post-emergence seedling disease (*Fusarium culmorum*, *Pythium* spp. and other fungi)

Seedlings of all grass species are susceptible to disease during the period following sowing, up to the production of tillers. The critical stage for infection by *F. culmorum* is between seed germination and seedling emergence. Under optimum conditions, ryegrasses have a rapid rate of seed germination and seedling emergence, but if this is slowed by lack of soil moisture or by sowing too deeply, *Fusarium culmorum* can greatly reduce seedling populations. Infected seedlings that are not killed have a reduced vigour. The result is an insufficient establishment of seedlings and a high risk of weed invasion. Those species with small seeds and/or slow seedling growth rates are most at risk. *Pythium* is associated with post-emergence death of seedlings, which is often known as 'damping-off'. Infection is favoured by cool, wet conditions. Patches of seedlings become yellow or red in colour, collapse and rot.

Chemical control can be obtained by the use of seed treatment with thiram. Good seedbed preparation and sowing technique will encourage rapid seedling emergence. Avoiding sowing when soil conditions are unfavourable is advisable.

However, the shallow depth of sowing of grass seed predisposes the seed to the rapid changes in soil moisture and temperature that can occur at the soil surface.

Rhynchosporium leaf spot (*Rhynchosporium secalis* and *R. orthosporum*)
This disease causes large, brown lesions with a lighter-coloured centre in ryegrasses and cocksfoot. Infection is most severe in Italian ryegrass during cool, moist weather in spring and autumn, when forage quality can be reduced. *R. secalis*, which exists in many specialized strains, also attacks barley, and one of the strains on Italian ryegrass has been shown to be pathogenic to barley. The other species, *R. orthosporum*, is also found on Italian ryegrass and on cocksfoot. Applications of nitrogen fertilizer of 300 kg/ha/year and above appear to increase the severity of infection.

Chemical control can be obtained by the use of sprays of propiconazole or triadimefon. The use of propiconazole is restricted to crops grown for silage or for seed. For silage crops, one application is permitted each year, before any cut. For seed crops, two applications are permitted each year, one in spring and one in autumn. Cultural control can be obtained by eliminating the humid microclimate within dense crops through reduced nitrogen inputs. Taking an early cut can reduce the impact of this disease in spring. The NIAB recommended list of varieties (cultivars) of Italian ryegrass gives ratings for resistance to leaf blotch. Tetraploid and hybrid ryegrasses are less susceptible to attack than diploids.

Rusts
Brown rust (*Puccinia recondita* f. sp. *lolii*), crown rust (*P. coronata*) and stem rust (*P. graminis*) are the most visually striking of the foliar fungal diseases of grasses. Infection reduces forage yield, quality, and competitive ability of plants, and can render forage unpalatable to livestock. Crown rust appears as distinctive orange pustules on the leaves, normally from late summer to early autumn. Epidemics occur during periods when it is warm and dry during daytime, which favours spore dispersal, and there is dew-fall at night, which favours spore germination and infection of leaves. Later, black pustules are produced on the underside of leaves. Epidemics are likely to be most severe in the south and south-west of England. Brown rust is similar in appearance to crown rust, but occurs in spring and early summer and in northern as well as southern England. Crops cut for conservation are at particular risk because there is no early defoliation. Italian ryegrass is particularly susceptible to brown rust. Stem rust can reach high levels of infection on perennial ryegrass after hot summers.

Chemical control can be obtained by the use of sprays of propiconazole or triadimefon. The use of propiconazole is restricted to crops grown for silage or for seed. For silage crops, one application is permitted each year, before any cut. For seed crops, two applications are permitted each year, one in spring and one in autumn.

Cultural control can be obtained by strategic grazing or cutting to prevent build-up of the disease, and by the use of resistant cultivars. The National Institute of Agricultural Botany (NIAB) recommended list of varieties (cultivars) gives indications of resistance to crown rust in perennial ryegrass and brown rust in Italian and hybrid ryegrasses. Crown rust infection tends to decrease with increasing nitrogen level, although this is contentious, and a strategic application of nitrogen in late summer may be of benefit in reducing infection levels.

Virus diseases

In contrast to foliar fungal diseases, virus diseases are systemic and infected plants remain infected even when the plant is defoliated. In fact, cutting and grazing can spread some virus diseases. In Europe, 26 viruses infecting grass species have been identified but only the two described below are likely to be of economic importance in the UK. Little is known of the extent and significance of virus infection in grassland.

Barley yellow dwarf virus (BYDV)

BYDV has been reported to infect many grass species in south-west England and the situation is probably the same for the rest of the UK. *Lolium* and *Festuca* species are particularly susceptible. Three strains of the virus have been identified in the UK. BYDV causes greater damage in perennial ryegrass than in Italian ryegrass. Infection is spread by several species of aphid and is mostly symptomless, although sometimes there may be a yellow, progressive discoloration of the leaves from the tip, turning to red or purple. The impact of BYDV on grass production in the UK has yet to be ascertained, although a potential for loss of yield and persistence has been demonstrated. BYDV is an important disease of cereal crops, and grass areas in the vicinity can constitute a reservoir of viruliferous aphids, although the risk of cross-infection may be small.

No chemical control of the virus is available in the UK; insecticides used to control aphids in grassland may reduce spread of BYDV infection, but the likely impact is unknown. Cultural control of BYDV is likely to depend on the development of resistant cultivars but none is available in the UK at present.

Ryegrass mosaic virus (RgMV)

Surveys made in the 1970s showed that RgMV is common in ryegrass crops in UK but no recent information is available. Unlike BYDV, RgMV causes greater damage in Italian ryegrass than in perennial ryegrass. There are a number of different strains of the virus, which differ in virulence. Mild strains produce mottling and streaking of the leaves, whereas severe strains cause a dark brown necrosis. Infection can spread rapidly in Italian ryegrass and yield can be reduced severely. Infected swards have a reduced response to nitrogen fertilizer, and have an increased susceptibility to environmental stress such as drought and cold. Infection is introduced to new sowings by the wind-dispersed mite vector (cereal rust mite, *Abacarus hystrix*), and mite populations increase rapidly in autumn. Thus, crops sown in the spring are more severely affected during the first full year of harvesting than crops directly sown in the autumn. This is because mites disperse and colonize new sowings during the summer, and therefore autumn sowings are not colonized until the year after sowing. The importance of sap transmission in the spread of infection within crops is uncertain, but mowing machinery does not appear to be an effective vector. Infection can spread rapidly in seed crops because they are infrequently defoliated, which allows the mites to disperse and multiply.

No chemical control of the virus or the vector is available in the UK. Cultural control can be obtained in Italian ryegrass by selecting cultivars with resistance, as indicated in the NIAB list of recommended varieties (cultivars). However, no cultivars have a high level of resistance. Some of the later-flowering cultivars of perennial ryegrass have a good degree of resistance. The introduction and subsequent spread of RgMV may be reduced by sowing in autumn rather than in spring, to delay ingress of the mite, and by defoliating in autumn to curtail the increase in mite populations. Early defoliation in the autumn reduces the mite population, decreasing virus infection in the following year.

Amenity and turf grass

For the purposes of this chapter, the term 'amenity grass' is used to describe grass that has a recreational, functional or aesthetic value and is subject to human trampling and at least some routine management. The term 'turf grass' is used to describe amenity grass that receives a high level of management, including regular close mowing. Thus, turf grass has a high 'value' and control measures for pests and diseases of amenity grass are mainly directed at turf grass. The total area of amenity grassland in the UK is 0.35 million ha, of which about 15% is turf grass.

In the preceding section the emphasis has been on perennial ryegrass because it is the mainstay of agricultural grassland. A similar situation exists with amenity grass, where perennial ryegrass is the major component of all but the finest turf. Perennial ryegrass is susceptible to the same pests and diseases whatever the situation in which it is grown, but this susceptibility is greatly influenced by the diverse grass management applied. This is true especially for diseases; the pests of importance are common to both situations. In essence, agricultural grass is grown for the maximum herbage production and amenity grass for the minimum. Thus, the repeated removal of leaves that is a feature of much amenity grass also removes foliar disease inoculum and foliar pests. Consequently, the foliar diseases, and possibly virus diseases, that are of importance in the agricultural situation are generally of little consequence in amenity grass. Breeding of perennial ryegrass has produced cultivars for amenity use that have smaller and finer leaves and have resistance to the main diseases.

Other species of grass used for amenity purposes are several subspecies of red fescue, a fine-leaved fescue, several species of *Agrostis*, commonly known as bent grasses, and several species of meadow grass. One species of meadow grass (annual meadow grass), is abundant in amenity grass, although it is not sown. It has long been regarded as a weed grass but, because it is difficult to eradicate, there is now a trend towards accepting its presence and altering management to accommodate it.

The continuous use of one fungicide to control a particular disease can lead to the development of resistant strains of the disease. Therefore, if there is a choice of fungicides, these should be used in rotation.

In the descriptions below of the major pests and diseases of amenity grasses, indications are given of the factors that are likely to predispose grass to attack. Strategies to avoid these factors playing a part will help to reduce risk of attack. Factors common to many diseases are the application of fertilizer at the wrong time and/or at the wrong rate, the presence of excess moisture on the leaves, and a high soil pH.

Pests

The insect pests that attack amenity grasses are the same as those attacking forage grasses.

Chafer grubs (e.g. *garden chafer, Phyllopertha horticola*)
See under Forage grasses, p. 89.

Domestic dogs (*Canis familiaris*)
Fouling by dogs is a common problem. They scorch the grass in patches by urinating and also deposit faeces; these depredations are unsightly and also pose a risk to human health.

No chemical control is available in the UK. Urine scorch can be avoided if immediate action is taken to apply copious amounts of water to dilute the toxic elements. Exclusion of dogs by fencing may be the most effective control.

Earthworms
There are many species of earthworm in the UK and their activities below amenity grass, where populations can reach 100 per m^2, are beneficial. However, three species (*Aporrectodea caliginosa, A. longa* and *Lumbricus terrestris*) produce casts on the surface that disfigure fine turf, affect the playing quality and become smeared over the surface during mowing. Earthworm activity is greatest during mild, wet weather.

Chemical control can be obtained in turf grass by the application of sprays of carbendazim, carbendazim + chlorothalonil, gamma-HCH + thiophanate-methyl, or thiophanate-methyl. Natural or artificial watering after application will increase efficacy of treatment. Best results are obtained when the chemicals are applied during periods of significant earthworm activity close to the surface, which is usually in spring or autumn after periods of wet weather. Cultural control is obtained by inhibiting earthworms from casting, by (a) maintaining the soil at a low pH, (b) removing grass clippings to reduce food supply, and (c) regular aeration and scarification of the grass.

Frit fly (*Oscinella frit*) and grass & cereal flies (*Geomyza* spp., *Opomyza* spp., *Oscinella vastator* and *Meromyza* spp.)
See under Forage grasses, p. 89.

Chemical control can be obtained in amenity and turf grass by the application of sprays of chlorpyrifos.

Leatherjackets
See under Forage grasses, p. 90, for a description of these pests.

Chemical control can be obtained in turf grass by the application of sprays of chlorpyrifos or gamma-HCH + thiophanate-methyl. Treatments should be applied from November to March and when high populations are detected or damage is seen. A non-chemical control for small areas can be obtained by thoroughly wetting the area and covering it with sheeting overnight. The leatherjackets present in the soil come to the surface and can be swept up or crushed when the sheet is removed the following day.

Moles (*Talpa europea*)
Moles are small mammals that are abundant in the UK. They burrow under grass areas seeking worms and insects for food, thereby causing damage to roots and smothering the grass with the spoil from their tunnels. The tunnels can collapse under the weight of people walking on the grass, and cause injury.

Chemical control can be obtained by the use of strychnine hydrochloride, which is incorporated into baits. Strychnine is subject to UK poison regulations and must be applied by trained operators only. Various non-chemical methods for deterring moles have been proposed but it is difficult to do rigorous testing of their efficacy.

Rabbits (*Oryctolagus cuniculus*)
Rabbits are abundant pests in the UK. They graze on grass, but this causes little damage, and they may be beneficial in certain areas of amenity grass by maintaining the habitat. The main damage by rabbits in turf grass is through digging, which causes direct damage of the grass and presents a hazard to people walking on the grass.

Chemical control can be obtained by the use of sodium cyanide powder, which is introduced to rabbit burrows. The burrows are sealed and the powder reacts with moisture in the soil to produce hydrogen cyanide, a lethal poison. Sodium cyanide is subject to UK poison regulations and must be applied by trained operators only. Various repellent chemicals are available as amateur products but these are likely to have only a short-term effect; they are active only when in a dry state. Shooting of rabbits provides effective control but can be carried out only under strict guidelines. Exclusion by fencing, where this is practical, is probably the best long-term solution.

Diseases

As a general rule, grass treated with fungicide should not be mown within 24–48 h following treatment. The product leaflet will provide specific information.

Anthracnose (*Colletotrichum graminicola*)
This disease is widespread in the UK, but occurs only on annual meadow grass; it is most severe on closely mown turf. Symptoms usually appear from late summer to late winter. Leaves of affected plants become red or yellow in colour and in severe cases show as patches up to 15 cm in diameter. Affected plants can be easily pulled up and a black rot is apparent at the stem base. Compaction, low fertility and prolonged soil wetness favour the disease.

Chemical control in turf grass is obtained by the application of sprays of carbendazim + chlorothalonil, carbendazim + iprodione and chlorothalonil. These chemicals are not effective as curative treatments and need to be applied as soon as infection is detected. Regular aeration and appropriate fertilizer application will reduce the risk of disease.

Brown patch (*Thanatephorus cucumeris* – anamorph: *Rhizoctonia solani*)
This disease appears as water-soaked or bleached brown spots, often with a grey halo; the spots spread to form patches up to 60 cm across. All amenity grass species are susceptible but *Agrostis* spp. are the worst affected. Risk of disease is greater during periods of hot weather when the grass is watered frequently, and when high levels of nitrogen fertilizer are applied.

Chemical control can be obtained in amenity and turf grass by the application of sprays of iprodione. Risk of disease can be reduced by 'switching' the grass to dislodge water droplets from the leaves and by appropriate applications of nitrogen fertilizer.

Dollar spot (*Sclerotinia homeocarpa*)
In the UK this disease is a problem only on subspecies of red fescue. The symptoms are distinct, circular spots up to 2 cm in diameter. Risk of disease is greatest during warm, humid conditions.

Chemical control can be obtained in amenity and turf grass by the application of sprays of iprodione or thiabendazole. Turf grass can also be treated with sprays of carbendazim, carbendazim + chlorothalonil, chlorothalonil, fenarimol, quintozene or thiophanate-methyl. Indications of varietal (cultivar) resistance to dollar spot are provided by the Sports Turf Research Institute (STRI).

Fairy rings (*Marasmius oreades* and other species of basidiomycete fungi)
The classic symptom of a fairy ring is a complete circle of dead or dying grass,

bordered on both margins by a zone of stimulated grass. At certain times, the fruiting bodies of the fungus ('toadstools') appear within the ring. The rings gradually increase in size and can be several metres in diameter; playing quality and appearance of the grass can be greatly affected. This type of ring is usually caused by *M. oreades*. Other species of fungus produce only fruiting bodies and are not damaging to the grass, although they affect its appearance and playing quality. Control measures are mainly directed at *M. oreades*, which is difficult to eradicate completely because of the depth to which the fungal mycelium is present in the soil and the water-repellent nature of this mycelium.

Chemical control can be obtained in turf grass by the application of triforine as a high-volume (HV) spray or drench. Efficacy is greatest when treatment is applied to actively growing rings. The use of spiking and a wetting agent prior to fungicide application is also beneficial.

Cultivation and re-sowing of affected areas appears to carry little risk of re-infection. Applications of iron darken the grass and help to disguise the presence of rings.

Fusarium patch (*Monographella nivalis* – anamorph: *Microdochium nivale*)
This is probably the most important disease occurring during the winter months in the UK. It affects *Agrostis* spp., *Festuca* spp. and perennial ryegrass, but annual meadow grass is the most susceptible grass. Symptoms are patches up to 5 cm in diameter that are orange-brown; the patches may increase in size and coalesce to form large, irregular patches. Under prolonged humid conditions, a white or pink mycelium appears around the perimeter of the patch. The disease is also favoured by high soil pH and by inappropriate nitrogen fertilization.

Chemical control is obtained by the application of sprays of iprodione and thiabendazole on amenity and turf grass, and carbendazim, carbendazim + chlorothalonil, carbendazim + iprodione, chlorothalonil, fenarimol and thiophanate-methyl on turf grass only. Cultural control can be obtained by alleviating the humid conditions that favour infection, through improving the airflow over the grass, 'switching' the grass to dislodge water droplets from leaves, and improving drainage. The removal of grass clippings will reduce disease inoculum.

Grey snow mould (*Typhula incarnata*)
In the UK this disease is restricted to regions where prolonged snow cover occurs. All amenity grass species are susceptible. The disease appears in winter as patches of yellow-brown grass up to 5 cm in diameter, even in the absence of snow cover. However, severe damage occurs only under prolonged snow cover, and when the snow melts, large areas of bleached, matted leaves are apparent. The leaves are covered with a greyish mycelium, and small, red-brown sclerotia (fungal resting bodies) can be seen within the leaves.

Chemical control can be obtained by the application of sprays of iprodione in amenity and turf grass. Chemicals need to be applied prior to snow cover, and weather forecasts should therefore be consulted during vulnerable periods.

Melting-out (*Bipolaris* spp., *Curvularia* spp. and *Drechslera* spp. e.g. *D. poae*)
Melting-out is a descriptive term for the severe effects of leaf spots caused by various fungi. Most amenity grass species are susceptible to leaf spots, which may occur as distinct spots or more diffuse streaks or mottles. Disease is favoured by warm, humid conditions. The fungal spores are spread by the impact of water droplets, so rainfall and irrigation are important factors.

Chemical control can be obtained by the application of sprays of iprodione on amenity and turf grass.

Pre- and post-emergence seedling diseases (*Cladochytrium caespitis*, *Fusarium culmorum*, *Pythium* spp., and other species)
All amenity grass species are susceptible to seedling diseases, but *Agrostis* spp., *Festuca* spp. and smooth meadow grass are particularly susceptible to post-emergence disease. Risk of infection with *F. culmorum* is greatest in warm, dry soils and with *Pythium* spp. and *C. caespitis* in cool, damp soils.

Chemical control can be obtained by seed treatment with thiram, or (off-label use only) a foliar spray with fosetyl-aluminium (SOLA 1971/98). Factors which promote rapid seed germination and seedling growth will reduce the risk of pre-emergence infection, but excessive fertilization can predispose seedlings to post-emergence infection. The dry soil conditions that favour infection by *F. culmorum* can be avoided by irrigation.

Red thread (*Laetisaria fuciformis*)
This fungus often occurs as a disease complex with the pink patch fungus, *Limonomyces roseipellis*. All turf-grass species are susceptible, but perennial ryegrass and *Festuca* spp. are the most affected. The symptoms are pink/red patches, usually seen in summer and autumn, in which red, coral-like structures are apparent on affected leaves. Damage is often superficial and has no lasting effect. The disease is favoured by light, sandy soils, mild temperatures, excess surface moisture and low soil fertility, especially low nitrogen.

Chemical control is obtained by the use of sprays of iprodione and thiabendazole on amenity and turf grass, and carbendazim + chlorothalonil, carbendazim + iprodione, chlorothalonil, dichlorophen, fenarimol, quintozene and thiophanate-methyl on turf grass only. Applications of fertilizer, either alone or in conjunction with fungicides, may be beneficial by stimulating grass growth, but may predispose the grass to fusarium patch (see above) if applied in late summer or autumn. Indications of resistance to red thread in cultivars of *Agrostis* spp., *Festuca* spp. and perennial ryegrass are provided by STRI.

Take-all patch (*Gaeumannomyces graminis*)
This is an important disease of turf grass, because it is destructive and difficult to

control. In the UK, take-all predominantly occurs on *Agrostis* spp. Symptoms, which usually appear during summer and progress into late autumn, are slightly depressed areas of bronze-coloured grass up to 30 cm in diameter. The fungus is ubiquitous in soil, where it is usually held in check by antagonistic micro-organisms. However, application of chemicals or use of sand-based turf can eliminate the antagonists and allow the pathogen to spread. The disease is favoured by high soil pH, low nutrient levels, excessive thatch and poor drainage. Nitrate forms of nitrogen fertilizer tend to increase disease, whereas ammonium forms decrease it.

No chemical control is available in the UK. Cultural control can be obtained by avoidance of the conditions favouring disease spread. Also, turf should not be established with *Agrostis* spp. alone. Management of the root zone to encourage the establishment of antagonists is beneficial.

Herbage legumes (excluding field beans)

Control measures available for pests and diseases of herbage legumes are very limited. This situation is partly due to the substantial decline in the use of red clover, lucerne and other herbage legumes in the UK, which has been matched by a decline in research on these crops. White clover remains the most important and widely grown of the herbage legumes but, again, there has been little research on pests and diseases of this crop. No chemical control specifically for pests and diseases of herbage legumes is available in the UK. However, chemicals that are applied to grass will control pests or diseases present on legumes growing in mixture with the grass. At present there is no published information on resistance to pests and diseases in herbage legumes.

Pests

Aphids

Herbage legumes are infested by various species of aphid, e.g. black bean aphid (*Aphis fabae*), pea aphid (*Acyrthosiphon pisum*), vetch aphid (*Megoura viciae*) and cowpea aphid (*Aphis craccivora*). These aphids vary in colour but are generally green to black. Eggs are laid in the autumn and these do not hatch until spring. Winged forms are eventually produced and these disperse to other crops. Large numbers of aphids present on herbage legumes can affect plant growth and vigour; in addition, they may transmit certain virus diseases.

Harvesting an infested crop will remove most of the aphids.

Clover leaf weevils (*Hypera nigrirostris* and *H. postica*)

The weevils overwinter as adults, which are brownish and up to 5.5 mm long. They emerge from hibernation in spring and feed on clover leaves, but damage is insignificant. The adults then lay eggs, and the resulting larvae, which are green to brown, apodous and up to 5 mm long, feed on the young leaves and flowers of clover, lucerne, sainfoin, trefoil and vetches. Seed production may be seriously affected.

In crops grown specifically for hay or silage, the first cut in the year should be taken when the plants are still in bud, to remove the larvae. If a seed crop is rendered unusable because of severe infestation, the crop can still be cut for hay, which again removes the larvae. Adult weevils are highly mobile, and therefore new sowings should not be sited near previously infested crops.

Clover seed weevils (*Apion apricans*, *A. dichroum* and *A. trifolii*)

Adult weevils are about 2 mm long and dark in colour, and have a pronounced snout. They emerge from hibernation in late spring and feed on clover leaves, making small holes that are of little importance. Eggs are laid in May and June in the developing flower heads, and the resulting larvae, which are whitish and up to 2 mm long, feed on the immature seeds. In red clover, larvae from the second generation of weevils extend the period of damage. Serious losses in seed crops can occur, particularly in red clover.

In seed crops, damage by seed weevils can be reduced by harvesting the crop for forage in early summer to remove larvae.

Leatherjackets

Seedlings and established crops of clovers and lucerne may be damaged and patches may develop which become invaded by weed grasses. See under Forage grasses (p. 90) for further details. There are no specific recommendations for chemical control of leatherjackets on herbage legumes.

Sitona weevils

Several species are associated with herbage legumes. These include common clover weevil (*Sitona hispidulus*), clover weevil (*S. lepidus*) and pea & bean weevil (*S. lineatus*). The adult weevils cause characteristic notching of the leaf margins of a wide range of leguminous crops. The three species vary in their preferences for feeding on different leguminous plants. Populations of up to 370 adult weevils per m^2 have been recorded, and weevils are more abundant during warm, dry periods. Sowings in August in eastern England are particularly susceptible. *Sitona* weevils can cause severe losses of seedlings through consumption of cotyledons and the first leaves. Also, weevil larvae feeding on root nodules reduce the growth of clover seedlings. Wounds caused by larval feeding can predispose legumes to various crown- and root-rotting fungi. Removal of leaf tissue by adult weevils probably has little impact on plants with a substantial amount of leaf, but may impede the growth of plants in spring when little leaf is present.

Sowing in spring rather than in autumn reduces the risk of severe infestation. In eastern England, the risk of *S. lineatus* damaging new sowings of clovers can be determined by examining nearby pea or bean crops for the presence of severe leaf notching.

Slugs
Slugs, such as field slug (*Deroceras reticulatum*), prefer legumes to grasses and, consequently, legumes suffer greater seedling and yield losses than grass. Leaf tissue is rasped away in strips between the veins. Extensive leaf damage to established plants and young seedlings may occur in prolonged wet conditions. Large populations are likely on soils that provide abundant crevices in which the slugs can hide. Such conditions are provided, for example, by heavy clay soils (which crack open in dry weather) and loosely tilled soils. Little is known of the effect of leaf damage on established swards, but newly established crops should be protected if slug grazing is extensive. Seedling morphology and age, and the presence of seedlings of other plant species, may be factors in determining the extent of slug feeding. White clover drilled into established grass is vulnerable to damage from slugs moving along the slits or rotavated strips. On established white clover plants, slug feeding on the leaves probably has little adverse effect but feeding on leaf buds may cause significant damage in spring, when growth commences. See under Forage grasses (p. 91) for further details.

Slug populations can be reduced if a fine, firm seedbed can be produced.

Stem nematode (*Ditylenchus dipsaci*)
This nematode infests the stem bases, nodes and petioles, causing swelling and distortion, and damage typically appears as patches of stunted plants. On slopes, these patches spread downhill, indicating that the nematode is spread by water. Most forage legumes are susceptible, but damage is particularly severe in red clover, white clover and lucerne, and on heavy soils. Infestations can be difficult to recognize when the legume is grown in mixture with grass. There are a number of distinct races of the nematode, and some legume species are host to more than one race. The nematode can survive for many years in the absence of the host, especially on heavy soils, and can be a contaminant of seed.

Indications of resistance in cultivars to stem nematode in red clover and lucerne are available in the NIAB list of recommended varieties (cultivars). Measures to control stem nematode in new sowings have resulted in three-fold increases in forage yield at the first cut. Cultural control measures include the avoidance of animal or machinery movement from old to new swards, and crop rotation, although breaks of eight or more years between clover crops may be necessary.

Diseases

Fungal diseases are commonly found on herbage legumes in the UK but, although they can reduce yield and quality of forage, the extent of loss is largely unknown. Pseudopeziza leaf spot and pepper spot increase the oestrogenic activity of white clover, which may affect the reproductive performance of grazing animals. For foliar diseases, probably the only action that can be taken is to harvest the crop and remove the disease inoculum. There are no recommended chemical control measures.

Anthracnose (*Colletotrichum trifolii*)
This seed-borne disease of lucerne and clover causes wilting and death of stems by forming a girdling lesion at the base of the shoots. The loss of foliage is only temporary, as the plants are not killed.

Black blotch (*Cymadothea trifolii* on clovers)
This disease appears as large, circular, shiny-black lesions on the underside of leaves in the autumn.

Black stem (*Ascochyta imperfecta*)
This seed-borne disease is common on lucerne, and also affects clovers and trefoil. Dark-brown, elongate lesions on stems and petioles often extend deeply into the tissues, causing cankering and death of the shoots. It is most severe in the early months and may affect the first cut.

Early cutting has a beneficial effect, and a 3-year interval between susceptible crops is advisable.

Clover rot (*Sclerotinia trifoliorum*)
This is a serious disease of red clover and lucerne, and has contributed to the sharp decline in red clover cultivation since the late 1980s. The disease also affects white clover, trefoil and sainfoin. Infection can be very damaging to lucerne in its establishment year, and may kill many plants, but established lucerne crops are generally resistant, temporary loss of crop occurring before the first cut. In white clover, forage yield is greatly reduced by clover rot in experimental plots at research stations, but the extent of damage in the field situation is not known. Infection begins in the autumn as necrotic spots on the leaves and, during mild weather, spreads within the stems into the crown, causing plant death. If mild, humid conditions persist, many plants can be killed and severe loss of stand may be evident at the end of the winter. Periods of frost reduce the infection process. Black sclerotia (fungal resting bodies) develop in diseased tissue, and can be found in the rotting crowns and stems. The sclerotia eventually fall on to the soil; these germinate in the following year to produce spores and continue the infection cycle. However, if conditions are unsuitable for germination, sclerotia can survive in the soil for many years before germinating. Infection can be introduced to new sowings by infected seed or seed contaminated by sclerotia.

Indications of resistance in cultivars to clover rot in red and white clover are available. Crop rotation will reduce the risk from infection by germinating sclerotia, but the longevity of sclerotia in soil necessitates an interval of 8 or more years. Crop rotation is generally recommended for the control of this disease, and it is now thought that a 4- to 5-year rotation between susceptible crops is adequate. Care should be taken to secure seed from disease-free crops, because sclerotia can be disseminated with the seed. Grazing or cutting the crops in the autumn may be beneficial, by reducing the amount of foliage available for infection by the aerial ascospores and reducing the humid microclimate within a leafy crop. The NIAB list of recommended varieties (cultivars) of red and white clover gives indications of varietal (cultivar) resistance to clover rot.

Crown wart (*Physoderma alfalfae*)
This disease of lucerne appears as characteristic warty galls at the crown of the plant, and may lead to wilting in hot weather and loss of yield. It is invariably associated with poor drainage, and is often found in fields that have been poached by winter grazing of cattle.

Control is best effected by ensuring good drainage and lengthening the crop rotation.

Downy mildew (*Peronospora trifoliorum* on clovers, lucerne and trefoil)
This disease produces chlorotic areas on the upper leaf surfaces, which are often puckered, and a purplish-grey weft of the fungus appears on the undersurfaces. It is most severe on white clover and lucerne, in which infection can become systemic.

Pepper spot (*Leptosphaerulina trifolii* on clovers and lucerne)
This disease appears as small, abundant brown lesions on both leaf surfaces, and on the petiole. It occurs particularly on white clover.

Powdery mildew (*Erysiphe trifolii* on clovers and sainfoin; *E. pisi* on lucerne, trefoil and vetches)
This disease appears as white, powdery infections on the upper leaf surfaces, and may cause extensive infection on red clover and sainfoin.

Pseudopeziza leaf spot (*Pseudopeziza trifolii* on clovers; *P. medicaginis* on lucerne and trefoil)
This disease appears as large, brown lesions with a star-shaped margin on the upper surface of clover and lucerne leaves. Mature lesions produce yellowish, glistening, dish-shaped fruiting bodies in the centre. Severe infection can occur in wet autumns, causing premature leaf shed.

Rust (*Uromyces* spp.)
Rusts infect the leaves and petioles of all the herbage legumes, and occasionally do serious damage. *U. fallens* is found on red clover and *U. trifolii* on white clover. *U. pisi* occurs on lucerne and trefoil and *U. onobrychidis* is very common on sainfoin.

Scorch (*Kabatiella caulivora* on red clover)
This seed-borne disease causes the death of leaves and shrivelling of flowers, by producing lesions that girdle the leaf and flower stalks. The blackened appearance of infected crops resembles scorching by fire. In a wet season, infection can cause a temporary, severe loss of foliage, and in a seed crop a total loss of seed production.

Verticillium wilt (*Verticillium albo-atrum*)
This is the most serious disease of lucerne and is one of the factors which has led to the reduction in the area grown. Infection is seed- and soil-borne, and the symptoms develop as a yellowing and wilting of the leaves and stems, followed by shrivelling of the whole plant from the base upwards. Symptoms appear after the first cut and increase throughout the season. Re-growth may be stunted, with shortened internodes. Eventually, the whole plant may be killed and stands rendered quite worthless in their third year.

Cultural control can be obtained by crop rotation. Crops should be ploughed after 2–3 years. The disease persists in the soil for many years, and therefore as long a break as possible should be allowed after an infected crop. To prevent inter-field infection via crop fragments, younger crops should be harvested before older ones. The NIAB recommended list of varieties (cultivars) of lucerne gives a rating for resistance to wilt.

Viruses
Numerous viruses have been detected in leguminous crops, many of them affecting a wide range of hosts, and infection of a single host by several viruses is commonplace. These viruses are spread by infected sap and insect vectors. Little is known of their effects in the UK, but they are associated with loss of production and lack of persistence in countries such as the US, Australia and New Zealand. Of the various viruses that infect white clover, there is unconfirmed evidence that white clover mosaic virus and clover yellow vein virus are the most frequently occurring in the UK. Red clover necrotic mosaic virus (RCNMV) was a serious virus disease of red clover in the UK in the 1970s, causing severe leaf mottle and distortion, browning and 'winter kill'. There is a renewed interest in red clover at present and RCNMV may again become a problem.

The only likely prospect of control of virus diseases is through the development of resistant cultivars, but none is available in the UK at present. In the US, virus-resistant white clover has yielded substantially more than virus-susceptible cultivars.

Fodder brassica crops (excluding oilseed rape)

Chemicals available for the control of individual pests and diseases in fodder brassicas vary according to the crop concerned. Therefore, the chemical controls listed below do not necessarily apply to all fodder brassica crops. Kale has the widest choice of chemicals.

Pests

See Table 4.3 for a summary of chemical control options.

Cabbage aphid (*Brevicoryne brassicae*)
This aphid is a blue-grey colour and has a powdery coating. Eggs are laid in the autumn and these hatch in the following spring. Winged forms are produced eventually and these disperse to other crops. It infests the leaves and shoots of kale, rape and swedes, but does not attack turnips. Heavy infestations may cause stunting and severe leaf distortion, particularly in swedes. Infestations on fodder swedes may cause a loss of yield, and control measures may be worth while. The aphid also transmits cauliflower mosaic virus and turnip mosaic virus. Other

Table 4.3 Chemicals available for control of insect pests of fodder brassicas

Pest	Kale	Swede	Turnip
cabbage aphid	alpha-cypermethrin	–	–
	–	carbofuran [g]	carbofuran [g]
	chlorpyrifos	chlorpyrifos	chlorpyrifos
	chlorpyrifos + dimethoate [g]	chlorpyrifos + dimethoate [g]	chlorpyrifos + dimethoate [g]
	cypermethrin	–	–
	deltamethrin + pirimicarb	deltamethrin + pirimicarb	deltamethrin + pirimicarb
	liquid soap*	liquid soap*	liquid soap*
	nicotine	nicotine	nicotine
	pirimicarb	pirimicarb	pirimicarb
	rotenone*	rotenone*	rotenone*
cabbage root fly	carbofuran [g]	carbofuran [g]	carbofuran [g]
cabbage stem weevil	carbofuran [g]	carbofuran [g]	carbofuran [g]
caterpillars	alpha-cypermethrin	–	–
	cypermethrin	–	–
	deltamethrin	deltamethrin	deltamethrin
	deltamethrin + pirimicarb	–	–
	nicotine	nicotine	nicotine
flea beetles	alpha-cypermethrin	–	–
	carbofuran [g]	carbofuran [g]	carbofuran [g]
	cypermethrin	–	–
turnip root fly	–	carbofuran [g]	carbofuran [g]

[g] = Granules; * see comments in text, p. 110.

aphids, e.g. peach/potato aphid (*Myzus persicae*), also occur on fodder brassicas but are of lesser importance.

See Table 4.3 for a summary of chemical control measures available. Rotenone and liquid soap can be used in organic systems but it is prudent to seek the approval of the Soil Association beforehand. Granules of carbofuran should be incorporated into the soil at sowing; granules of chlorpyrifos + dimethoate should be applied as a surface or sub-surface band treatment at drilling or by mid-April. Cultural control is aimed at reducing the numbers of overwintering eggs by eliminating crop plants left behind after harvest and some brassicaceous weeds. In spring, avoid sowing near to a seed crop that has overwintered, unless it has been sprayed with an aphicide.

Cabbage leaf miner (*Phytomyza rufipes*)

The larvae tunnel within the leaf petioles and stems of rape and kale, and infested plants with wilting lower leaves may be seen in seedbeds. The larvae are white, smooth, shiny and up to 6 mm long, and can be mistaken for those of cabbage root fly (see below). They are produced from eggs laid from May to October and there can be up to three generations of the pest during a year. Larvae of another leaf-mining species (*Scaptomyza flava*) also infest brassica plants, forming blotch mines in the leaves.

Chemical control in seed crops can be obtained by the application of a spray of dimethoate.

Cabbage root fly (*Delia radicum*)

The larvae are white, apodous and up to 10 mm long. Egg laying begins in April, reaches a peak in May and continues until November. The resulting larvae burrow through the soil to attack the roots of kale, swede and turnip. There can be several generations in a year. Damage is most severe in late April and May, when small seedlings wilt and die, and damage to tap roots at an early stage causes stunting and reduced yields.

See Table 4.3, p. 109, for a summary of chemical control measures available. Granules of carbofuran should be incorporated into the soil at sowing. Control is most likely to be cost-effective in crops drilled before the end of April, when seedling emergence coincides with the peak of egg laying.

Cabbage stem flea beetle (*Psylloides chrysocephala*)

The adults are about 5 mm long and a shiny green, blue or bronze in colour. After a period of summer aestivation, adults appear in September and disperse to feed on newly sown brassicas. Severe damage may occur, especially if plant growth is slow. Eggs are laid in the soil, mainly from September to November. The resulting larvae invade plants and feed from autumn to spring, tunnelling within the stem and causing extensive damage.

No chemical control on fodder brassica crops is available in the UK.

Cabbage stem weevil (*Ceutorhynchus pallidactylus*)
The larvae are creamish-white, apodous and up to 6 mm long. They are produced from eggs laid in spring and they tunnel in the stem during May and June, causing the seedlings to become stunted and 'spongy'. Damage is most severe in crops sown in late April to late May.

Chemical control can be obtained by the application of carbofuran granules, incorporated into the soil at sowing.

Caterpillars
Caterpillars of several pests, e.g. cabbage moth (*Mamestra brassicae*), diamond-back moth (*Plutella xylostella*), garden pebble moth (*Evergestis forficalis*) and cabbage white butterflies (*Pieris* spp.), may feed on the foliage.

See Table 4.3, p. 109, for a summary of chemical control measures available.

Cutworms
Caterpillars of turnip moth (*Agrotis segetum*), and certain other species, feed as cutworms. The name 'cutworm' is given to a moth caterpillar that damages plants at or just below soil level; as the name suggests, cutworms sever roots or stems near ground level, resulting in the death of the plant. Such larvae are up to 50 mm long and typically dull greyish-brown in colour. They are produced from eggs laid in spring/early summer and are usually fully fed by the autumn. In years of 'high' cutworm activity (usually associated with hot, dry summers), extensive damage to fodder crops has been reported, especially in July.

See Table 4.3, p. 109, for a summary of chemical control measures. The use of additional wetter is generally recommended when spraying brassicas with insecticides. Sprays should be timed to control the young larvae feeding above ground; they will have little or no impact on older larvae feeding below ground. Cutworms feed on many non-crop species; therefore, fallow ground should be kept clean to eliminate food sources and reduce the likelihood of egg laying by the adult moths.

Flea beetles (*Phyllotreta* spp.)
Various species of flea beetle attack fodder brassicas, and different species dominate in different areas of the UK. In general, they are all dark coloured, although some have a distinct yellow band running longitudinally along each elytron (wing case); the beetles are typically up to 1.5–3.0 mm in length. They leap into the air when disturbed – hence their common name. The adult beetles overwinter and become active in warm weather in the spring. The cotyledons and stems of young seedlings are holed and often destroyed in April and May. Damage is accentuated when dry soil conditions slow growth of spring-sown crops. Egg laying takes place in the spring, and the new generation of adults appears in late summer before eventually overwintering. Large numbers of

young adults may feed on the mature crop, sometimes resulting in serious damage.

See Table 4.3, p. 109, for a summary of available chemical control measures. If used, carbofuran granules should be incorporated into the soil at sowing. In order to maximize seedling growth before beetle attack occurs, spring crops should be sown as early as possible on a fine tilth with adequate fertilizer in the seedbed.

Leatherjackets

Leatherjackets, the larvae of crane flies (e.g. *Tipula paludosa*), may sever the stems of plants, but on fodder crops they are troublesome only occasionally at the seedling stage.

No chemical control on fodder brassica crops is available in the UK. Crops sown after the larvae have pupated in spring or before egg laying in autumn should avoid damage.

Turnip root fly (*Delia floralis*)

This pest is closely related to the cabbage root fly (*Delia radicum*), and in the northern parts of Britain the larvae cause damage to bulbs of swedes and turnips from late September onwards.

Chemical control can be obtained on swede or turnip by the application of carbofuran granules, incorporated into the soil at sowing.

Wireworms

Many fodder brassica crops are highly resistant to wireworm attack and can be grown safely after ploughing infested grassland. No chemical control is available on fodder brassica crops in the UK.

Diseases

Although fodder brassicas are attacked by the same diseases as vegetable brassicas, the impact of disease is of less economic importance in fodder crops because they have a lower cash value. The carry-over of some diseases can be reduced by chopping and ploughing crop stubble and debris as soon as possible after harvest.

Alternaria (*Alternaria* spp.)

This seed-borne disease appears as dark spots on the leaves of all brassica crops, but especially turnips. Crops in the south and south-west of England are particularly prone to infection. Wet weather promotes infection.

Chemical control can be obtained by the application of iprodione as a seed treatment to swede and turnip, and as a spray to stubble turnip.

Canker (*Leptosphaeria maculans*)

Canker is now becoming important as a disease of oilseed rape (see Chapter 3,

p. 62). It has long been known as a serious disease of swedes and turnips, sometimes completely rotting the roots.

No chemical control is available in the UK. In forage crops, an adequate rotation of at least 4 years should be used between susceptible crops.

Clubroot (*Plasmodiophora brassicae*)
This soil-borne disease will attack all brassica crops; swedes and turnips may be damaged severely but kale is rarely affected. Damage is most frequent in crops on acid soils in the north and west of England. The fungus can survive in the soil for many years in the absence of host plants.

No chemical control is available in the UK. Cultural control may be obtained by the use of lime to produce an alkaline reaction in the soil, improvement of drainage and crop rotation. An interval of at least 5 years between susceptible crops is sufficient for the disease to diminish. However, the disease can be maintained in susceptible weeds, thus nullifying the crop rotation.

Downy mildew (*Peronospora parasitica*)
This pathogen attacks the undersurfaces of the leaves, producing chlorotic areas on the upper surfaces, and loss of leaves. It is most troublesome at the seedling stage, particularly on swede crops. Infection is favoured by autumn sowing and wet weather.

Chemical control in kale only can be obtained by the application of a spray of chlorothalonil. As a general recommendation, avoid sowing crops in low-lying, cold, wet soils.

Powdery mildew (*Erysiphe cruciferarum*)
This disease gives a white, powdery appearance to the leaves. It attacks all brassica crops, but is most important on swedes and turnips. Hybrid catch crops, crosses between stubble turnips and Chinese cabbage, appear to be very susceptible to this disease. The disease is favoured by dry conditions. Yield can be reduced if the crop is infected at an early stage.

Chemical control can be obtained by the application of a spray of copper sulfate + sulfur. This treatment can be used in organic systems but it is prudent to seek the approval of the Soil Association beforehand.

Root rot (*Phytophthora megasperma*)
This disease causes the roots and bases of the plants to rot completely and is one of the most destructive diseases of kale. Waterlogged conditions predispose the crop to infection.

Correction of drainage is the only means of control.

Seed and seedling diseases (*Pythium* spp.)
Chemical control can be obtained by the application of a seed treatment of thiram alone or in combination with gamma-HCH.

Viruses

Cauliflower mosaic virus and turnip mosaic virus (both aphid-borne) affect kale, swedes and turnips. In addition, on swedes and turnips, there are three other viruses transmitted by flea beetles.

No direct chemical control of viruses is available in the UK, but the aphid and flea beetle vectors can be controlled (see pp. 109 and 111, respectively).

Wirestem (*Thanatephorus cucumeris* – anamorph: *Rhizoctonia solani*)

This disease causes a severe foot rot and damping-off in seedlings of all brassicas.

No chemical control measures are available in the UK.

Fodder beet and mangolds

The pests and diseases that attack sugar beet (see Chapter 6, p. 168) also attack fodder beet and mangolds, but have a lesser status in fodder crops because these crops are of lower cash value. Some of the insecticides cited below will also control other pests that occur on fodder crops but are of far greater significance in sugar beet.

Pests

Aphids

The two main species infesting fodder beet and mangold are peach/potato aphid (*Myzus persicae*) and black bean aphid (*Aphis fabae*). *M. persicae* is a greenish colour, whereas *A. fabae* is mainly black and commonly known as 'blackfly'. *M. persicae* can overwinter in crops as adults or nymphs, but *A. fabae* overwinters as eggs on spindle (*Euonymus europaeus*). *A. fabae* may occur on beet in large numbers, particularly in hot, dry seasons, and may cause considerable damage. Both aphid species are vectors of virus yellows.

Chemical control can be obtained by applying a spray of dimethoate, demeton-S-methyl, liquid soap or rotenone, or by applying granules of carbosulfan or disulfoton incorporated into the soil at drilling. Use of dimethoate excludes control of *M. persicae* because some strains are resistant to this chemical. Dimethoate should not be applied during hot, dry conditions. Liquid soap and rotenone can be used in organic systems but it is prudent to seek the approval of the Soil Association beforehand.

Mangold flea beetle (*Chaetocnema concinna*)

The larvae of this pest are white and up to 6 mm long and the adults are a metallic-green or bronzy-black and up to 2.3 mm long. Eggs are laid in the soil in spring and the resulting larvae feed on roots for up to 6 weeks before pupating.

The adults feed on the foliage. The pest is commonly found on fodder crops but is only occasionally serious. The most severe damage occurs in warm, dry conditions in spring when seedlings in the cotyledon stage may be killed by heavy attacks.

Chemical control can be obtained by the application of granules of carbofuran or carbosulfan incorporated into the soil at drilling.

Mangold fly (*Pegomya hyoscyami*)
Mangold fly lays white eggs, singly or in groups, on the undersides of leaves. The larvae (sometimes referred to as 'beet leaf miners') are whitish and up to to 8 mm long; they burrow into the leaf, producing large, blister-like mines. The first-generation occurs in May and early June, and the growth of very young plants may be retarded when the attack is severe, particularly if the growing point is damaged. The risk is greatest when plant growth is slow. Beet and mangolds are able to tolerate considerable damage, and often recover completely if the growing point is undamaged. There are typically two or three generations of the pest in a year.

Chemical control can be obtained by applying a spray of demeton-S-methyl or trichlorfon or by applying granules of benfuracarb, carbofuran, carbosulfan or disulfoton. Granules are incorporated into the soil at drilling. A wetting agent should be added to a spray of trichlorfon. Early drilling is advisable, to allow plants to be sufficiently developed and so resist attack.

Pygmy mangold beetle (*Atomaria linearis*)
The adult beetles are red-brown to blackish and less than 2 mm long. They overwinter on the previous year's host crop and migrate in April to lay eggs. The resulting larvae feed on roots for up to 6 weeks before pupating. The adults emerge soon afterwards and feed on the hypocotyl of young seedlings, making small blackened pits in the tissues and sometimes causing death. Damage is most likely in areas where sugar beet, fodder beet or mangolds are grown regularly, and is almost inevitable if host crops are grown in succession. Adults arrive in April from the previous year's host crop and damage is likely only in areas where sugar beet, fodder beet or mangolds are grown regularly. Damage is almost inevitable if a host crop follows a previous host crop in the rotation.

Chemical control in fodder beet can be obtained only by the application of tefluthrin to the seed, or by granules of benfuracarb, carbofuran or carbosulfan incorporated into the soil at drilling. Crop rotation will prevent a build-up of pest populations.

Diseases

Black leg (*Pleospora bjoerlingii*)
This seed-borne fungus is the most important of the several species of fungi that can kill seedlings of fodder beet and mangolds. Infection takes the form of a black

lesion that kills the hypocotyl as the seedlings emerge and is known as black leg. The disease continues to infect all parts of the plant during growth, and it is spread by splash during wet weather. The rot continues in stored roots. In seed crops, the seed may be infected. Plants deficient in boron are more susceptible to damage.

Chemical control can be obtained by the application of thiram to the seed. Boron deficiency in the soil should be corrected by suitable applications of sodium borate.

Clamp rot (*Botryotinia fuckeliana* – anamorph: *Botrytis cinerea*)
This is the commonest root rot found in fodder beet and mangolds, and forms a greyish, furry mould on affected roots. The disease usually attacks only mechanically damaged roots (e.g. the damage caused by machinery at harvest). Such damage should therefore be kept to a minimum.

Downy mildew (*Peronospora farinosa* f. sp. *betae*)
This fungus attacks the lower surfaces of the leaves, and particularly the younger leaves, in the growing point, and can severely check the growth of plants. The disease is favoured by cool, wet seasons.

No chemical control is available in the UK.

Powdery mildew (*Erysiphe betae*)
This fungus forms a white, powdery covering on the upper leaf surface and is favoured by hot, dry summers.

Chemical control in fodder beet can be obtained only by the application of a spray of triadimefon or triadimenol. Indications of resistance to mildew are given in the NIAB recommended list of fodder beet varieties (cultivars).

Ramularia leaf spot (*Ramularia beticola*)
This disease forms greyish-brown spots, and in cool, humid conditions the spots spread and coalesce, causing leaf senescence.

No chemical control is available in the UK. Ratings for resistance to leaf spot are given in the NIAB recommended list of fodder beet varieties (cultivars).

Rust (*Uromyces betae*)
This disease appears as reddish-brown pustules on the leaves in late summer, causing premature leaf senescence. Infection is favoured by hot, dry weather.

No chemical control is available in the UK. Ratings for varietal (cultivar) resistance are given by NIAB.

Violet root rot (*Helicobasidium purpureum*)
This soil-borne disease appears as a purple, superficial mycelium on the roots of a wide range of plants, including non-crop species. Infection is most common on light, alkaline soils and is favoured by warm soil conditions. Fungal infection is often followed by secondary bacterial infection that causes extensive decay when

the roots are stored in clamps. The fungus can remain viable in the soil as sclerotia (resting bodies) for many years in the absence of host plants.

No chemical control is available in the UK. The removal of affected roots from the field will prevent the disease building up in the soil. Further, improved drainage will reduce survival of sclerotia in wet patches. The elimination of non-crop host plants will reduce the risk of carry-over of the disease.

Virus yellows

Virus yellows – beet mild yellowing virus (BMYV) and beet yellows virus (BYV) – is a serious disease of sugar beet. It also affects mangolds and fodder beet, which may be a source of infection for sugar beet crops. Both viruses are spread by aphids, especially peach/potato aphid (*Myzus persicae*). Symptoms consist of a severe yellowing of the leaves, which become brittle and bronzed, or develop fine necrotic spots. Yield is affected severely. The disease usually occurs in patches in the field.

No direct chemical control of the viruses is available in the UK. Chemical control is aimed at the aphid vectors (see p. 114). To prevent this disease being carried over into stored mangolds by infected aphids, the roots should always be topped of any green tissue before storing. Clamps should be cleared before the emergence of the sugar beet crop in the following season, to prevent the spread of infective aphids from the stored roots to the new sowing.

Forage maize

Pests

Cutworms (e.g. larvae of turnip moth, *Agrotis segetum*)

Chemical control of cutworms on forage maize can be obtained by the application of gamma-HCH spray to the seedbed. See under Fodder brassica crops (p. 111) for further details.

Frit fly (*Oscinella frit*)

Frit fly is a major pest, and it is the spring generation that causes the damage to newly sown maize. See under forage grasses (p. 89) for further details. Damage to young maize seedlings is revealed as stunting and distortion, often resulting in death of the central shoot. Plants develop with multiple, small tillers and produce poor cobs. When seedling growth is rapid, the larvae are pushed away from the apical meristem and damage is restricted to some twisting and raggedness of the leaves, causing a temporary check in growth.

Chemical control can be obtained by the application of granules of carbofuran or phorate to the seedbed, or sprays of chlorpyrifos or fenitrothion. Maize seed imported from other countries is often supplied treated with methiocarb, ostensibly for control of frit fly.

Leatherjackets
See under Forage grasses, p. 90, for a description of these pests.

Chemical control of leatherjackets on maize can be obtained by spraying fenitrothion or gamma-HCH on to the seedbed.

Wireworms (e.g. larvae of *Agriotes lineatus*, *A. obscurus*, *A. sputator*, *Athous haemorrhoidalis* and *Ctenicera* spp.)
Maize sown into ploughed grassland is at greatest risk of damage. See under Forage grasses, p. 92, for further details of wireworms.

Chemical control can be obtained by a spray of gamma-HCH, applied to the seedbed.

Diseases

Seedling diseases (*Pythium* spp. and other fungi)
Seedlings are attacked mainly after emergence and may be killed. Cool, wet soil conditions favour infection.

Chemical control can be obtained by treating the seed with thiram. Most maize seed imported from other countries is supplied pre-treated with thiram. Avoid early sowing in cold, wet soils.

Smut (*Ustilago maydis*)
This disease is easily recognized by the large, white galls formed mainly on the cobs, although other parts of the plant can be infected. The galls break up and release black spores that can accumulate in the soil, and it is from this source that most outbreaks rise. Thus, smut is more frequent where maize is grown continually in the same field.

No chemical control is available in the UK. Crop rotation will reduce the build-up of disease inoculum in the soil.

Stalk rot (*Fusarium* spp.)
This soil- and seed-borne disease attacks the roots and lower parts of the stem; this leads to lodging, which renders harvesting difficult. The disease accumulates in the soil.

No chemical control is available in the UK. Cultural control is aimed at preventing the build-up of disease inoculum in the soil. This is achieved by crop rotation and by encouraging rapid breakdown of crop debris after harvest, through fine chopping and incorporation into the soil.

Cereals for forage

Cereals, including wheat, rye, oats, triticale and barley, are grown widely for forage, as whole-crop silage. These cereals suffer from the same range of pests and diseases as they do when grown for grain (see Chapter 2). One major difference is that cereals are harvested for forage before the inflorescence is fully developed; therefore, pests and diseases specifically attacking the inflorescence do not attain the importance that they have in grain crops. The chemical control measures for whole-crop cereals and grain crops are also the same, providing that there is no withholding period for chemicals applied to crops to be used as animal feed.

A novel cultural control has been developed for autumn-sown whole-crop cereals, in which the crop is sown into a perennial understorey of white clover (*Trifolium repens*). In this system, the damage by pests and diseases (and weeds) has been reduced markedly, with a consequent reduction in the need for agrochemicals.

List of pests cited in the text*

Abacarus hystrix (Prostigmata: Eriophyidae)	cereal rust mite
Acyrthosiphon pisum (Hemiptera: Aphididae)	pea aphid
Agriotes lineatus (Coleoptera: Elateridae)	a common click beetle
Agriotes obscurus (Coleoptera: Elateridae)	a common click beetle
Agriotes sputator (Coleoptera: Elateridae)	a common click beetle
Agrotis segetum (Lepidoptera: Noctuidae)	turnip moth
Aphis craccivora (Hemiptera: Aphididae)	cowpea aphid
Aphis fabae (Hemiptera: Aphididae)	black bean aphid
Apion apricans (Coleoptera: Apionidae)	a clover seed weevil
Apion dichroum (Coleoptera: Apionidae)	a clover seed weevil
Apion trifolii (Coleoptera: Apionidae)	a clover seed weevil
Aporrectodea caliginosa (Opisthopora: Lumbricidae)	grey worm
Aporrectodea longa (Opisthopora: Lumbricidae)	long worm
Athous haemorrhoidalis (Coleoptera: Elateridae)	garden click beetle
Atomaria linearis (Coleoptera: Cryptophagidae)	pygmy mangold beetle
Bibio marci (Diptera: Bibionidae)	St. Mark's fly
Brevicoryne brassicae (Hemiptera: Aphididae)	cabbage aphid
Canis familiaris (Carnivora: Canidae)	domestic dog
Cerapteryx graminis (Lepidoptera: Noctuidae)	antler moth
Ceutorhynchus pallidactylus (Coleoptera: Curculionidae)	cabbage stem weevil
Chaetocnema concinna (Coleoptera: Chrysomelidae)	mangold flea beetle
Chrysoteuchia culmella (Lepidoptera: Pyralidae)	garden grass veneer moth
Ctenicera spp. (Coleoptera: Elateridae)	upland click beetles
Delia floralis (Diptera: Anthomyiidae)	turnip root fly
Delia radicum (Diptera: Anthomyiidae)	cabbage root fly
Deroceras reticulatum (Stylommatophora: Limacidae)	field slug
Dilophus febrilis (Diptera: Bibionidae)	fever fly
Ditylenchus dipsaci (Tylenchida: Tylenchidae)	stem nematode
Evergestis forficalis (Lepidoptera: Pyralidae)	garden pebble moth
Geomyza spp. (Diptera: Opomyzidae)	grass and cereal flies
Hepialus humuli (Lepidoptera: Hepialidae)	ghost swift moth
Hepialus lupulinus (Lepidoptera: Hepialidae)	garden swift moth
Hypera nigrirostris (Coleoptera: Curculionidae)	a clover leaf weevil
Hypera postica (Coleoptera: Curculionidae)	a clover leaf weevil

Lumbricus terrestris (Opisthopora: Lumbricidae)	lob worm
Mamestra brassicae (Lepidoptera: Noctuidae)	cabbage moth
Megoura viciae (Hemiptera: Aphididae)	vetch aphid
Melolontha melolontha (Coleoptera: Scarabaeidae)	cockchafer
Meromyza spp. (Diptera: Opomyzidae)	grass & cereal flies
Metopolophium festucae (Hemiptera: Aphididae)	fescue aphid
Myzus persicae (Hemiptera: Aphididae)	peach/potato aphid
Opomyza spp. (Diptera: Opomyzidae)	grass and cereal flies
Oryctolagus cuniculus (Lagomorpha: Leporidae)	rabbit
Oscinella frit (Diptera: Chloropidae)	frit fly
Oscinella vastator (Diptera: Chloropidae)	*larva* = a grass stem borer
Pegomya hyoscyami (Diptera: Anthomyiidae)	mangold fly
Penthaleus major (Prostigmata: Eupododae)	red-legged earth mite
Phyllobius pyri (Coleoptera: Curculionidae)	common leaf weevil
Phyllopertha horticola (Coleoptera: Scarabaeidae)	garden chafer
Phyllotreta spp. (Coleoptera: Chrysomelidae)	flea beetles
Phytomyza rufipes (Diptera: Agromyzidae)	*larva* = cabbage leaf miner
Pieris spp. (Lepidoptera: Pieridae)	cabbage white butterflies
Plutella xylostella (Lepidoptera: Yponomeutidae)	diamond-back moth
Psylloides chrysocephala (Coleoptera: Chrysomelidae)	cabbage stem flea beetle
Rhopalosiphum padi (Hemiptera: Aphididae)	bird-cherry aphid
Scaptomyza flava (Diptera: Drosophilidae)	*larva* = a brassica leaf miner
Sitobion avenae (Hemiptera: Aphididae)	grain aphid
Sitona hispidulus (Coleoptera: Curculionidae)	common clover weevil
Sitona lepidus (Coleoptera: Curculionidae)	clover weevil
Sitona lineatus (Coleoptera: Curculionidae)	pea & bean weevil
Talpa europea (Insectivora: Talpidae)	mole
Tipula paludosa (Diptera: Tipulidae)	a common crane fly

* The classification in parentheses refers to order and family.

List of pathogens/diseases (other than viruses) cited in the text*

Alternaria spp. (Hyphomycetes)	dark leaf spot of brassicas
Ascochyta imperfecta (Coelomycetes)	black stem of lucerne
Bipolaris spp. (Hyphomycetes)	melting-out of amenity grasses
Blumeria graminis (Ascomycota)	powdery mildew of grasses
Botryotinia fuckeliana (Ascomycota)	clamp rot of beet and mangolds
Botrytis cinerea (Hyphomycetes)	– anamorph of *Botryotinia fuckeliana*
Cladochytrium caespitis (Coelomycetes)	damping-off of amenity grasses
Claviceps purpurea (Ascomycota)	ergot of grasses
Colletotrichum graminicola (Coelomycetes)	anthracnose of amenity grasses
Colletotrichum trifolii (Coelomycetes)	anthracnose of lucerne
Curvularia spp. (Hyphomycetes)	melting-out of amenity grass
Cymadothea trifolii (Ascomycota)	black blotch of clover
Drechslera andersenii (Hyphomycetes)	leaf spot of ryegrass
Drechslera dictyoides (Hyphomycetes)	– anamorph of *Pyrenophora dictyoides*
Drechslera festucae (Hyphomycetes)	leaf spot of tall fescue
Drechslera phlei (Hyphomycetes)	leaf spot of timothy and cocksfoot, and melting-out of amenity grasses

Drechslera poae (Hyphomycetes)	melting-out of amenity grasses
Drechslera siccans (Hyphomycetes)	– anamorph of *Pyrenophora lolii*
Erysiphe betae (Ascomycota)	powdery mildew of fodder beet and mangold
Erysiphe cruciferarum (Ascomycota)	powdery mildew of brassicas
Erysiphe pisi (Ascomycota)	powdery mildew of lucerne
Eryisphe trifolii (Ascomycota)	powdery mildew of clover
Fusarium spp. (Hyphomycetes)	stalk rot of maize
Fusarium culmorum (Hyphomycetes)	pre-emergence disease of grass seedlings
Gaeumannomyces graminis (Ascomycota)	take-all of grasses
Helicobasidium purpureum (Basidiomycetes)	violet root rot
Kabatiella caulivora (Hyphomycetes)	scorch of clover
Laetisaria fuciformis (Basidiomycetes)	red thread of amenity grasses
Leptosphaeria maculans (Ascomycota)	canker of brassicas
Leptosphaerulina trifolii (Ascomycota)	pepper spot of clover
Limonomyces roseipellis (Basidiomycetes)	pink patch of amenity grasses
Marasmius oreades (Basidiomycetes)	fairy rings of amenity grass
Microdochium nivale (Hyphomycetes)	– anamorph of *Monographella nivalis*
Monographella nivalis (Ascomycota)	fusarium patch and snow mould of grasses
Peronospora farinosa f. sp. *betae* (Oomycetes)	downy mildew of fodder beet and mangold
Peronospora parasitica (Oomycetes)	downy mildew of brassicas
Peronospora trifoliorum (Oomycetes)	downy mildew of clover and lucerne
Physoderma alfalfae (Chytridiomycetes)	crown wart of lucerne
Phytophthora megasperma (Oomycetes)	root rot of kale
Plasmodiophora brassicae (Plasmodiophoromycetes)	clubroot of brassicas
Pleospora bjoerlingii (Ascomycota)	black leg of fodder beet and mangold
Pseudopeziza medicaginis (Ascomycota)	leaf spot of lucerne
Pseudopeziza trifolii (Ascomycota)	leaf spot of clover
Puccinia coronata (Teliomycetes)	crown rust of ryegrass
Puccinia graminis (Teliomycetes)	stem rust of ryegrass
Puccinia recondita f. sp. *lolii* (Teliomycetes)	brown rust of ryegrass
Pyrenophora dictyoides (Ascomycota)	leaf spot of meadow fescue and melting-out of amenity grasses
Pyrenophora lolii (Ascomycota)	leaf spot and foot rot of ryegrass and melting-out of amenity grasses
Pythium spp. (Oomycetes)	damping-off of grasses, maize and fodder crops
Ramularia beticola (Hyphomycetes)	leaf spot of fodder beet and mangold
Rhizoctonia solani (Hyphomycetes)	– anamorph of *Thanatephorus cucumeris*
Rhynchosporium orthosporum (Hyphomycetes)	leaf blotch of grasses
Rhynchosporium secalis (Hyphomycetes)	leaf blotch of grasses
Sclerotinia homeocarpa (Ascomycota)	dollar spot of amenity grasses
Sclerotinia trifoliorum (Ascomycota)	clover rot
Thanatephorus cucumeris (Basidiomycetes)	wirestem of brassicas and brown patch of amenity grasses
Typhula incarnata (Basidiomycetes)	grey snow mould (snow rot)
Uromyces betae (Teliomycetes)	rust of fodder beet and mangold
Uromyces fallens (Teliomycetes)	rust of red clover

Uromyces onobrychidis (Teliomycetes)	rust of sainfoin
Uromyces pisi (Teliomycetes)	rust of lucerne
Uromyces trifolii (Teliomycetes)	rust of white clover
Ustilago maydis (Ustomycetes)	smut of maize
Verticillium albo-atrum (Hyphomycetes)	wilt of lucerne

* For fungi, the classification in parentheses refers to class, although this is not possible within the phylum Ascomycota where classes have yet to be satisfactorily defined (see *Mycological Research*, February 2000). Oomycetes are now classified in Chromista with the brown algae, rather than as true fungi. Plasmodiophoromycetes are now classified as Protozoa rather than as true fungi. Some fungi have an asexual (anamorph) and a sexual (teleomorph) state, and the convention is to refer to them by their teleomorph name. However, where anamorph names are still in common use these are listed and cross-referenced to the teleomorph name. Strictly, fungi classified as Coelomycetes and Hyphomycetes should be known as 'hyphomycetous anamorphs' and 'coelomycetous anamorphs' of the relevant teleomorph taxon (e.g. hyphomycetous anamorphic Sclerotiniaceae, for *Botrytis fabae*), respectively. These problems highlight the continual changes in the classification of the fungi.

Chapter 5
Pests and Diseases of Potatoes

A. Lane
Independent Consultant, Church Aston, Shropshire

N.J. Bradshaw
ADAS Consulting Ltd, Cardiff, South Glamorgan

D. Buckley
ADAS Wolverhampton, West Midlands

Potatoes

Introduction

The area of potatoes grown in the UK, together with the number of producers, has been declining slowly over the last 10 years. However, over the same period, yields have increased to an average of 45 t/ha in 1998. This has been due largely to widespread adoption of irrigation and improvements in soil management aided by more effective machinery for seedbed preparation. Potato production is now in the hands of fewer, but more specialized, growers.

The UK crop is grown on approximately 160 000 ha, of which 18 000 ha are used for seed production. Of this seed area, 14 000 ha are in Scotland, 2500 ha in England and Wales and 1500 ha in Northern Ireland. The distribution of area cropped with early, second-early and maincrop potatoes is now much less distinct, as the market seeks all-year-round supplies of an ever-increasing variety of potato products, including salad, pre-pack, baking, chipping and crisping potatoes. Demand for potatoes and potato products is estimated at 6.5 million tonnes, of which 1.5 million tonnes are imported.

Trends in purchase and consumption of potatoes continue to change. The increasing importance of supermarkets, currently responsible for in excess of 65% of fresh potato sales, has been accompanied by a decline in the traditional wholesale outlets. In addition, the processing sector continues to grow at the expense of the fresh-market sector as demand for convenience food increases.

UK producers are having to meet ever increasing demands for high-quality potatoes, especially those grown for direct retail sales as pre-pack, where appearance is very important. At the same time, however, and in common with most other farm crop enterprises, there is increasing pressure for potatoes to be grown using integrated crop management (ICM) principles. This has led to the introduction and adoption of crop protocols and quality assurance/verification schemes for potato production. Most supermarkets, and other buyers of potatoes, now insist that growers comply with specific production protocols. These require an integrated approach to pest and disease control, taking into account pre-planting planning, crop rotation, nutrition, cultivar choice and use of pesticides.

Potato cultivars are selected and managed to meet specific market outlets. Although their suitability for certain soil types and locations may affect cultivar choice, market outlet is the main influence. This inevitably makes it more difficult to manage crop-protection programmes, especially those for potato cyst nematodes (PCN), soil-borne *Thanatephorus cucumeris* (anamorph: *Rhizoctonia solani*) and late blight. Many of the recently introduced potato cultivars exhibit some resistance/tolerance to certain pests and diseases. However, unless they have outstanding quality characteristics for a particular market outlet, they will not be grown. For example, cultivars with some resistance to the white potato cyst nematode (*Globodera pallida*) are readily available, but are not widely grown because of limited demand. Much of the industry, therefore, still depends on cultivars introduced many years ago, e.g. Maris Piper (introduced in 1962) and Estima (introduced in 1973), with their inherent susceptibility to pests and diseases and demandingly high levels of crop-protection inputs.

Appearance of potatoes grown for direct retail sales is very important. Therefore, growers are increasingly seeking to avoid tuber blemishes caused by diseases, and damage during harvest and handling which can contribute to the development of tuber diseases during storage. The main disease affecting potatoes in store is silver scurf. Although fungicides are available for application post-harvest, increased usage of refrigerated storage has greatly improved control of silver scurf and reduced infection by other storage diseases.

The fact that potatoes are planted, harvested, stored and consumed in the vegetative state (as the tuber) imposes rather different crop-protection problems from those applying to most other arable crops. In particular, the number of diseases caused by bacteria, fungi and viruses that can be transmitted in or on the potato tuber demands a high level of seed regulation to ensure reliable tuber and crop health. In the past, most effort was directed towards minimizing virus infection and spread. Whilst this is still paramount, the importance of bacterial and fungal diseases affecting the storability and appearance of potatoes has led to a broadening of focus in the production of healthy seed. A short growing season plus timely haulm destruction and early harvest under dry soil conditions are now accepted as being necessary to minimize disease levels in seed potatoes.

Sound crop-protection programmes depend on forward planning, good basic husbandry and appropriate rotations, because only in the case of a few pests and diseases can full remedial measures be taken once the crop is planted. This applies especially to soil-borne pests, such as potato cyst nematodes and wireworms, the latter becoming more troublesome in all arable rotations.

Despite the recent decline in area grown, the use of pesticides on potatoes is substantial and considered essential to maintain yield and quality. In 1998, ware potatoes in Great Britain received an average of 12 pesticide applications, accounted for largely by fungicides applied to control late blight; ware crops were treated an average of nine times with a blight fungicide. Insecticide usage was most extensive on crops grown for seed, illustrating the priority given to minimizing levels of aphid-borne virus diseases. The effectiveness of many aphicides used to control the main virus vector, *Myzus persicae*, is now limited, owing to the occurrence in this aphid of several mechanisms conferring resistance to insecticides (see below).

PCN continues to be the most damaging pest of potatoes, with an annual cost to the industry estimated at £50 million. Effective control depends on the adoption of a management programme integrating rotation, cultivar choice and nematicides. However, because of the limited availablity of irrigation and the high cost of potato-growing land, rotations tend to be shorter than is desirable for keeping PCN in check. This, and the practice in some areas of double cropping, which gives two crops in one season from the same field, has resulted in increased PCN infestations, especially of *Globodera pallida*, and has also led to an increase in soil-borne infections of stem canker caused by *Thanatephorus cucumeris* (anamorph: *Rhizoctonia solani*).

Late blight remains the greatest single threat to the potato crop, as it has done since the Irish famine of the mid-nineteenth century. This is not helped by the market's dependence on older, more susceptible cultivars. Because blight is potentially so serious a threat to potato production, the majority of maincrops are routinely treated with fungicides applied in a spray programme adjusted according to cultivar, risk of infection and cost of treatment.

Pests

Aphids

These pests can cause yield loss to potato crops as a result of their feeding on sap and by the transmission of viruses. Damage from aphid feeding is significant only when large numbers are present. In general, yield losses due to virus diseases are more serious than those caused directly by aphids.

Four species of aphid occur commonly on potato foliage. These are peach/potato aphid (*Myzus persicae*), potato aphid (*Macrosiphum euphorbiae*), glasshouse & potato aphid (*Aulacorthum solani*) and buckthorn/potato aphid (*Aphis nasturtii*). Other species occasionally breed on potato foliage, including shallot aphid (*Myzus ascalonicus*) and violet aphid (*Myzus ornatus*). Black bean aphid (*Aphis fabae*), and other migratory species which alight and feed on potato foliage may contribute to the spread of certain virus diseases.

In the potato store or chitting house, bulb & potato aphid (*Rhopalosiphonius latysiphon*), *A. solani*, *M. euphorbiae* and *M. persicae* are sometimes found colonizing the sprouts of seed potatoes. In addition to the direct damage they cause, all are capable of spreading virus diseases, the potato being particularly susceptible to infection at this stage. To control aphid infestations on seed potatoes, fumigation with nicotine shreds is recommended.

In the growing crop, numbers of each species vary considerably between seasons and localities. *A. nasturtii, A. solani* and *M. euphorbiae* are of little importance as field vectors of potato virus diseases but may cause physical damage to the foliage and yield loss. *M. persicae* is sometimes present in sufficient numbers to cause direct feeding damage, but is more important as the main vector of potato leaf roll virus (PLRV) and potato virus Y (PVY), the most damaging aphid-borne viruses of potatoes (see p. 162).

Virus can be spread by aphids in two ways. For PVY and some others, the aphid quickly acquires virus as it feeds on infected plants and can then rapidly transmit it on moving to a healthy plant – the process often taking only a few minutes. This kind of virus spread is called stylet-borne or, since the virus is quickly lost, non-persistent transmission. By contrast, some viruses such as PVY are acquired only after long feeds by the aphid and transmission can take several hours. This is called circulative transmission or, since the virus once acquired by the aphid is retained for life, persistent transmission.

Chemical treatments available for aphid control in the field are summarized in Table 5.1. The effectiveness of these treatments against *M. persicae* is limited owing to the occurrence of strains with varying levels of resistance to many of the compounds listed.

Three mechanisms conferring resistance to insecticides in *M. persicae* have been identified.

Table 5.1 Insecticides recommended for control of aphids on potato

Active ingredient	Chemical group
Granules applied at planting	
aldicarb*	monomethyl carbamate
disulfoton	OP
oxamyl*	monomethyl carbamate
phorate	OP
Sprays applied to growing crop	
deltamethrin + heptenophos	pyrethroid + OP
deltamethrin + pirimicarb	pyrethroid + dimethyl carbamate
demeton-S-methyl	OP
dimethoate**	OP
lambda-cyhalothrin	pyrethroid
lambda-cyhalothrin + pirimicarb	pyrethroid + dimethyl carbamate
malathion	OP
nicotine	nicotinyl
pirimicarb	dimethyl carbamate

* When applied overall and incorporated for PCN control, will give some control of early aphid infestations.

** Not recommended for control of *Myzus persicae*.

The well-established carboxylesterase-based resistance, discovered in the mid-1970s, is widespread. Resistant aphids are classified according to the levels of esterase enzymes they contain: S (susceptible); R1 (mildly resistant); R2 (moderately resistant); and R3 (highly resistant). Most of the main groups of insecticides, i.e. organophosphates (OPs), pyrethroids and carbamates, are affected to a greater or lesser extent by this resistance mechanism. OPs and pyrethroids are the groups mostly affected; carbamates such as pirimicarb and aldicarb are least affected, whilst nicotine-containing products are not affected at all. Results from recent surveys have shown that most populations of *M. persicae* now contain R1 and R2 aphids; R3 aphids are uncommon in the field and are found only at the end of the growing season.

The resistance problem in *M. persicae* has recently been accentuated by an additional mechanism in which the aphids also contain an insecticide-insensitive form of acetylcholinesterase, the target for organophosphorus and carbamate insecticides. This new mechanism, termed MACE (Modified Acetyl-CholinEsterase), which confers resistance specifically to pirimicarb and triazamate, was first discovered in field crops in Lincolnshire in 1996 and has now been found in aphid populations from other regions.

The presence of these two resistance mechanisms in this species, together with a third but as yet less common mechanism, kdr (knock-down resistance) specific to pyrethroids, has important consequences for *M. persicae* control in potatoes. The advent of MACE resistance is a very recent phenomenon and consequently there has been insufficient time to test control strategies experimentally. Table 5.2 summarizes the likely relative performance of the commonly used aphicides against different resistant strains of *M. persicae*. These guidelines are based on the best information available and will need to be refined in the light of experience and further research.

Where crops are grown for seed (certified or home-saved), effective control of *M. persicae* is essential to minimize the risk of virus infection. Whilst insecticides cannot prevent the entry of PVY into the potato crop, products based on pyrethroids (which exert an anti-feedant and/or repellent action on the aphid), may reduce in-crop spread of the virus. After a mild winter, or when forecasts predict early aphid infestations, the use of a granular aphicide at planting is recommended, to limit introduction and within-crop spread of PLRV. The alternative is to apply a number of sprays, starting at 80% crop emergence and continued until 10–14 days before haulm destruction. The choice of insecticide will depend on some knowledge of the success of previous aphid control strategies. For crops grown in areas where MACE resistance is present, the first two foliar sprays should be based on a pyrethroid or on an OP + pyrethroid mixture. If aphid numbers build up rapidly following the two-spray programme, a nicotine spray is recommended. For crops grown in non-MACE areas, spray programmes should be based on pirimicarb or on pirimicarb/pyrethroid mixtures. Whatever programme is used, it is important to monitor the effectiveness of the treatments and to seek specialist advice if aphids are not being controlled.

Table 5.2 Likely level of control of resistant *Myzus persicae* by insecticides approved for use on potato (+ + + = good, + + = fair, + = poor)

Active ingredient	Resistance type				
	S	R1	R2	R3	MACE*
Dimethyl carbamate					
pirimicarb	+ + +	+ + +	+ +	+	none
Monomethyl carbamate					
aldicarb	+ + +	+ + +	+ +	+	+ + +
oxamyl	+ + +	+ + +	+ +	+	+ + +
Organophosphate (OP)					
demeton-S-methyl	+ + +	+ +	+	none	+ + +
disulfoton	+ + +	+ +	+	none	+ + +
phorate	+ + +	+ +	+	none	+ + +
Nicotinyl					
nicotine	+ +	+ +	+ +	+ +	+ +
Pyrethroid					
lambda-cyhalothrin	+ + +	+ +	+	none	+ + +
Dimethyl carbamate + pyrethroid					
pirimicarb + deltamethrin	+ + +	+ + +	+ +	+	+ + +
pirimicarb + lambda-cyhalothrin	+ + +	+ + +	+ +	+	+ + +
OP + pyrethroid					
heptenophos + deltamethrin	+ + +	+ +	+	none	+ + +

* Control rating for MACE assumes it is the only resistance mechanism present. When present with carboxylesterase resistance, control with active ingredients rated as + + + will be reduced.

For ware crops there is minimal risk of yield and quality reduction by virus infection acquired in the same year. Insecticide treatment is recommended only for those cultivars susceptible to aphid-induced false top roll (e.g. Desirée and Record) and when aphid numbers are increasing rapidly. A single spray of the specific aphicide pirimicarb, timed to coincide with aphid population increase (usually late June/early July) should be sufficient in most seasons. Crops treated with a soil-applied pesticide for PCN control (see p. 131) may benefit from a suppression of early aphid infestation.

Crops intended for home-saved seed should be isolated from other potato crops. Only healthy seed should be planted, and this should be followed up by careful removal of virus-infected plants during the early part of the growing season. The build-up of aphid natural enemies may be encouraged by using specific aphicides, such as pirimicarb. Use of a virus testing service to establish the health of the harvested crop is recommended. Destruction of the foliage of seed crops should be done as early as possible.

Capsids

The three main potato-infesting species of capsid are common green capsid

(*Lygocoris pabulinus*), potato capsid (*Calocoris norvegicus*) and tarnished plant bug (*Lygus rugulipennis*). These, together with occasional feeders such as slender grey capsid (*Dicyphus errans*), feed on many herbaceous and woody plants. Extensive brown necrotic spots may be caused on potato foliage, the brown tissue later collapsing to leave holes. Young shoots and foliage may die or become distorted under heavy attacks. Damage is usually confined to plants on headlands.

Chemical control measures are seldom if ever required for whole fields. Granules of phorate, applied at or before planting to control aphids, also control capsids.

Caterpillars

The caterpillars of a number of moth species occasionally damage the foliage, stems or tubers of potato plants. Attacks are often sporadic and chemical control measures for most species are rarely necessary.

Cutworms

These are the caterpillars of a number of noctuid moths, particularly turnip moth (*Agrotis segetum*). Attacks are worse in hot, dry summers in light textured soils. The stems and roots of potato plants are bitten and severed near to ground level; more importantly, cutworms tunnel into the tubers.

Other cutworms, with similar habits and life-cycle to the turnip moth and which can be found damaging potato tubers include caterpillars of large yellow underwing moth (*Noctua pronuba*) and garden dart moth (*Euxoa nigricans*).

Insecticide sprays are effective only if applied while the cutworms are small and are feeding above ground on the foliage, usually during late June/early July. Larger cutworms remain at or below ground level, where they are very difficult to control. Spray warnings, based on predictive models using temperature and rainfall data, are available to the industry and these indicate when treatment of susceptible crops is necessary. Spray treatments include chlorpyrifos, cypermethrin and lambda-cyhalothrin + pirimicarb – all applied at high volume.

Land should be kept as free from weeds as possible to deter egg laying. Because young cutworms cannot survive in wet soil, frequent irrigation should also help to prevent the development of damaging infestations.

There are several other species of noctuid moth caterpillars that are occasionally found on potato. Although closely related to cutworms, they tend to feed on the aerial parts of the plant and are not usually found in the soil. The commonest is the caterpillar of rosy rustic moth (*Hydraecia micacea*), known as 'potato stem borer'. Stems of attacked plants are hollowed from the base upwards; in badly attacked plants, the foliage wilts and finally collapses. Others which damage potato foliage include caterpillars of silver y moth (*Autographa gamma*), tomato moth (*Lacanobia oleracea*) and angle-shades moth (*Phlogophora meticulosa*). Specific control measures are rarely necessary; sprays applied to control cutworms will give incidental control of other caterpillars.

Swift moths

Caterpillars of ghost swift moth (*Hepialus humuli*) and garden swift moth (*H. lupulinus*) may feed on the roots of potatoes grown after pasture or cereal stubbles infested with grass weeds. Occasionally, the caterpillars tunnel into developing tubers.

Soil cultivations made before planting potatoes help to keep swift moth caterpillars in check by disturbing and injuring them and exposing them to birds. Chemical control measures are not usually necessary or practical.

Chafer grubs

These pests (larvae of chafers) occasionally attack potatoes planted after old pasture, damaging the roots and the tubers. The cockchafer (*Melolontha melolontha*) and the garden chafer (*Phyllopertha horticola*) are most commonly found attacking potato. Attacks are sporadic and difficult to control chemically.

Avoid planting potatoes immediately after old pasture in areas where chafer grubs are frequently seen. Thorough cultivation of the soil before planting destroys eggs and grubs.

Leafhoppers

Green leafhoppers (*Edwardsiana flavescens* and *Empoasca decipiens*) and potato leafhoppers (*Eupterycyba jucunda* and *Eupteryx aurata*) frequently suck the sap from potato foliage, sometimes causing speckling, browning and wilting of leaves. The damage is never serious and, unlike aphids, leafhoppers do not transmit potato viruses in the UK.

Many of the foliar sprays applied to control aphids on potatoes will give some reduction in leafhopper damage.

Leatherjackets

These are the larvae of crane flies (daddy longlegs) and occur commonly in grassland. The species most commonly encountered are *Tipula oleracea* and *T. paludosa*. Potato may be damaged by leatherjackets when a crop is grown after grassland. The shoots emerging from the planted tuber are sometimes eaten, but damage is usually negligible on a field scale.

Control measures are rarely necessary, but methiocarb pellets, as used for slug control (see p. 136), and high-volume sprays of chlorpyrifos, will give some control of leatherjackets feeding close to the soil surface.

Millepedes

The spotted snake millepede (*Blaniulus guttulatus*), one species of black millepede (*Cylindroiulus londinensis*) and a flat millepede (*Polydesmus angustus*) commonly feed in holes in potato tubers already attacked by other pests such as slugs and wireworms. When abundant, however, millepedes can act as primary feeders, scarring the tuber surface or tunnelling into the flesh to form shallow cavities.

Millepedes are common in many fields but are most numerous on heavier, wetter soils, especially those with a high organic content.

Millepedes are difficult to control. Granular carbamate pesticides, such as aldicarb, carbofuran or oxamyl used against nematodes (see below), will give some control of millepedes.

Nematodes

A number of nematode species feed on and damage potatoes. Those that live exclusively in the soil and feed externally on roots and tubers are known as ectoparasites, e.g. stubby-root nematodes. Others, which spend most of their life-cycle within plant tissue, are endoparasites; these may be relatively immobile or sedentary, e.g. PCN (Table 5.3).

Table 5.3 Nematode pests of potato

Type of attack	Nematode species
Below ground on roots	
(a) Endoparasitic – sedentary	Potato cyst nematodes (*Globodera* spp.)
(b) Endoparasitic – mobile	Root-lesion nematode (*Pratylenchus penetrans*)
(c) Ectoparasitic – mobile	Needle nematodes (*Longidorus* spp.)
	Stubby-root nematodes (e.g. *Trichodorus* spp.)
Below ground on tubers	
(d) Endoparasitic – mobile	Potato tuber nematode (*Ditylenchus destructor*)
	Stem nematode (*Ditylenchus dipsaci*)
On stems and leaves	
(e) Endoparasitic – mobile	Stem nematode (*Ditylenchus dipsaci*)

Needle nematodes

Needle nematodes (*Longidorus* spp.) are commonly occurring ectoparasites. One species, *Longidorus leptocephalus*, feeds on potato roots growing below cultivation depth and may reduce yield when present in large numbers. Chemical control measures directed specifically against needle nematodes are not usually necessary or practical. Soil fumigants applied for PCN control (see below) will give some reductions in numbers.

Potato cyst nematodes (PCN)

Two species, yellow potato cyst nematode (*Globodera rostochiensis*) and white potato cyst nematode (*G. pallida*), are present in the UK, and they are by far the most important pests of potatoes. Recent surveys have shown that approximately 42% of land currently cropped with potatoes in the UK is infested with PCN, resulting in annual losses to the industry in the order of £50 million. In addition to their effects on yield, PCN are subject to stringent plant health regulations. Seed potatoes cannot be sold within the UK unless grown in land shown to be free from PCN. The pests also have important implications for the export of potatoes and other plant material (e.g. bulbs, nursery stock) from the UK, as many importing countries prohibit entry of such material unless it was grown in land in which no PCN has been found.

Both PCN species exist as biological races or pathotypes, which can be distinguished only by their ability to multiply on certain resistant potato cultivars. Three pathotypes, one of *G. rostochiensis* (Ro1) and two of *G. pallida* (Pa1 and Pa2/3), are known to be present in the UK. However, only Ro1 and Pa1 can be regarded as distinct pathotypes. Surveys have shown that over the last 20 years *G. pallida* has become the most dominant species in the main potato-producing areas and that, currently, infestation levels are increasing. Control of *G. pallida* has proved to be more difficult than for *G. rostochiensis*, because of the former's slower rate of decline in the absence of a potato crop and longer period of egg hatch, together with lack of availability of potato cultivars with total resistance to this species.

Effective management of PCN depends on the integration of some or all of the available control options, especially crop rotation, resistant cultivars and chemical controls (use of nematicides). The overall aim is to limit yield loss in the current potato crop and to reduce PCN populations likely to damage future crops.

Over the last 10 years, potato cropping has become more intensive, resulting in closer rotations. Growing potatoes one year in four or less will not allow PCN to decline to safe levels, even with chemical controls. As a basis for rotational planning, regular and intensive soil sampling to track PCN population levels and species composition is recommended.

The use of PCN-resistant potato cultivars to reduce nematode infestations can be very effective. Many cultivars now incorporate resistance to *G. rostochiensis* (Ro1), e.g. Cara, Maris Piper and Nadine. However, their use is limited, as populations of this species have declined to very low levels. Resistance to the predominant *G. pallida* is present in only a few commercially available cultivars, e.g. Midas, Sante and Sierra, but the resistance is only partial (80% or less). These partial resisters, therefore, only reduce PCN multiplication rather than prevent it, and some are intolerant of nematode attack, so requiring nematicide treatment to maintain yield. The NIAB recommended list of potato varieties (cultivars) identifies those with Ro1 resistance and gives a rating for resistance to *G. pallida*.

Nematicides suitable for use in the management of PCN are listed in Table 5.4. Essentially, granular nematicides are used to minimize damage caused by nematode attack, whilst also giving some measure of population control. Their effectiveness depends on PCN infestation level, method of application and soil type. They are generally less effective against *G. pallida*, whose eggs hatch over a longer period. Fumigant nematicides are recommended to reduce very large PCN infestations to levels where they become more manageable by other methods. Under ideal conditions, up to 80% control of PCN can be achieved. Fumigant nematicides are equally effective against both species of PCN. They are only

Table 5.4 Nematicides for control of potato cyst nematodes (PCN)

Nematicide	Application method	Comments
Granular nematicides		
aldicarb, carbofuran, ethoprophos, fosthiazate oxamyl	Broadcast overall by means of suitable granular applicator just before planting. Incorporate thoroughly to depth of 10–15 cm, preferably with rotary cultivator or power harrows.	Effective only when used on the growing crop. More effective at controlling PCN damage than reducing PCN population increase. Poor control may result when granules applied during stone/clod separation.
aldicarb		Controls early-session aphid infestations and reduces incidence of spraing.
carbofuran		Avoid wet or waterlogged soils.
ethoprophos		Pest control reduced on organic soils.
fosthiazate		Do not apply more than once every 4 years on same area of ground.
oxamyl		Some control of early aphid infestations and reduces incidence of spraing.
Fumigant nematicides		
1,3-dichloropropene	Use in early autumn of any year in rotation when soil is warm (at least 6°C), friable and moist. Inject to a depth of 20–30 cm. Soil should be sealed immediately after injection.	Do not use in heavy clay or in high organic soils or in soils with a high proportion of large stones. Treatment may be more effective in the autumn immediately after potato crop. Also controls other soil nematodes and can reduce incidence of spraing.

suitable, however, for use on medium-to-light soil types that can be sealed after nematicide application.

An alternative non-chemical means of reducing high *G. pallida* infestations is by use of trap cropping. This involves planting closely spaced potato tubers, which encourages the PCN juvenile nematodes to 'hatch' and invade the potato roots. The plants are then lifted a few weeks after planting and destroyed. This prevents the nematode population from increasing and substantially reduces the soil nematode infestation. Recent research has demonstrated that trap cropping can reduce infestations by up to 80%. However, the technique is costly and requires very careful management. Also, failure to destroy the trap crop in time can increase, rather than decrease, the PCN population.

Successful management of PCN, especially *G. pallida*, depends on adopting a long-term strategy integrating the various control methods described above until cultivars which are totally resistant and also tolerant of attack become available. On the basis of regular soil sampling for PCN, decisions can be made as to the approach to be taken (see Table 5.5).

Table 5.5 Interpretation of soil sampling results for planning management of *Globodera pallida*

	Nematode infestation category		
Not found/very low (no viable cysts found)	Low (1–10 eggs/g)	Moderate (11–60 eggs/g)	High (> 60 eggs/g)
Safe to grow potatoes without chemical treatment, but continue to soil-sample regularly.	Apply a granular nematicide at 5 eggs/g or more if growing an intolerant variety on very light soils.	Apply a granular nematicide with a partially resistant cultivar, to reduce crop damage and limit PCN increase.	Use a fumigant nematicide or trap crop to reduce PCN infestation.
	Apply a granular nematicide if cropping closer than 1 year in 6, to limit PCN increase.	Increase crop rotation to at least 1 year in 6.	Resample to check success of controls.
	Cropping with a partially resistant cultivar will limit PCN increase.		Apply a granular nematicide with a partially resistant cultivar, to reduce crop damage and limit PCN increase.
			Increase crop rotation to at least one crop in six.

Potato tuber nematode

Potato tuber nematode (*Ditylenchus destructor*) is an endoparasitic species which attacks potato, bulbous *Iris* and corms or tubers of *Dahlia* (and has a number of weed hosts) but infestations are uncommon. Infested potato plants usually show symptoms of damage but only in the tubers.

There are no effective chemical control measures available for potatoes. Infested potatoes should not be used for seed. Effective weed control helps to eliminate potato tuber nematode from field soils.

Root-lesion nematode

Root-lesion nematode (*Pratylenchus penetrans*), a mobile endoparasitic species, feeds on the roots of potato, causing patchy growth. *P. penetrans* is troublesome only locally and is found chiefly in south-west England on light sandy soils; damage to potatoes is uncommon.

Nematicide granules used for PCN control (see above) will give some control of root-lesion nematodes and may reduce damage.

Stem nematode

Stem nematode (*Ditylenchus dipsaci*), like the closely related potato tuber nematode, is an endoparasitic species and a destructive pest of many crops. Although potatoes are a host of stem nematode, and damage to the stems and leaves is often recorded in continental Europe, attacks are rare in the UK.

Stubby-root nematodes

Nematodes of the genera *Trichodorus* and *Paratrichodorus* feed externally (ectoparasitically) on potato roots and developing tubers. They are restricted to light, open-textured soils such as the sandy soils of the Vale of York, Norfolk and parts of the West Midlands. Stubby-root nematodes are capable of causing direct feeding injury; however, more importantly, they can transmit tobacco rattle virus (TRV), which produces an internal disorder of the potato tuber called spraing (see p. 160). Virus infection does not affect crop yield but reduces tuber quality in susceptible cultivars. In some seasons, crops can suffer severe loss, affected tubers being unacceptable for sale yet impossible to grade out. Less frequently, potato mop top virus (PMTV), which is transmitted by the powdery scab fungus, also causes spraing-like symptoms (see p. 161).

Potato cultivars differ in their susceptibility to TRV, e.g. Pentland Dell is highly susceptible whereas Record is one of the least susceptible to damage. The NIAB recommended list of potato varieties (cultivars) gives a rating for spraing-TRV susceptibility and should be consulted when choosing a cultivar to be grown on fields with a history of the disorder.

Soil nematicide treatments designed primarily for PCN control (see Table 5.4, p. 133) and/or aphid control will give some control of the nematode vectors and reduce the incidence of spraing symptoms. The granular nematicides aldicarb and oxamyl are currently recommended as in-furrow treatments to reduce spraing on susceptible cultivars. These, and other granular nematicides applied overall before planting for PCN control, as well as the fumigant nematicide 1,3-dichloropropene, may also give a reduction in spraing symptoms.

Soils can be tested for the presence of the nematode vectors, but it is not practicable to test these for TRV. However, spraing soils are usually known from previous experience and in these situations highly susceptible cultivars should be avoided. TRV can persist in populations of the vector nematode for several years, because the virus also infects some common weeds including common chickweed (*Stellaria media*) and shepherd's purse (*Capsella bursa-pastoris*). Effective weed control will reduce the persistence of the virus. Some crops grown in rotation with potatoes can help to reduce the incidence of spraing. For example, barley, although a good host of *Trichodorus* spp., is not a host of TRV and so the risk of spraing in potatoes declines after a series of barley crops.

Potato flea beetle (*Psylliodes affinis*)

This flea beetle feeds on potato leaves, producing the characteristic 'shot holing' effect. Damage is of little consequence unless beetle infestations are high and plants are small. Many of the foliar sprays used to control aphids on potatoes, especially those containing a pyrethroid, will give incidental control of this pest.

Slugs

Damage to newly formed potato tubers may be serious when wet autumns follow mild, wet summers, especially when harvesting is delayed. Crops grown in heavier, more-moisture-retaining soils are most at risk. Garden slug (*Arion hortensis*) and two species of keeled slug (*Milax gigantes* and *Tandonia budapestensis*) cause the most serious damage to potatoes. Field slug (*Deroceras reticulatum*), which is the most common slug species and largely a surface-feeding pest that frequently damages cereal crops, is less important on potatoes than the subterranean species cited above. However, it is often found in holed tubers and can cause considerable secondary damage.

Although slugs are often secondary feeders (enlarging holes made in tubers by other pests), they can penetrate the tuber skin as primary feeders. The entry hole is usually small and circular, leading to large tunnels eaten deep into the flesh of the tuber. This contrasts with cutworm damage (see p. 129), in which large, uneven holes are produced in the skin and shallow galleries are eaten into the tuber flesh. Even low levels of slug damage can affect the marketability of crops grown for pre-packing.

Potato cultivars vary in their susceptibility to slug attack. This is known to be associated with the starch content of the tuber and the concentration of secondary compounds such as glyco-alkaloids in the skin. The NIAB recommended list of potato varieties (cultivars) gives a rating for susceptibility to slug damage. The following maincrop cultivars are listed in order of increasing susceptibility: Pentland Dell (the least susceptible cultivar), Romano, Pentland Squire, Russet Burbank, Record, Desirée, Cara, Valor, Maris Piper (the most susceptible cultivar).

A worthwhile reduction of tuber damage can be achieved by applying pellets of metaldehyde, methiocarb or thiodicarb, broadcast over the potato ridges in late July and again in early August. It is important to make the applications when slugs are active on the soil surface. This can be gauged by test baiting with small quantities of pellets placed under tile traps distributed throughout the crop. When treating, the soil should be moist together with a good cover of foliage, so creating a humid environment in which slugs will be active. Applications made later, when slug damage becomes obvious, are unlikely to be effective.

Potatoes, especially maincrop cultivars, should be lifted as soon as possible after tubers become mature. Also, avoid planting very susceptible cultivars on fields with a history of slug damage.

Wireworms

Wireworms are the larvae of certain click beetles, of which *Agriotes lineatus*, *A. obscurus* and *A. sputator* are the most common and responsible for the majority of attacks on potatoes. Damage may also be caused by species of the genera *Athous* and *Ctenicera*, which can be found in mixed populations with *Agriotes*.

Wireworms are abundant in pasture and they frequently attack arable crops, including potatoes, in the first few years after ploughing up old grassland. However, even populations as low as 75 000/ha (sometimes found in arable fields without a history of grassland in the rotation) can damage a potato crop. The larvae tunnel deeply into potato tubers, leaving small round holes on the surface. Early-season attacks on seed tubers and sprouts are not usually of much consequence. Later in the season, holing of newly formed tubers, although not affecting yield, causes a serious loss in quality and provides access for slugs, millepedes and other soil organisms. Damage increases the longer the crop remains in the ground; early-maturing crops are little affected.

Damage caused by wireworms used to be regarded as a sporadic problem, mainly of concern to growers in mixed farming areas in northern and western areas. In recent years, however, wireworm damage has become increasingly widespread. This may be due to a number of factors, including: (a) more stringent quality demands from retailers which have lowered the tolerance for wireworm and other pest damage; (b) an increase in the use of rented land for growing potatoes, often involving ploughing up old grassland, and (c) a marked increase in wireworm damage in all-arable rotations. Recent surveys have shown that wireworm populations can build up on set-aside land.

It is becoming increasingly important to apply integrated management strategies for managing wireworms. Pre-crop sampling to detect wireworm infestations, using either soil sampling or the recently introduced bait trapping method, should be regarded only as tests for the presence or absence of the pest. This is because of the poor relationship between sampling or bait trap catches and subsequent crop damage. Where chemical control measures are considered necessary on crops at risk, applications of granular formulations of ethoprophos or phorate applied into the furrow close to the seed tuber will reduce but not prevent wireworm damage. Gamma-HCH (which must not be used if potatoes are to be planted within 18 months) can be applied to a preceding winter cereal crop within the rotation to reduce wireworm infestations. However, it does not necessarily eliminate the need for a further insecticide treatment to the potato crop.

Wireworm populations decrease quickly under arable rotations. After ploughing-out grassland, growing crops less likely to be damaged by wireworms (such as peas and beans) is advised. Where damage is expected, early-maturing potato cultivars are recommended or the lifting of maincrop cultivars as soon as tubers mature and skins set. Cultivate the soil thoroughly before ridging; this is likely to have most effect on populations when done in the autumn, when wireworms are active in the upper layers of the soil profile.

Diseases

Black dot (*Colletotrichum coccodes*)
Black dot is caused by the saprophytic (weakly parasitic) fungus, *Colletotrichum coccodes*. It is particularly common on the haulm, roots and stolons of dying and senescing potato plants at the end of the season. Numerous pinhead-sized resting bodies (sclerotia) are produced on the dead skin or on the underlying tissues. Infection of the ware tubers is becoming increasingly important and black dot is regarded as one of the more important tuber blemish diseases. This may be because symptoms are very similar to those caused by silver scurf (see below) and the industry is becoming more aware of an existing problem owing to an increasing demand for better skin finish.

Black dot can be distinguished from silver scurf (with the aid of a hand lens), by the presence of black sclerotia (0.5 mm diameter). Setae (bristles) are usually present on the sclerotia. Stress factors such as drought or over-irrigation, diseases (such as *Fusarium* and *Verticillium*) and pests (such as potato cyst nematode) may predispose crops to infection. Warm, moist conditions are reported to favour the development of black dot in store, so that dry curing (used to reduce bacterial soft rots and silver scurf), may also suppress the disease.

C. coccodes remains viable in the soil for a considerable period (and much longer than the 4–5 years between normal potato rotations), possibly for at least eight years. The contribution of soil-borne inoculum to the incidence of black dot on progeny tubers is at least twice that of tuber-borne inoculum, even when tubers are severely infected. Irrigation has been shown to increase the incidence and severity of black dot, especially later in the season. Any delay between senescence or defoliation and harvest increases infection levels of black dot.

No resistant cultivars are available and no fungicides have a current label recommendation for the control of black dot.

Black heart
This physiological disorder occurs in storage, mostly in processing crops, but is now rare. It is caused by asphyxiation of the centre of the potato tuber under conditions of low oxygen and high carbon dioxide concentrations in the affected tissues, leading to cell death. The condition is accentuated by high temperatures, as tuber respiration increases with increasing temperature. There are no external tuber symptoms.

Black heart is thought to be associated with the use of very effective materials for insulating stores, which restricts ventilation. This, together with the practice of re-circulating the same air within the store, may lead to decreased oxygen concentration and eventual asphyxiation of the tubers. Warnings of possible problems are loss of breath by operators within the store or difficulty in maintaining a lighted match or cigarette.

To reduce the risk of black heart, carbon dioxide levels should be maintained at approximately 0.05%. Assuming a respiration rate of 6 mg carbon dioxide/kg/h (i.e. normal steady-state storage conditions), the ventilation requirement would be in the range 30–60 m^3/t/day, depending on how much natural store leakage occurs.

Blackleg and tuber soft rots (*Erwinia carotovora* ssp. atroseptica and *E. carotovora* ssp. *carotovora*)
Blackleg is the haulm disease caused by the bacterium *Erwinia carotovora* ssp. *atroseptica*. This bacterium, and the closely related *E. carotovora* ssp. *carotovora*, is a major cause of tuber soft rotting, both in the ground and in store. In warmer climates, *E. carotovora* ssp. *carotovora* is reported to cause blackleg, and in the UK this pathogen is sometimes responsible in wet seasons for a bacterial rot in the mature haulm. In plants affected by blackleg the shoots are stunted, with pale green to yellowish foliage which has a tendency to wilt. The underground stem is slightly discoloured at the point of attachment to the seed tuber or, in the later stages of infection, a brown or black rot extends up the stem well above soil level. In wet conditions, blackleg appears as a soft wet rot of the stem, sometimes liquefying the internal tissues or spreading in black streaks higher up the plant. If affected stems are cut across at soil level, or above the blackened rotted area, a black to brown discoloration of the vascular tissue can be seen. As the season progresses, affected shoots usually wilt and die. Blackleg symptoms on the tuber are usually a soft rot at the heel end. The rotted tuber flesh is often bounded by a darker brown or black margin.

Blackleg is one of the commonest potato diseases, and a few affected plants may be found in most crops. It is occasionally serious, particularly in wet seasons, in irrigated crops and in those grown on poorly drained land, when progeny tubers may become infected and rot in the ridges before harvest. Infected seed tubers rot during establishment of the haulm, and bacteria released into the soil contaminate and infect the progeny tubers. In poorly drained, waterlogged soils the seed tuber breaks down earlier and the risk of tuber infection is greater. Although soil populations of *Erwinia* spp. rapidly decrease after potatoes, the blackleg bacterium and other *Erwinia* spp. persist in this way from season to season, without necessarily causing recognizable blackleg symptoms. This source of inoculum could lead to bacterial rotting in store, particularly if tubers are lifted in wet conditions and are stored wet and in poorly managed stores.

Although blackleg in the growing crop is mostly due to bacteria carried in or on the seed tubers, there are other ways by which healthy crops can become infected. For example, bacteria can be brought from nearby potato crops or dumps by insects, via contaminated irrigation water or in aerosol mists formed by rain-splash. Airborne distribution of blackleg bacteria can also occur when affected stems are pulverized mechanically during haulm destruction.

Store management
Whilst blackleg can reduce yield through premature plant death and loss of shoots, the main losses are through rotting during bulk storage (although severe losses are unusual nowadays). In box stores, severely rotting tubers can easily be removed and, therefore, do not pose a threat to tubers stored beneath them. Bacteria multiply in water on the surface of tubers and spread in store through drips and contact. Ventilation of stores should be such that moisture released during respiration will be removed and the occasional diseased tuber will be dried off. Air moving through a bulk of potatoes gathers moisture. As this air cools at the top of the store, moisture condenses out; if this runs back over the tubers, it will create conditions ideal for bacterial activity. Good roof insulation will prevent condensation and, together with a covering of straw over the stack of potatoes, should be able to contain the problem.

There are no potato cultivars resistant to blackleg, although some are more susceptible than others. Some control of blackleg is achieved through the certification scheme by the rejection of infected seed crops. However, inspection clearly cannot exclude tuber infections which occur in the absence of above-ground symptoms. A number of organisations offer a laboratory testing service to determine the level of contamination by blackleg bacteria in seed stocks. Although this test provides a valuable means of identifying high-risk stocks, the results require careful interpretation.

The use of disinfectants is currently being investigated as a means of improving store hygiene. Hot-water dip treatment (45°C for 30 minutes) developed as a technique to reduce *Erwinia* contamination of seed tubers is rarely used commercially. Where wet rots are noted at harvest, such tubers should not be stored. However, if wet rots develop in store, the tubers should receive short blasts of cool, dry air (to dry them) and the temperature should be reduced to 7–10°C for crisping potatoes and to 5–7°C for ware potatoes. The store should be closely monitored and if the temperature continues to rise, or if there is any smell or sinkage, then the store should be cleared as soon as possible.

Black scurf and stem canker (*Thanatephorus cucumeris* – anamorph: *Rhizoctonia solani*)
The black scurf/stem canker complex is both seed- and soil-borne. Black scurf is the seed-borne phase, and is so named because of the presence of conspicuous black, tar-like sclerotia of the fungus which appear on tuber skin. These sclerotia are often variable in size and are easily removed by the finger nail but, characteristically, do not cause any rotting of the tuber. In exceptional circumstances, black scurf reduces the marketability of pre-washed tubers, i.e. pre-pack and baker-trade ware potatoes. However, black scurf is not generally considered a major tuber blemish disease.

More importantly, the black sclerotia are an important source of inoculum in the initiation of stem canker (which is the field phase of the disease). Symptoms of stem canker appear on the young shoots below ground level as black/brown sunken lesions on the white stem tissue. In severe cases, stem canker infections can completely girdle the stems and cause 'pruning', which leads to death of the shoots (resulting in delayed emergence and gappy and uneven plant stands). Even in the most severe cases, secondary shoots are produced and infected plants survive to produce ware tubers. However, in early-planted crops this can lead to a significant delay in reaching maturity and, consequently, in attaining optimum yield and returns in sensitive market conditions.

Sometimes, on maincrop potatoes, the perfect stage of the fungus can be seen towards the end of the growing season as a white or fawn-coloured collar of fungal growth on the stems at soil level. This fungal growth can be easily rubbed off to reveal healthy stem tissue underneath. Inoculum of the stem canker fungus may be present in the soil and may build up where rotations are too close. In addition to inoculum on the seed as sclerotia, the fungus may also be present as fragments of mycelium on the tubers.

Factors which lead to a delay in crop emergence also predispose to stem canker infection. Cold, dry springs (when crop emergence is slow) will encourage stem canker on first-early crops in particular.

Fungicide treatment of seed with products containing iprodione, pencycuron or tolclofos-methyl has given excellent protection of developing sprouts from both seed- and soil-borne infection but rarely (except in first-early cultivars) a corresponding increase in yield. Iprodione, pencycuron and tolclofos-methyl also reduce black scurf on the ware crop. Routine fungicide treatment is probably not worthwhile, except where crops are particularly at risk (close rotations and/or a previous field history of the disease; first-early production) or where freedom from tuber blemishes is an important market requirement.

Blight

See Late blight.

Brown rot (*Ralstonia solanacearum*)

Brown rot is a potentially serious disease of potatoes, previously thought of as a tropical or warm-country disease unlikely to survive under UK conditions. However, the first case of brown rot was confirmed on ware potatoes in 1993 in the Thames Valley. Since then, there have been further outbreaks in England and the disease is subject to a government eradication policy. The disease has also been found in some northern and southern European countries, where similar policies have been adopted.

In the growing crop, symptoms are likely to be seen only in warm conditions, and they begin with a transient wilting of the upper leaves during the heat of the day (often with recovery at night). Permanent wilting usually follows and, eventually, the plant dies. In severe cases, internal brown streaking of the stem occurs, starting at soil level and extending upwards. Wilted stems usually exude a white bacterial slime. When an infected tuber is cut in half, the initial symptom is a brownish staining around the bundles in the vascular ring, and it is often possible to squeeze a pale bacterial slime out of the discoloured vascular tissue. As the disease progresses, extensive rotting of the vascular tissues is common. In advanced cases, bacterial ooze may exude from the eyes and the heel end of the tubers, which often have soil attached. Brown rot also affects tomatoes and, to date, two outbreaks have been confirmed in Bedfordshire.

In the UK, the outbreaks in both potatoes and tomatoes have been linked to contaminated irrigation water from rivers where the bacterium persists and multiplies by infecting a wild host: woody nightshade (*Solanum dulcamara*). This wild host is subject to control measures in areas where it has been shown to be carrying *Ralstonia solanacearum*.

Brown rot is subject to statutory control under both EC and UK legislation.

Common scab (*Streptomyces scabies*)
Common scab is caused by several closely related soil bacteria, usually grouped under the name *Streptomyces scabies*. It is one of the most widespread and common tuber blemish diseases and can severely reduce ware quality; it is of particular importance to the pre-pack trade. On affected tubers, loose corky tissue is formed on the tuber skin. These lesions are usually superficial, and range from small angular lesions to round scabs. They are sometimes raised on mounds or may penetrate several millimetres, causing deep cracking or pitting of the tuber surface.

Although the disease is seed-borne, this is of relatively little significance compared with soil-borne inoculum. However, severely scabbed seed should not be used, as the eyes may be affected and this may lead to poor crop emergence. The incidence and severity of scab is strongly influenced by soil type and soil moisture. Severe infections are commonest in dry seasons on light alkaline sandy soils low in organic matter or where lime has recently been applied. Sometimes the disease is severe on potatoes grown immediately after a permanent grass ley. Common scab can be confused with powdery scab, which is more prevalent on heavier soils in wet seasons (see below).

Some cultivars are relatively resistant to common scab and susceptible cultivars should not be grown on land with a history of scab problems. For the latest disease ratings of currently available potato cultivars, the *NIAB Potato Variety Handbook* should be consulted. The only consistently effective method of controlling common scab on prone soils is to avoid a Soil Moisture Deficit (SMD) during the 6 weeks from the first appearance of tuber initiation. In practice this means applying irrigation (12 mm) when the SMD reaches 15 mm during a period of 6 weeks from the time tubers have started to form. Maintaining soil moisture at or near to field capacity following tuber initiation could increase the risk of other diseases occurring, particularly powdery scab. The amount and duration of irrigation is thus a compromise between the need to control common scab and minimizing the risk from other diseases.

Dry rot (*Fusarium solani* var. *caeruleum*, *F. avenaceum* and *F. sulphureum*)
Dry rot is caused by several species of the fungus *Fusarium*, all of which are normal inhabitants of arable soils. For a long time dry rot was considered to be the most serious cause of storage losses and gappiness in emerging crops. In recent years, however, its importance has declined, possibly owing to greater care in handling tubers to avoid mechanical damage, chemical treatment of tubers and cooler storage conditions.

Infection occurs through minor mechanical wounds, made usually at lifting or during riddling. Symptoms take several weeks to produce small, brown lesions on the tuber surface which in time gradually increase in size. Internally, the tuber flesh becomes extensively rotted and cavities develop within the rotted tissue. These cavities are often lined with a fluffy, white, pink or pale-blue fungal growth.

In dry storage conditions, and usually in seed trays, tubers dry out as they rot and, as the skin shrinks, concentric rings are formed. In these conditions, pustules of *Fusarium* burst through the skin and, eventually, the tuber mummifies. In moist conditions there is less shrinkage and, instead of becoming mummified, the tubers become wet and pulpy, with large, white, gelatinous spore pustules scattered over the whole surface of the tuber.

Infection and development of dry rot is favoured by higher temperatures, and tubers become more susceptible during the later periods of storage, particularly when seed tubers are riddled out of bulk store late in the storage season. Infected seed tubers rot away rapidly if planted, and even slightly affected tubers will fail to produce a viable plant.

Fungicides recommended for the control or reduction of dry rot (i.e. imazalil, imazalil + thiabendazole and thiabendazole) are best applied as soon as possible after lifting or at first grading, and should be applied as LV or ULV treatments. There are no resistant cultivars.

Gangrene (*Phoma exigua* var. *foveata*)
Gangrene is a serious disease of stored potatoes. It is particularly serious in seed delivered from the cooler seed-producing areas of the UK, especially when lifted late in cold, wet conditions when inoculum levels of the fungus are higher. Gangrene symptoms do not become visible until at least one month after lifting, and often appear much later in store. Initially, small, black, circular lesions appear on the tuber surface and these may develop into larger irregularly shaped, thumb-mark depressions. Internally, affected tissue forms a dark rot with a clear distinction between diseased and healthy tissue. The extent of the surface lesion is little guide to the severity of the rot internally. Small surface lesions are often associated with extensive internal rotting, with deep cavities inside the tuber. Internal cavities may be lined with a mass of brown or purple fungal growth and, as the rot develops, small pinhead-sized spore-producing bodies of the fungus (called pycnidia) appear on the surface of the tuber.

Gangrene infection takes place either from spores produced on infected mother tubers or from diseased haulm. The fungus readily colonizes dying haulm, so that soil adhering to the tubers after lifting is often heavily contaminated. Infection then occurs through wounds following mechanical damage at lifting, particularly in wet soil conditions or later in the storage period during riddling, when seed tubers are separated from ware. Low temperatures at this time delay the wound healing process and allow the gangrene pathogen to infect more easily.

Gangrene is typically a disease of seed tubers delivered from the cooler seed-producing areas during the winter. Compared with dry rot, it is less likely to result in rapid breakdown of the seed after planting, and lightly affected seed often produces normal plants.

No cultivars are fully resistant to the disease but some are particularly susceptible. For the latest disease ratings for currently available potato cultivars, consult the *NIAB Potato Variety Handbook*. Early lifting leads to less gangrene, because temperatures are generally higher and the shorter the interval between haulm death and lifting the lower the potential for infection. However, lifting too early conflicts with the requirements for late blight and dry rot control, so care is needed.

Tubers should be handled carefully at all times to reduce mechanical damage. Wound healing can be promoted by a curing period (10 days at 13–16°C, with high relative humidity) given after any process which is likely to damage the tubers, especially lifting, riddling or boxing of seed. The high temperature accelerates wound healing and checks the advance of pathogens. This curing process not only reduces the incidence of gangrene, but causes the sprout shoots to break dormancy and be less susceptible to skin spot infection.

A number of fungicide treatments are recommended for the control or reduction of gangrene. Fumigation with 2-aminobutane is usually done 'post-harvest – in store' by contractor. LV or ULV treatments (e.g. imazalil, imazalil + thiabendazole or thiabendazole) are best applied as soon as possible after lifting and before storing or at first grading.

Glassiness and jelly end rot

Warm wet conditions in the autumn following a protracted dry period sometimes stimulate tubers into 'second growth' (see p. 158). In some situations, tubers swell beyond the carbohydrate resources available. The tuber tissue then becomes partially starch depleted so that the cut surface of the tuber looks glassy.

Jelly end rot is a specific manifestation of second growth, and is also a physiological disorder caused by the mobilization and movement of starch from cells at the heel end of tubers towards those at the rose end. The heel end becomes glassy and, in extreme examples, degenerates into jelly end rot. The condition occurs during a period of second growth or re-growth when a drought period, which has stopped growth, is broken by heavy rain or by irrigation. Affected tubers lose turgidity at the heel end and may leak fluid, so attracting secondary bacterial infection. The internal tissues do not stain blue in the presence of iodine, indicating a lack of starch. The problem can be particularly serious during the long-term storage of potatoes in clamps or bulk stores where compression hastens leakage.

Adequate irrigation is the key to prevention of jelly end rot, as this regulates water supply. Tubers which are stored and which show symptoms of second growth should be examined regularly; the stored tubers should also be examined where any rise in temperature is noted. If the temperature cannot be controlled by blowing in cold air, then the store should be cleared as soon as possible.

Hollow heart

Hollow heart is thought to result from tissue tension during rapid tuber enlargement. Manipulation of agronomic factors to favour steady growth rates can provide some control. Excessive nitrogen availability and low soil calcium tend to induce symptoms.

Internal rust spot (IRS)

IRS is a quality-related disorder affecting the flesh of potatoes. It is known as 'internal brown spot' or 'internal heat necrosis' in the US and as 'fleck' in Australia. Symptoms are seen in the perimedullary zone between the pith and the vascular ring, and can range from individual flecks no more than 1–2 mm in diameter, to necrotic blotches up to several centimetres across with cavities developing within them. IRS has been related to lack of calcium during tuber bulking, and cultivars differ in their susceptibility to the disorder. IRS occurs most frequently on light soils, and is induced by environmental and agronomic conditions that favour irregular rates of tuber growth. Fluctuations in temperature and soil moisture, high temperature stress and low calcium availability are associated with symptom development.

Many commonly grown cultivars regularly develop IRS, including Cara, Estima, Maris Piper, Pentland Squire and Saturna. The most susceptible commercially available cultivar is probably Cultra.

Various treatments have been shown to reduce the incidence of IRS in crops, including (a) soil applications of calcium sulfate and calcium carbonate, (b) aldicarb, when applied against the nematode vectors of spraing, and (c) maleic hydrazide applied to the growing crop. It is not known why aldicarb should reduce the incidence of IRS. Maleic hydrazide stops cell division and may act by reducing the demand for calcium required for the formation of cell walls.

N.B. Brown centre is a similar condition to IRS, except that the symptoms appear in the central pith of the tuber. There is also evidence of a role for calcium in the development of brown centre, and there are differences in cultivar susceptibility. It has been reported that brown centre precedes hollow heart (see above).

Late blight (*Phytophthora infestans*)

It is over 150 years since late blight devastated the potato crop in Ireland and caused the Irish Potato Famine. Nevertheless, blight still remains the greatest potential disease threat, not only to UK potato crops but to potato crops world wide. Despite considerable research effort and breeding of cultivars with improved levels of resistance, in the UK, as in many other developed countries, routine use of fungicides is still the most effective means of control.

Blight can destroy the haulm extremely rapidly, leading to reduced photosynthetic area and consequent yield reduction. In 38 replicated field experiments between 1978 and 1992, the yield responses to fungicide treatment for the control of blight compared with unsprayed controls ranged from zero to 30.8 t/ha, reflecting the timing of the epidemic in relation to the tuber-bulking phase of crop development. The mean yield response was 12.92 t/ha. Observations over many years from fungicide trials have shown just how fast a blight epidemic can develop. In untreated plots, foliage blight has often increased from 5% to 75% haulm destroyed in less than 10 days.

Blight can also infect the tubers and in doing so directly reduces marketable yield. Tuber infection may also lead to breakdown in store as a result of secondary infection with soft-rotting bacteria. Because blight is potentially so devastating, growers need to apply fungicides prophylactically as routine sprays in programmes, and well before the disease becomes established in the crop or locality. The choice of fungicide, and the frequency of use, will depend on (a) the cost, (b) weather conditions, and (c) the perceived risk of blight in the locality. Cultivar resistance in terms of foliar vs. tuber blight may also influence the intensity of spray applications but, often, cultivar choice is directed by market requirements rather than disease resistance characteristics.

Symptoms

Viewed from above, blight on the foliage typically produces brown spots, each surrounded by a yellowish green margin. This margin is where the fungus is most active in the leaf tissue, and in warm ($>10°C$) and wet or humid conditions ($>90\%$ RH) will produce a delicate white halo of spore-bearing structures known as sporangiophores. In optimum conditions sporangiophores are usually produced within 5–7 days after initial infection. The spores themselves, known as sporangia, are produced within as little as 12 hours on mature sporangiophores in warm, wet weather and particularly in the humid microclimate conditions that are often prevalent within the crop canopy.

Sporangia are the sole means by which blight spreads, and epidemics develop during the growing period. The sporangia are airborne but the exact distance that they can travel is not known precisely. Circumstantial evidence would suggest many kilometres, providing that the sporangia do not desiccate in the wind. Much will also depend on the conditions on the leaf surface once the spores have landed, as to whether infection takes place. Clearly, the whole process of epidemic development is a matter of chance but is also a function of the sheer numbers of spores that are produced.

Sporangia require a film of water on the leaf surface for at least 12 hours for infection to occur. At temperatures higher than 15°C they directly infect the leaf/stem tissues but at lower temperatures they will be stimulated to produce up to 10–12 motile (swimming) spores known as zoospores, each one of which is capable of causing infection. This is known as indirect infection. However, in dry conditions, sporangia/zoospores may die before infection can occur, and in such conditions the only place within the foliage where moisture is retained long enough for infection to occur is at the leaf axils.

Infection at this position on the plant will result in the development of stem lesions and may be one explanation why so-called 'stem blight' is more prevalent in some seasons than others. Stem blight is **not**, therefore, any different from leaf blight but is a result of the particular conditions within the foliage at the time when blight spores arrive and land on the canopy. Stem blight may also occur when the fungus, having infected a leaf, progresses rapidly down the petioles towards the stem. Sporulation is usually more profuse on leaf lesions than on the stems but both sources of spores can result in tuber infection, i.e. tuber blight.

Tubers become infected when spores produced on the foliage are (a) washed down through the soil profile, (b) washed down the stems themselves from stem lesions, or (c) come into contact with tubers at lifting. Symptoms of tuber blight are very characteristic, but where soil is adhering to the tubers (and particularly in wet conditions) they can be extremely difficult to detect. Externally, young lesions on the surface of the tuber appear as small leaden-grey areas through normal-looking skin. At this stage there may be little evidence on the surface of the quite extensive penetration by the fungus into the tuber flesh. Blight in the tuber spreads initially through the surface tissues outside the vascular ring and appears as a firm, 'foxy' (reddish-brown) granular rot before penetrating towards the centre of the tuber. In wet conditions in the ground or in bulk stores, moisture released from the infected tissue encourages bacterial wet rots to develop, and it is these secondary organisms which have the potential to cause extensive losses in store.

Sources of blight

Infected tubers from the previous year's crop are still the main source of blight, either on dumps of discarded potatoes or via the seed. More recently, resting spores of the blight pathogen (oospores) have been discovered but their practical significance in the initiation of blight outbreaks is not fully understood (see p. 149).

Dumps are particularly dangerous because they often include blighted tubers and also tend to be situated in damp and sheltered sites such as ditches or wherever rubbish gathered after grading the previous year's potatoes has already been dumped. Here, sporulation is likely to take place earlier than in the open field, because the dense growth of potato shoots holds a microclimate which encourages blight development. Foliage should not be allowed to develop on dumps, as this often proves to be an important source of primary inoculum to nearby crops.

Control of haulm on potato dumps

The surest way of preventing haulm growth developing on dumps is to bury the waste potatoes and cover with soil to a depth of at least 0.5 m. This should be done before foliage has started to appear in the spring and is probably the most practical option in many cases but would require suitable heavy machinery such as bulldozers.

Dumps should be sited in some easily accessible place on the farm where measures can be taken to destroy the haulm growth in the spring. Growth can be burnt off with a desiccant herbicide as soon as it appears, but this has the disadvantage of allowing some potential for blight to develop before control is achieved. A quick-acting herbicide (such as diquat or paraquat), although very effective within a few days, may require frequent application to ensure complete kill of later-emerging shoots. An alternative is to use glyphosate, which would ensure control of the tubers but would take 1–2 weeks to be completely effective, during which time blight could still develop. Similarly, more than one application may be necessary to control late-emerging shoots. Whichever option is chosen, it is important to check treated dump sites regularly and to re-treat if necessary.

Another herbicide option is the use of the persistent herbicide dichlobenil, which should be applied before the tubers have sprouted. Once treated, dump sites should be immediately covered with soil. At the rate required for control of potatoes on dumps, dichlobenil has residual activity for 12 months and so should not be used on sites intended for cropping during that period.

Another method (particularly suitable for smaller dumps) is to spray them with water and cover with a plastic sheet held down at the edges with soil. Under these conditions the potatoes quickly rot away.

Seed infection

Blight in seed tubers is almost impossible to detect. Only a very low proportion of blighted tubers, as little as one in 200, succeeds in setting up an above-ground lesion, yet 1% infection in the seed could provide two 'primary infectors' per hectare. In warm, moist conditions, this is more than enough to initiate an epidemic, as spores from 'primary infectors' spread locally and initiate 'primary foci'. Grading-out infected tubers from an infected stock is unlikely to be entirely successful, as lightly infected tubers (those most likely to survive a storage period) are also the most difficult to detect.

Seed-borne blight could be a significant problem in home-saved seed stocks, particularly after a blight year. Ideally, seed should not be saved from a stock that has been infected with blight. Tolerances set for certified seed represent the best attainable in practice. Even so, the practical limitations of grading and inspection mean that certification cannot completely exclude the possibility of blight in the seed. It is essential that the management of seed-producing crops is such that they are desiccated before blight (and other diseases) develop.

Volunteer potatoes

The importance of volunteers as a source of primary inoculum is not known but could be on a par with other sources, and certainly cannot be discounted. As with seed-borne infection, only a small proportion of infected volunteer tubers are likely to survive the winter to produce an above-ground lesion, but in the right conditions could be sufficient to initiate an early outbreak. Volunteer potatoes invariably appear in other crops that cannot be treated with a suitable blight fungicide. In years when blight is rife and weather conditions are conducive to infection, volunteers will pick up blight and thereby act as an important local source of inoculum. In these situations, volunteers are picking up infection from the same sources to which potato crops themselves are also being exposed (such as nearby affected crops and uncontrolled dumps). Volunteers can therefore act as a 'sink' for blight but in favourable weather could rapidly become a very important 'source' of infection. Whole-farm hygiene therefore remains a vital component in a blight control strategy.

Sexual reproduction and oospores

Since the discovery of the two mating types of the blight pathogen in the UK and Europe, known as the A1 and the A2 strains, the possibility that the fungus is able to reproduce sexually has been recognized. Sexual reproduction increases the gene flow within the blight population and with it carries implications for both varietal (cultivar) and fungicide resistance. Sexual reproduction gives rise to resting spores, called oospores, which have the potential to survive in soils in the absence of potatoes. Experimentally, oospores introduced into soil have been shown to cause stem and leaf lesions on potatoes; their significance as a source of blight in commercial potato production is not known but is the subject of current research. The proportion of A2 types in England and Wales is low (less than 5% in 1997). Therefore, whilst the opportunity for sexual reproduction and oospore formation exists, in practice the frequency is likely to be low. In relation to other inoculum sources the role of oospores must be considered less important.

'New blight'

Researchers in the UK, the Netherlands and the US (using sophisticated techniques akin to DNA fingerprinting) have monitored changes in genetic structure of the blight pathogen over the last 15–20 years. This has shown that there have been several migrations of the blight fungus from Mexico (which is considered to be the epicentre of blight populations) since the mid-1970s. In the US some of these new strains have been called 'superblight', which is a rather emotive phrase used to describe their increased aggressiveness in laboratory studies. The 'old' A1 population has been displaced by a 'new' population comprising both A1 and A2 mating types.

These 'new' isolates have been shown to be more aggressive than the 'old' isolates and are also insensitive to phenylamide fungicides, e.g. metalaxyl. Indeed, it has been suggested that the population displacement may have been driven by the use of phenylamide fungicides. The 'new' blight isolates present in the UK since the early 1980s are from different migrations (arising out of Mexico) from those currently causing problems on both potatoes and tomatoes in the US, but they do share some of the properties of the new US blight populations. New blight strains have been present in Europe for almost 20 years and so, over this time, blight fungicides have been tested against them (and used effectively to control them). There is no evidence to support the claim that 'new' blight has increased the incidence of stem blight.

The characteristics of new blight strains are as follows:

- They are more 'aggressive' than 'old' strains, i.e. they produce more spores when tested on detached potato leaves.
- They are less sensitive to phenylamide fungicides.
- They are genetically diverse – at least 16 different strains, compared with a single 'old' strain.
- They may be either A1 or A2 mating type (the old strain was A1 only).
- When A1 and A2 strains are present together, they have the potential to form oospores in the foliage, stems and tubers.

Role of forecasting

In the UK, two forecasting schemes (developed empirically) for potato blight have been in widespread use – the Beaumont Period (used since 1950) which was superseded by the Smith Period in 1975. Smith Periods are defined as *'two consecutive 24- hour periods ending at 09.00 GMT in which the minimum temperature is 10°C or above and in each of which there are at least 11 hours with a relative humidity above 90%'*. Forecasts are based on temperature and humidity data received from a network of synoptic weather stations. However, these are frequently sited at airfields which are not necessarily close to major potato-growing areas. Other systems are in use in Europe (e.g. Guntz Divoux, NegFry, ProPhy, Symphyt and Televis), and in parts of the US (Blitecast™).

The ready availability of portable weather stations has raised considerable interest in using meteorological data from in-field weather stations to utilize blight forecasting models developed for the purpose. These models include such schemes as Smith, NegFry and Blitecast™. Ideally, a forecast should give a warning 14 days in advance of blight occurring, to provide sufficient time for the application of fungicides and the start of a routine spray programme. However, evaluation of a number of these models in England and Wales has shown great temporal and spatial variation in their performance, with some models indicating blight too far in advance of infection (including instances where the disease did not occur), or failing to give sufficient warning for growers to protect their crops adequately.

The variability in the performance of the different models is a cause for concern. The microclimate within a canopy is likely to play an important part in such variation, as would other factors such as damp hollows in fields, tree shading and differential rates of foliage growth. These would all influence the in-field conditions, and even in-crop sensors could not account for such variability. Additionally, information regarding the presence/absence and quantity of blight inoculum is not taken into account by any of the forecasting schemes. So far, blight forecasting has not improved the precision of spray applications; nor has it resulted in a reduction in fungicide use in years of low blight risk, when the opportunity should be greatest.

At present, therefore, forecasting does not offer the precision necessary for individual crops. However, it may be of value in warning of blight risk in a broader geographical area before routine spray programmes have started, or for areas where blight is an infrequent problem and routine programmes are not the rule. Forecasts that rely on one station regardless of its distance from the crop cannot be considered safe, for a number of reasons – not least the accuracy of the instrumentation. More generalized but robust schemes are likely to be of practical value to growers. In short, forecasting schemes are not a panacea but rather an aid to effective blight control, as part of an overall decision support system.

Fungicides

Blight control nowadays is certainly more challenging, with wider planting dates and less segregation of early crops. Early crops under polythene are always at a higher risk until they have been treated with a fungicide. They can be carrying blight whilst nearby second earlies/maincrops are emerging. Second crops are at particular blight risk in most seasons, and nearby organically grown crops are also a potential threat if not managed correctly with regard to blight control. It is helpful, therefore, to have good local intelligence of what is being grown in the area.

Because of the lack of robustness of forecasting systems, fungicides will continue to be used routinely in conventional production, at least for the foreseeable future. Spray programmes should be underway well before blight becomes established in the area. Protective spraying is essential for effective blight control. Fungicides are effective in the early stages of an epidemic, before blight can readily be found, but usually have little effect once blight is well established. It is not always appreciated that in 'blighty weather' when the disease is very active, sprays will never provide 100% control, but at best will delay the onset of the epidemic by 3–4 weeks.

As a strategy for fungicide use, the first precautionary spray should be applied just before the haulm meets along the rows and followed by a regular spray programme. The intervals between sprays should be no longer than 14 days, reducing to 7-day intervals depending on factors such as weather (blight risk) conditions, cultivar resistance, rate and stage of canopy development, whether irrigation is being used, presence of blight in the locality and, of course, fungicide product. Maintaining short spray intervals in high-risk conditions is essential. In these situations, recent research has shown that the interval between applications is more important than product choice, and effective control of foliar blight is achievable with cheaper protectant materials when applied at 7-day intervals. However, as the intervals are extended the more sophisticated mixtures currently available give better control.

The following suggestions may be useful in judging blight risk but they should be modified according to local experience and market requirements. Blight risk is high when:

- weather conditions are warm ($> 10^\circ C$) and wet or have satisfied the Smith criteria;
- blight is present in a locality on dumps or on volunteers;
- blight is present in the crop;
- home-saved seed is being used from a crop which carried blight in the previous season.

In areas of early potato production where blight occurs in most seasons (e.g. south-west England, south and west Wales), many crops are now grown under plastic cover. A fungicide spray should be applied as soon as the cover is removed, and this should be the start of a routine spray programme. Intervals between applications should not exceed 10 days and, as blight risk in these areas is invariably high, they should be maintained at the closest recommended interval for the chosen product. Early crops are usually lifted 'green top' (i.e. in the presence of green haulm). As even low levels of blight in the foliage dramatically increase the risk of tuber infection during lifting, fungicides should continue to be applied as close to lifting as the product label permits.

In intensive maincrop potato-growing areas such as the Cambridgeshire Fens, especially where irrigation is being used or where blight-susceptible cultivars predominate, routine spraying with a recommended fungicide should commence just before the haulm meets along the rows, or earlier if any of the above blight risk criteria are met. Sprays should also be applied at intervals not exceeding 10 days and this interval should be reduced to 7 days immediately blight risk increases and providing the product label permits. Second crops, which are usually grown after the early crop has been lifted, also fall into this category.

In the less-intensive production areas, the first precautionary spray in a programme should be applied when the tops are well met along the rows, but before meeting across the rows. Even in these areas the start of the spray programme should be brought forward if any of the above-mentioned risk criteria are met. In most seasons 10-day intervals between sprays are usually adequate, according to the manufacturers' recommendations.

High volume (HV) (1000–3000 litres/ha) spraying is rarely used these days, and most manufacturers recommend low volume (LV) (200–450 litres/ha) applications. A number of blight fungicides may be applied from the air as ultra low volume (ULV) applications (20–60 litres/ha) but this use has declined dramatically in recent years. Less than 1% of the potato crop in England and Wales was treated from the air in 1998. However, aerial application is useful when ground conditions prevent conventional spraying.

When considering the choice of fungicide for blight control, it is useful to be aware of their relative effectiveness and mode of action characteristics. These are given for a selected range of fungicide active ingredients in Table 5.6.

The application intervals indicated in Table 5.6 are **not** intended as a guide as to how frequently a particular fungicide should be used but have been chosen for comparative purposes only. Where disease pressure is low, intervals between applications may be extended and, in some instances, fungicide applications may be made in response to nationally or locally issued spray warnings. It is essential to follow the appropriate instructions for use given on the approved label before handling, storing or using a blight fungicide.

Fungicide resistance – phenylamide fungicides

Resistance to the phenylamide fungicide metalaxyl was first reported in the Netherlands, Ireland and Switzerland in 1980, when it was being used alone without the usual mancozeb component. In the UK, where phenylamide fungicides were introduced as co-formulated mixtures with mancozeb, strains of the blight pathogen resistant to phenylamide fungicides were identified in 1981. At the time, phenylamide fungicide mixtures were being used season long, and often in curative situations where the disease was well established and difficult to control. Proprietary phenylamide fungicides have been available in the UK only as 'two-way' mixtures with mancozeb or as 'three-way' mixtures with mancozeb and cymoxanil. This approach is an integral component of a resistance management strategy.

Since the identification of phenylamide resistance, levels of resistance have been monitored annually in the UK and Europe by testing samples of *P. infestans* for the presence or absence of resistant spores. To address the resistance problem manufacturers of phenylamide fungicides formed the Fungicide Resistance Action Committee (FRAC) and established industry guidelines for an anti-resistance strategy.

The FRAC guidelines (here reworded) are as follows:

- Use phenylamide-containing fungicides but protectively only.
- Use early in the season, during the period of active growth.
- Adopt a maximum spray interval of 14 days.
- Applications can be made up to the end of active crop growth, which usually finishes by the middle of August.
- Up to five applications may be made to any one crop; however, application in the period of rapid growth will often mean that, in practice, only two or three sprays are used.

There is no evidence of resistance to any of the other major blight fungicides in use at the time of writing. These fungicides are dithiocarbamates (e.g. mancozeb, maneb), chlorothalonil, cymoxanil, dimethomorph, fluazinam, propamocarb hydrochloride and the fentin (tin) products. Many proprietary blight fungicides also contain two different active ingredients, and are available only as co-

Table 5.6 The effectiveness and characteristics of selected fungicides for the control of *Phytophthora infestans* on potato

Active ingredient	Spray interval (days)	Effectiveness				Activity			Rainfastness	Mode of action
		Leaf blight	New growth	Stem blight	Tuber blight	Protectant	Curative	Eradicant		
chlorothalonil	7	+ +	0	(+)	0	+ +	0	0	+ +(+)	contact
copper	7	+	0	+	+	+(+)	0	0	+	contact
cymoxanil	7	+ +(+)	0	+(+)	0	+ +(+)	+ +	+	+ +	translaminar
dimethomorph	7	+ +(+)	0	+(+)	+ +	+ +(+)	+(+)	+ +	+ +(+)	translaminar
fentin acetate	7	+ +	0	+	+ +(+)	+ +	0	0	+ +	contact
fentin hydroxide	7	+ +	0	+	+ +(+)	+ +	0	0	+ +	contact
fluazinam	7	+ + +	0	+	+ +(+)	+ + +	0	0	+ +(+)	contact
mancozeb or maneb	7	+ +	0	+	0	+ +	0	0	+(+)	contact
metalaxyl*	10	+ +(+)	+ +	+ +	N/A	+ +(+)	+ +(+)	+ +(+)	+ + +	systemic
oxadixyl*	10	+ +(+)	+ +	+ +	N/A	+ +(+)	+ +(+)	+ +(+)	+ + +	systemic
propamocarb hydrochloride	7	+ +(+)	+(+)	+ +	+ +	+ +(+)	+ +	+ +	+ + +	systemic

Source: Schepers & Bouma (2000).
* Assumes a phenylamide-sensitive blight population.
Key to ratings: 0 = no effect; + = reasonable effect; + + = good effect; + + + very good effect; N/A = not recommended for control of tuber blight.
Protectant activity: Spores killed before or upon germination/penetration. The fungicide has to be present on/in the leaf/stem surface before spore germination/penetration occurs.
Curative activity: The fungicide is active against *P. infestans* during the immediate post-infection period but before symptoms become visible, i.e. during the latent period.
Eradicant activity: *P. infestans* is killed within sporulating lesions, thereby preventing further lesion development. This mode of action prevents sporangiophore formation and, therefore, anti-sporulant activity is included within the definition of eradicant activity.
Stem blight control: Effective for the control of stem infection, either by direct contact or via systemic activity.
Tuber blight control: Activity against tuber infection as a result of mid-/late-season post-infection fungicide application and has a direct effect on the tuber infection process.

formulated mixtures. This in itself is a useful resistance management strategy, as the different modes of action of mixture partners are being fully exploited. In this way, the blight pathogen is being controlled at different metabolic sites or at different stages in its life cycle. A further safeguard against the possibility of resistance developing is either to alternate products or to use different materials at different stages in the crop's development.

Systemic products containing either a phenylamide fungicide or propamocarb hydrochloride are ideally suited for use during the early phase of rapid haulm development. Protectant fungicides (such as chlorothalonil, fluazinam and mancozeb) or materials with translaminar and curative properties (e.g. cymoxanil or dimethomorph) are particularly suited during canopy stabilization. Towards the end of the growing period, when tuber protection is required, fluazinam or tin-based products should be used.

Haulm destruction

This reduces the risk of tuber blight by removing the source of infection; it also facilitates crop lifting by destroying weed growth. To achieve good control of tuber blight, haulm destruction should take place, ideally, as soon as possible after blight is seen in the field (at approximately the 5% level – up to one leaflet in 10 per plant infected). In practice, where infection has occurred during the tuber-bulking phase of growth this would often result in an unacceptably heavy loss of crop yield. On the other hand, it is rarely worth destroying haulm already half dead with blight, unless the crop is shortly to be lifted. Factors to be considered before deciding whether to desiccate the haulm to mitigate the effects of blight are:

- the amount of blight on the leaves and stems;
- the amount of crop already formed – it is not worth risking an already good crop for the sake of a little extra weight;
- the rate of bulking – it is useful to do weekly sample lifting to get some idea of this;
- the nature of the soil and the state of the ridges – some ridges seem to encourage tuber blight, either because they crack or because they tend to retain water;
- susceptibility to tuber blight – it is rarely necessary to desiccate tuber-resistant cultivars specifically for the control of tuber blight, although it may be agronomically desirable to do so to prevent second growth or to deal with weeds;
- the crop should not be lifted for at least 14–21 days after the haulm is completely dead – this will reduce the viability of blight spores on the soil surface. Also, a crop to be stored should never be lifted while there is any green tissue at all on the leaves or stem bases as this could be harbouring blight spores.

The materials currently available for haulm desiccation are shown in Table 5.7.

Table 5.7 Chemicals available for potato haulm desiccation

Product	Comments
diquat	Application during or shortly after a dry period may result in damage to the tubers. It is important to check the label for details of maximum allowable soil moisture deficit and cultivar drought resistance score. Only one application can be made per crop. Harvest interval 14 days.
glufosinate-ammonium	Apply to listed cultivars only and do **not** use on seed crops. Two applications for desiccation can be made per crop from the onset of senescence, 14–21 days before harvest. Harvest interval 7 days.
sulfuric acid	Commodity substance. Requires specialized machinery – usually applied by contractor. Quickest-acting desiccant. Up to three applications may be made per crop.

Pink rot (*Phytophthora erythroseptica*)

Pink rot is caused by the soil-borne fungus *Phytophthora erythroseptica*, and derives its name from the tuber symptoms. A section through partially rotted tubers reveals a rubbery-textured flesh with an 'off white' colour, which turns pink within a few minutes and eventually black. The tissue may have a pungent, vinegary or slightly alcoholic smell. Affected tubers leak fluid and, because of this, usually have soil attached to them when harvested.

Pink rot is usually confined to patches in crops where the drainage is poor, and in mid-summer affected plants will wilt. Infection of the underground stems, tubers and roots is favoured by high soil moisture and above-average temperatures. The disease is therefore usually more common in hotter seasons, and on heavier soils when the spring and summer are particularly wet.

There are no resistant cultivars and control relies on maintaining good drainage and a wide rotation, although it is important to note that *P. erythroseptica* is able to produce resting spores (oospores) which are capable of remaining viable for the 4–5 years between potato crops. Potatoes should not be grown in fields with a known history of pink rot, and tubers known to be infected should not be destined for long-term storage. If pink rot is recognized in the growing crop, then the crop should be left as long as possible before lifting, to allow infected tubers to rot before harvest.

Powdery scab (*Spongospora subterranea*)

Powdery scab is both seed- and soil-borne, and infects tubers through the lenticels, eyes and small wounds. Infection occurs under cool, wet, growing conditions. Alternating periods of soil saturation and non-saturation result in the most severe disease. The disease can occur in all soil types. Previously more common in the north and west of the UK, it has become more widespread in recent years and has been recorded annually in ADAS Crop Intelligence Reports since records began. Powdery scab came to prominence in the late 1970s, when the highly susceptible cv. Pentland Crown was widely grown. The disease has persisted as a problem because of the predominance of susceptible cultivars and the widespread use of irrigation. There are many alternate hosts but few allow it to complete its life-cycle in the absence of potatoes.

Powdery scab symptoms are of two kinds: (a) open scabs, with a brown powdery surface, and (b) cankerous outgrowths. The scabs start as watery pimples which, as they mature, darken and shrivel, the tissue inside breaking down into 'spore balls' of the fungus. The skin over the scabs usually breaks with a ragged edge to expose the powdery mass of 'spore balls'. The scabs vary in size and appearance, sometimes resembling symptoms of common scab or of skin spot. Microscopic examination of the tissues is necessary for a reliable diagnosis. The canker phase of the disease occurs when the pathogen causes malformations of the tuber, usually at the rose-end. These cankers are associated with infection through the eyes, stimulating tissue to grow abnormally. The unsuberized tissue of the canker is often invaded by the fungus and then develops surface scabs. In some cases cankers resemble wart disease.

The 'spore balls' of powdery scab are the resting stage of the fungus and contain hundreds of spores. The 'spore balls' are very hardy and long-lived, being capable of surviving for up to 18 years. In wet soil conditions, the spores within 'spore balls' release swimming spores (zoospores). These zoospores are short-lived but can infect roots or tubers. Powdery scab spores are also introduced on infected seed and, possibly, on dung from stock fed with diseased tubers.

Tubers are most susceptible to infection around the time of tuber initiation. There is conflicting evidence concerning the effect of soil pH on infectivity, although there is a suggestion that acid soils may be slightly less conducive. There is also some evidence to suggest that potatoes growing in soils with high levels of zinc are at less risk from powdery scab. The relationship between seed infection and progeny infection is unclear, although more infection and reduced vigour have been recorded with very high levels of seed infection.

In the absence of any effective chemical control measures in the UK, disease avoidance by planting healthy seed into uncontaminated land is the most effective control method. However, lesion-free tubers do not necessarily mean freedom from the fungus, for cross-contamination with 'spore balls' can occur during grading. Grader hygiene is therefore important. Further, field selection is crucial, although contamination of soils is widespread. The risk of powdery scab infection depends on factors such as: (a) soil type (moisture retention), (b) drainage (avoid poorly drained fields), (c) previous history of powdery scab (if this is not known, assume a risk if potatoes have been grown in the last 15 years), (d) avoiding fields where contaminated slurry may have been applied, and (e) choosing a resistant cultivar for high-risk sites. Where soil contamination occurs, planting infected seed will add little to the risk. The incorporation of zinc, often included with the fertilizer, to a maximum of 15 kg/ha, has given modest reductions in disease severity in experiments.

Rubbery rot (*Geotrichum candidum*)
Tubers affected by rubbery rot show symptoms of irregular brown patches on the skin with dark margins. On a cut surface, the affected flesh is only slightly discoloured but discolours further on exposure to the air (similar to pink rot but over several hours rather than minutes). Affected tubers have a rubbery texture and a tendency to weep. When incubated at 20°C for 48 hours in a sealed, damp container, small greyish-white tufts of mycelium and spores grow out of the skin and lenticels. This is even more evident on cut surfaces.

Rubbery rot is an infrequent disease and is rarely severe. Usually, only one daughter tuber per plant is affected. No haulm symptoms are evident and the disease is not noticeable until harvest. Infected tubers do have the potential to degenerate with secondary bacterial infection and may result in wet pockets in bulk store.

Outbreaks of rubbery rot are invariably associated with heavy irrigation or following rain within 3 weeks of lifting, especially in warm weather in compacted, poorly drained soils. Ensure that soil pans are broken and that fields are adequately drained. It is also sound practice not to store ware potatoes from headlands as these areas are likely to be more compacted and, therefore, to carry an increased risk of infection.

Second growth
Second growth is the term applied to a group of abnormal tuber conditions, associated with fluctuations in crop growth and brought about by drought stress followed by a sudden abundance of water – following either rainfall or irrigation. One manifestation of this physiological condition is chain tuberization, where primary tuber growth ceases but stolons emerge which give rise to small daughter tubers. This process is repeated so that, eventually, a chain of small, worthless tubers is formed. In serious cases of second growth, starch depletion of the primary tuber occurs and affects cooking quality. The starch-depleted area becomes 'glassy' after boiling. This condition is called glassiness or, where complete breakdown occurs, jelly end rot (see p. 144).

Silver scurf (*Helminthosporium solani*)
This disease causes a very common skin blemish, which is particularly important on long-term-stored ware potatoes for the washed pre-pack trade. Infection causes a slight blemish of the skin, which is darkened or becomes slightly silvery owing to the separation of the outer cell layers as the lesions age. In very humid conditions, the fungus produces dark spores that look like minute flecks of soot at the edges of the lesions. Symptoms are not usually noticeable at lifting but develop further during storage, particularly in warm, humid conditions. Severely affected tubers lose moisture and the effect of this disease is a weight loss, as well as a loss in quality. Silver scurf is a superficial disease and does not affect the eyes or a tuber's ability to sprout. It is easily confused with black dot (see above).

Silver scurf is not thought to be soil-borne, although this aspect is the subject of ongoing research. The main source of infection is thought to be seed-borne inoculum but the means by which spores reach the progeny tubers is not understood. Attention to seed health is essential, as even low levels of infection can result in significant infection in the ware crop. Store hygiene and management are also important, as spores of the fungus may reside in dust in the store and it is important to avoid high humidity during storage as this will encourage sporulation and re-infection during the storage period.

A number of fungicide treatments are recommended for the control of silver scurf, most of which should be applied as soon after lifting as possible. Fumigation with 2-aminolontane is usually done by contractor. Treatment of seed tubers reduces the amount of disease on the progeny at lifting and during subsequent storage, although additional fungicide treatment of the ware may also be necessary. LV or ULV treatments (e.g. imazalil, imazalil + penycuron, imazalil + thiabendazole or thiabendazole) are best applied as soon as possible after lifting and before storing or at first grading. A pre-planting dust treatment (imazalil + penycuron) is also available.

Since 1993, when isolates of *H. solani* resistant to thiabendazole were identified, the monitoring of populations in England and Wales continues to show that resistant isolates are present in almost all stocks of ware potatoes, irrespective of fungicide use. Fungicides are rarely applied to potatoes solely for the control of silver scurf, but a prudent resistance management strategy would be to use fungicide mixtures and to alternate with different fungicide active ingredients on the same stock of potatoes.

Skin spot (*Polyscylatum pustulans*)

Skin spot is an important blemish disease of tubers in store, where superficial pimple-like spots develop surrounded by a dark sunken ring of tissue. These spots are produced either singly or in groups, and may not become obvious until February or March. Severe spotting detracts from the appearance of the tubers and the fungus also infects the eyes, resulting in damage or death which leads to delayed emergence, gappy crops or, in extreme circumstances, crop failure. Where chlorpropham has been used as a treatment for sprout suppression, this also may delay healing which, in turn, leads to deeper penetration and more-severe skin spot lesions.

Although skin spot is soil-borne, the main source of infection is the seed. In the growing crop, superficial light-brown to rusty-coloured lesions develop on the shoots, roots and stolons of young plants and these provide inoculum for the progeny tubers. There are no resistant cultivars, although severe skin spot is rarely recorded on cvs Arran Consul, Estima, Home Guard or Pentland Squire. Skin susceptibility is not necessarily associated with eye damage, e.g. the cv. King Edward is resistant to skin spotting but susceptible to eye damage. The worst effects of skin spot, from eye damage and failure to sprout, occur in non-chitted seed planted into cold soils.

A number of fungicide treatments are recommended for the control of skin spot, most of which should be applied as soon after lifting as possible. Fumigation with 2-aminobutane is usually done by contractor. Treatment of seed tubers reduces the amount of disease on the progeny at lifting and during subsequent storage, although additional fungicide treatment of the ware crop may also be necessary. LV or ULV treatments include imazalil, imazalil + penycuron, imazalil + thiabendazole and thiabendazole.

Verticillium wilt (*Verticillium dahliae*)
This disease is caused by a soil-borne fungus. It is somewhat difficult to identify in the field as the haulm symptoms could be thought to be due to a number of other factors. The American term for the disease, 'potato early dying' (PED), describes the premature senescence that it causes. Infection usually takes place at the time of tuber initiation, and is favoured by wet soil conditions. Irrigation to minimize common scab infection, which is done at this time, is therefore likely to encourage infection. The haulm symptoms rarely appear before late July, and the distribution of affected areas in a crop may be patchy. Earliest symptoms are a reversible wilting which occurs in hot weather, but this eventually becomes permanent. Leaf chlorosis follows, often confined initially to one side of the leaflets; later, necrosis occurs. Meanwhile, the stems remain erect, to give the affected plant a staring habit; sometimes, only part of a plant shows these symptoms. Dry soil conditions later in the season will aggravate symptoms. Tubers are affected via the stolons; they show no external symptoms but when cut across near to the heel-end they may show a browning of the vascular ring.

Verticillium wilt may severely reduce yield, and all UK cultivars can be affected to varying degrees. The cvs Estima, Record and Saturna are very susceptible, whereas cv. Cara is possibly the least affected by the disease. The pathogen has been detected within the vascular tissue of seed tubers, and may be introduced into previously uninfested land by this means. No chemical control is known but a soil assay can determine the level of *V. dahliae* infestation, enabling farmers to avoid seriously affected fields.

Virus diseases – spraing
Spraing is the term given to the tuber symptoms resulting from infection either by tobacco rattle virus (TRV) (nematode transmitted) or, less commonly, by potato mop top virus (PMTV) (transmitted by the powdery scab pathogen). It is not possible to distinguish between these two viruses by visual examination of the tubers.

Tobacco rattle virus (TRV)
Tubers infected with TRV during the growing season (primary infection) show symptoms of single or concentric brown streaks, arcs or circles on the cut surface. These symptoms vary widely in different potato cultivars, including thick brown lines, fine streaks or broken arcs. These internal symptoms are invariably evident at lifting but do not progress further during storage. TRV-infected tubers usually produce healthy plants, although occasionally individual stems show symptoms

called 'stem mottle'. This depends on the cultivar and strain of TRV, and affected stems are stunted and easily overlooked. Tubers formed on 'stem mottle' stems may be misshapen and show secondary symptoms of spraing, i.e. internal necrotic flecks and, occasionally, small rings. Tubers with secondary spraing invariably produce plants with 'stem mottle' symptoms.

TRV is transmitted by migratory nematodes of the genera *Trichodorus* and *Paratrichodorus* but is not transmitted by potato cyst nematodes. The nematode vectors are more commonly found in light, sandy soils, where they feed on the root hairs of potatoes and various weeds. TRV is very persistent in the nematodes, and transmission of the virus occurs from an infective nematode during the feeding process. Weeds infected with TRV act as a reservoir of the virus and in some species, e.g. field pansy (*Viola arvensis*), can be seed transmitted.

The incidence of TRV-induced spraing varies from year to year, reflecting the suitability of soil moisture conditions for mobility of the nematodes. Increased soil moisture favours nematode activity, and irrigation to prevent common scab infection could increase spraing if TRV and nematodes are present. Potato cultivars differ in their susceptibility to spraing and some are highly susceptible, e.g. Arran Comet, Maris Bard and Pentland Dell.

As there are no haulm symptoms from current season infection, TRV cannot be detected by field inspection. However, secondary infections from infected seed are very rare and spraing problems relating to seed quality seldom arise. Where fields are known to have a history of spraing they should either be avoided or, alternatively, cropped with a more resistant potato cultivar. Nematicides applied specifically for spraing may be worth while in some circumstances, although they will not eradicate infective nematodes but merely decrease their numbers (see p. 135).

Potato mop top virus (PMTV)

PMTV symptoms in the tuber are similar to those caused by TRV, although they may vary with cultivar. As with TRV there are no foliar symptoms during the year of infection. Secondary symptoms from seed infected in the previous year show either a 'mop top' (i.e. short stems with crowded leaves forming a cushion of foliage) or aucuba leaf markings. Aucuba symptoms are bright yellow blotches, rings and chevrons on normally vigorous foliage. Symptoms are usually more likely to appear following low temperatures during early growth.

PMTV is transmitted by the motile zoospores of *Spongospora subterranea*, the powdery scab pathogen (see above), as they infect the root hairs. PMTV survives in the oospores of *S. subterranea* within the soil for 10–15 years or more.

Some cultivars infected with PMTV develop spraing symptoms (e.g. Arran Pilot, Pentland Crown, Ulster Sceptre); others do not (e.g. Desirée, King Edward, Maris Peer, Maris Piper, Record).

Virus diseases other than spraing

Potatoes are susceptible to a wide range of virus diseases, but consideration here will be limited to the more important or 'severe' viruses. Severe viruses are so called because of their impact on crop yield. This yield effect is made up of two components: (a) the reduction in tuber number and size of plants which have been grown from infected seed (secondary infection), and (b) the yield loss which results when infection occurs during the current growing season (primary infection). Severe virus diseases in the UK are caused by: severe mosaic virus or potato virus Y (PVY), tobacco veinal necrosis virus (PVY^N) and potato leaf roll virus (PLRV).

Potato virus Y (PVY)

PVY is common and widespread wherever potatoes are grown, and is the most important virus disease of potatoes in the UK. It is known as 'severe mosaic' because it can produce a characteristic severe mosaic symptom on plants grown from infected seed. Infected plants have a much reduced vigour and are dwarfed, and the foliage becomes a roughened or crinkled, pale, mottled green. Different cultivars vary in their expression of PVY symptoms. In some, the reaction is mild, whereas in others the reaction is strongly necrotic, leading to leaf browning and early death of the plant.

PVY is transmitted by aphids, mainly peach/potato aphid (*Myzus persicae*) (see p. 126) in the non-persistent manner, i.e. a short acquisition feed and short transmission feed. Feeding aphids can acquire PVY from infected plants and transmit it to healthy ones within minutes, before they can be killed by insecticides. When a healthy plant is infected with PVY in this way during the growing season, symptoms may take up to 4 weeks to appear and are called 'leaf drop streak'. 'Leaf drop streak' is where brown spots or streaks develop on the infected leaf, and this necrosis spreads up and down the stem, initially on one side only. Dead leaves remain hanging on the stems, and on some cultivars (e.g. King Edward) can lead to premature defoliation. Not all reactions are so severe, and some cultivars (e.g. Estima and Record), react only with a mild mottle, although yield losses may still be high.

Varietal resistance is available and is an important factor, particularly when considering whether to save home-produced seed. The PVY resistance ratings usually produced for cultivars refer to the ease with which aphid transmission occurs in the field and not the disease severity. Because PVY is transmitted in a non-persistent manner, insecticides do not reliably prevent virus spread.

Tobacco veinal necrosis virus (PVY^N)

PYV^N is the next most common strain of PVY and induces a mild mottle in most cultivars, following both primary and secondary infection. Occasionally, as with cv. Record, symptoms are much more marked. In some cultivars, e.g. Maris Peer, yield losses can be heavy following both primary and secondary infection of the plants.

Potato leaf roll virus (PLRV)

In the UK, PLRV is second only in importance to PVY. Symptom expression is also dependent on whether infection originates from the seed or is transmitted by an aphid vector. Primary infection from the aphid vector may not result in symptoms during the growing season, unless infection occurs very early in the season in which case symptoms appear on later-produced foliage. The upper leaves of such plants roll upwards at the edges and become pale green in colour, often tinged with pink or purple.

Symptoms of 'secondary' leaf roll are much more marked once the plants have reached a height of approximately 30 cm. The lower leaves begin to show an upward and inward rolling of the margins, usually more pronounced at the leaflet base than at the tip. As the plant grows, the characteristic rolling also affects leaves on the upper part of the plant and eventually the whole plant is affected. The leaflets are not only rolled but become thickened and feel crisp, owing to the accumulation of starch.

In the field, 'false top roll' caused by aphid feeding on the growing point is often indistinguishable from virus leaf roll, except that 'top roll' plants occur in patches. Symptoms similar to virus leaf roll may be caused by other diseases, including stem canker and black leg, or by drought.

The severe viruses PVY and PLRV cause loss in ware yield in two ways.

- Loss resulting from plants grown from infected seed (up to 80% yield loss per plant). In some cases, at low disease incidence, overall crop yield losses can be reduced by compensatory growth and (in cultivars where the virus markedly reduces the growth of the infected plant) by the yield of adjacent healthy plants.
- Loss from infection during the season (up to 50% yield loss per plant), particularly where 'leaf drop streak' occurs, and this depends on:
    - varietal (cultivar) susceptibility to the virus;
    - frequency of source of infection (adjacent crops, volunteer plants, seed-borne infection in the crop);
    - extent and earliness of infection due to aphid (vector) activity.

Losses resulting from virus infection can be minimized by:

- choosing a resistant cultivar;
- starting with healthy seed (certified seed is inspected to ensure a minimum level of virus – home-saved seed should be tested to determine virus levels);
- keeping aphids out of the chitting house;
- avoiding local sources of virus, e.g. potato dumps, infected volunteers and nearby crops of lower health status (isolation reduces the risk of severe virus infection);
- spraying regularly with aphicides, to prevent the spread of virus and direct aphid damage;
- early burning-off of the crop if a seed fraction is to be riddled out.

Watery wound rot (*Pythium ultimum*)

Infection by this soil-borne fungus (a disease favoured by high temperatures) occurs only following bruising, mechanical damage or where the skins have been scuffed. The affected flesh is initially only slightly discoloured but when infected tubers are cut and exposed to the air, the cut surface turns grey and finally black. The rotted flesh is wet and pulpy with cavities, but the texture is never rubbery as with pink rot. The internal tissue of affected tubers rapidly develops into a watery mass, leaving the skins intact; affected tubers having a 'fishy' odour.

It is important to avoid growing potatoes in fields with a history of this disease and to improve drainage. Haulm should be destroyed at least 2 weeks before lifting, to enable tuber skins to set, and damage to tubers should be minimized at lifting and during store loading. Lifting during hot weather should be avoided. Chemical control is not available.

List of pests cited in the text*

Agriotes lineatus (Coleoptera: Elateridae)	a common click beetle
Agriotes obscurus (Coleoptera: Elateridae)	a common click beetle
Agriotes sputator (Coleoptera: Elateridae)	a common click beetle
Agrotis segetum (Lepidoptera: Noctuidae)	turnip moth
Aphis fabae (Hemiptera: Aphididae)	black bean aphid
Aphis nasturtii (Hemiptera: Aphididae)	buckthorn/potato aphid
Arion hortensis (Stylommatophora: Arionidae)	garden slug
Athous spp. (Coleoptera: Elateridae)	garden click beetles
Aulacorthum solani (Hemiptera: Aphididae)	glasshouse & potato aphid
Autographa gamma (Lepidoptera: Noctuidae)	silver y moth
Blaniulus guttulatus (Diplopoda: Blaniulidae)†	spotted snake millepede
Calocoris norvegicus (Hemiptera: Miridae)	potato capsid
Ctenicera spp. (Coleoptera: Elateridae)	upland click beetles
Cylindroiulus londinensis (Diplopoda: Iulidae)†	a black millepede
Deroceras reticulatum (Stylommatophora: Limacidae)	field slug
Dicyphus errans (Hemiptera: Miridae)	slender grey capsid
Ditylenchus destructor (Tylenchida: Tylenchidae)	potato tuber nematode
Ditylenchus dipsaci (Tylenchida: Tylenchidae)	stem nematode
Edwardsiana flavescens (Hemiptera: Cicadellidae)	a green leafhopper
Empoasca decipiens (Hemiptera: Cicadellidae)	a green leafhopper
Eupterycyba jucunda (Hemiptera: Cicadellidae)	a potato leafhopper
Eupteryx aurata (Hemiptera: Cicadellidae)	a potato leafhopper
Euxoa nigricans (Lepidoptera: Noctuidae)	garden dart moth
Globodera pallida (Tylenchida: Heteroderidae)	white potato cyst nematode
Globodera rostochiensis (Tylenchida: Heteroderidae)	yellow potato cyst nematode
Hepialus humuli (Lepidoptera: Hepialidae)	ghost swift moth
Hepialus lupulinus (Lepidoptera: Hepialidae)	garden swift moth
Hydraecia micacea (Lepidoptera: Noctuidae)	rosy rustic moth
Lacanobia oleracea (Lepidoptera: Noctuidae)	tomato moth
Longidorus leptocephalus (Dorylaimida: Longidoridae)	a needle nematode
Longidorus spp. (Dorylaimida: Longidoridae)	needle nematodes
Lygocoris pabulinus (Hemiptera: Miridae)	common green capsid
Lygus rugulipennis (Hemiptera: Miridae)	tarnished plant bug
Macrosiphum euphorbiae (Hemiptera: Aphididae)	potato aphid
Melolontha melolontha (Coleoptera: Scarabaeidae)	cockchafer
Milax gigantes (Stylommatophora: Limacidae)	a keeled slug

Myzus ascalonicus (Hemiptera: Aphididae)	shallot aphid
Myzus ornatus (Hemiptera: Aphididae)	violet aphid
Myzus persicae (Hemiptera: Aphididae)	peach/potato aphid
Noctua pronuba (Lepidoptera: Noctuidae)	large yellow underwing moth
Paratrichodorus spp. (Dorylaimida: Trichodoridae)	stubby-root nematodes
Phyllopertha horticola (Coleoptera: Scarabaeidae)	garden chafer
Phlogophora meticulosa (Lepidoptera: Noctuidae)	angle-shades moth
Polydesmus angustus (Diplopoda: Polydesmidae)†	a flat millepede
Pratylenchus penetrans (Tylenchida: Pratylenchidae)	a root-lesion nematode
Psylliodes affinis (Coleoptera: Chrysomelidae)	potato flea beetle
Rhopalosiphoninus latysiphon (Hemiptera: Aphididae)	bulb & potato aphid
Tandonia budapestensis (Stylommatophora: Limacidae)	a keeled slug
Tipula oleracea (Diptera: Tipulidae)	a common crane fly
Tipula paludosa (Diptera: Tipulidae)	a common crane fly
Trichodorus spp. (Dorylaimida: Trichodoridae)	stubby-root nematodes

* The classification in parentheses refers to order and family, except (†) where order is replaced by class.

List of pathogens/diseases (other than viruses) cited in the text*

Colletotrichum coccodes (Coelomycetes)	black dot
Erwinia carotovora ssp. *atroseptica* (Gracilicutes: Proteobacteria)†	blackleg and tuber soft rots
Erwinia carotovora ssp. *carotovora* (Gracilicutes: Proteobacteria)†	blackleg and tuber soft rots
Fusarium avenaceum (Hyphomycetes)	dry rot
Fusarium solani var. *caeruleum* (Hyphomycetes)	dry rot
Fusarium sulphureum (Hyphomycetes)	dry rot
Geotrichum candidum (Hyphomycetes)	rubbery rot
Helminthosporium solani (Hyphomycetes)	silver scurf
Phoma exigua var. *foveata* (Coelomycetes)	gangrene
Phytophthora erythroseptica (Oomycetes)	pink rot
Phytophthora infestans (Oomycetes)	late blight
Polyscylatum pustulans (Hyphomycetes)	skin spot
Pythium ultimum (Oomycetes)	watery wound rot
Ralstonia solanacearum (Pseudomonadales)	brown rot
Rhizoctonia solani (Hyphomycetes)	– anamorph of *Thanatephorus cucumeris*
Spongospora subterranea (Plasmodiophoromycetes)	powdery scab
Streptomyces scabies (affinity uncertain)†	common scab
Thanatephorus cucumeris (Basidiomycetes)	black scurf/stem canker
Verticillium dahliae (Hyphomycetes)	verticillium wilt

* For fungi, the classification in parentheses refers to class, although this is not possible within the phylum Ascomycota where classes have yet to be satisfactorily defined (see *Mycological Research*, February 2000). Oomycetes are now classified in Chromista with the brown algae, rather than as true fungi. Plasmodiophoromycetes are now classified as Protozoa rather than as true fungi. Some fungi have an asexual (anamorph) and a sexual (teleomorph) state, and the convention is to refer to them by their teleomorph name. However, where anamorph names are still in common use, these are listed and cross-referenced to the teleomorph name. Strictly, fungi classified as Coelomycetes and Hyphomycetes should be known as 'hyphomycetous anamorphs' and 'coelomycetous anamorphs' of the relevant teleomorph taxon (e.g. hyphomycetous anamorphic Sclerotiniaceae, for *Botrytis fabae*), respectively. These problems highlight the continual changes in the classification of the fungi.

† Bacteria – the classification in parentheses refers to division and class.

Chapter 6
Pests and Diseases of Sugar Beet

A. Lane
Independent Consultant, Church Aston, Shropshire

P. Gladders
ADAS Boxworth, Cambridgeshire

D. Buckley
ADAS Wolverhampton, West Midlands

Sugar beet

Introduction

In 1999, the area of sugar beet grown under contract to British Sugar plc was 185 000 ha and this produced the UK white sugar quota of 1.144 million tonnes, representing just over half the total tonnage of sugar consumed in the UK. The areas in which sugar beet is grown are dictated by the location of the nine remaining British Sugar factories, which are centred mainly in eastern England (from Yorkshire to Essex – seven factories), and in western counties (mainly in Shropshire and Herefordshire – two factories).

The 5-year average yield of sugar beet is now 50 t/ha at 16% sugar content; this compares with the 5-year average up to 1987 of 43 t/ha at 17% sugar. Yields have increased because of better cultivars, improvements in the pelleting of seed, more effective seed treatments for pest and disease control, a trend to earlier sowings and more use of irrigation.

About 90% of the crop follows cereals in the rotation, beet being a good break crop for cereals. The current contract with British Sugar stipulates that sugar beet must not be grown in fields which have grown sugar beet or other *Beta* species (e.g. fodder beet, mangold, red beet) in either of the two preceding years (i.e. one year in three). This restriction is aimed primarily at preventing the build-up and spread of rhizomania, but it also helps with the control of beet cyst nematode (*Heterodera schachtii*). The incidence of rhizomania is increasing steadily year on year and whilst it remains a statutory disease, which restricts the cropping of sugar beet, some farmers may have to abandon beet growing.

All beet seed, supplied exclusively by British Sugar, is treated with thiram to control the seed-borne fungus *Phoma betae*, and with hymexazol to control the soil-borne fungus *Aphanomyces cochlioides*. These treatments are aimed at protecting the germinating seedlings against blackleg and damping-off. Other seed treatment options are available to growers, including the insecticides imidacloprid and tefluthrin for use against soil pests. Imidacloprid also provides early-season control of many foliar pests and aphid vectors of virus yellows; approximately 70% of seed was treated with this insecticide in 1998. Its use has had a significant and positive impact on the management of virus yellows by providing effective control of the principal virus vector, peach/potato aphid (*Myzus persicae*), strains of which have become resistant to other aphicides. Imidacloprid is also an effective management tool, as growers do not normally have to spray to control aphids at a time of year when the farm sprayer is already busy. Pest and disease control, using seed treatments rather than granules or sprays, is widely regarded as being environmentally desirable as the quantity of active ingredient used is very small and is placed exactly where needed. An additional non-pesticide seed treatment ('Advantage') is available, often used in combination with other seed treatments, which acts as a primer to encourage early germination and boost uniformity of plant stand.

With improved crop establishment methods, time of sowing has become earlier and is now done from early March to mid-April. Early and rapid crop establishment is essential to achieve optimum yield. Pest and disease attacks can reduce leaf cover of the soil, due to plant loss and/or leaf damage, and decrease the photosynthetic efficiency of leaves. In both cases, less of the available sunlight energy is intercepted and converted into sugar.

Crops are drilled to a stand using pelleted seed, which aids precision drilling. Rows are usually 46–53 cm apart, and growers aim for a plant population of 80 000–90 000/ha; a target of 80 % establishment is achieved in most cases. Crop failure, necessitating re-drilling, is now rare. However, effective pest and disease control is essential to achieve target plant populations. The now widespread use of insecticide-treated seed ensures that damage to seedlings caused by soil insect pests is minimized. The use of granular pesticides applied in the seed furrow at sowing has declined rapidly with the availability of insecticide seed treatments. However, neither of the seed treatments is effective against the ectoparasitic nematodes (*Longidorus* spp., *Paratrichodorus* spp. and *Trichodorus* spp.) which, on light soils, cause Docking disorder. To control these nematodes, a granular nematicide is applied at sowing. These and other pesticide granules were used on 15% of UK sugar beet crops in 1998.

Later in the season, the crop may be threatened by several diseases (e.g. virus yellows, powdery mildew, rust and ramularia) and by some pests. Growers are advised on how best to control damage by optimizing the timing of sprays if needed, e.g. via the virus yellows and powdery mildew warning schemes; equally important, growers are discouraged from making unnecessary applications. Sugar beet crops are monitored regularly during the growing season by independent agronomists and area managers from British Sugar, and collation of their reports at IACR-Broom's Barn ensures that beet growers are kept well informed of pest and disease outbreaks.

The crop is harvested from the end of September until mid-January. Most sugar beet is delivered to the factories on a permit system within 2–3 weeks of lifting, but some may be clamped for 8 weeks or more before delivery.

Pests or diseases are often blamed as the prime cause of crop loss, when they are merely adding to losses due to other causes (e.g. poor seedbed preparation, inaccurate drilling, and damage resulting from careless herbicide application and soil acidity). Correct identification of pests and diseases, together with accurate diagnosis of damage, is essential if appropriate control measures are to be taken to ensure profitable production.

Pests

Beet cyst nematode (*Heterodera schachtii*)
This nematode can cause severe yield losses and is the most important nematode pest of the crop in most beet-growing countries. In England it is well established in the Fens of East Anglia, some adjacent mineral soils and other small, localized areas, especially where brassicas (including oilseed rape) are grown frequently. Affected plants usually occur in patches, their leaves wilt readily in the sunshine and their root systems are small but with excessive development of lateral (hunger) roots. From about late June onwards, white, lemon-shaped female nematodes can be seen protruding from infested roots; these females (each containing up to 600 eggs) later turn brown, forming the protective cysts which remain in the soil for several years. The nematode has a restricted host range, and under non-host crops (e.g. cereals) about 50% of eggs hatch each year and the juveniles which emerge soon die; therefore, wide rotations of host crops will prevent rapid population increases. Under sugar beet crops, two generations can be completed in a season in England, but up to five may occur during the longer, hotter growing seasons of other more southerly European countries.

Severe damage (sometimes known as 'beet sickness') is uncommon in England, possibly because for many years wide rotations of sugar beet crops were enforced. From 1943 to 1976, the Beet Eelworm Order restricted the frequency of growing host crops of the nematode in a 'scheduled area' of the Fens and in all other fields known to be infested. Until 1983 a clause in the contract between British Sugar and the grower stipulated that sugar beet should not be grown on land that had grown a host crop of beet cyst nematode in either of the two preceding years. From 1983 to 1986, however, there were no restrictions on the frequency of cropping. Between 1977 and 1988, in an annual survey of 300 randomly selected sugar beet crops in the Fens, the proportion of detectable infestations of beet cyst nematode increased from 8% in 1977 to 34% in 1985, stabilizing at around 30% thereafter. At present, a rotation clause in the British Sugar contract forbids the growing of sugar beet in fields where a *Beta* species (i.e. sugar beet, fodder beet, mangold or red beet) has been grown in either of the two preceding years.

In some countries, chemical control by soil fumigants or by granular, carbamate pesticides is necessary to achieve economic yields in infested fields.

However, in England the benefit of chemical control is not proven and no recommendations are made. At present, the only economic method of control is by crop rotation. However, research continues into the development of nematode-resistant beet cultivars and the use of nematode-resistant brassicaceous (cruciferous), green-manure catch crops to increase the decline rate of the nematode. Control of this pest in England, in the immediate future, will continue to be based upon a sound policy of crop rotation.

Black bean aphid (*Aphis fabae*)

'Blackfly', as this aphid species is often known, is one of the most common pests of sugar beet. Very dense colonies can develop on beet plants during the summer months, causing wilting and poor growth. This aphid does not introduce yellowing viruses into the crop, but it can spread those introduced by other species, e.g. peach/potato aphid (*Myzus persicae*), albeit less efficiently. Outbreaks vary in severity between years and also between regions within the beet-growing areas. Forecasts of their likely abundance in field beans (see Chapter 3, p. 77) are co-ordinated by researchers at Imperial College at Silwood Park. These forecasts are based on egg counts on the aphid's primary host, spindle (*Euonymus europaeus*). Further information is available during the growing season from suction traps operated by IACR-Broom's Barn and IACR-Rothamsted. To some extent these forecasts can be applicable to sugar beet in areas where both crops are grown.

Migration of black bean aphid to sugar beet from spindle can occur as early as May in some years but, more commonly, the aphids migrate to beet from various other sources (mostly secondary host plants), in early July.

Early infestations will be suppressed by in-furrow applications of aldicarb, carbofuran, carbosulfan or oxamyl and by the use of imidacloprid seed treatment. Where preventive treatments have not been applied, the aphids can be controlled by spraying with either pirimicarb or triazamate; both compounds will kill fewer parasitoids and predators than other, more broad-spectrum, aphicides. Control is recommended when 10% of plants are infested with aphid colonies and aphid numbers are increasing.

Control of black bean aphid after mid-July is rarely worth while since parasitoids, predators and pathogens frequently attack the colonies and reduce aphid numbers. Moreover, the then larger plants can compensate for damage caused by aphids so long as water is not limiting. If plants are under drought stress, severe damage may occur and control, using HV sprays of the specific aphicides pirimicarb or triazamate, may be worthwhile.

Capsids

Potato capsid (*Calocoris norvegicus*) and tarnished plant bug (*Lygus rugulipennis*) damage sugar beet occasionally.

Damage by potato capsid is confined to crop edges, close to hedgerows and woods from where the nymphs migrate to feed on beet plants. Damage symptoms are necrotic spotting, puckering of the leaf laminae and general distortion and yellowing of the leaves, especially at the tips. Damage is rarely severe and, at most, only the field margins need treatment. Pyrethroid insecticides used for aphid control will give some control of capsids.

Tarnished plant bugs migrate, as adults, into sugar beet fields very early in the season. They feed on the growing point of the young seedlings, causing blindness. Although damaged seedlings are not killed, several growing points subsequently develop from axillary buds, leading to a multi-crowned plant. Damaged plants, however, are rarely numerous and control measures seldom justified. Aldicarb, when applied as an in-furrow treatment to control nematodes or other soil pests, may give some control of these bugs.

Caterpillars

The caterpillars of a number of moth species occasionally damage the foliage, stems or roots of sugar beet plants. Attacks are often sporadic and chemical control measures for most species are rarely necessary.

Cutworms

The caterpillars of several noctuid moths, commonly known as cutworms, damage young sugar beet plants. The commonest, but not the most serious, are the caterpillars of turnip moth (*Agrotis segetum*). These occur from mid-summer onwards and, after feeding on the foliage for a time, feed below soil level for the rest of the season. Superficial root damage is sometimes extensive, but does not justify control measures. Less common, but more damaging, are the caterpillars of dart moths, including garden dart moth (*Euxoa nigricans*) and white-line dart moth (*E. tritici*). These feed from early spring to mid-June, attacking plants at or just below soil level; where damage is caused to beet seedlings this can lead to thin stands. Attacks of dart moth caterpillars cannot be forecast, so preventive measures are not usually possible. Where damage is occurring, applications of cypermethrin or lambda-cyhalothrin as HV sprays are recommended, preferably under moist conditions which favour surface feeding by the cutworms. Irrigation applied when cutworms are small may also provide useful control by washing the caterpillars off the plant.

Foliar-feeding caterpillars

Caterpillars of several other species of noctuid moth are sometimes found on sugar beet, feeding on the aerial parts of the plant. These include caterpillars of silver y moth (*Autographa gamma*) (this migratory species was a widespread and damaging pest in 1994, and in 1996 when it also affected various vegetable crops), tomato moth (*Lacanobia oleracea*) and angle-shades moth (*Phlogophora meticulosa*). Specific control measures are rarely necessary; sprays of a pyrethroid insecticide, such as cypermethrin, deltamethrin or lambda-cyhalothrin when used for other pests, will give incidental control of foliar-feeding caterpillars.

Tortrix moth caterpillars (usually those of flax tortrix moth, *Cnephasia asseclana*) bind parts of a leaf or leaves together and feed on the leaf surface within.

Only occasionally are numbers sufficient for damage to be noticeable and, even then, yield is unlikely to be affected. There are no specific recommendations for controlling tortrix moth caterpillars on sugar beet, but foliar sprays used to control beet leaf miner (see under Mangold fly, below) are also likely to be effective against them.

Stem-boring caterpillars

Caterpillars of rosy rustic moth (*Hydraecia micacea*) burrow inside the swelling roots and stems of beet plants from late May onwards and may kill them. Damage is usually negligible and control neither necessary nor possible. Attacks by this pest tend to occur most frequently on weedy sites.

Chafer grubs

Larvae of cockchafer (*Melolontha melolontha*) and sometimes of summer chafer (*Amphimallon solstitialis*) very occasionally damage sugar beet, especially in well-wooded areas where the soil is light. The grubs feed entirely below the soil surface, down to depths of 30 cm. Control in the growing crop is impossible, and the application of preventive treatments at sowing impractical because of the difficulty of incorporating the treatment to a sufficiently deep level.

Leatherjackets

These pests, the larvae of crane flies (most commonly *Tipula paludosa*) are a sporadic but increasing problem in sugar beet, especially in western counties, possibly owing to the increase in grass weeds and volunteers in cereal stubbles. The grey maggots feed on young beet seedlings, biting the stems at ground level. Insecticide granules, when applied at sowing to control other soil pests, may give some control of leatherjackets as will methiocarb pellets when used at the rate recommended for slug control. Most effective control is given by spraying and incorporating chlorpyrifos or gamma-HCH prior to sowing. Annual forecasts of the likely threat of leatherjacket damage, based on autumn and early-spring soil sampling by ADAS, should be used as a guide to the need for pre-sowing treatments. Post-emergence sprays of chlorpyrifos, made after the two-leaf stage if damage is detected in crops, can give some control.

Mangold flea beetle (*Chaetocnema concinna*)

This small, bronzy-metallic beetle is common in beet crops, but rarely causes economic damage. The beetles produce small, irregular pits on either the upper or the lower surface of the cotyledons, leaves and petioles. The pits break open as the leaves expand, so that holes develop. Severe damage occurs only in the drier parts of the country, especially in sheltered fields during a cold, dry spring. Damage caused by larvae feeding on the roots is of no significance.

Control can be obtained up to the four-true-leaf stage by the use of imidacloprid seed treatment or by a granular pesticide applied at sowing for the control of soil pests (see under millepedes, below); granular treatments are not justified for the control of flea beetle alone.

Spraying with the pyrethroid insecticides deltamethrin or lambda-cyhalothrin will also control the pest. However, unnecessary treatments should be avoided not only to reduce the impact on beneficial insects but also to minimize the selection for resistant peach/potato aphids and, therefore, to increase the risk of subsequent aphid infestations and virus yellows infection.

Mangold fly (*Pegomya hyoscyami*)

Attacks on sugar beet vary from year to year and from district to district, but recently damage has been less common generally, as the use of imidacloprid seed treatment has increased; 1993 was the last year of significant damage on sugar beet in the UK. White eggs are laid singly or in groups on the underside of the cotyledons and true leaves; they hatch about 5 days later. The larvae (commonly known as 'beet leaf miners') at first produce linear mines in the leaves; later, large blotch mines are formed between the upper and lower leaf surfaces. Heavily attacked plants have a scorched appearance. There are two or three generations each year, but only the first generation is worth controlling (during May), and only then when attacks are severe.

Imidacloprid seed treatment and furrow-applied granular pesticides, when used to control other pests, will prevent damage at the critical seedling stages. Where preventive treatments have not been applied, sprays of dimethoate, lambda-cyhalothrin or pirimiphos-methyl are recommended for control of beet leaf miner. These compounds, however, should be applied only when necessary, to avoid selecting for insecticide-resistant peach/potato aphids. As an approximate guide, spraying is worthwhile when the number of fresh eggs plus living maggots per plant exceeds the square of the number of true leaves (e.g. if, on plants with four true leaves, an average of 16 eggs and larvae are present).

Migratory nematodes

In sandy soils, including sandy peats, needle nematodes (*Longidorus* spp.) and stubby-root nematodes (*Paratrichodorus* and *Trichodorus* spp.) can damage the roots of beet seedlings, causing a condition known as Docking disorder (named after the parish of Docking, in Norfolk, where the problem was first recognized). Potentially, about 15% of the national crop is at risk from such damage. The nematodes aggregate around seedlings soon after germination, their feeding causing stunted growth, usually in patches or along lengths of row. Damaged seedlings may have characteristic, stubby, lateral roots (caused by stubby-root nematodes) or terminal root galls (caused by needle nematodes). Affected plants remain stunted, often have misshapen taproots and usually show symptoms of nitrogen or magnesium deficiency. The nematodes are most active in wet soils, so that damage is more likely in seasons with prolonged rainfall after germination. Irrigation during a dry summer can prolong nematode activity and increase damage in crops that have not been treated with a nematicide.

Changing the soil structure, increasing the amount of fertilizer or reducing the frequency of herbicide application may improve growth slightly on affected fields, but the only practical method of control is to apply a granular nematicide at sowing. Aldicarb, benfuracarb, carbofuran, carbosulfan and oxamyl are recommended for the control of migratory nematodes on sugar beet.

Millepedes

Millepedes, springtails and symphylids tend to occur together in organic and silty soils, and are regarded as the most important of the soil pest complex affecting the establishment of sugar beet. All three pests can be controlled with the same pesticides.

The most common species of millepede attacking beet is the spotted snake millepede (*Blaniulus guttulatus*), but the flat millepede (*Brachydesmus superus*) occurs more widely and can cause damage occasionally. Both species feed on the seedling roots and stems below soil level, damaged areas turning brown or black. Damage can sometimes be severe, especially when the spring is cold and wet, and is usually associated with that caused by springtails and symphylids.

Until the availability of seed treatments, control was by the use of in-furrow granular pesticides (aldicarb, benfuracarb, carbofuran, carbosulfan and oxamyl) or by sprays of gamma-HCH applied pre-sowing and lightly incorporated. Imidacloprid and tefluthrin seed treatments are now widely used for the control of millepedes and associated soil pests. However, in exceptional cases, where the pest population is very high, control by seed treatments may be only partial.

Peach/potato aphid (*Myzus persicae*)

This aphid rarely reaches population levels which cause direct damage to sugar beet plants, but it is the most important vector of the viruses causing virus yellows. Control measures are described below under virus yellows (p. 181).

Details are given under potatoes (Chapter 5, p. 127) about the insecticide resistance status of *M. persicae* and strategies for its control. Unlike potatoes, where there are no insecticides approved that will control all of the known resistant strains of this aphid, imidacloprid (as an approved seed treatment for sugar beet) has given very effective control of resistant *M. persicae*, especially MACE variants (see p. 127). Since it first became available for use on sugar beet in 1994, imidacloprid is now used on about 70% of sugar beet crops and, as a result, the incidence of virus yellows over this period has been low. To date, there is no known resistance to imidacloprid in populations of *M. persicae* in the UK, but very low levels of resistance to this pesticide have been found elsewhere in the world on other crops. It is important, therefore, with the continued widespread use of this active ingredient on sugar beet (together with new recommendations for its use on other crops, including cereals and some horticultural crops), that control programmes for aphids are monitored very closely for any signs of resistance developing.

Potato aphid (*Macrosiphum euphorbiae*)
This aphid occurs commonly on sugar beet but is not regarded as a serious pest or as an efficient vector of the viruses causing virus yellows. However, the young nymphs of this species can easily be confused with those of *Myzus persicae*, which may lead growers to spray crops unnecessarily. Potato aphid is large and elongate (up to 4 mm long), with long, thin siphunculi and a long, finger-shaped cauda; peach/potato aphid is oval (up to 2.6 mm long), with moderately long, noticeably swollen, siphunculi and a more-or-less triangular cauda.

Pygmy mangold beetle (*Atomaria linearis*)
Damage is characterized by small, blackened pits in the hypocotyl of young beet plants. Small seedlings may be killed, either directly or indirectly by secondary fungi invading through the wounds. However, once the stem starts to thicken, this pest does little harm. From May onwards, the small, dark-brown beetles migrate in fine weather from the previous year's beet fields to the current crop. Damage is most common in intensive beet-growing areas.

Damage can be controlled by using imidacloprid and tefluthrin seed treatments or an in-furrow application of a granular pesticide (see under Millepedes, p. 173). Where damage is serious and these treatments have not been applied, spraying with chlorpyrifos is recommended.

Root-knot nematodes
Two species of root-knot nematode occasionally attack sugar beet in England; infected plants are usually stunted and have a tendency to wilt during hot weather. The northern root-knot nematode (*Meloidogyne hapla*) has been found on sugar beet and other crops (such as carrots and potatoes) on light, sandy soil in East Anglia. The cereal root-knot nematode (*M. naasi*) occurs in Wales and in western counties of England, where it is most often found on the roots of cereals and grasses; however, it also attacks sugar beet roots on which it induces the formation of small, often elongated galls (unlike those induced by *M. hapla* which are also small, but usually round). Control measures have not been evaluated in England, where damage occurs only infrequently.

Sand weevil (*Philopedon plagiatus*)
Damage by sand weevils is confined to crops grown in sandy soils and is most prevalent in the Breckland regions of Norfolk and Suffolk. The adults feed on the foliage from late April to early June, producing a characteristic notched appearance. The pest, however, seldom occurs in sufficient numbers to cause economic damage and no insecticide treatment is recommended.

Slugs
Damage, usually by field slugs (*Deroceras reticulatum*) but sometimes by species of *Arion*, especially garden slug (*A. hortensis*), is a problem only on heavier soils, when plant establishment can be affected. Damage is most serious when the

previous autumn has been mild and wet, and where straw has been incompletely incorporated into the soil after the previous harvest. Where slugs are found damaging seedling stems above or below ground, applications of methiocarb or metaldehyde pellets are recommended. Once crops are established and plants have at least four true leaves, slug attacks are rarely severe enough to justify treatment.

Springtails (*Onychiurus* spp.)
Along with millepedes and symphylids, springtails are serious pests of sugar beet seedlings, especially in organic and silty soils. They feed on young seedlings before they emerge, causing blackened pits on the stem and roots. Seedlings may be killed before or at emergence, or their growth may be stunted resulting in irregular plant establishment. Seed treatments containing imidacloprid or tefluthrin will give effective control of springtails; pre-sowing soil treatment with gamma-HCH or in-furrow applications of the granular pesticides listed under millepedes also give good control.

Stem nematode (*Ditylenchus dipsaci*)
Stem nematodes invade seedlings, causing galling, bloating and distortion of the petioles and midribs, and sometimes death of the growing point. Obvious symptoms of infestation are absent during the summer but they appear in the autumn as a canker (a dry, corky canker in the region of the lower leaf scars) which develops rapidly and eventually invades the whole crown. There are many hosts of stem nematode (especially oats and onions), and sugar beet should not immediately follow these crops if they were infested. Damage is rarely, if ever, severe enough in England to warrant preventive measures, although this nematode has been recorded as a serious sugar beet pest in other European countries.

Symphylids (*Scutigerella immaculata*)
Symphylids are particularly common in soils with a fissured soil structure, such as chalks and in silts, where they can be found along with millepedes and springtails. Their feeding on the roots either kills the seedlings or decreases their vigour but damage tends to be patchy within the field. Control is as for millepedes (see above).

Thrips
Thrips occasionally damage sugar beet seedlings. Field thrips (*Thrips angusticeps*), which causes most damage to beet, overwinters in the soil as wingless adults and is found on beet seedlings in April and May, feeding mainly on the curled heart leaves. When these leaves expand they are elongated and even strap-like, roughened with irregular and partially reddened or blackened margins and tips. Small, silvery lesions on the leaf surfaces are also usually present. Damage, which was locally severe in Norfolk in 1995, is more important in cold, dry weather; attacked seedlings are rarely killed but their growth is retarded.

There are no specific insecticides recommended for the control of thrips on sugar beet, but some control may be achieved with imidacloprid seed treatment or with pyrethroid sprays applied to control aphids and flea beetles.

Two-spotted spider mite (*Tetranychus urticae*)
These mites (common pests of protected crops and outdoor strawberry), have become more noticeable in beet crops in recent years, possibly encouraged by warmer, drier summers. The mites feed on the beet foliage, causing yellow/grey flecking on the upper surface of the foliage which turns prematurely yellow and withers with a severe attack. Damage is confined mostly to headland plants, but can be more widespread within the crop if the attack is serious. It is not known what effect the pest has on yield, but reports from Belgium and France suggest that significant losses can occur with severe attacks and a prolonged drought. There are no control treatments recommended specifically for the control of spider mites on sugar beet in the UK, although it is likely that some of the spray treatments used for aphid control may reduce mite infestations.

Wireworms
Wireworms, the larvae of click beetles (*Agriotes* spp.), damage beet in rotations containing grass. They are most troublesome in the second year after grass, or where there were many grass weeds in stubble following a cereal crop. Wireworms have two periods of active feeding: one in the spring, coinciding with the critical beet seedling stage, and another in the autumn; the latter is of no importance. They feed on the seedling stem below ground and attacked plants usually wilt and die. There are indications that wireworm damage is increasing in completely arable rotations when beet is grown after several cereal crops. This so-called 'arable wireworm' problem is commonest on chalky soils.

Seed treatments of imidacloprid or tefluthrin may give some protection against wireworm attack; however, where the risk of serious damage is high, more effective control can be achieved with pre-sowing soil application of gamma-HCH worked into the seedbed, or by in-furrow applications of the granular pesticides carbofuran and carbosulfan. There are no control measures available after sowing.

Wood mouse (*Apodemus sylvaticus*)
Mice dig out newly sown pelleted seeds, and then extract and eat the true seed within the pellet. The worst damage occurs after mild winters in fields where sowing has been shallow into a dry seedbed. Wood mice can be found in all fields, where they feed mainly on weed seeds and insects. Damage to beet is most likely to occur if mice find pellets that have not been covered properly after sowing and discover that they contain a seed. Measures taken to encourage birds of prey on farmland will help to reduce wood mouse populations. At sowing, it is important to ensure that all beet seeds are covered by soil. Crops should be inspected daily, soon after sowing, and if mouse damage is seen, in-field trapping or a spray of

aluminium ammonium sulfate (a bird and animal repellant) may reduce further losses. Anticoagulant rodenticides are recommended for use only around farm buildings and yards; they must **not** be used in the field.

Diseases

Bacterial leaf spot (*Pseudomonas syringae* pv. *aptata*)
This is a minor leaf-spot disease, which became quite common on sugar beet in 1997 following high rainfall in June. Symptoms are usually circular leaf spots with dark, water-soaked margins and pale centres. The pathogen can be carried on seed as a contaminant and is capable of causing vascular blackening and root necrosis. Control measures are not usually required in the UK, and there are no available chemical treatments.

Barney patch
This disorder is caused mainly by the soil-borne fungus *Thanatephorus cucumeris* (anamorph: *Rhizoctonia solani*), but *Rhizoctonia oryzae* may be involved at some sites. Both of these fungi attack the roots of seedlings, resulting in patches of stunted growth and root proliferation after root tips are killed. *R. solani* occurs on light sandy soils and affects other arable crops, particularly barley, in a similar way. Patches of damaged plants can appear suddenly for 2 or 3 years and then disappear. These vary from small, oval patches to large, kite-shaped areas and enlarge or spread to new areas of the field in successive years. The name Barney patch was first attributed to a kite-shaped patch found in the parish of Barney in Norfolk. The problem occurs world wide, particularly in cereals, and is now referred to as 'bare patch'. The area of the crop affected is usually very small and yield losses are significant only within the patches. The problem is aggravated by compaction and lack of soil disturbance, so thorough cultivation shortly before drilling may reduce its impact.

Blackleg
Several fungi can attack seeds or young seedlings, producing similar symptoms (known as blackleg) when roots and hypocotyl are shrivelled or carry sunken lesions. Germinating seeds may be killed below ground or the young seedlings may 'damp-off' soon after emergence. Those surviving the early infection may then die from stem girdling. Blackleg rarely causes complete crop failure, but leads to thin and gappy stands of plants, with consequent loss of yield.

The pathogen *Pleospora bjoerlingii* (anamorph: *Phoma betae*) is the most prevalent and important seed-borne disease of sugar beet and a common cause of blackleg. It is currently controlled by steeping seed in thiram prior to the pelleting process. Prior to 1989, diethyl mercuric phosphate (EMP) was the standard treatment but this was replaced by thiram because of restrictions on the use of mercury imposed by the EC – the original Directive (79/117/EEC) was dated

December 1978 and this remains in place through amended EC Directives.

The soil-borne fungus *Aphanomyces cochlioides* has been detected in about one third of soils and can kill seedlings in late-sown crops when the soil remains wet, sometimes also infecting the fine roots of sugar beet later in the season. Most crops are normally sown when the soil is too cold for *A. cochlioides* to attack seedlings and, therefore, escape the disease. Typically, the fungus spreads up the stem from below ground level to the base of the cotyledons, leaving a thin and shrivelled stem. In the event of crop failure, resulting from *A. cochlioides*, the field should not be re-sown with beet because this crop would almost certainly also fail. Under wet conditions, *Pythium* spp. (which have also been found in about 30% of soils) may be responsible for pre-emergence losses. Control of both these pathogens can be achieved by using hymexazol incorporated in the seed pellet. This treatment improves seedling establishment and has been used widely since 1988.

Cercospora leaf spot (*Cercospora beticola*)

This is a major disease in warmer Mediterranean and central European climates, but is seldom seen in England and control is not needed. The first symptoms are small, reddish dots that later develop into circular, depressed lesions with a red-brown halo. Its spores are produced on black conidiophores, which are a useful distinguishing feature from the brown leaf spots caused by *Ramularia*, which bear white conidiophores.

Downy mildew (*Peronospora farinosa* f. sp. *betae*)

Downy mildew is prevalent in some years, in areas where a cycle of infection is maintained between seed crops and root crops. Contracts for growing seed crops specify minimum distances between seed crops and root crops, and this separation helps to check the spread of the disease.

Seed crops sown in the summer and early autumn are particularly susceptible. The disease can also overwinter on self-seeded beet in previous beet fields, on old cleaner-loader sites etc., and so good farm hygiene is an essential preventive control measure.

The disease also affects root crops, and is of greatest concern when the youngest leaves become severely affected and turn grey with intense fungal sporulation early in the season. However, the area affected is usually small and, therefore, routine spraying of crops is unnecessary. Although differences in cultivar susceptibility can occur, there is no current published information to guide growers. Plants infected in June may have their root yield more than halved, but infection in September has little effect on yield. Severe attacks greatly decrease the purity of the root juice and adversely affect sugar extraction in the factory. There are no fungicides recommended for use against downy mildew on sugar beet.

Powdery mildew (*Erysiphe betae*)

Powdery mildew is common on foliage in dry weather in late summer, when it may reduce sugar yield by up to 20%. The first signs of infection are found from July onwards, when leaves show small numbers of white pustules with radiating hyphae (this is typical of powdery mildews). Warnings are issued for powdery mildew each year by IACR-Broom's Barn, based on the number of frosts in February and March. Powdery mildew usually starts near the East Anglian coast and spreads gradually northwards and westwards during the season.

In southern counties of England, when powdery mildew has appeared before the end of August, a single spray of wettable sulfur (applied within a week of finding the disease) has given, on average, a 7% yield response. Powdery mildew control is still worth while up to mid-September, provided there are still at least 4 weeks before harvest. A copper sulfate + sulfur formulation is also available. Alternatively, the triazole fungicides cyproconazole, flusilazole + carbendazim, propiconazole, triadimefon or triadimenol may be used, and these products have been more widely used in recent years because they give long-lasting control and are also effective against rust. Carbendazim + prochloraz has off-label approval for general disease control in sugar beet seed crops (SOLA 1862/96), as does prochloraz alone (SOLA 1241/97).

Ramularia leaf spot (*Ramularia beticola*)

This occurs in most years in south-west England, where it defoliates the crop in a wet summer. In other regions, the disease is only occasionally of economic importance, though spotting can be found on a few plants in most years. The symptoms are pale-brown, circular leaf spots, which bear chains of spores on tiny, white cushions. Although differences in cultivar susceptibility can occur, there is no current published information to guide growers. Some seed crops are defoliated in July by *R. beticola*. The fungicides cyproconazole and propiconazole carry recommendations for the control of ramularia leaf spots, whilst carbendazim + prochloraz is available under off-label arrangements for disease control in sugar beet seed crops (SOLA 1862/96).

Rhizomania

This disease, caused by a root-infecting virus (beet necrotic yellow vein virus, BNYVV) and transmitted by a soil-borne 'fungus' (*Polymyxa betae*), was first detected in England in 1987. The first symptoms often appear from late summer onwards, as patches of pale, yellow-green plants. The affected plants may have erect foliage with narrow leaves and long petioles or may wilt. In some cases, warm, wet conditions can lead to the appearance of yellowing along the veins. However, the disease takes its name from the pronounced proliferation of fibrous roots, a symptom known as 'bearding'. A vertical section through the root reveals a constricted shape and vascular browning at the root tip. Late infection may produce symptoms on the lateral roots rather than on the tap root.

Rhizomania is subject to statutory control and suspected cases must be reported to a MAFF Plant Health Inspector who will advise on future action. The UK currently has Rhizomania Free Zone status and statutory action is aimed at containing outbreaks, thereby limiting its spread to other fields. Infected crops are usually destroyed, and the grower compensated from a grower levy fund.

By 1998, rhizomania had been found in 346 fields on 97 farms (affecting over 4000 ha – about 2% of the production area), mainly in Norfolk and Suffolk but with a few cases in other counties. However, the fungal vector (a relatively harmless root parasite) has been found in soils throughout Britain, wherever there is a history of beet growing. Rhizomania therefore remains a major threat to the UK sugar beet industry. Current models for the spread of rhizomania predict that the number of cases will increase gradually for another 4–5 years, and then increase rapidly. The disease is widespread in continental Europe and causes severe losses in France, Germany and Italy (each with over 100 000 ha affected each year), as well as in California and Japan. Although the most severe yield losses are associated with warmer climates, rhizomania has spread into Belgium and Holland, where, like other European countries, use of resistant cultivars enables production to continue with only slightly reduced yields.

Measures to prevent its spread in this country are aimed primarily at minimizing the transfer of soil between farms, especially on machinery and root crops. At present there is no method of controlling the disease once it has become established, though resistant and partially resistant cultivars are being developed in Europe and in the US. Some evaluation of resistant cultivars has been undertaken in the UK, and cultivar selection will become increasingly important in future management of rhizomania. At present, use of resistant cultivars is restricted to non-infected fields on farms with rhizomania. Extended rotations (some growers on affected farms have opted to give up sugar beet cropping in the short-term and transferred quota to unaffected farms), the maintenance of good soil structure and drainage, and the avoidance of excessive irrigation are ameliorative measures. To prevent further introductions of the disease, imports of plant material that may be contaminated with rhizomania-infested soil (e.g. unprocessed sugar beet, seed and ware potatoes) are subject to strict legislative control.

Rust (*Uromyces betae*)

Reddish-brown pustules of beet rust usually appear from July onwards and may become numerous in late summer, killing some of the older leaves. The disease has become more prevalent in recent years. Most current cultivars are susceptible and have shown favourable 10% yield responses when severe rust was controlled with fungicides.

Sprays of the triazole fungicides cyproconazole and flusilazole (formulated with carbendazim) have given good control of rust in recent field experiments and were rather more effective than propiconazole. Sprays are most beneficial when applied to crops as rust starts to spread throughout the crop in July and August. There is variation between products in the maximum number of applications and the harvest interval, and details are specified on the product label. There are off-label approvals for the use of carbendazim + prochloraz (SOLA 1862/96) and fenpropimorph (SOLA 1807/96) in sugar beet seed crops.

Control of late infections of rust may be worthwhile if the tops are used for stockfeed.

Violet root rot (*Helicobasidium purpureum*)

This rot is common but rarely causes serious losses. The roots of affected plants have a purplish fungal growth on the surface, below which the root tissues decay. In recent years, mild, wet autumns have led to an increase in violet root rot problems. The fungus survives in the soil as resting sclerotia, and grows on the roots of many crops and weeds. Problems are usually associated with close rotations of sugar beet with other root crops, such as carrots and potatoes. Control is by crop rotation (at least a four-year break from root crops), crop hygiene, keeping the land free from weeds, deep and thorough cultivation, and, where the disease is recognized in a crop, early harvesting and rapid delivery to the factory. There are no available fungicide treatments for controlling this disease.

Virus yellows

This disease is caused by either or both of the following viruses: beet yellows virus (BYV) and beet mild yellowing virus (BMYV). The two viruses have several different properties which affect the rate at which each spreads within beet crops, and also, possibly, the strategy for control.

BYV is a semi-persistent virus with a relatively limited host range confined to members of the genus *Beta*. The principal sources of the virus are fodder-beet and mangold clamps, ground-keeper sugar beet (common where set-aside follows beet), fodder-beet and red-beet seed crops, and maritime beet (*Beta maritima*). Segregation, in recent years, of seed crops from root crops and the reduction in number of mangold clamps have resulted in a decrease in the incidence of this virus, which is now found principally in the south-east of the beet-growing area within the Bury St. Edmunds and Ipswich sugar factory areas.

BMYV is a persistent virus, being retained by the aphid vector for its lifetime. This virus has a much wider host range than BYV, including members of the genus *Beta* and numerous common weeds. Consequently, it is much more widespread and occurs over the whole UK beet-growing area.

Yield losses caused by these viruses can be severe (up to 50% with BYV and up to 35% with BMYV, depending on the time of infection). The main vector is peach/potato aphid (*Myzus persicae*) but other aphids, e.g. black bean aphid (*Aphis fabae*), can also spread the viruses, albeit less efficiently.

The most effective way of limiting introduction to, and spread of the viruses within, the sugar beet crop, is to ensure that overwintering sources, such as mangold clamps and beet remnants from cleaner-loader sites, are destroyed before the onset of the spring migration of aphids. However, there is little that can be done to prevent migration from the overwintering weed hosts of these vectors.

Some areas are at greater risk of infection than others, mainly owing to the number and size of local aphid overwintering sites, e.g. vegetable brassicas or oilseed rape. However, that risk is modified each year by the severity of the winter. Virus yellows is most common after a mild winter (fewer than 20 ground frosts in January and February).

The current forecasts of the likely incidence of virus yellows at the end of August, made by IACR-Broom's Barn in early spring, allow growers who have not sown seed treated with imidacloprid time to decide whether or not to use a granular aphicide at sowing. When an early aphid migration is expected (i.e. after a mild winter and a warm spring), the risk of a virus yellows epidemic is increased, and growers are advised to apply the granular aphicide aldicarb. This treatment provides some protection up to the six-true-leaf stage, depending on rainfall after application. Thereafter, control (where necessary) is given by sprays of deltamethrin + heptenophos, deltamethrin + pirimicarb, lambda-cyhalothrin + pirimicarb, pirimicarb or triazamate. However, these sprays may give poor control if many of the aphids are resistant types. For this reason, the use of organophosphate aphicides to control *M. persicae* is no longer advised. Relying solely on post-emergence sprays is unlikely to be as effective in high-risk areas in seasons when aphid migration into beet crops is early.

Timing of sprays is important, and warnings are sent out by British Sugar to farmers when it is appropriate for treatments to start. Warnings are issued on the basis of local and general information about the development of aphid populations, including that gathered from the network of IACR-Rothamsted aphid suction traps; a revised forecast is usually made by IACR-Broom's Barn in early May, based on aphid activity. Aphid bulletins are issued regularly by Broom's Barn during the growing season and are available electronically and in hardcopy. The forecasting scheme was revised in 1998, based on a new modelling system which takes more account of the relationship between the spread of virus yellows and the population dynamics of the aphid vector. This will provide farmers with more local and detailed information on the potential threat of virus, with advice on options for control.

When plants have fewer than 12 leaves, the presence of an average of one or more wingless green aphids per four plants justifies a spray; this can occur as early as the two-true-leaf stage following a mild winter and early migration of aphids. Plants with more than 12 leaves do not need spraying until they have an average of one wingless green aphid per plant.

The current forecasting system for virus yellows is of limited benefit to farmers sowing imidacloprid-treated seed, a product that gives effective aphid control for up to 10 weeks after sowing and so protects plants from virus infection during the most susceptible period. Moreover, this compound controls those insecticide-resistant strains of *M. persicae* that are commonly found in the crop. Therefore, imidacloprid is being used as an insurance, rather than as an essential strategy, and is probably being applied to beet seed to be sown in areas not likely to be at risk from virus yellows; this is due to the need to order seed, and the insecticide treatment applied to it, in July of the previous year. This now widespread use of one active ingredient carries the risk of aphids becoming resistant to it.

Research is now in progress to further develop the forecasts for virus yellows incidence, with recommended pest management to include new aphid control options, including seed treatments.

List of pests cited in the text*

Agriotes spp. (Coleoptera: Elateridae)	click beetles
Agrotis segetum (Lepidoptera: Noctuidae)	turnip moth
Amphimallon solstitialis (Coleoptera: Scarabaeidae)	summer chafer
Aphis fabae (Hemiptera: Aphididae)	black bean aphid
Apodemus sylvaticus (Rodentia: Muridae)	wood mouse
Arion hortensis (Stylommatophora: Arionidae)	garden slug
Atomaria linearis (Coleoptera: Cryptophagidae)	pygmy mangold beetle
Autographa gamma (Lepidoptera: Noctuidae)	silver y moth
Blaniulus guttulatus (Diplopoda: Blaniulidae)†	spotted snake millepede
Brachydesmus superus (Diplopoda: Polydesmidae)†	a flat millepede
Calocoris norvegicus (Hemiptera: Miridae)	potato capsid
Chaetocnema concinna (Coleoptera: Chrysomelidae)	mangold flea beetle
Cnephasia asseclana (Lepidoptera: Tortricidae)	flax tortrix moth
Deroceras reticulatum (Stylommatophora: Limacidae)	field slug
Ditylenchus dipsaci (Tylenchida: Tylenchidae)	stem nematode
Euxoa nigricans (Lepidoptera: Noctuidae)	garden dart moth
Euxoa tritici (Lepidoptera: Noctuidae)	white-line dart moth
Heterodera schachtii (Tylenchida: Heteroderidae)	beet cyst nematode
Hydraecia micacea (Lepidoptera: Noctuidae)	rosy rustic moth
Lacanobia oleracea (Lepidoptera: Noctuidae)	tomato moth
Longidorus spp. (Dorylaimida: Longidoridae)	needle nematodes
Lygus rugulipennis (Hemiptera: Miridae)	tarnished plant bug
Macrosiphum euphorbiae (Hemiptera: Aphididae)	potato aphid
Meloidogyne hapla (Tylenchida: Heteroderidae)	northern root-knot nematode
Meloidogyne naasi (Tylenchida: Heteroderidae)	cereal root-knot nematode
Melolontha melolontha (Coleoptera: Scarabaeidae)	cockchafer
Myzus persicae (Hemiptera: Aphididae)	peach/potato aphid
Onychiurus spp. (Collembola: Onychiuridae)	white blind springtails
Paratrichodorus spp. (Dorylaimida: Trichodoridae)	stubby-root nematodes
Pegomya hyoscyami (Diptera: Anthomyiidae)	mangold fly
Philopedon plagiatus (Coleoptera: Curculionidae)	sand weevil
Phlogophora meticulosa (Lepidoptera: Noctuidae)	angle-shades moth
Scutigerella immaculata (Symphyla: Scutigerellidae)†	glasshouse symphylid
Tetranychus urticae (Prostigmata: Tetranychidae)	two-spotted spider mite
Thrips angusticeps (Thysanoptera: Thripidae)	field thrips
Tipula paludosa (Diptera: Tipulidae)	common crane fly
Trichodorus spp. (Dorylaimida: Trichodoridae)	stubby-root nematodes

* The classification in parentheses refers to order and family, except (†) where order is replaced by class.

List of pathogens/diseases (other than viruses) cited in the text*

Aphanomyces cochlioides (Oomycetes)	black wirestem of sugar beet
Cercospora beticola (Hyphomycetes)	cercospora leaf spot
Erysiphe betae (Ascomycota)	powdery mildew of sugar beet
Helicobasidium purpureum (Basidiomycetes)	violet root rot
Peronospora farinosa f. sp. *betae* (Oomycetes)	downy mildew of sugar beet
Phoma betae (Coelomycetes)	– anamorph of *Pleospora bjoerlingii*
Pleospora bjoerlingii (Ascomycota)	blackleg
Polymyxa betae (Plasmodiophoromycetes)	vector of rhizomania
Pseudomonas syringae pv. *aptata* (Gracilicutes: Proteobacteria)†	bacterial leaf spot
Pythium spp. (Oomycetes)	damping-off
Ramularia beticola (Hyphomycetes)	ramularia leaf spot
Rhizoctonia solani (Hyphomycetes)	– anamorph of *Thanatephorus cucumeris*
Thanatephorus cucumeris (Basidiomycetes)	Barney patch or bare patch
Uromyces betae (Ustomycetes)	beet rust

* For fungi, the classification in parentheses refers to class, although this is not possible within the phylum Ascomycota where classes have yet to be satisfactorily defined (see *Mycological Research*, February 2000). Oomycetes are now classified in Chromista with the brown algae, rather than as true fungi. Plasmodiophoromycetes are now classified as Protozoa rather than as true fungi. Some fungi have an asexual (anamorph) and a sexual (teleomorph) state, and the convention is to refer to them by their teleomorph name. However, where anamorph names are still in common use these are listed and cross-referenced to the teleomorph name. Strictly, fungi classified as Coelomycetes and Hyphomycetes should be known as 'hyphomycetous anamorphs' and 'coelomycetous anamorphs' of the relevant teleomorph taxon (e.g. hyphomycetous anamorphic Sclerotiniaceae, for *Botrytis fabae*), respectively. These problems highlight the continual changes in the classification of the fungi.

† Bacteria – the classification in parentheses refers to division and class.

Chapter 7
Pests and Diseases of Field Vegetables

R. Kennedy and Rosemary Collier
Horticulture Research International, Wellesbourne, Warwickshire

Introduction

In 1998, *c.* 150 000 ha of vegetables were grown in the open in the UK, peas and beans occupying nearly 70 000 ha, brassicas nearly 40 000 ha, and roots and onions *c.* 30 000 ha. The remaining area supported minor vegetable crops such as asparagus, celery, leek, lettuce, rhubarb and watercress.

Most vegetables were produced for human consumption, as either fresh or stored produce or preserved by freezing, dehydration or canning. High-quality produce is essential for marketability, especially for those vegetables that are pre-packed, canned or frozen. These vegetables and salads require particularly high-quality standards as they are assumed to be ready to eat. Therefore, they must be free of all contaminants.

Pesticide usage on field vegetables is aimed at improving yield and quality by preventing damage. In the UK, a relatively small number of target pest species (e.g. cabbage aphid, cabbage root fly, carrot fly, lettuce aphid, onion thrips) account for a large proportion of the total insecticide usage. However, for many diseases on many vegetable crops in the UK there is no effective chemical control. The pesticides used, and their time of application and dose, must not result in unacceptable chemical residues or taint in the produce. The period stipulated between the time of pesticide application and crop harvest (the harvest interval) must always be observed (as should all other statutory restrictions). The development and the extension of usage of new fungicides are at present deemed uneconomic for many vegetable crop/pathogen interactions.

Pest and disease control on field vegetable crops is entering a new phase and the next few years are likely to include many changes in the range of active ingredients used. Until recently, carbamate insecticides, organophosphorus insecticides and carbendazim fungicides were applied to control many of the insect pests and diseases of field vegetables. However, the MAFF Review of Anticholinesterase Compounds, initiated in 1998, led to the withdrawal of several active ingredients, some of which were used previously on large areas of crop. There is a use-up period for most of these insecticides, but after that they will no longer be approved. This has stimulated a search for alternative effective active ingredients. In addition, many of the key insecticides and fungicides now used on vegetable crops do not have on-label approval, but are applied at growers' own risk under the system of specific off-label approvals (SOLAs). As cropping areas of many vegetables have declined, agrochemical companies have not always actively supported original uses, owing to economic factors. For this reason, many of the current approvals have been obtained by the Horticultural Development Council (HDC) on behalf of growers. Because of this continual and rapid change in approvals (including SOLAs), it is important that growers check regularly to ensure that their intended use of any particular pesticide is legal.

In the past, growers have relied on routine chemical treatments to maintain high quality vegetable crops. However, vegetable growers in the UK are now under increasing pressure to justify and to reduce the use of pesticides. Coincidental with these changes has been an increased intensification of production that has resulted in greater risk of disease build-up and transfer between crops. With no break in cultivation of many vegetable crops in the UK this has provided a 'green bridge', enabling many diseases to be active at any stage of the year. Many crops are grown according to the standards set out in integrated crop management (ICM) protocols produced by the NFU and major retailers. Greater emphasis is now being placed on the advantages to be gained from more efficient application of pesticides by, for example, incorporating pesticides into polymers for film-coated application to seeds and into the media used for raising vegetable plants in modules. Another way of reducing insecticide use is to avoid routine spray applications, an approach that has stimulated the development of pest and disease forecasting and monitoring systems, so that pesticides are applied only when necessary. As disease forecasting and monitoring systems become available for particular crops, major retailers are increasingly insisting that their suppliers should adopt these if such growers wish to continue to market produce through their retail outlets. In addition, non-insecticidal and fungicidal methods of control are being considered whenever possible. For example, commercial cultivars resistant to pests and diseases are beginning to have a role in the production of certain crops. The widespread use of biological control agents in field vegetable crops is still a long way off, with *Bacillus thuringiensis* being the only 'biological' product used widely in the UK. However, MAFF is currently funding considerable research in this area. Organic vegetable production is also becoming increasingly important, although it still accounts for only a very small percentage of the overall market.

Information is available on growing vegetables and their pests and diseases through the normal advisory channels; these include research organisations, crop consultants, pesticide distributors and the HDC. In addition, computer software and the Internet are becoming increasingly important methods of disseminating crop protection information.

Asparagus

Pests

Asparagus beetle (*Crioceris asparagi*)
This beetle feeds only on members of the genus *Asparagus*, including ornamental species such as *A. plumosus*, and most vegetable asparagus crops become infested. In severe attacks, the feeding of adults and larvae may skeletonize the fronds. The crown is weakened and yield may be reduced. Feeding of adults and larvae continues until October, when the pest enters the soil to hibernate during the winter. The removal of crop debris helps to reduce the hibernating population. Adults and larvae can be controlled with foliar sprays of cypermethrin (off-label) (SOLAs 3134/98, 3135/98).

Slugs
Biological details and recommended treatments for all edible vegetable crops are given under Lettuce, p. 228.

Diseases

Grey mould (*Botryotinia fuckeliana* – anamorph: *Botrytis cinerea*)
This disease is of only minor importance in the UK, causing die-back in the field and some storage losses. Crops affected by the disease can be sprayed with iprodione (off-label) (SOLA 2772/96). Ten to 14 days must be allowed between sprays.

Rust (*Puccinia asparagi*)
The pathogen *Puccinia asparagi* can infect plants in the genera *Asparagus* and *Allium*. Rust is noticeable in the late summer when rusty-brown patches (uredinial stage) occur on the fronds; weather conditions at this time are much more favourable for disease development. Heavy dew periods are more favourable than rain for infection. In the winter, teliospores (which cause dark streaks on the stems and needles) are produced. Teliospores on fallen stem and leaf debris germinate in the spring to produce sporidia which infect the emerging buds, producing yellow pustules (aecial stage). These spores complete the disease cycle, infecting the fronds to produce the uredinial stage of the rust.

Sanitation of infected material in early autumn may be effective in reducing the potential for the pathogen to overwinter. Debris from young asparagus beds and volunteer plants should be removed and burned. Applications of difenconazole (off-label) (SOLA 1539/98) or thiabendazole (off-label) (SOLA 0525/95) can be used on crops affected by leaf spots of asparagus – this includes lesions caused by pathogens such as *Botrytis cinerea, P. asparagi* or *Stemphyllium* (only a very minor problem). Although difenoconazole may have activity against, for example, *Stemphyllium*, the off-label approval is for use only on crops affected by rust (i.e. *P. asparagi*).

Violet root rot (*Helicobasidium purpureum*)
This fungal pathogen is serious in some commercial areas, causing root rot, yellowing of fronds and die-back. It appears in patches, which increase in size, and spreads from plant to plant in late summer. Cultivated crops, such as carrot, parsnip, potato, red beet and sugar beet, and common weeds in asparagus beds, such as bindweeds (e.g. *Convolvulus arvensis*), dandelion (*Taraxacum officinale*) and docks (*Rumex* spp.), are all susceptible. Infected material may become covered with fungal resting structures (sclerotia). There are no chemicals that currently hold approval for control of this disease in the UK.

Cultural control practices include burning diseased roots and some of the surrounding healthy roots and the removal of weeds. Where the disease is severe, a non-susceptible crop, such as a brassica, should be grown and the land kept free from asparagus for several years. The fungal resting bodies can remain viable in the soil for several years.

Watery soft rot (*Sclerotinia minor*)
This disease causes some losses in UK production, and occurs as a post-harvest storage rot. Applications of iprodione (off-label) (SOLA 2772/96) can be used on crops affected by the disease.

Wilt (*Fusarium moniliforme* and *F. oxysporum* f. sp. *asparagi*)
Both species of *Fusarium* are seed-borne and specific to asparagus. Infection results in a vascular staining of roots, stems and crowns of plants, with the appearance of yellowing of the foliage. Both fungi are also common in soil, and can survive for long periods as resting structures called chlamydospores. Diseased parts have reddish-brown lesions, and root numbers are often reduced. These symptoms combine to produce a slow decline in the productivity of affected plants.

Wilt is a prevalent disease of sandy soils, especially where these are compacted or badly drained. It occurs in traditional asparagus-growing areas and is present in the newer asparagus-growing areas of the eastern region of the UK.

There are no known resistant cultivars of asparagus. Elsewhere, benzimidazole-based fungicides have been applied as seed and plant treatments, but these are not approved for use in the UK. In experiments in Canada, some cross-protection has been afforded by spraying asparagus with spores of a non-viable isolate of *F. oxysporum* derived from bean.

Beetroot

See Red beet.

Brassica crops

Because brassicaceous (cruciferous) crops often have the same pests and diseases, radish, swede and turnip have been included with the cole crops (botanical cultivars of *Brassica oleracea*) under brassicas. The area devoted to brassicas in the UK has decreased during the 1990s and is now *c.* 36 663 ha, but still exceeds that of any other vegetable crop. Although most types of brassica crop have declined, production of calabrese has increased by approximately 50% over the last ten years to 7226 ha. Sequential production guarantees a supply of produce for the fresh market throughout the year.

Pests

Aphids (*Brevicoryne brassicae* and *Myzus persicae*)
Cabbage aphid (*Brevicoryne brassicae*) is a mealy grey aphid, which infests the leaves and shoots of many brassicaceous crops and frequently causes severe damage. It remains on herbaceous Brassicaceae (Cruciferae) throughout its life-cycle. In most northern areas of its distribution, cabbage aphid overwinters as an egg on the stems of plants that remain in the field throughout the winter (e.g. oilseed rape and overwintering horticultural brassica crops). The eggs hatch in February/March and the resulting aphids colonize seed crops or vegetable crops. However, in the last 20 years or so, cabbage aphid adults and nymphs have formed a large proportion of the overwintering population.

During late spring and early summer, winged cabbage aphids disperse from their overwintering sites to colonize new host plants. They are one of the species captured regularly in the network of suction traps run by the IACR-Rothamsted Insect Survey and, in general, the first winged cabbage aphid is captured earlier following a mild winter. Once the winged aphids migrate from their overwintering hosts on to horticultural brassica crops, cabbage aphid numbers increase rapidly. This initial increase is followed, usually, by a mid-season population 'crash'. It is not known which natural control agents contribute most to this decline. Aphid numbers then begin to increase once more in early autumn, before finally declining in late autumn/early winter. The early summer peak in abundance occurs usually between mid-July and mid-August, and the late peak in abundance between mid-September and mid-December. There is considerable variation in the pattern of infestation from year to year.

Although, in the past, cabbage aphid has been the major aphid pest of brassicas, during 1996 large infestations of peach/potato aphid (*Myzus persicae*) were found in many crops. Both cabbage aphid and peach/potato aphid transmit turnip mosaic virus and cauliflower mosaic virus.

Researchers have been developing a management system for brassica aphids, particularly on Brussels sprout. This has involved the development of a crop-walking system, to estimate the size of aphid infestations, and the use of treatment thresholds based on the percentage of plants infested. Details are available from Horticulture Research International (HRI) and ADAS. One of the main findings is that growers should sample the edges of crops, rather than the complete crop, to estimate the size of aphid infestations. However, a different crop-walking strategy may be required for disease monitoring.

It is difficult to control aphids during the autumn, particularly on Brussels sprout, so it is important to suppress infestations by late summer. Several of the insecticides approved previously for aphid control on brassicas have been withdrawn recently, and will no longer be available to growers once the use-up period has expired. The active ingredients that can be used to control aphids on the various brassica crops are shown in Table 7.1. Granular insecticide treatments, using either carbosulfan, chlorpyrifos + dimethoate or phorate, can be applied at sowing or planting. The remaining insecticides may be applied as foliar sprays. Generally, sprays should be applied in just enough water to wet the foliage, and an additional wetting agent may be required to ensure wetting of the waxy aphids and foliage. Nozzles fitted to drop-legs on the spray boom may help to achieve adequate coverage when applying insecticides for the control of aphid infestations on the lower leaves and buttons of Brussels sprout in the late summer/autumn.

There is no evidence that UK populations of cabbage aphid have become resistant to insecticides. However, care should be taken if crops are infested also with peach/potato aphids, as this species may be resistant to several insecticides (see under Lettuce, p. 226, for more details). Experimental studies have shown that cabbage aphids are susceptible to some of the newer active ingredients, such as imidacloprid and triazamate. However, currently, these are not approved for use on brassica crops.

Cabbage leaf miner (*Phytomyza rufipes*)

The larvae, which are locally common, mine young shoots and petioles, and are particularly damaging on calabrese. The adults lay eggs from spring until autumn. Recently, this insect has become a more important pest in southern England.

Foliar sprays of nicotine are approved for control of leaf miners on borecole and kale, broccoli, Brussels sprout, cabbage, calabrese, cauliflower, radish, swede and turnip. There is a 2-day harvest interval.

Cabbage root fly (*Delia radicum*)

Cabbage root fly occurs throughout the UK and is a serious pest of brassica crops. The larvae feed on the roots, which may be destroyed completely. Infested plants are stunted, and may collapse and die. The larvae also tunnel into Brussels sprout buttons and the fleshy roots of radish, swede and turnip, reducing yield and quality.

Cabbage root flies overwinter as pupae in the soil. The adults emerge from the soil from March to May, the precise timing being dependent on spring temperatures. There are some areas in the UK where a proportion of cabbage root

Table 7.1 Insecticides for control of aphids on brassica crops

Compound	Formulation	Harvest interval (days)	Borecole	Broccoli	Brussels sprout	Cabbage	Calabrese	Cauliflower	Chinese cabbage	Kale	Kohl rabi	Radish	Swede	Turnip
alpha-cypermethrin	spray	7	+	+	+	+	+	+	–	+	–	–	–	–
carbosulfan	granules	see*	–	+	+	+	+	+	–	–	–	–	+	+
chlorpyrifos	spray	21	–	+	–	+	+	+	+	–	–	–	–	–
chlorpyrfos + dimethoate	granules	28	–	+	–	+	–	+	–	–	–	–	–	–
cypermethrin	spray	–	–	+	+	+	+	+	†	–	–	–	–	–
deltamethrin	spray	–	+	+	+	+	+	+	‡	+	–	–	+	+
deltamethrin + pirimicarb	spray	3	+	+	+	+	+	+	–	+	–	–	+	+
dimethoate	spray	7	–	+	+	+	+	+	††	††	††	–	–	–
fatty acids	spray	0	–	–	+	+	–	–	–	–	–	–	–	–
lambda-cyhalothrin + pirimicarb	spray	3	–	+	+	+	+	+	–	–	–	–	–	–
nicotine	spray	2	+	+	+	+	+	+	–	+	–	+	+	+
phorate	granules	see**	–	+	+	+	+	+	–	–	–	–	–	–
pirimicarb	spray	3	+	+	+	+	+	+	+	+	‡‡	‡‡‡	+	+

* 100 days on swede and turnip; 56 days on other brassicas.
** At sowing or planting only.
† Off-label (SOLA 3133/98).
†† Off-label (SOLA 0389/94).
‡ Off-label (SOLA 1691/96).
‡‡ Off-label (SOLA 0328/96).
‡‡‡ Off-label (SOLA 1634/95, SOLA 1626/95).

flies emerge from the soil later in the year than expected. These 'late-emerging' flies are genetically different from 'early-emerging' ones. Significant numbers of late-emerging flies occur in Devon, South Wales and south-west Lancashire, and mixed populations of early- and late-emerging flies may produce almost continuous fly pressure from May to September.

The rate of cabbage root fly development is dependent on temperature, so that fly activity will occur earlier during a warm year. There are two full generations each year and there is usually a partial third generation at sites in the south and in the Midlands. However, even at sites in the south, some of the progeny of third-generation flies may be unable to complete their development and will not cause damage. This is because there are insufficient heat units in the autumn for them to complete all of the larval stages before the onset of winter. Egg-laying by early-emerging flies occurs usually during May, with egg-laying by the second and third generations occurring in July and September, respectively. However, the timing of each cabbage root fly generation varies in different parts of the UK. It can also vary by as much as 3–5 weeks between years.

Adult cabbage root flies can be captured using water traps or sticky traps (usually coloured yellow), with or without a volatile chemical attractant. The eggs can be sampled from the soil around the base of host plants. Both monitoring methods show when cabbage root fly numbers are increasing or decreasing. However, reliable treatment thresholds are not available. A weather-based forecast of the timing of cabbage root fly activity has been developed at HRI, and is now available through the HDC. This can be used to time the application of treatments to established crops, such as Brussels sprout and swede, where mid-season treatments are required to control the later generations.

Cabbage root flies can be excluded from host crops using fleece or fine-mesh crop covers. Crop covers are a particularly effective way of controlling cabbage root fly on swede. However, crop covers may increase costs considerably and cause other problems (such as weeds) for growers. Covers would not be viable in areas where crucifers are grown intensively, with little or no rotation, as pupae from the previous generation would produce flies under the covers. Some commercial cultivars of brassica crops are partially resistant to cabbage root fly. However, levels of resistance are not sufficiently high to obviate the need for insecticides.

Cabbage root fly has a range of predators and parasitoids, and the biological control of this pest (using various arthropods, nematodes and fungi) is being investigated in the UK and elsewhere. However, the commercial use of such techniques is still several years away. There may be financial and other constraints to the commercialization of biological control, such as the difficulties of mass production of natural enemies for inundative release. However, such techniques may be feasible in the future in small areas of high-value crops, or in seedbeds.

Methods of insecticidal control vary according to the type of crop and stage of development. Treatments are confined almost exclusively to the use of carbamate and organophosphorus compounds. Several of the insecticides approved previously for cabbage root fly control have been withdrawn recently, and will no longer be available to growers once the use-up period has expired. Insecticides are applied as granules, drenches or sprays or as a seed treatment. Insecticides that can be used currently on one or more horticultural brassica crops are shown in Table 7.2. In intensive vegetable-growing areas, where carbosulfan or other soil insecticides from the same chemical group have been applied annually, enhanced biodegradation by soil organisms may lead to reduced levels of control. To minimize the risk of enhanced biodegradation, carbosulfan should not be applied more frequently than once every 2 years.

Application of a pre-planting drench (chlorpyrifos) to plant modules is an effective method of cabbage root fly control, with accurate placement of the active ingredient. Film-coating seed with insecticide offers another method of applying even smaller amounts of insecticide and some growers use seed-film coated with chlorpyrifos. However, this treatment is not approved in the UK, and is available only on imported seed.

On leafy brassica crops, cabbage root fly control is usually essential only whilst the plants are small. Larger plants can tolerate relatively large numbers of cabbage root fly larvae without suffering economic damage. However, on crops where the marketable part of the plant is likely to be damaged by subsequent generations (e.g. Brussels sprout buttons, swede and turnip roots) mid-season treatments may be necessary. In future, effective insecticide treatments may not be available as mid-season treatments for use on established crops in the UK. Alternative active ingredients are being sought.

Cabbage seed weevil (*Ceutorhynchus assimilis*)

This is a potentially important pest of all brassica seed crops, except white mustard. See Chapter 3, p. 57, for further details. No insecticides are approved specifically for control of cabbage seed weevil on horticultural crops.

Cabbage stem flea beetle (*Psylliodes chrysocephala*)

Most brassica crops are attacked, the damage being caused by larvae tunnelling into the leaves and stems in the late summer and early autumn, reducing plant vigour considerably. Some plants may be killed; seed-yield can be reduced by even slight attacks. See Chapter 3, p. 58, for further details.

Foliar sprays of cypermethrin (no harvest interval is specified) can be used specifically to control cabbage stem flea beetle infestations on cabbage. However, pyrethroids used for other pests usually keep infestations under control. A range of insecticides can be used to control adult flea beetles (see under Flea beetles, below).

Cabbage stem weevil (*Ceutorhynchus pallidactylus*)

This weevil infests spring-sown brassica plants, especially in seedbeds. Adults emerge from their overwintering sites in April and lay their eggs. The attacks are

Table 7.2 Insecticides for control of cabbage root fly on brassica crops

Compound	Formulation	Harvest interval	Broccoli	Brussels sprout	Cabbage	Calabrese	Cauliflower	Chinese cabbage	Radish + mooli	Swede	Turnip
carbosulfan	granules	see*	+	+	+	+	+	–	–	+	+
chlorpyrifos	spray	21	+	–	+	+	+	–	†	–	–
chlorpyrifos	drench	see**	+	+	+	+	+	+	–	–	–
chlorpyrifos + dimethoate	granules	28	+	–	+	–	+	–	–	–	–
phorate	granules	see***	+	+	+	–	+	–	–	–	–

* 100 days on swede and turnip; 56 days on other brassicas.
** Applied 4 days after transplanting or at seedling emergence.
*** Applied at sowing or at planting only; 42-day harvest interval.
† Off-label (SOLA 2935/99).
†† Off-label (SOLA 0507/99).

most serious in April–July when the larvae mine in the petioles and stems. The stems of infested plants may be spongy and snap readily during transplanting. Damage caused by the larvae can decrease the quality, as well as the yield, of mature plants. Fully grown larvae pupate in the soil and the new generation of adults begins to emerge from July onwards. At this stage, adults reared on oilseed rape crops can invade vegetable brassicas and cause considerable damage to the stems and the underside of the major veins of leaf brassicas. There is one generation each year.

Carbosulfan granules can be applied at drilling or transplanting to control cabbage stem weevil on broccoli, Brussels sprout, cabbage, calabrese, cauliflower, swede and turnip. There is a 56-day harvest interval on broccoli, Brussels sprout, cabbage and cauliflower, and a 100-day harvest interval on swede and turnip. Only one treatment may be applied per crop.

Cabbage whitefly (*Aleyrodes proletella*)

This pest infests vegetable brassicas, especially broccoli, Brussels sprout and cabbage. Feeding of the scale-like nymphs on the underside of the leaves from the end of June until late autumn causes white or yellow patches to develop. The honeydew excreted by the pest, together with the fungal growth (sooty mould) that it supports, can cause loss in quality of infested plants.

The destruction of crop remains in the winter restricts the development of sources from which the insects can spread in the spring and early summer, since adults overwinter on the undersides of leaves of brassica crops.

At present, cabbage whitefly is a minor pest and causes few problems for most growers. This may be because some insecticides used for the control of cabbage aphid and caterpillars also provide incidental control of cabbage whitefly. Foliar sprays of the insecticides shown in Table 7.3 can be applied specifically to control whiteflies.

Caterpillars (foliar-feeding)

Brassica crops may be attacked by several caterpillar species, of which those of diamond-back moth (*Plutella xylostella*) and small white butterfly (*Pieris rapae*) are the most common. At least in gardens and allotments, caterpillars of cabbage moth (*Mamestra brassicae*), garden pebble moth (*Evergestis forficalis*) and large white butterfly (*Pieris brassicae*) can also be significant pests. Although caterpillar attacks can be severe, they do not occur in every crop in every year, and routine insecticide treatments may be wasted.

In recent years, diamond-back moth has been the most damaging caterpillar pest of commercial brassicas in the UK. Although diamond-back moths may overwinter in sheltered locations in the UK, large infestations are due usually to the migration of moths across the Channel, which may occur at any time during the summer. Diamond-back moth caterpillars develop rapidly and it is important to control large infestations quickly. Pheromone trap captures of male moths give

Table 7.3 Insecticide sprays for control of cabbage whitefly on brassica crops

Compound	Harvest interval (days)	Borecole	Broccoli	Brussels sprouts	Cabbage	Calabrese	Cauliflower	Chinese cabbage	Kale
chlorpyrifos	21	–	+	–	+	+	+	+	–
cypermethrin	–	+	+	+	+	+	+	–	+
deltamethrin	–	+	+	+	+	+	+	–	+
fatty acids	0	–	–	+	+	–	–	–	–
lambda-cyhalothrin	–	–	+	+	+	+	+	–	–
lambda-cyhalothrin + primicarb	3	–	+	+	+	+	+	–	–

a good indication of the start of egg-laying, but not the size of the subsequent caterpillar infestation. Attacks are particularly severe in warm, dry summers, when these pests are able to develop rapidly.

Diamond-back moth has become resistant to a range of insecticides worldwide, and repeated use of one group of insecticides may select for resistance in this pest in the UK. However, the UK climate is relatively cool and, consequently, diamond-back moth is unable to complete more than four (usually no more than two or three) generations each season. Therefore, compared with abroad, under UK conditions it is not exposed to such a high selection pressure. The numbers of caterpillars found on plants usually decline after the first generation, possibly as a result of mortality due to natural enemies.

Caterpillars of large white and small white butterflies feed on many types of brassicaceous plants, including weeds, but attacks on vegetable brassica crops are sporadic. The large white butterfly lays its eggs in batches, and groups of caterpillars may feed on isolated plants, skeletonizing the leaves. Treatment against them on field crops is usually unnecessary. In contrast, the small cabbage white butterfly lays its eggs singly and the caterpillars feed in the centres (hearts) of plants, fouling them with excrement. Infestations by this species often require insecticide treatment. There are usually two generations of both species of butterfly each year.

Cabbage moth and garden pebble moth are sporadic and localized pests of crucifers. Cabbage moth caterpillars may cause damage to crops from June to October. Cabbage plants suffer most because the caterpillars eat into the heart. Garden pebble moth caterpillars feed on the leaves of older plants and sometimes mine into the hearts. There are generally two generations each year, and moths are active during May/June and August/September. Male moths of both species can be captured in pheromone traps. However, recent studies have shown that the pheromone lures supplied for cabbage moth are relatively non-specific and may capture only small numbers of the target species, even when subsequent caterpillar infestations are large. In both cases, trap captures do not provide a good indication of the size of an infestation.

In recent years, caterpillars of silver y moth (*Autographa gamma*) (a notorious migrant species) have been occasional, but minor, pests of brassica crops. Infestations were particularly large in 1996.

Researchers have been developing a management system for caterpillars, particularly on Brussels sprout. This has involved the development of a weather-based forecasting system and the use of treatment thresholds based on the percentage of plants infested. Details are available from HRI and ADAS. One of the main findings is that growers should sample the edges of crops, rather than the complete crop, to estimate the size of caterpillar infestations. However, a different crop-walking strategy may be required for disease monitoring. Using such a management system it is possible to target spray applications accurately. In some years, very small numbers of sprays may be required to maintain good caterpillar control.

Foliar sprays are most effective when applied to control young caterpillars. Insecticides that can be used for caterpillar control on one or more vegetable brassica crops are shown in Table 7.4. Spray volumes should be sufficient to ensure adequate crop coverage, and additional wetters should be used as necessary.

Recent work has shown that sprays of *Bacillus thuringiensis* (*Bt*) can be used to produce levels of caterpillar control (diamond-back moth, small white butterfly) which are no worse than those achieved with sprays of the pyrethroid deltamethrin. However, cabbage moth and garden pebble moth caterpillars are less susceptible to the strains of *Bt* used in products currently available in the UK.

Cutworms

Cutworms are occasional pests of brassicas. See under Red beet, p. 246, for further details. Sprays of chlorpyrifos can be used to control cutworms on broccoli, cabbage, calabrese, cauliflower and Chinese cabbage. These treatments all have a 21-day harvest interval.

Cyst nematodes

Beet cyst nematode (*Heterodera schachtii*) and brassica cyst nematode (*H. cruciferae*) can attack all brassica crops, and may slow the growth of plants in nursery seedbeds and in the field. Attacked plants are susceptible to nutrient deficiencies and water stress, and are usually stunted, but yield is reduced only rarely.

Damage in the field is best avoided by a rotation of one host crop in 4 or more years.

Flea beetles (*Phyllotreta* spp.)

Flea beetles are becoming an increasing problem, and often cause serious damage to newly emerged brassica seedlings. They are a particular problem on speciality salad vegetables (e.g. mizuna and roquette), because they cause cosmetic damage, and on drilled brassicas (such as swedes). Adults form holes in the leaves and stems, and their feeding may check or even destroy young plants.

There are several species of flea beetle, but all have a similar life-cycle. Adults hibernate in sheltered sites and move out as soon as temperatures rise in the spring to feed on weed seedlings. As temperatures become higher, most beetles disperse. When they find a suitable host crop they start to feed, often on seedling tissues below ground. Towards the end of May the beetles mate and begin to lay their eggs. The larvae feed on plant roots or, sometimes, as leaf miners in the cotyledons. The larvae then pupate and the resulting adults emerge in late July/August, to feed and build up their reserves prior to hibernation.

Most damage is caused in April and May; crops sown before early April or after the end of May usually suffer only slight damage.

Carbosulfan applied as a granule treatment is approved for control of flea beetles. Only one treatment may be applied/crop. Alternatively, foliar sprays of

Table 7.4 Insecticide sprays for control of caterpillars on brassica crops

Compound	Harvest interval (days)	Borecole	Broccoli	Brussels sprout	Cabbage	Calabrese	Cauliflower	Chinese cabbage	Kale	Kohl rabi	Radish	Swede	Turnip
alpha-cypermethrin	7	+	+	+	+	+	+	–	+	–	–	–	–
Bacillus thuringiensis	0	–	+	+	+	+	+	–	–	–	–	–	–
chlorpyrifos	21	–	+	–	+	+	+	+	–	–	–	–	–
cypermethrin	–	+	+	+	+	+	+	–	+	–	–	–	–
deltamethrin	–	+	+	+	+	+	+	†	+	–	–	+	+
deltamethrin + pirimicarb	3	+	+	+	+	+	+	–	+	–	–	+	+
diflubenzuron	14	–	+	+	+	+	+	–	–	–	–	–	–
lambda-cyhalothrin	–	–	+	+	+	+	+	–	–	–	–	–	–
lambda-cyhalothrin + pirimicarb	3	–	+	+	+	+	+	–	–	–	–	–	–
nicotine	2	–	+	+	+	+	+	+	+	–	–	+	+

† Off-label (SOLA 1691/96).

several insecticides may be applied (Table 7.5). Sprays of pyrethroid insecticides are not persistent, and the crop can be re-invaded rapidly after spraying; even repeated insecticide treatment does not always give adequate control. Alternative insecticides and methods of control are being sought.

Pollen beetle (*Meligethes aeneus*)

These insects (also known as blossom beetles) occur in large numbers in the flower heads of brassicaceous (cruciferous) crops. In recent years, feeding by adults on the florets of calabrese and cauliflower has reduced the marketability of many crops. Apart from direct damage done by the grazing beetles, individuals often conceal themselves within calabrese heads, so that when plastic-wrapped packs of produce are transferred from cold stores to shop temperatures, the beetles become active inside the packs. They may also damage brassicaceous crops grown for seed.

Adult pollen beetles overwinter in the soil. They become active in early spring and later fly to host crops (notably winter oilseed rape, see Chapter 3, p. 60), where the beetles feed on the buds and flowers and lay their eggs. The larvae feed on pollen. Fully grown larvae drop to the soil where they pupate in earthen cells. Young beetles emerge usually in late June/July and the majority leave to feed on other plants. This is to accumulate reserves that maintain them through the winter period, there being just one generation each year. Although the young beetles have a strong preference for brassicaceous plants, including calabrese and cauliflower, many fail to locate such crops and, instead, feed on the pollen of weeds and garden flowers.

Large infestations of beetles on vegetable brassicas do not occur at all sites every year, and the numbers of beetles migrating will depend on the size of the local population (affected by the proximity of oilseed rape crops) and weather conditions during the migration period. Pollen beetles are particularly active when it is warm and humid. To highlight periods when they are active, the adults can be captured on yellow sticky traps. However, there is no reliable treatment threshold for vegetable brassicas. A weather-based forecast of the timing of the pollen beetle migration has been developed at HRI and is available from the HDC. This information can be used to target crop walking.

Foliar sprays of alpha-cypermethrin (off-label) (SOLA 1750/96) (on broccoli, calabrese and cauliflower), deltamethrin + pirimicarb (on borecole and kale, broccoli, Brussels sprout, cabbage, calabrese and cauliflower) or lambda-cyhalothrin + pirimicarb (on cauliflower and calabrese) can be applied to control pollen beetle infestations. These treatments have harvest intervals of 7, 3 and 3 days, respectively. Sprays should be applied as soon as beetles are seen feeding on the florets.

Slugs

Biological details and recommended treatments for all edible vegetable crops are given under Lettuce, p. 228.

Table 7.5 Insecticide sprays for control of flea beetles on brassica crops

Compound	Harvest interval (days)	Borecole	Broccoli	Brussels sprout	Cabbage	Calabrese	Cauliflower	Kale	Swede	Turnip
alpha-cypermethrin	7	+	+	+	+	+	+	+	–	–
carbosulfan	see*	–	+	+	+	+	+	–	+	+
cypermethrin	–	+	–	+	+	–	+	+	–	–
deltamethrin	–	+	+	+	+	–	+	+	+	+
deltamethrin + pirimicarb	3	+	+	+	+	+	+	+	–	–

* 100 days on swede and turnip; 56 days on other brassicas.

Swede midge (*Contarinia nasturtii*)
This is predominantly a pest of swedes but it does attack other brassicas. It is uncommon in the UK. The characteristic damage caused by larvae feeding in the growing points is known as 'many-neck'. Bacterial rots frequently follow the damage caused initially by the midge larvae. No insecticide treatment is currently approved for its control.

Turnip gall weevil (*Ceutorhynchus pleurostigma*)
The rounded galls produced by this weevil may occur on all cultivated brassicaceous crops, and on wild hosts such as charlock (*Sinapis arvensis*), especially in south-west England. They occur on the root just below soil level and can be distinguished from the swellings caused by clubroot as each gall contains a larva, or a cavity with an exit hole through which the fully fed larva has emerged. Most of the damage arises in the seedbed, and the growth of badly galled young plants may be checked severely. Well established plants usually suffer little but the quality of culinary swede and turnip is reduced if they are galled badly.

No insecticide treatment is approved. Where feasible, any galled seedlings found at transplanting should be discarded. Also, where feasible, crop rotation should be practised.

Turnip root fly (*Delia floralis*)
This pest, which is very similar to cabbage root fly, is common in Scotland and the North of England (including south-west Lancashire), especially on light soils. There is one generation each year and the adults emerge in June/July. Turnip root fly causes direct damage to swedes and turnips through the mines made by the larvae, and this also enables secondary pathogens to infect attacked roots.

No insecticide is currently approved specifically for use against turnip root fly. However, brassica crops attacked by turnip root fly are usually attacked also by cabbage root fly (although the generation times are different) and it is likely that treatments applied to control the latter pest will give incidental control of turnip root fly.

Diseases

Bacterial leaf spot (*Pseudomonas syringae* pv. *maculicola*)
The pathogen is seed-borne. In general, the disease is rare in the UK, but in recent years outbreaks on cauliflower seedlings raised in modules may have resulted from the use of infected seed stocks.

Symptoms comprise small (up to 3 mm), brown or purplish spots on leaves and brownish spots on cauliflower curds. Severely affected leaves may become distorted, and may turn yellow and fall. Crop rotation and hot water treatment (HWT) of seeds should reduce the incidence of this disease. However, HWT may depress the germination of less vigorous seed stocks. There are at present no approved chemicals which can be used to control this disease. However, sprays of copper oxychloride hold specific off-label approval (SOLA 0993/92) for control of spear rot on calabrese in the UK, and this may be effective against other bacterial leaf spots on calabrese if applied protectively. Spear rot is caused by *Pseudomonas syringae* but largely of unknown type.

Black rot (*Xanthomonas campestris*)
The bacterium is seed-borne, attacking horticultural brassicas, including broccoli, cabbage and kale. The disease has become increasingly common in the UK and is economically important on cauliflower in Cornwall and other western areas of brassica production in the UK. Typical external symptoms include leaf yellowing from the tips inwards, accompanied by blackening of the veins. Internally, vascular necrosis is a typical symptom and a black ring is present in cross-sections of stems and roots. HWT can reduce seed-borne inoculum but may have an adverse effect on the germination of certain seed lots. A break of 2–3 years between brassica crops can reduce soil-borne inoculum and the disease. There are no approved chemicals for control of black rot in the UK, although soaking seeds in sodium hypochlorite controls *Xanthomonas* infection in heavily infected seed lots.

Canker (*Leptosphaeria maculans* – anamorph: *Phoma lingam*)
This is a seed-borne disease, the asexual stage (*Phoma lingam*) causing dry rot of swede and turnip, and sometimes damping-off and stem canker. It also affects broccoli and Brussels sprout crops produced from home-saved seeds. The disease is extensive in winter oilseed rape (*Brassica napus*). The sexual state (*Leptosphaeria maculans*) develops on the field stubble of harvested crops, and air-borne spores from this substrate may disseminate the pathogen over considerable distances. There are no approved chemicals for the control of canker on horticultural brassicas. However, applications of tebuconazole for the control of ring spot and other leaf diseases may be effective. Tebuconazole is approved for control of canker in winter oilseed rape.

Clubroot (*Plasmodiophora brassicae*)
This pathogen is economically important in the UK in Cornwall, Lancashire and Scotland. It is not commonly found in Lincolnshire, and the reasons for this are, at present, unknown. The disease appears as swellings on the roots, causing wilting and death of plants. It is particularly severe on summer crops and first appears in patches within the crop. Many chemicals, such as mercurous chloride, have lost approval for control of this disease during the last 5–10 years, and cultural practices are often the only realistic means of control. Crop rotation can reduce the build-up of disease, although resting spores can remain active in the soil for many years. Controlling soil pH can be an effective way of preventing disease development. Addition of lime, to retain soil pH at 7.2 and above, will reduce the disease or control it completely. Application of calcium cyanamide to the soil has also been effective in controlling the disease.

Application of boron and agral (at 15 ppm and 0.2%, respectively), applied as a pre-planting drench, has also been shown to be an effective control treatment, although this does not hold approval for control of clubroot.

Damping-off and wirestem (*Thanatephorus cucumeris* – anamorph: *Rhizoctonia solani*)
This pathogen causes a dark-brown or black rot on the stem base of seedlings, especially cauliflower, usually resulting in a pronounced constriction of the stem. The seedlings often die, but sometimes they may survive as stunted plants with a typical 'wirestem' appearance. The pathogen persists in the soil as sclerotia or on infected debris. The sclerotia germinate to release basidiospores, which are wind dispersed. The fungus is more commonly found on seedlings. The disease can be controlled by applying quintozene or tolclofos-methyl, either at seed sowing or to established seedlings.

Dark leaf spot (*Alternaria brassicae* and *A. brassicicola*)
Dark leaf spot is a severe disease of brassica seed crops, especially in wet seasons when the fungus readily invades the developing pods and causes considerable loss of seed. The disease is seed-borne and gives rise to a damping-off of young seedlings. In recent years, dark leaf spot has increased in horticultural brassica seed samples, and the disease is now one of the most important problems in Brussels sprout production. The appearance of the disease on the buttons downgrades their value, in some cases rendering them unmarketable. The disease is not economically important on broccoli, calabrese or cauliflower as it affects only the leaves of these crops.

Alternaria on leaves of horticultural brassicas can be controlled by applications of chlorothalonil. Sprays containing chlorothalonil + metalaxyl are also approved for control of *Alternaria* on Brussels sprout, calabrese and cauliflower. In addition, difenoconazole tebuconazole and triadimenol are approved for control of *Alternaria* on Brussels sprout and cabbage. Sprays of tebuconazole, applied against ring spot on broccoli and cauliflower (SOLAs 2048/97, 2495/96), will be effective on these crops. Difenoconazole also holds approval for control of *Alternaria* on calabrese and cauliflower. Sprays should be applied at intervals of 14–21 days, up to a maximum permitted dosage per crop. Finally, iprodione is approved for control of dark leaf spot on broccoli, Brussels sprout, cabbage and cauliflower; also, iprodione can be applied effectively as a seed treatment on both swede and turnip.

Downy mildew (*Peronospora parasitica*)
This may be troublesome in the seedling stage, especially on cauliflower in Dutch lights, causing stunting or death of young plants. The disease can be economically important in the field. Lesions are first seen on brassica leaves as pale-green to yellowish spots, which are angular in shape and usually bounded by leaf veins. The pathogen readily sporulates on the underside of the infected leaves. There are

many races of downy mildew, some of which attack brassicaceous (cruciferous) weeds. However, these do not usually infect horticultural brassicas.

The disease can be controlled using sprays of chlorothalonil on all horticultural brassica plants and seedlings (except calabrese); sprays must be applied at the first sign of disease. Propamocarb hydrochloride is approved for control of downy mildew on all horticultural brassicas where it should be applied as a drench in plant propagation. Chlorothalonil + metalaxyl is approved for control of downy mildew in the field in Brussels sprout, calabrese and cauliflower crops; a maximum of two sprays are permitted to each crop and should be applied immediately disease is observed, with a 14- to 21-day interval between sprays. Mixtures of maneb and zinc have on-label approval for control of the pathogen on broccoli, calabrese and cauliflower. Fosetyl-aluminium (off-label) (SOLA 1533/95) and copper oxychloride + metalaxyl (off-label) (SOLA 1383/97) can be used on broccoli, Brussels sprout, cabbage, calabrese and cauliflower crops affected by downy mildew. Fosetyl-aluminium and mixtures of copper oxychloride + metalaxyl are usually applied to affected seedlings. Mancozeb + metalaxyl-M (off-label) (SOLA 0937/99) is available for use on cabbage crops infected with downy mildew. Sprays should be applied in the field at first sign of disease.

Light leaf spot (*Pyrenopeziza brassicae*)

This disease is of sporadic importance in horticultural brassicas in the UK, where there is strong evidence that agricultural brassicas, particularly oilseed rape, represent a major source of ascosporic inoculum. In Scotland, the disease is problematic on Brussels sprout crops. The older leaves of broccoli, Brussels sprout, cabbage and cauliflower are attacked. Both surfaces are invaded, and pink to white lesions develop in concentric rings beneath the cuticle of the interveinal tissues of the leaf. The cuticle ruptures to reveal the clumps of spores, which are spread by rain splash. Lesions darken with age. Light leaf spot also blemishes Brussels sprout buttons and may cause them to rot. There is little evidence that the ascosporic stage produced on horticultural brassicas plays any role in the epidemiology of the disease on this crop.

Leaf and stem debris should be ploughed in to reduce this source of inoculum, and brassica crops should be rotated on a 4-year cycle. Difenoconazole is approved for control of this disease on Brussels sprout, cabbage, calabrese and cauliflower. A three-spray programme with this chemical is effective when commenced at the first sign of disease. Tebuconazole, which has approval for use on horticultural brassica crops infected by *Mycosphaerella brassicicola*, may also be effective against light leaf spot, although it is not approved for this use.

Powdery mildew (*Erysiphe cruciferarum*)

This disease causes severe leaf infections of swede and turnip, particularly in dry summers. Infection by powdery mildew on *B. oleracea* types (e.g. Brussels sprout, cabbage, cauliflower) is increasingly important. The fungus forms greyish patches that become white on the upper surfaces of leaves; in severe attacks, these may coalesce to cover the entire leaf and result in defoliation. Free (unbound) water has been shown to reduce brassica powdery mildew infection on the leaf.

Up to three sprays of triadimenol have approval for control of powdery mildew on Brussels sprouts and cabbage. Sprays should be applied at first sign of disease. On swede and turnip, two sprays of triadimenol are permitted, with a 14-day interval between each spray. Tebuconazole and sulfur (2–3 applications per crop) hold approval for control of powdery mildew on swede. On turnip, only tebuconazole holds approval (up to a maximum permitted dosage per crop) for powdery mildew control. Tebuconazole can be used on Brussels sprout and cabbage crops infected with powdery mildew, but should not be applied to these crops at either 'button formation' in Brussels sprout or at 'heart formation' in cabbage. Only three sprays are permitted per crop; there is also a maximum dosage per crop. Tebuconazole may also be applied to control powdery mildew on swede and turnip. Sprays should be applied to these crops at root diameters of < 2.5 cm. Alternatively, to control powdery mildew, infected swede and turnip crops can be treated with sulfur. Sulfur should be applied at first sign of disease in the crop, with repeat sprays 2 or 3 weeks later.

Ring spot (*Mycosphaerella brassicicola*)

The pathogen frequently affects Brussels sprout, cabbage and cauliflower, and more rarely horseradish, kale and oilseed rape. Although previously considered to be of economic importance only in south-west England, the fungus is now widely distributed in all the main brassica-growing areas. The leaves are normally diseased but the fungus can attack all the living plant parts. Seed pods can become diseased and Brussels sprout buttons may be so badly affected as to render them unsaleable. Initially, small spots appear between the veins of leaves. These have a characteristic, clearly defined edge; however, at the early stage in their development, ring spot and dark leaf spot lesions are difficult to differentiate. Lesions increase in size to produce concentric rings of pseudothecia, which give it a characteristic appearance and its name. The pseudothecia give rise to ascospores which are airborne and are the only spore form recorded for this pathogen in the UK. Ring spot lesions may cover the surface of sprout buttons. Badly affected leaves fall, and from the debris ascospores of *M. brassicicola* are discharged into the air. These may be carried downwind over considerable distances, to cause new infections on other brassica crops. Seeds may become infected but their role in the transmission of the disease is not clear, and infected debris probably constitutes the main source of the disease. Rotation of crops and burning of debris will help to reduce inoculum levels.

Two applications of chlorothalonil can be applied to infected broccoli, Brussels sprout, cabbage, calabrese, and cauliflower crops, with a 10- to 21-day spray interval. This chemical is often used to treat Brussels sprout crops affected by many foliar pathogens for which it has approval because it has a short (7-day) harvest interval after spray application. Other chemicals, such as tebuconazole, are also approved for control of ring spot on infected Brussels sprout and cabbage crops. Tebuconazole (off-label) (SOLA 2048/97) may also be used on broccoli, calabrese and cauliflower crops infected with ring spot. A maximum permitted dosage per crop is allowed. However, the harvest interval after application is 21 days. Tebuconazole should be applied at first sign of disease. Care should be taken that the dosage used is appropriate for the growth stage of the crop. Triadimenol (on-label approval), like tebuconazole, has eradicant activity against ring spot on Brussels sprout and cabbage, and up to three applications of this fungicide are allowed per crop to control ring spot infection. Mixtures of chlorothalonil + metalaxyl are approved for control of ring spot on Brussels sprout, calabrese and cauliflower. Difenoconazole is approved for control of ring spot on Brussels sprout, cabbage, calabrese and cauliflower. Up to three sprays are permitted per crop (up to a maximum dosage) with a 14- to 21-day interval between sprays. The harvest interval after using this eradicant is 21 days. Sprays should be commenced at first sign of disease. Benomyl is also approved for control of ring spot, when applied as field spray, with a 21- to 28-day spray interval.

Soft rot (*Erwinia carotovora* ssp. *carotovora*)
The bacterium causes internal decay of the stems of broccoli, cauliflower, kale, etc., and affected plants become putrid and collapse.

The soft rot organism is soil-borne and invades plant tissue when this has been mechanically damaged or stressed (use of excess farmyard manure, poor drainage, etc). Disposal of crop debris by burning or ploughing and the use of 2- to 3-year rotations should reduce inoculum and decrease the likelihood of attack. There are no approved chemicals that can be used to control the disease.

Spear rot
See under Bacterial leaf spot, p. 202.

Storage rots of cabbage
Bacteria, fungi and viruses cause deterioration of Dutch white cabbage heads in refrigerated storage. The most extensive damage is caused by fungi, particularly *Botryotinia fuckeliana* (anamorph: *Botrytis cinerea*) and, to a lesser extent, *Alternaria* spp. (*A. brassicae* and *A. brassicicola*), *Mycosphaerella brassicicola* and *Phytophthora porri*. Mixtures of iprodione applied with carbendazim + metalaxyl are approved as a post-harvest dip for control of these diseases. The harvest interval after treatment is 2 months. Two viruses (turnip mosaic and, to a lesser extent, cauliflower mosaic) are particularly important in causing internal necrosis of stored Dutch white cabbage. A few commercially available cultivars are highly resistant, however, and these do not develop necrosis.

White blister (*Albugo candida*)
White blister infects brassicaceous (cruciferous) seedlings and plants. The disease is of increasing importance, having occurred more frequently on horticultural

brassicas in recent years. The pathogen produces leaf distortion, often resulting in raised lesions which break open to liberate zoosporangia. Lesions are a distinctive white colour and, on breaking, often superficially resemble blisters. Zoosporangia liberate motile zoospores; these require free (unbound) water for their liberation. The zoospores infect brassica tissues rapidly (4–6 hours at optimal temperatures).

Infection of Brussels sprout buttons by the pathogen produces raised lesions, which can reduce the quality of the crop. Lesions occur mainly on the lower surfaces of leaves. Most crucifers are susceptible but, as several races of *A. candida* have been identified, it is unlikely that related weeds, e.g. shepherd's purse (*Capsella bursa-pastoris*), act as infection sources to cultivated brassicas. Chlorothalonil + metalaxyl is approved for control of white blister on infected Brussels sprout and cauliflower crops. Sprays must be applied at first sign of disease, at intervals of 14–21 days. However, only two sprays are recommended (up to a maximum permitted dosage) per crop. Mancozeb + metalaxyl-M (off-label) (SOLA 0937/99) can be used on cabbage crops affected by white blister.

Broad bean

Approximately 2500 ha of broad beans are grown in the UK with much of it spring-sown.

Pests

Bean beetle (*Bruchus rufimanus*)

The incidence of damage caused by the larvae of this beetle as they feed inside the developing seeds is low generally, although local infestations (which have become more widespread in recent years) may be severe in hot, dry summers. Although subsequent development of seedlings from infested seed may not be impaired, infested crops may be rejected if grown for processing.

Sprays of deltamethrin can be applied to control this pest on beans, but not specifically on broad beans. Current advice from PGRO is to spray mid-flower as soon as the beetle is found and before eggs are laid on the lowest pods. A second spray may be needed.

Black bean aphid (*Aphis fabae*)

This aphid is a common and often serious pest of field and broad beans, but heavy infestations may occur also on French and runner beans, mangold, red beet, spinach and sugar beet. Wild summer hosts include fat-hen (*Chenopodium album*) and thistles (e.g. *Sonchus* spp.). The eggs overwinter on spindle (*Euonymus europaeus*). On beans, heavy infestations on the leaves and stems cause stunting, leading to a marked reduction in yield. Autumn or early-spring sowing of beans allows the plants to become established and usually to flower before the attack starts. Removal of the plant tips when an infestation has just begun generally decreases the subsequent attack. If 5% or more of the plants on the south-west headland of a field are colonized by mid-June, sprays should be applied.

Foliar sprays of nicotine or pirimicarb are approved for control of aphids on broad beans, and have harvest intervals of 2 and 3 days, respectively. Insecticides should be applied when colonies first appear and treatments repeated if necessary. Aphid colonies should not be allowed to establish in the crop.

Capsids (*Lygocoris pabulinus* and *Lygus rugulipennis*)
Biological information is given under French beans, p. 220.

Nicotine is approved for capsid control on broad beans, with a harvest interval of 2 days.

Pea & bean weevil (*Sitona lineatus*)
This weevil commonly infests broad bean crops, the adults causing characteristic notching of leaves whilst feeding in the spring and early summer. Overwintered crops are damaged less seriously than less-established crops, the growth of which may be stunted when weevils feed on young shoots.

Adult weevils overwinter in sheltered locations. When spring temperatures rise, they migrate to pea and bean crops, where they feed on the foliage, typically notching the edges of the leaves. Eggs are eventually laid on the leaves or in the soil. Following egg hatch, the larvae make their way to the root nodules, on which they feed for several weeks before pupating in the soil. Members of the new generation of adults that emerges from these pupae feed on foliage before migrating to overwintering sites. Destruction of the nodules by larvae has more influence on seed weight than leaf notching caused by adults.

Spraying crops as soon as the first leaf damage is seen will disrupt the egg-laying period. A monitoring system is available which indicates the risk of high levels of damage. The system comprises five funnel traps containing the weevil aggregation pheromone. Traps are placed on the grassy edges of the previous year's pea or bean crop in mid-February, and monitored three times a week until an average of more than 30 weevils per trap is exceeded. High-risk crops are those that are within 10 days of emergence.

Foliar sprays of alpha-cypermethrin or deltamethrin are approved for control of pea and bean weevils on broad bean. Alpha-cypermethrin has a harvest interval of 11 days; no harvest interval is specified for deltamethrin. Sprays of these insecticides will not kill larvae in the root nodules.

Diseases

Several diseases of *Vicia* crops (including Fusarium root and stem rots, sclerotinia rot, viruses and certain leaf diseases) are described under Field beans (see Chapter 3, p. 76).

Chocolate spot (*Botryotinia fuckeliana* – anamorph: *Botrytis cinerea*, and *Botrytis fabae*)
Chocolate spot can cause severe damage to sowings of winter beans. In wet conditions the aggressive phase develops, causing extensive blackening, wilting and almost complete destruction of bean foliage. More commonly, the non-aggressive phase produces a brown spotting or flecking of leaves, stems and pods. Both pathogens are seed-borne, and also survive on debris in the soil. Development of *Botrytis fabae* on the pods can be particularly rapid.

Good drainage, sufficient supplies of potassium and phosphorus, and adequate spacing between plants are factors which decrease the likelihood of attack. Chlorothalonil + metalaxyl and vinclozolin are approved for control of chocolate spot pathogens on broad beans in the UK. Sprays must be applied from mid-flower to first pod set, or as soon as disease appears in the field. Chlorothalonil + metalaxyl may also have some effect against other leaf spots of broad bean.

Leaf and pod spot (*Didymella fabae* – anamorph: *Ascochyta fabae*)
In the past, this seed-borne fungus was common in samples of field bean seeds. Diseased seeds produce infected plants, which have clearly defined, circular, sunken lesions on leaves, stems and pods. Pod infection results in poor-quality seeds of low viability. Thiabendazole + thiram is approved as a seed treatment for control of this disease. Antagonism between *A. fabae* and *Botrytis fabae* on beans has been reported.

Rust (*Uromyces viciae-fabae*)
This disease appears as typical brown lesions on the pods. Cypress spurge (*Euphorbia cyparissias*) is an alternate host, on which the pathogen produces other life-cycle stages. In the UK, the pathogen probably survives as teliospores on debris or as infected tissue. Teliospores germinate to produce basidiospores, which infect the crop at the beginning of the season. The typical brown pustules, which release urediniospores, develop during the spring and summer.

Broccoli

See under Brassica crops, p. 189.

Brussels sprout

See under Brassica crops, p. 189.

Cabbage

See under Brassica crops, p. 189.

Carrot

About 11 000 ha of carrots are grown annually in the UK, which is more than any vegetable crop other than brassicas and peas.

Pests

Carrot fly (*Psila rosae*)
This widespread and often serious pest of carrot also damages celeriac, celery, parsley and parsnip. The larvae feed on roots, and young plants may be stunted or killed. Tunnelling by the larvae may make the mature roots of carrot and parsnip unmarketable.

Carrot flies overwinter in the soil, either as larvae or as diapausing pupae. Late-developing insects remain as larvae and continue to feed on carrot roots throughout the winter before pupating in the spring. In the UK, carrot flies complete two generations each year; in particularly warm locations there may be a partial third generation. Depending on the weather, first-generation fly emergence occurs during April/June and second-generation fly emergence during July/September. The timing of peak emergence by the first and second generations of carrot fly may vary by as much as 3–5 weeks from one year to the next.

Some carrot cultivars are partially resistant to carrot fly attack. At present, the most resistant commercial cultivars have levels of partial resistance of approximately 50% compared with a susceptible one. Breeding lines held by commercial companies now have levels of partial resistance as high as 75%. Information on the relative susceptibility of many commercial cultivars can be obtained from HRI. Experiments at HRI have shown that the effects of a partially resistant cultivar and use of a granular soil insecticide are additive.

Crop covers can be used to exclude adult carrot flies from carrot crops and are being used by some organic growers. Crop rotation, and isolation from crops infested with carrot fly, can reduce carrot fly numbers. Recent work at HRI showed that there was a strong inverse relationship between the numbers of flies captured on sticky traps in carrot plots and the distance of the plot from the pest's overwintering site. Few flies infested plots more than 1 km away from the emergence site.

Carrot fly damage can be reduced by sowing crops late, to avoid the period of peak egg-laying by the first generation, and by lifting carrots early to avoid the development of larval feeding damage. However, this is not always feasible when growers wish to harvest crops almost year-round. Carrots at the edges of the field are generally more severely damaged than those in the middle and, therefore, should be lifted first. In countries such as Denmark, where low winter temperatures prevent growers storing their crops in the field, carrots are lifted in the autumn and kept in cold stores. This reduces the incidence of carrot fly damage considerably.

Until 1995, carrot fly was controlled using organophosphorus and carbamate insecticides. In 1995, as a result of research on residue levels in individual carrot roots, MAFF PSD announced a limit of three organophosphorus insecticide applications per carrot crop, with a concession of four applications on soils with more than 10% organic matter. At about the same time, off-label approvals were granted for the pyrethroids tefluthrin as a seed treatment (SOLAs 0873/00, 0874/00) and lambda-cyhalothrin as a foliar spray (SOLAs 1737/96, 1738/96, 0283/2000; harvest interval 14 days), for carrot fly control on carrots and parsnips. A third off-label approval was granted later for foliar sprays of deltamethrin for general insect control on carrots (SOLA 1265/95; harvest interval 21 days). Most of the organophosphorus and carbamate insecticides approved previously for carrot fly control on carrots have now been withdrawn. Once their use-up periods have expired, growers will have very little choice of insecticides. Alternatives are being sought.

The move to replace organophosphorus insecticides with pyrethroids means that growers need to adopt a new strategy for carrot fly control. This is because the two groups of insecticide affect completely different stages in the life-cycle of the pest. The organophosphorus compounds are effective mainly against the neonate larvae, whereas the pyrethroid compounds kill the fly adults. HV sprays should not be used to apply pyrethroids. The target is the leaf surfaces. Because sprays of lambda-cyhalothrin become rain-fast straight away, they can be applied even if rain is expected. The best time to apply pyrethroid sprays to kill flies by direct contact is between 4 pm and 6 pm, when maximum numbers of female flies are in the crop.

Adult carrot flies can be captured on orange/yellow sticky traps, and in the past trap captures have been used to time the application of treatments against the larvae (using organophosphorus insecticides). In contrast, pyrethroid insecticides should be targeted against the adults, and a warning based on trap captures may come too late. A weather-based forecast of the timing of carrot fly generations (adult emergence and egg laying) has been developed at HRI (with validation data from ADAS) and is available through the HDC. The optimum timing for the first pyrethroid spray against the second generation of adults appears to be one week ahead of the forecast for 10% egg laying.

Although six sprays of lambda-cyhalothrin and three sprays of deltamethrin are permitted on each crop, it should be possible to control carrot fly with fewer sprays. Experimental work done by HRI and ADAS has shown that, usually, insecticidal sprays are not required after the end of September, even on crops to be harvested as late as May. It is larvae resulting from uncontrolled flies from the beginning of the second generation that cause the damage that then increases during the winter months.

Caterpillars, including cutworms

Foliar-feeding caterpillars, e.g. those of silver y moth (*Autographa gamma*) (see under Lettuce, p. 228, for details), and cutworms, e.g. caterpillars of turnip moth (*Agrotis segetum*) (see under Red beet, p. 246, for details), are sporadic pests of carrots.

Foliar sprays of chlorpyrifos, lambda-cyhalothrin or lambda-cyhalothrin + pirimicarb, or cypermethrin (off-label) (SOLA 2184/98) can be applied to control cutworms; these treatments have harvest intervals of 14, 63, 63 and 0 days.

Nematodes (migratory)
Several species of migratory nematodes (*Longidorus* spp. and *Trichodorus* spp.) feed on carrots, especially in eastern England, causing symptoms such as stunting and fanging. Granules of aldicarb and carbosulfan are approved for control of migratory nematodes on carrot. Treatment, if required, should be applied at drilling. Aldicarb and carbosulfan have 84- and 100-day harvest intervals, respectively. Soil samples can be taken and tested prior to drilling to determine the risk from nematode attack.

Nematodes (other)
Carrots are also attacked by carrot cyst nematode (*Heterodera carotae*) and by northern root-knot nematode (*Meloidogyne hapla*), both local endoparasites. Granular nematicides do not give effective control of these species.

Slugs
Biological details and recommended treatments for all edible vegetable crops are given under Lettuce, p. 228.

Willow/carrot aphid (*Cavariella aegopodii*)
These aphids overwinter as eggs on willow (*Salix*) trees, or as aphid colonies on umbelliferous plants. Eggs on willow hatch in February or March. In May, winged aphids migrate to carrot crops over a period of 5–6 weeks, reaching a peak in numbers in early June. Heavy infestations of this aphid often occur from late May to early July, and cause considerable loss of yield in early and mid-season crops. The aphids infest carrots at the cotyledon stage, and also invade older plants. They are vectors of the carrot motley dwarf virus complex, which produces a yellow mottling of the leaves and stunts the plants. They also transmit parsnip yellow fleck virus, which can cause severe damage, stunting plants and blackening the central core. Celery and parsnip are also attacked.

Aphids may be controlled using aldicarb or carbosulfan granules applied at drilling. Aldicarb and carbosulfan have 84- and 100-day harvest intervals, respectively. Alternatively, foliar sprays of lambda-cyhalothrin + pirimicarb, nicotine and pirimicarb may be applied; harvest intervals are 63, 2 and 3 days, respectively.

Diseases

Black rot (*Alternaria radicina*)
This fungus causes black sunken lesions on the mature roots and can cause storage losses. It can also cause a serious foliage blight, similar to that caused by

Alternaria dauci (see below), but is more often found on the crown of the carrot. Black rot infection predisposes stored carrots to grey mould (*Botryotinia fuckeliana* – anamorph: *Botrytis cinerea*). *Alternaria radicina* is seed-borne, and can cause loss of seed in the seed crop and damping-off of seedlings. Thiram applied as a seed treatment is approved for control of *A. radicina* and other fungi that cause damping-off diseases in carrots. In addition, fenpropimorph (off-label) (SOLA 2483/96), iprodione + thiophanate-methyl (off-label) (SOLA 1868/99) and tebuconazole (off-label) (SOLA 1588/98) are available for use on crops affected by *A. radicina*. Triadimenol (off-label) (SOLA 0836/95) is available for general disease control in carrot crops in the UK.

Cavity spot (*Pythium* spp.)
Cavity spot, identified as a carrot disorder in 1961, is now known to be a disease caused by pathogens of the genus *Pythium*. This disease has become one of the most important occurring on carrots in the UK. The first signs of disease are elliptical depressions, up to 6 mm wide, on the sides of the roots. The skin (periderm) remains intact but the tissue underneath collapses. As the roots mature the cavity spot lesions enlarge and the periderm breaks. The exposed lesions are dark, discoloured and corky in texture. Lesions may extend up to half way around roots. Carrots grown on peaty soils are most frequently affected but cavity spot has also infected crops grown in sandy conditions.

Many fast-growing *Pythium* spp. affect the periderm of carrots but two slow-growing species, *P. violae* and *P. sulcatum*, singly or in combination, mainly cause the cavities beneath the skin. Metalaxyl-M is approved for control of cavity spot in carrots in the UK.

Leaf blight (*Alternaria dauci*)
Although this fungus can cause lesions on the leaves in wet seasons, it is more important (especially in precision-drilled crops) as a cause of seedling damping-off. Crop losses result from the use of untreated infected seeds. Eradication of infection from the seed by a thiram soak treatment gives a clean stand of seedlings. Thiabendazole + thiram is also approved for control of damping-off in carrot. Sprays of fenpropimorph (off-label) (SOLA 2483/96) and iprodione + thiophanate-methyl (SOLA 1869/99) are also available for use on carrot crops infected with *Alternaria dauci*.

Liquorice rot (*Mycocentrospora acerina*)
This disease occurs on carrots grown in organic soils and produces distinctive, sunken, black lesions, with brown water-soaked margins, on the crowns and shanks of the roots. There are no approved chemicals that can be used to control this disease in the UK.

Scab (*Streptomyces scabies*)
On carrot, *Streptomyces scabies* affects only the roots, producing corky, raised or sunken scab-like lesions. Parsnip roots may also be affected. The only methods of control are cultural and include crop rotation, maintaining acid soil conditions and irrigation. The disease is of limited importance on carrot and the main description is given under Red beet, p. 248.

Sclerotinia rot (*Sclerotinia sclerotiorum*)
This is a soil-borne fungus, which survives by means of black, hard-walled resting bodies (sclerotia). It attacks a wide range of vegetable species, including artichoke, beans, carrot and celery. Generally, the fungus produces a watery-brown rot on the petiole bases of carrot, infected foliage turning brown and dying. White mould grows on the surface of infected carrots during storage. Very similar symptoms are produced on celery petioles, which appear water-soaked and pink in colour. Primary symptoms are first observed on the crown of the carrot plant.

In dense crops of carrot or celery, the fungus may spread rapidly. In early summer, the perfect stage may be produced from soil-borne sclerotia, and air-borne spores (ascospores) are released; these are wind-borne and transmit the disease to newly established crops. Because the sclerotia persist for long periods, rotations of more than 5 years may be necessary to reduce this source of the disease. In addition, the pathogenicity of *S. sclerotiorum* to a wide crop range makes it difficult to control the disease by husbandry methods, e.g. by planting of non-susceptible crops. Field sprays of iprodione, applied for control of foliar pathogens on non-carrot crops grown on the same land in the previous season, may reduce the initial incidence of *Sclerotinia*. However, there are currently no approved chemical controls for *Sclerotinia* on carrots in the UK. Recent research has shown that chitosan is effective in reducing the incidence of this disease on carrots but no approval for this chemical currently exists in the UK.

Storage rots
Several fungal pathogens may cause deterioration of stored carrots. Of these, *Botryotinia fuckeliana* (anamorph: *Botrytis cinerea*) (grey mould), *Rhizoctonia carotae* (crater rot) and *Thielaviopsis basicola* occur occasionally in stored carrots in the UK. Cultural control practices must be followed if problems with these pathogens are to be avoided.

Violet root rot (*Helicobasidium purpureum*)
See under Asparagus, p. 188.

Cauliflower

See under Brassica crops, p. 189.

Celeriac and celery

Celeriac is a minor crop, the produce of which often goes for soup manufacture. Fresh market production is now increasing. Celery is grown on 650 ha in the UK, which is considerably less than the other umbelliferous crops: carrot and parsnip. However, the farm-gate value of the crop is high.

Pests

Carrot fly (*Psila rosae*)
This pest can attack both celeriac and celery. Biological details of this pest are given under Carrot, p. 211.

No insecticides are approved specifically for control of carrot fly on celeriac. Phorate granules have on-label approval for the control of carrot fly on celery. However, growers are under pressure not to use this organophosphorus insecticide and, on celery, foliar sprays of lambda-cyhalothrin (off-label) (SOLAs 2066/97, 2067/97, 0289/2000) can be used as an alternative, for which there is a 7-day harvest interval. See under Carrot, p. 212, for more details.

Caterpillars, including cutworms
Foliar-feeding caterpillars, e.g. those of silver y moth (*Autographa gamma*) (see under Lettuce, p. 228, for details), and cutworms, e.g. caterpillars of turnip moth (*Agrotis segetum*) (see under Red beet, p. 246, for details), are sporadic pests of celery.

Foliar sprays of lambda-cyhalothrin (off-label) (SOLAs 2066/97, 2067/97) can be used to control caterpillars of silver y moth (7-day harvest interval). Up to four sprays of deltamethrin (off-label) (SOLA 0125/99) can be used for general insect control, for which no harvest interval is stipulated. No insecticides are approved specifically for cutworm control on celery.

Celery fly (*Euleia heraclei*)
The larvae feed in the leaves of celery and parsnip, causing large blisters. The damage is most severe on celery, especially, when the plants are small, although large plants may also be attacked quite heavily. Attacks occur from May to October. No insecticide is approved specifically for control of celery fly.

Slugs
Biological details and recommended treatments for all edible vegetable crops are given under Lettuce, p. 228.

Willow/carrot aphid (*Cavariella aegopodii*)
This aphid, discussed in greater detail under Carrot, p. 213, rarely causes severe problems on celeriac or on celery.

On celeriac, foliar sprays of nicotine (2-day harvest interval) or pirimicarb (off-label) (SOLA 0328/96), for which there is a 3-day harvest interval, can be applied to control aphids. Foliar sprays of dimethoate (off-label) (SOLA 0389/94) may be applied for general insect control, before plants have seven true leaves.

Foliar sprays of nicotine and pirimicarb can be applied for aphid control on celery. These have harvest intervals of 2 and 3 days, respectively. Alternatively, up to four sprays of deltamethrin (off-label) (SOLA 0125/99) can be used to control aphids on celery, for which no harvest interval is stipulated. Foliar sprays of dimethoate (off-label) (SOLA 0778/96) can be used for general insect control (7-day harvest interval).

Diseases

Black rot (*Alternaria radicina*)
This disease is of very minor importance in the UK. Chlorothalonil applied to control other leaf diseases may be effective against this pathogen though it does not hold approval for control of black rot.

Crown rot (*Mycocentrospora acerina*)
Crown rot is the main limiting factor restricting storage of celery in the UK. Infected celery plants, although apparently healthy when harvested, develop a basal rot after 7–8 weeks in store. Dark-green to black lesions develop at the base of the crown, sometimes extending to the upper parts of petioles. Badly affected plants are unmarketable. The fungus is soil-borne, particularly in organic soils where it persists as resting spores (chlamydospores). There are no chemicals currently approved for control of this disease in the UK.

Damping-off diseases
See under Pea, p. 243.

Grey mould (*Botryotinia fuckeliana*)
See under Carrot, storage rots, p. 215.

Leaf spot (*Septoria apiicola*)
This fungus causes brown spots on leaves and stems of celery plants, particularly in cool, damp weather. It is important in celery production in the UK. The optimal temperature for infection is between 20 and 25°C, in combination with wetness durations of 48–72 hours. The fungus is often deep-seated in the celery seed but also persists in debris from previous celery crops. Seeds can be treated by immersing them in water at 50°C for 25 minutes. Alternatively, seeds can be treated with thiram. Chlorothalonil, copper oxychloride and cupric ammonium carbonate are all approved for field-treatment of the disease.

Root rot (*Phoma apiicola*)
Root rot occasionally causes severe losses of celery seedlings in nursery beds. Infected seed is an important source of the disease. The butt of the plant becomes progressively brown, then black, before decay appears on the stalks. Seed-borne infection can be controlled by using thiram as a seed treatment.

Sclerotinia rot (*Sclerotinia sclerotiorum*)
See under Carrot, p. 215.

Chicory

Chicory is grown on a limited scale in the UK, as a root crop mostly for the fresh market.

Pests

Aphids
Some of the species found on lettuce may occur on chicory. Foliar sprays of pirimicarb (off-label) (SOLA 1078/98) may be applied to control aphids on chicory. Sprays should be applied if infestations start to develop.

Diseases

Violet root rot (*Helicobasidium purpureum*)
This disease may attack roots grown under unfavourable conditions in the field, i.e. where rotations are restricted and soil conditions poor. There is currently no approved chemical control for violet root rot on chicory. For cultural control see under Asparagus (p. 188).

Courgette

Pests

Aphids
Aphids are vectors of viruses, of which zucchini yellows mosaic virus presents a particular threat to the UK crop. It is a limiting factor to production in southern Europe. The viruses are mainly non-persistent, so insecticidal control may have relatively little effect. The main aphid pest is peach/potato aphid (*Myzus persicae*), so there is a need to consider an insecticide resistance management strategy (see under Lettuce, p. 227, for details).

Foliar sprays of pirimicarb (off-label) (SOLA 1626/95) can be applied to control aphids on courgettes. The harvest interval is 3 days. Sprays should be applied as soon as aphid infestations are seen.

Bean seed flies (*Delia florilega* and *D. platura*)
The most serious damage to cucurbits occurs on plants raised in peat blocks, soon after planting out. The plants collapse completely, often within a week of planting. Later attacks result in plants wilting during dry weather. The plants may die, often because of secondary infections. No insecticides are approved for bean seed fly control on courgette.

Diseases

Grey mould (*Botryotinia fuckeliana* – anamorph: *Botrytis cinerea*)
See under Cucumber, below.

Powdery mildew (*Erysiphe cichoracearum*)
Powdery mildew is common on leaves and stems of courgette in the autumn, turning them white and causing them to wither prematurely. Both species of pathogen have probably slightly different climatic requirements. *E. cichoracearum* is more common earlier in the growing season. Bupirimate (on-label approval) and imazalil (off-label) (SOLA 1492/99) can be used to control the disease and should be applied at intervals of 10–14 days from first sign of disease.

Cucumber (outdoor)

Diseases

Grey mould (*Botryotinia fuckeliana* – anamorph: *Botrytis cinerea*)
Grey mould is an important disease on cucumbers (and on courgettes). The disease survives on plant debris, and occurs on leaves and fruit. Good hygiene is important in maintaining control. There are currently no chemical controls which can be applied against *Botrytis* on outdoor cucumbers.

Gummosis (*Cladosporium cucumerinum*)
This disease infects ridge cucumbers grown in the open, where it causes scab-like depressions from which sap is exuded, which later turns into an amber gum. The disease is common in colder, wetter areas of cucumber production and is often associated with poorly drained land. The pathogen can directly penetrate young fruit, which may be distorted if infected at an early stage. There are no chemicals with approval for control of gummosis in outdoor crops of cucumber in the UK.

Powdery mildew (*Erysiphe cichoracearum*)
See under Courgette (p. 219) for details. The following chemicals are approved for control of powdery mildew on cucumbers: bupirimate, cupric ammonium carbonate and fenarimol. These can be applied at 7- to 14-day intervals as protective sprays or at first sign of disease. Triforine (off-label) (SOLA 1730/96) has eradicant activity and may also be used for control of powdery mildew.

Stem and fruit rot (*Didymella bryoniae*)
This disease, sometimes known as 'black rot', has been reported in the field but, in the UK, is more important in protected cucumber crops (see Chapter 9, p. 344). The pathogen is both seed-borne and soil-borne. Disease symptoms usually occur on the stems and fruits. No chemicals are approved for control of this disease on outdoor cucumber.

French bean and runner bean

Approximately 3500 ha of French and runner beans are grown in the UK.

Pests

Bean seed flies (*Delia florilega* and *D. platura*)
The larvae of these insects tunnel into the germinating seeds and young stems of French and runner beans, and may cause serious crop losses by killing or stunting the seedlings. The damage is usually worst on crops sown early and when germination is slow. The flies are attracted to decaying plant material to lay their eggs and damage is often worst in soils that have a high organic-matter content. In the UK there are usually three or four overlapping generations each year. The final generation overwinters as pupae in the soil or as larvae that may continue feeding on plant residues. Damage to runner beans can be reduced markedly by raising plants in pots and planting out. No insecticides are approved specifically for control of bean seed flies on beans.

Black bean aphid (*Aphis fabae*)
Heavy infestations often develop on French and runner beans during July and August. Large colonies cause stunting, flower drop and malformation of the pods.

Crops should be treated with insecticides before they become heavily infested but, to avoid harm to bees, should not be sprayed when in flower. Foliar sprays of nicotine or pirimicarb can be applied to control aphids on French and runner beans. These insecticides have harvest intervals of 2 and 3 days, respectively.

Capsids (*Lygocoris pabulinus* and *Lygus rugulipennis*)
Adult bugs and nymphs inject toxic saliva into punctures made in shoot tips and unfolding leaves, causing deformed leaves and damage to seeds.

Nicotine (off-label) (SOLA 0080/92) can be used for general insect control on French beans. It has a harvest interval of 2 days. Nicotine also has on-label approval for control of capsids on runner beans.

Caterpillars

Caterpillars of silver y moth (*Autographa gamma*) and occasionally those of other species are sporadic pests of French and runner beans. Male silver y moths can be monitored by means of a pheromone trapping system.

Caterpillars of silver y moth, and other foliar pests of French and runner beans, can be controlled with foliar sprays of lambda-cyhalothrin (off-label) (SOLA 1247/95). *Bacillus thuringiensis* (off-label) (SOLA 0029/92) can be used, specifically, to control caterpillars on French beans. These treatments have harvest intervals of 7 and 0 days, respectively.

Two-spotted spider mite (*Tetranychus urticae*)

This can be a serious problem on dwarf bean and runner bean. Biological control agents (predatory mites) can be used to control the pest, in which case care must be taken to use compatible pesticides (such as pirimicarb) for control of other pests. No chemical pesticides are approved for spider mite control on beans.

Diseases

Anthracnose (*Colletotrichum lindemuthianum*)

The pathogen causes cankers on stems, brownish lesions associated with the leaf veins, and circular lesions on pods. The disease is not economically important in the UK. However, higher quality requirements in food processing mean that small amounts of disease require control.

Seeds are an important source of disease, and methods that reduce seed-borne infection could be an effective control strategy. Plants are susceptible to infection at all stages of growth; in particular, young tissue is very susceptible. However, experimentally, the disease has been reduced on cotyledons when the hypocotyls of plants were previously inoculated with non-pathogenic strains of binucleate *Rhizoctonia* spp. There are no approved fungicidal seed treatments and no approved fungicides for control of this pathogen in the field.

Halo blight (*Pseudomonas syringae* pv. *phaseolicola*)

This bacterium is seed-borne and affects both French and runner beans. Symptoms vary with age of plant and environmental conditions. Classic symptoms consist of small lesions, surrounded by yellow 'halos' on the leaves, and water-soaked lesions on the pods (grease spot). The bacterium is spread by rain and the disease is damaging in wet seasons, sometimes reducing yield but mainly affecting the marketable quality of the pods. Crop rotation practices, excluding beans for at least 2 years, should be used to control the disease. At present, there are no approved fungicides for control of this pathogen (although copper sprays applied from emergence onwards have been shown to be effective).

Pod rot (*Botryotinia fuckeliana* – anamorph: *Botrytis cinerea*)
This disease is most severe in wet seasons, and causes grey, water-soaked lesions on the pods, which may render them unmarketable. Sclerotia (black fungal resting bodies) may form on the pods. One important source of disease may be the colonization of dying flowers by *B. cinerea*, which may then transmit the disease to the young bean pods. The disease can be controlled by applying up to two applications of vinclozolin (on-label approval) or iprodione (off-label) (SOLA 1565/98) from first pod set or earlier if disease is detected. The possibility of resistance problems occurring in the pathogen population, through use of the latter chemical in particular, cannot be ruled out.

Rust (*Uromyces appendiculatus*)
The disease causes brown lesions on pods and leaves. Infection requires warm, moist conditions, and severe attacks can adversely affect yields by reducing the marketability of the pods. The pathogen has many stages in its life-cycle and has been shown to be seed-borne. No chemicals currently hold on-label approval for control of rust on French bean. However, the disease can be controlled using tebuconazole (off-label) (SOLA 0073/00). This has an eradicant mode of activity and spray applications must be applied when rust is first detected in the crop.

White mould (*Sclerotinia* spp.)
This disease usually occurs on pods as water-soaked lesions, which usually give rise to white mould. However, it can also cause plants to wilt and die at earlier growth stages. Sclerotia appear on infected stems and pods. These normally give rise to ascospores, which transmit the disease over longer distances. Several species of *Sclerotinia* can infect French bean, but all produce the same symptoms. They can result in heavy losses in the field. At present, apart from iprodione (off-label) (SOLA 1565/98), there are no approved treatments for use on French bean. However, application of tebuconazole (for control of rust) and vinclozolin (for control of pod rot) may have beneficial effects in controlling white mould.

Globe artichoke

This crop is grown commercially on a small scale, in the south of England.

Pests

Aphids
Aphids can be damaging, especially black bean aphid (*Aphis fabae*). Nicotine can be used for control aphids.

Diseases

Powdery mildew (*Leveillula taurica*), and also the pathogen *Ascochyta caynarae*, have been recorded in UK crops, but there are currently no chemicals which can be used to control these diseases of globe artichoke.

Jerusalem artichoke

Production of this crop is limited chiefly to gardens and allotments, but the plant is grown in other places for wind breaks and pheasant cover.

Pests

Aphids
Aphids, although sometimes present, are rarely numerous enough to impair plant growth.

Diseases

White mould (*Sclerotinia minor* and *S. sclerotiorum*)
This soil-borne disease can cause collapse and loss of plants. Roguing of affected plants will decrease disease spread. Wherever possible, crops should be raised on clean land. Infected seed tubers are also a significant source of disease. There are no approved fungicides that can be used to control the disease.

Leek

The crop area has increased slightly in the UK in the past few years and leeks are now grown on about 2470 ha. The area of this crop has remained remarkably stable over the past 10 years.

Pests

Bean seed flies (*Delia florilega* and *D. platura*)
These flies often damage leeks in the seedling stage (see under Onion, p. 233, for more details).

Leek seed may be film-coated with tefluthrin (off-label) (SOLA 1748/00) to prevent bean seed fly damage.

Cutworms
Cutworms, e.g. caterpillars of turnip moth (*Agrotis segetum*) (see under Red beet,

p. 246, for more details), are sporadic pests of leeks. The leaves and stems of leek plants damaged by cutworms become characteristically twisted and malformed. Even plants with only slight damage become unmarketable.

No insecticide is approved for cutworm control on leeks, although a product containing deltamethrin (off-label) (SOLA 1273/99) can be used as a foliar spray for general insect control. Such treatment, however, would not be effective once the caterpillars adopt their typical soil-inhabiting cutworm habit.

Leek moth (*Acrolepiopsis assectella*)

This is sometimes a minor pest in the coastal regions of south-east and eastern England. It can also attack onions, but here the larvae are less damaging than when burrowing into leeks. First eggs are laid in early May and there are three or four generations/year. In countries where leek moth is a major problem, male moths are monitored using pheromone traps, to provide early warning of attack.

Once leek moth infestations have developed, the application of insecticides is usually of limited value.

Onion fly (*Delia antiqua*)

This pest may sometimes cause damage, especially in central and eastern England (see under Onion, p. 233, for details). Small plants are affected most seriously, and these wilt and die rapidly as a result of the larvae feeding just below the soil surface. Larvae may move along the row from one plant to another. On larger plants, the larvae feed in the shank of the leek.

Leek seed may be film-coated with tefluthrin (off-label) (SOLA 1748/00) to prevent onion fly damage.

Onion thrips (*Thrips tabaci*)

Onion thrips is the most important pest of leeks in the UK. Thrips can damage leeks badly, particularly in periods of drought. Feeding injury gives attacked plants a whitish, silvery appearance and their growth may be checked. Thrips are susceptible to rainfall and irrigation. Regular overhead irrigation, therefore, may reduce the severity of thrips infestations.

Foliar sprays of dimethoate (off-label) (SOLA 0778/96) or deltamethrin (off-label) (SOLA 1273/99) can be used to control thrips on leeks. These treatments have harvest intervals of 7 and 3 days, respectively. Sprays should be applied as soon as crops become infested. Despite the fact that several sprays may be applied, growers in many areas obtain very poor levels of control. This may be because the insecticides do not make contact with the thrips, which are hidden in the plant, or because treatments are poorly timed. Possible ways of improving the timing of treatments may include the use of a day-degree model developed in the US to predict periods of immigration or treatment thresholds to time spray applications. Several research groups outside the UK have developed management systems for onion thrips, based on crop sampling and the use of treatment thresholds. Large samples are required to estimate infestation levels accurately, and destructive samples are far more accurate than visual assessments of thrips numbers or damage. However, such sampling is time-consuming and may be done too infrequently to detect the initial increase in thrips numbers. In France, researchers have developed an action threshold, based on adult trapping, to initiate a programme of insecticide sprays.

Alternative and more effective insecticide treatments are being sought. Seed treatments using either fipronil or imidacloprid are being considered.

Diseases

Downy mildew (*Peronospora destructor*)
The disease is occasionally problematic on seedlings (see under Onion, p. 234, for description). Downy mildew on leek can be controlled by applying propamocarb hydrochloride as a field drench.

Leaf blotch disease (*Cladosporium allii*)
This pathogen infects the leaves of leeks. Individual lesions are eye-shaped and white, often with brown centres as a result of sporulation. Where there is extensive infection, numerous white lesions can occur in the crop, causing it to appear as if it had been sprayed with a herbicide. The presence of lesions on the leaves reduces the marketability of the leeks, and severe attacks may result in complete crop loss. The disease has become important in the UK in the last few years, particularly in areas where the crop is grown intensively. Propiconazole is approved for control of leaf blotch on leeks; it will also control leek rust. Sprays should be applied immediately disease appears in the crop.

Leek rust (*Puccinia allii*)
Leek rust has become more prevalent in the last decade, mainly because of the general intensification in leek production leading to sequential planting of crops on some farms. Overlapping crops enable the transfer and increase of the rust pathogen. Fungal perennation may have been aided to some extent by the recent succession of mild winters. Long, bright-orange lesions (uredosori) develop on the outer leaves of leeks, disfiguring them and reducing their marketability. The pathogen requires only very short periods of leaf surface wetness over a wide temperature range to infect host tissues. Rust spreads and develops within the crop at a maximum rate during most field conditions. The optimal temperature for infection is 15°C. Therefore, fungicidal control is important, for which cyproconazole fenpropimorph, propiconazole, tebuconazole and triadimefon have approval. Sprays must be applied from first appearance of disease in the crop. There may be some potential for using reduced dosages of chemical but only where incidences of rust in the crop are low.

White tip (*Phytophthora porri*)
This disease is common in the Evesham area of Worcestershire and is present in all other leek production areas. It appears in August and early September, causing the leaves to turn yellow. Infected leaves later become crisp and bleached, and eventually die. Young and old leek plants are affected, and severe infection causes the leaves to rot at soil level. Oospores of the fungus overwinter in infected debris in the soil, and may persist for as long as 3 years. Therefore, long rotations are necessary to prevent re-infection of crops, but removal and destruction of plant debris will reduce inoculum. High soil temperatures (45–55°C) may reduce the viability of oospores but these conditions do not occur often in the field. Latent/incubation periods are rapid at 11°C (4–11 days). Chlorothalonil + metalaxyl is approved for control of the disease, and has been effective in reducing disease severity in locations where the disease is problematical. Propamocarb hydrochloride (off-label) (SOLA 2447/98) is also currently available for use on leek crops infected with white tip.

Lettuce

Outdoor lettuce is the third most valuable vegetable crop currently grown in the UK and is produced commercially on approximately 6290 ha.

Pests

Aphids (foliar-feeding)
Three species of aphid, currant/lettuce aphid (*Nasonovia ribisnigri*), peach/potato aphid (*Myzus persicae*) and potato aphid (*Macrosiphum euphorbiae*), are the major foliar pests on outdoor lettuce. They cause damage by stunting, malforming and contaminating the leaves with their cast skins and honeydew. Certain of these aphids (notably peach/potato aphid), besides directly damaging the plants, spread viruses, such as lettuce mosaic virus, which cause severe stunting.

In the UK, both peach/potato aphid and potato aphid overwinter mainly as adults and nymphs on host crops and weeds. Both species are captured in suction traps run by the Rothamsted Insect Survey and, in general, the first winged aphid is captured earlier following a mild winter. In contrast, currant/lettuce aphid overwinters in the egg stage on currant or gooseberry bushes. These eggs usually hatch in March or April, nymphs then infesting the tips of the young shoots. Colonies are formed on the developing leaves, and in May or June winged aphids migrate to lettuce and other Asteraceae (Compositae), their summer hosts, on which successive generations are produced until September or October. During October and November, winged aphids migrate to the winter hosts, where eggs are laid.

The relative abundance of the three aphid species varies during the growing season. For example, currant/lettuce aphid appears to be the most predominant species in late summer. In addition, the timing of the initial aphid immigration varies from year to year, as does the timing of the mid-season decline in aphid numbers or aphid 'crash'. Weather-based forecasts of the timing of aphid immigration and the mid-season 'crash' are being developed at HRI and IACR-Rothamsted.

Cultivars resistant to currant/lettuce aphid have been developed and released recently. In addition, there are commercial lettuce cultivars with resistance to peach/potato aphid and potato aphid. Widespread use of such cultivars will depend on how appropriate they are for the commercial market.

In the UK, populations of all three aphid species may contain individuals that are resistant to insecticides. Three forms of resistance have been identified in peach/potato aphid, conferring resistance to a range of carbamate, organophosphorus and pyrethroid insecticides. In addition, some populations of peach/potato aphid possess low-level tolerance to imidacloprid, correlated with a decreased susceptibility to nicotine. At present, this does not seem to affect the field performance of these insecticides. Insecticide resistance to pirimicarb and pyrethroids in currant/lettuce aphid has been reported in the last few years, and there is now also evidence that populations of potato aphid may contain individuals with insecticide resistance to these chemicals. Therefore, there is a need to develop a control strategy to avoid increasing insecticide resistance within local populations. A UK Insecticide Resistance Action Group has been formed, and is publishing resistance-management guidelines for peach/potato aphid.

Control measures should be applied early, not only to prevent the crop being infested when it begins to heart, but also to decrease the spread of virus when the plants are young. Foliar sprays of cypermethrin, fatty acids, lambda-cyhalothrin + pirimicarb, nicotine, pirimicarb and deltamethrin (off-label) (SOLA 1691/96) can be applied to control aphids on lettuce. They have harvest intervals of 0, 0, 3, 2, 3 and 0 days, respectively. As several applications may be required, specialist advice should be sought on the sequence of active ingredients that should be used to minimize the development of insecticide resistance. Much of the UK crop is now treated with imidacloprid (off-label) (SOLA 1041/96); the lettuce seed is film-coated with this insecticide. However, it is important that this treatment is used carefully, as part of an overall management strategy, to avoid the development of resistance to imidacloprid in aphid populations in the UK. The use of imidacloprid-treated seed should not be necessary for every crop.

Aphids (root)

Lettuce root aphid (*Pemphigus bursarius*) feeds on roots, causing a yellowing and stunting of the foliage; when the soil is dry, infested plants will die. The pest overwinters on black poplar (*Populus nigra*), lombardy poplar (*Populus nigra* var. *italica*) and Manchester poplar (*Populus nigra* var. *betulifolia*) and, in June and July, winged migrants fly to lettuce where their progeny infest the roots. Here, aphid numbers increase rapidly, especially in hot weather. After mid-August, most of the aphids return to poplar, but a few remain in the soil to infest subsequent crops of lettuce planted on the same ground in the autumn and following spring. A day-degree forecast of the timing of the migration from poplar to lettuce has been developed at HRI, and validated with data collected by ADAS.

Lettuce cultivars resistant to lettuce root aphid have been available to growers for some time. Lettuce seed may be film-coated with imidacloprid (off-label) (SOLA 1041/96) to control lettuce root aphid. Alternatively, phorate granules may be incorporated into the soil. However, the long harvest interval for phorate (42 days) means that this treatment can be used only on drilled crops. The foliar spray treatments used to control foliar-feeding aphids (see p. 226) will have only a limited impact on lettuce root aphid infestations.

Caterpillars (foliar-feeding)

Foliar-feeding caterpillars are sporadic pests of lettuce. Silver y moth (*Autographa gamma*) is a migrant species and can be a major pest in some years. Caterpillars of flax tortrix moth (*Cnephasia asseclana*) may also occur.

Crops should be inspected regularly for foliar-feeding caterpillars. Pheromone traps can be used to capture male silver y moths, and trap captures provide early warning of egg-laying. Foliar sprays of cypermethrin or deltamethrin (off-label) (SOLA 1691/96) can be used to control caterpillars on lettuce, with a harvest interval of 0 days. Sprays should be applied as soon as damaging infestations are seen, since small caterpillars are more susceptible to insecticides than large ones.

Cutworms

Cutworms (see under Red beet, p. 246, for more details), e.g. caterpillars of turnip moth (*Agrotis segetum*), are sporadic but sometimes damaging pests of lettuce, a particularly susceptible crop. The first two larval instars feed on the aerial parts of the plants but older individuals remain in the soil and feed on the plant stems at about ground level. Following attack, plants are either severed at ground level or large holes are bitten into the stems, so these are weakened. Most damage occurs to crops just after transplanting, or in the seedling stage when direct drilled.

Sprays of cypermethrin, deltamethrin, lambda-cyhalothrin or lambda-cyhalothrin + pirimicarb can be used to control cutworms on lettuce, with harvest intervals of 0, 0, 3 and 3 days, respectively. These should be applied in response to warnings from ADAS or other advisors or when damage is first seen. However, many lettuce crops receive regular irrigation, obviating the need for insecticide treatments.

Slugs

Slugs, e.g. field slug (*Deroceras reticulatum*), feed on a wide range of vegetable crops, their presence being encouraged by heavy soils and plant residues in the soil. In lettuce, they can provide contamination problems, particularly in crops for processing. Slugs are less active in hot or cold conditions, when they move down into the soil. However, in the spring and summer their activity increases considerably, especially in warm, moist conditions such as those under a crop canopy.

Recommended methods of control are based on baits broadcast on the soil surface, but growers should check for slug activity before treatment. Simple shelter traps, using hardboard squares, polythene sacks, etc., with a little food underneath (any vegetable material is suitable but chicken food is particularly effective), should be left in the field and checked for slugs in the early morning. The presence of one or more slugs in most of the traps indicates a potential problem population.

Metaldehyde- or methiocarb-based baits, either broadcast or incorporated into the soil, can be used to control slugs on all edible crops, but methiocarb must not be applied to crops within 7 days of harvest. Best results will be achieved from an application a few days before drilling or planting. For emerged or established crops, pellets should be applied immediately damage appears. Pellets are most effective if applied during mild, damp weather. However, molluscicide baits may be spoiled by rain and re-application may then be necessary. The application of slug bait to growing crops can lead to pellets lodging in the foliage and contaminating harvested produce.

Diseases

Beet western yellows virus (BWYV)

Serious outbreaks of this disease first occurred on mid- to late-summer lettuce field crops in the early 1970s. BWYV has recurred commonly since then, causing intense inter-veinal yellowing of the outer leaves of maturing butterhead lettuce, often necessitating trimming of these leaves or causing yield reductions. In cos and crisp lettuce, symptoms are slight, and plant growth is not affected adversely.

BWYV is transmitted by peach/potato aphid *Myzus persicae*. The isolates of the virus present in the UK, unlike American isolates, do not infect sugar beet or red beet but are frequently present in weed species including cleavers (*Galium aparine*), groundsel (*Senecio vulgaris*), hairy bittercress (*Cardamine hirsuta*), shepherd's purse (*Capsella bursa-pastoris*) and wild radish (*Raphanus raphanistrum*). Effective weed and insect (i.e. aphid-vector) control measures (see above) are important, therefore, in preventing the introduction of the virus.

Big-vein

This disease affects summer and winter field-grown and greenhouse crops of lettuce, expressing itself as a pale yellow or blanched vein-banding, particularly evident near the base of the outer leaves of diseased plants. Affected leaves may also have a puckered appearance. Symptoms appear in maturing plants and reduce the market value more than directly affecting yield. The virus-like causal agent is transmitted by zoospores of the fungus *Olpidium brassicae* to the roots of plants growing in contaminated soil. Resting spores of *Olpidium* also carry the virus, enabling it to survive from crop to crop.

Control is best obtained by growing lettuce away from affected areas of land. In experiments, soil sterilization with methyl bromide reduced the disease and increased the harvest weights of plants. The disease may be controlled satisfactorily in peat-block-raised plants by the addition of carbendazim (off-label) (SOLA 1751/96) directly to the peat block.

Bottom rot (*Thanatephorus cucumeris* – anamorph: *Rhizoctonia solani*)
This disease is more important on lettuce produced under glass or plastic but, occasionally, can occur in the field. Lettuce plants become vulnerable to attack when the leaves come into direct contact with the soil. Moist conditions at temperatures of 20°C are optimal for disease development. Disease appears first as sunken, rust-coloured lesions on the midribs of the outermost leaves. Sprays of tolclofos-methyl, applied before planting, are approved for control of this disease on lettuce.

Downy mildew (*Bremia lactucae*)
Downy mildew is one of the most important diseases of outdoor lettuce in the UK. The disease appears as pale-green or yellow, angular areas on the older leaves, which usually have whitish spores on the lower leaf surface. The infected areas become brown and die. Under cool, moist conditions, copious sporulation may occur. The disease is most important in the early autumn on outdoor lettuce and in late autumn on frame lettuce, but it also attacks overwintering lettuce. Disease development is dependent on the number of hours (duration) of leaf wetness in the morning, when there is concurrent spore release. Tissues infected by downy mildew may be colonized by *Botrytis* or by soft rots.

Mixtures of metalaxyl and thiram are approved for use on outdoor lettuce crops infected with downy mildew in the UK. Additionally, mancozeb, thiram (alone) and zineb also hold approval for control of downy mildew. Sprays should be applied at first sign of disease. Also, thiram can be applied as a seed treatment to lettuce seeds for control of damping-off fungi, including *Bremia lactucae*. Alternatively, fosetyl-aluminium (off-label) (SOLA 0255/97) or propamocarb hydrochloride (off-label) (SOLA 1971/99) may be applied to infected seedlings as a drench treatment.

In some lettuce-production areas resistance to metalaxyl has arisen in *B. lactucae*. The use of cultivars carrying genes resistant to metalaxyl-resistant pathotypes can overcome this problem. Careful use of chemical control sprays may also be useful in dealing with this problem.

Grey mould (*Botryotinia fuckeliana* – anamorph: *Botrytis cinerea*)
This pathogen is most serious under cool, damp weather conditions and is, therefore, most prevalent in the spring on overwintered lettuce. The first sign of infection is often a complete collapse of the plant, caused by a basal stem rot, which may occur at any stage of growth. On young plants this condition is known

as 'red leg', but sometimes the most severe attack occurs as the plants approach maturity. The fungus produces copious grey spores on decaying leaves and stems, and also forms sclerotia which persist in the soil. The disease may start on the seedlings, sometimes following an attack of bottom rot, damping-off or infection by *Bremia lactucae*, and remains quiescent for many weeks before causing serious damage. Therefore, protective treatments should be applied from the seedling stage onwards, and careful attention should be paid to efficient culture. Iprodione holds approval for control of *Botrytis* on outdoor lettuce. Up to 7 applications are permitted, with 14-day intervals between sprays. Thiram also holds approval for control of *Botrytis*. Further, both propamocarb hydrochloride (SOLA 1971/99) and iprodione (SOLA 0715/95) hold specific off-label approval for use on lettuce crops affected by this disease. Seedlings and young plants should be sprayed at intervals of 7–14 days. Sprays of carbendazim, which has specific off-label approval (SOLA 1751/96) for use on crops affected by big-vein virus, may also help reduce the incidence and development of *Botrytis* in the crop.

Lettuce mosaic virus

Lettuce mosaic virus is seed-borne and can affect all lettuce cultivars. Samples containing more than 0.1% infected seeds can cause significant primary field outbreaks of the disease, which are then spread further by aphids. Affected seedlings show vein clearing and rosette symptoms. Plants affected at later growth stages become stunted, and develop yellowed leaves with poorly formed hearts. Using virus-free seed is an important method of controlling the disease. See also (above), comments concerning control of the aphid vectors on lettuce.

Miscellaneous minor diseases

Lettuce can be infected by a range of minor diseases, such as septoria spot (*Septoria lactucae*), stemphyllium spot (*Pleospora herbarum* f. sp. *lactucum*) and bacterial rots (*Erwinia* spp., *Pseudomonas* spp., *Xanthomonas* spp.), but there are no approved chemical controls.

Powdery mildew (*Erysiphe cichoracearum* f. sp. *lactucae*)

This disease is a very minor problem in UK lettuce production. Symptoms are similar to those of powdery mildew on other vegetables (see under Courgette, p. 219). There are currently no chemical controls approved for controlling powdery mildew on lettuce in the UK.

Ring spot (*Microdochium panattonianum*)

This fungus forms circular spots (3–7 mm in diameter) on the outer leaves and elongated spots (resembling slug injury) on the leaf midribs. The disease is rarely serious, although rather disfiguring and, as such, it downgrades the value of the crop. The fungus is carried on seed and on plant debris as microsclerotia. It has been observed that thiram gives some control of the disease if applied from the seedling stage onwards. Therefore, sprays containing thiram applied to control downy mildew on lettuce may have an effect on *Microdochium panattonianum*. Prochloraz holds off-label approval (SOLA 2002/99) for use on crops infected with this disease.

Watery soft rot (*Sclerotinia minor* and *S. sclerotiorum*)
Sclerotinia can be a major problem on outdoor lettuce. Symptoms are similar to those described for these pathogens on other crops. The occurrence of the disease on many hosts makes crop rotation a less effective control measure. The resting bodies produced by *S. sclerotiorum* are larger (10 mm in diameter) than those of *S. minor* (2 mm in diameter). *Sclerotia* may germinate to form apothecia that release ascospores, the airborne spore form. For control, iprodione holds full and off-label approval (SOLA 0715/95) for use on infected lettuce crops.

Marrow

This is a minor crop which, together with courgettes, is being grown more often on small-holdings and for the 'pick your own' market.

Pests

Aphids (*Myzus persicae*, etc.)
Aphids are also pests of marrow (see under Courgette, p. 218, for details).

Foliar sprays of nicotine or pirimicarb (off-label) (SOLA 1626/95) can be applied to control aphids on marrow; they have harvest intervals of 2 and 3 days, respectively. Up to four sprays of dimethoate (off-label) (SOLA 0778/96) can be applied for general insect control; this insecticide has a 7-day harvest interval.

Diseases

Many of the pathogens which affect marrow also occur on the cucumber crop grown under glass (see Chapter 9, p. 339).

Onion

The area of bulb onion production has increased from 8364 ha to over 9522 ha in the last 10 years. Harvesting methods have changed and over 95% of crops now have their foliage removed mechanically in the field and the bulbs placed immediately in store and dried. The remainder are lifted with their foliage intact, field dried for 7–14 days and then removed to store where drying is completed.

Production of salad (green) onions has increased steadily over the last 10 years to 2298 ha.

Pests

Bean seed flies (*Delia florilega* and *D. platura*)

Bean seed fly larvae often damage salad and bulb onion crops in the seedling stage. There may be three or four overlapping generations during the summer. Damage is usually worst on soils which are freshly disturbed and where there is a high organic-matter content. Damage to onions may appear merely as poor emergence, since the larvae often attack seedlings between germination and emergence. Damage to emerged onions is indistinguishable from that caused by onion fly larvae – plants wilting suddenly and collapsing. Where very small plants are attacked, the larvae may move from plant to plant along a row. In some cases, mature onion bulbs have been infested with bean seed fly larvae.

Bulb and salad onion seed may be film-coated with tefluthrin (off-label) (SOLA 1748/00) to avoid bean seed fly damage.

Cutworms (*Agrotis segetum* and other species)

Cutworms (see under Red beet, p. 246, for more details) are sporadic pests of onions.

Treatments should be applied only when there is a risk of cutworm damage (see p. 246). Up to two sprays of chlorpyrifos can be applied to control cutworms on bulb and salad onions. Sprays should be applied when warnings are given by ADAS or other advisors, or when damage is first seen. There is a 21-day harvest interval. In addition, foliar sprays of deltamethrin (off-label) (SOLA 0506/96) can be used for general insect control on bulb onion crops. Up to five treatments/crop may be applied and there is a 0-day harvest interval.

Onion fly (*Delia antiqua*)

Damage caused by larvae of this fly feeding on the leaf bases of salad and bulb onion crops may be severe locally, especially in eastern England. There are generally two generations each year, but a third may occur in warm locations. The worst damage occurs in June and July, but damage can also occur in August and early September. Small plants are most seriously affected, and they wilt rapidly and die as a result of the larvae feeding just below the soil surface. Larvae may move along the row from one plant to another. On larger plants, the larvae feed in the bulb of the onion (or in the shank of the leek). The most important cultural control is to avoid physical damage to onion crops, which would increase the likelihood of attacks from onion fly. Crop rotation may also reduce onion fly damage considerably. The worst examples of onion fly damage occur on small farms with close rotations.

To avoid onion fly damage, bulb and salad onion seed may be film-coated with tefluthrin (off-label) (SOLA 1748/00).

Onion thrips (*Thrips tabaci*)

Onion thrips can damage onions badly, particularly in drought periods. Salad onions are most susceptible. Feeding injury gives attacked plants a whitish, silvery appearance and their growth may be checked severely (see under Leek, p. 224, for more details).

Foliar sprays of deltamethrin (off-label) (SOLAs 0506/96, 0125/99) or dimethoate (off-label) (SOLA 0778/96) can be used to control thrips on both bulb and salad onions. These treatments have harvest intervals of 0 and 7 days, respectively. Sprays should be applied as soon as crops become infested. Despite the fact that several sprays may be applied, growers in many areas obtain very poor levels of control. This may be because the insecticides do not make contact with the thrips, which are hidden within the plant, or because treatments are poorly timed (see p. 224). Alternative and more effective treatments are being sought.

Stem nematode (*Ditylenchus dipsaci*)

This nematode may be introduced to a field on onion seed, infested soil or debris. Since much seed is fumigated or routinely examined for nematodes, the numbers of nematodes introduced will not damage the crop in the same year, but they may damage subsequent crops. The nematode invades bulbs and leaves, causing distortion (bloat) and rotting. Early attacks cause seedling losses; later attacks rot the bulbs and often separate the bulb from the roots. Distortion curls the leaves, bending them at the top of the bulb, and the bulb may also split. Slightly infected bulbs, apparently sound at harvest, may rot in store.

To control nematodes on bulb onions, aldicarb granules may be applied at drilling or planting and incorporated into the soil. The recommendations for rates of use depend on the type of crop and the length of time it is to be stored (consult the pesticide manufacturer's label for details).

Diseases

Downy mildew (*Peronospora destructor*)

This has become the most important fungal disease on bulb onions in the UK. The pathogen produces pale, oval areas on the leaves, or causes leaf tips to become pale and to die back. The leaves often fold downwards at the infected area, upon which grey, later brown-purple, spores of the fungus may develop; these spread extensively in cool, moist conditions. Other leaf moulds also grow on the lesions and may obscure the pathogen. The fungus grows within infected leaf tissues and can infect the bulbs, which, if kept for seed raising, produce stunted leaves. The flower stalks of seedling plants may bear a conspicuous, pale, oval lesion, and the stalk often breaks at this point. Both the asexual stage (sporangia) and the oospore can be wind dispersed. Oospores of the fungus can persist in the soil and infect new crops. The disease is also carried over in perennial onions and shallots, and possibly also in wild *Allium* spp.

Control measures include avoiding contaminated land and growing onions, where possible, on warm, well-drained soils in sites with good air circulation.

Plant density, poor orientation and irrigation may also be important in control of the pathogen, particularly on salad (green) onions. Forecasting systems (DOWNCAST) have been developed for the control of this disease, based on the prediction of fungal infection and sporulation. Overwintering hosts should be eradicated.

Mixtures of metalaxyl with chlorothalonil are approved for control of downy mildew in bulb and salad onions. A number of field sprays are permitted, up to a maximum dosage per crop. These should be applied immediately disease is detected in the field. Propamocarb hydrochloride (off-label) (SOLA 1971/99) can be applied as a drench to affected bulbs or salad onions.

Leaf blight (*Botrytis squamosa*) ***and collar rot*** (*Botryotinia fuckeliana* – anamorph: *Botrytis cinerea*)

These pathogens are particularly destructive to densely sown onions, such as salad and pickling onions. They are less important in bulb-onion production. The leaf rot fungus (*B. squamosa*) produces roughly circular to elliptical, white lesions on the leaves and die-back of the leaf tips. White spots may also be produced by hail, but these occur on the weather side of leaves and are variable in shape and alignment relating to variation in force of impact and direction of the hail.

Leaf rot affects emerging salad onions in the autumn, and overwintering and newly established crops in the spring, when temperatures are greater than 10°C. Under moist conditions the fungus may spread from crop to crop, so that where successional sowings during spring and summer are sited close to one another, inoculum levels increase with time. In the final plantings, considerable numbers of lesions on leaves may occur with leaf die-back. If prophylactic measures are not applied in time, affected plant parts have to be trimmed to make onion bunches presentable for market, thus incurring extra costs to the grower.

Collar rot (*B. cinerea*) invades seedlings at emergence, and thereafter the disease is spread through developing crops by conidia produced on dead plant tissue. This disease is difficult to see; no obvious symptoms are produced, the fungus growing downwards within leaves to attack the stem base at or just above soil level. The pathogen develops progressively, and plants may take up to a month to die. Plant losses are unobtrusive in densely sown crops. The disease is destructive only in overwintered salad onions, and reaches maximum incidence in January when low temperatures inhibit growth of onion leaves but not that of the fungus. Most plants are killed at this time, but become less susceptible when temperatures rise (> 10°C) and leaf growth resumes. Infected seeds are a significant source of both diseases.

Seed treatment with thiram may be effective in controlling the disease on seedlings; and chlorothalonil and iprodione have approval for control of both diseases in the field, on both bulb and salad onions. Therefore, use of chlorothalonil + metalaxyl for control of downy mildew may be effective in reducing *Botrytis* infection levels in the crop. Biological control agents that affect sporulation have been shown to control the disease but, at present, these do not have approval for use in onion crops. Disease forecasting systems (BOTCAST) have been developed for use on bulb onions in the US. As both diseases are of minor economic importance on bulb onions in the UK, the uptake of these systems in this crop has been limited.

Leaf blotch (*Cladosporium allii-cepae*)
This disease can occur on the leaves of overwintered and spring-sown bulb onion crops. The symptoms are similar to those of leaf blotch of leek. Propiconazole (off-label) (SOLA 2279/97) is available for use on bulb and salad onion crops affected by leaf blotch but this off-label use for some formulations of propiconazole (SOLA 2280/97) is due to end shortly. There are no other approved chemicals which can be used to control the disease.

Neck rot (*Botrytis allii*)
In recent years, severe outbreaks of neck rot have occurred in stored bulb onions in the UK. The disease, which becomes evident only in store, causes onions to soften and rot internally. The black sclerotia of the fungus are produced on the necks of the bulbs, and the disease was thought to infect plants in the field at harvest. Recent research has shown, however, that the pathogen is carried internally in onion seeds and that it infects the green tissues of the emerging seedling cotyledons. Conidiophores produced on necrotic parts of leaves release spores, which spread infection through developing onion crops. The disease is symptomless, healthy and infected onion plants being indistinguishable. The pathogen occurs on the infected bases of older leaves before invading the neck tissues of the developing bulbs and growing downwards within them, causing neck rot. Neck rot affects only bulb onion crops. Conidia of *B. allii* produced at low temperatures (-2°C) cause greater disease incidence and severity than those produced at higher temperatures.

The disease can be controlled by applying chlorothalonil. In the absence of disease symptoms up to five sprays must be applied protectively at 14-day intervals. The disease can be managed on harvested bulb onions by using diagnostic tests to predict their storage potential.

Smut (*Urocystis cepulae*)
Smut is a soil-borne disease. Germinating mycelium from spore balls in the soil infects the bases of the leaves of young seedlings, causing dark, lead-coloured spots or streaks. These leaves later become thickened, and twist or curl backwards; further leaves or bulb scales may bear dark areas which later split, exposing black spores of the fungus. Soil that has carried an infected crop should not be used for onion growing for as long as possible, as the spores (ustilospores) of the fungus may survive for 20 years. Stringent precautions should be taken against spreading infested soil on boots, implements, etc. The disease also occurs on chive, garlic, leek and shallot. There are no currently approved treatments for control of the disease in the UK.

Storage rots of onions

The main temperate storage rot of dry bulb onions is neck rot (*Botrytis allii*). Seed treatment with benomyl virtually eliminated the disease in the UK. However, this fungicide no longer holds approval for use on onions. Seed treatment with thiram may have some effect on the disease.

In recent years, fungi and bacteria which grow at high temperatures have caused rotting in some stores. Their occurrence is related to the introduction of direct harvesting methods by which onions are mechanically 'topped' and, without field drying, taken into store. This necessitates the use of high-temperature drying regimes, to seal the necks of onions, remove excess moisture from the bulbs and prepare them for storage. Where storage drying temperatures and humidities exceed 30°C and 80% RH for a period of *c.* 7 days, fungi (including *Aspergillus fumigatus*, *A. niger* and *Penicillium* spp.) and bacteria (including *Erwinia herbicola* and *Lactobacillus*-like spp.) develop, causing blemish and rot problems.

The occurrence of bacterial rots may result from 'topping' the bulbs too close to the neck of the onion. These problems may be overcome where the correct drying procedures (30°C air at 425 m^3/h/t onions for 3–5 days, followed by secondary drying with recirculated air at 170 m^3/h/t onions maintained at 70–75% RH for up to 2 weeks) are employed, although this is expensive. The fungi do not develop at these humidities but the bacteria are not as well controlled and these will rot bulbs in some stores in some years. Affected crops can be treated with copper oxychloride (off-label) (SOLA 1127/99).

White rot (*Sclerotium cepivorum*)

White rot is a serious disease, mainly affecting salad onion. Affected plants are stunted, with yellow leaves, and the base of the plant is rotten and often covered with a white fungal growth in which the black resting bodies (sclerotia) of the pathogen may be embedded. The fungus is soil-borne and its sclerotia may persist for many years. Infected plants may show disease symptoms after harvest, resulting in considerable storage losses. In the UK, some production areas remain largely free of the disease. In affected areas, land not previously used for onion production has been cultivated, reducing the incidence of the disease. Affected salad onion crops can be treated with tebuconazole (off-label) (SOLA 1447/99). Shallots, garlic and bulb onions can also be treated with tebuconazole (off-label) (SOLA 2062/97). Soil solarization may be effective for the control of the disease in some production systems.

Parsley

Pests

Carrot fly (*Psila rosae*)

The biology of the fly is described under Carrot, p. 211. Larvae may damage the tap and lateral roots of parsley.

Tefluthrin seed treatment (off-label) (SOLA 0234/99), applied to control bean seed flies, may give incidental control of carrot fly.

Willow/carrot aphid (*Cavariella aegopodii*)
As in carrot, this aphid may cause direct damage to plants when feeding but it may also, and usually more importantly, transmit the carrot motley dwarf virus complex.

Foliar sprays of pirimicarb (off-label) (SOLA 1303/96) may be applied to control aphids on parsley. There is a 3-day harvest interval.

Diseases

Leaf spot (*Septoria petroselini*)
The pathogen is seed-borne, and samples of parsley seeds sometimes have the fungal pycnidia associated with them. In many cases, the pycnidiospores from these pycnidia fail to germinate and may not be viable. When these spores are viable, however, they transmit the disease to the emergent seedlings. Symptoms are similar to those of celery leaf blight, with tan lesions bearing pycnidia occurring on the leaves. The disease can increase rapidly on leaves of parsley, causing defoliation and loss of crop. Leaf spot disease can be of economic importance where the crop is produced for drying.

There are no approved chemicals for control of this disease on parsley. Affected debris should be well ploughed in and a 2-year rotation practised.

Parsnip

About 3600 ha of parsnip are grown annually in the UK. The cropped area for parsnips has increased over the last 10 years. Many of the diseases found on carrots can also occur on parsnips. However, they are less common on this crop owing to the smaller areas of parsnips grown in comparison with carrots.

Pests

Aphids (*Cavariella* spp.)
Parsnip aphid (*Cavariella pastinacae*), willow/carrot aphid (*Cavariella aegopodii*) and willow/parsnip aphid (*C. theobaldi*) all infest parsnip. As with carrot, severe infestations can stunt or even kill young seedlings. The aphids can also transmit viruses that occasionally cause severe symptoms.

Aldicarb and carbosulfan granules can be applied at sowing to control aphids on parsnip. Aldicarb has an 84-day harvest interval and carbosulfan a 100-day harvest interval. Sprays of nicotine or pirimicarb can be applied to control aphids. The harvest intervals are 2 and 3 days, respectively.

Carrot fly (*Psila rosae*)
The larvae often damage parsnips – see under Carrot, p. 211, for further details.

No insecticides will have approval for carrot fly control on parsnip once the use-up periods of the carbamate and organophosphorus compounds previously approved have expired. However, parsnip seed can be film-coated with tefluthrin (off-label) (SOLAs 0873/00, 0874/00) to reduce the risk of carrot fly damage. Up to six foliar sprays of lambda-cyhalothrin (off-label) (SOLAs 1737/96, 1738/96, 0283/2000) may be applied also, with a 14-day harvest interval (see under Carrot, p. 212, for more details).

Caterpillars, including cutworms
Foliar-feeding caterpillars and cutworms may damage parsnip roots, especially in hot, dry summers – see under Red beet, p. 246, for control options and other details.

Foliar sprays of cypermethrin (off-label) (SOLA 2184/98) can be applied to control cutworms if there is a risk of damage (see p. 246).

Celery fly (*Euleia heraclei*)
See under Celeriac and celery, p. 216, for details. Although the larvae may cause blisters on the leaves of parsnip, the damage rarely warrants the use of insecticide.

Nematodes
Needle nematodes (*Longidorus* spp.), stubby-root nematodes (*Trichodorus* spp.) and northern root-knot nematode (*Meloidogyne hapla*) are key nematode pests of parsnip. The last-mentioned species causes fanged and galled roots and is not controlled by granular nematicides.

Carbosulfan and aldicarb granules are approved for use at drilling for nematode control on parsnip. Aldicarb has an 84-day harvest interval and carbosulfan a 100-day harvest interval.

Diseases

Canker (*Itersonilia pastinacea*)
Canker causes the shoulder and crown of parsnip to rot during autumn and winter. The two main types of canker are either black or orange-brown in colour. Black canker can be caused, separately, by *Itersonilia pastinacea*, *Phoma* sp. or *Mycocentrospora acerina*. *I. pastinacea* is the common cause of the disease in the UK, though *M. acerina* is prevalent on black fen soils. *I. pastinacea* releases airborne spores (ballistospores) which cause leaf spots on the foliage. Spores from leaf spots may get washed on to the crown depression of the parsnip, resulting in new canker lesions. Orange-brown canker is fairly common, and probably results from root infection by weak pathogens. Reduction in both types

of canker can be obtained by decreasing the root size by using close plant spacing. Earthing-up of roots in the summer will decrease *Itersonilia* canker. There are currently no chemical controls for this disease that can be applied to parsnip crops in the UK.

Violet root rot (*Helicobasidium purpureum*)
In addition to parsnip, this pathogen occurs on several other vegetable crops, such as carrot and red beet; it is also important on asparagus. The symptoms are similar to those on carrot (see under Asparagus, p. 188). No chemical holds approval for control of this disease on parsnips.

Pea

Nearly 41 181 ha are grown each year in the UK for green processing and a further 908 ha (including mange-tout) for the fresh market trade. Dry pea seed production for human consumption covers an area of approximately 20 450 ha.

Pests

Caterpillars
Caterpillars of flax tortrix moth (*Cnephasia asseclana*) appear early in the season (after overwintering in their first instar), and feed on the foliage before pupating in early summer but rarely cause economic damage. Caterpillars of silver y moth (*Autographa gamma*), a sporadic, migrant pest, also feed on pea foliage, typically feeding from the end of June until the beginning of September. The quality of crops, especially those for processing, may be reduced by severe infestations.

A pheromone-based monitoring system is available to determine the infestation levels of silver y moths. If a cumulative total of 50 or more moths is caught in a trap by the time that the first pods have been formed, then an insecticide spray should be applied 10 days later.

Cypermethrin is approved for caterpillar control on mange-tout and has a 7-day harvest interval. Insecticides approved for pea moth control will also control caterpillars of silver y moth.

Field thrips (*Thrips angusticeps*)
This pest is similar to pea thrips (*Kakothrips pisivorus*) (see p. 242) but damages young pea plants in April and May, causing stunting, leaf malformation and discoloration (yellow blotching). Peas following brassica seed crops are especially liable to damage.

Serious infestations can be treated with foliar sprays of dimethoate. There is a 14-day harvest interval.

Pea & bean weevil (*Sitona lineatus*)
The adults feed on the leaves and the larvae on the root nodules of a wide range of leguminous crops (see under Broad bean, p. 209, for more details). The adults cause greatest damage and produce semi-circular notches in the leaf margins. Adults are active in April and May and do most damage when conditions for plant growth are poor, as in cloddy soil or in cold, dry weather.

A monitoring system is available which will identify crops at most risk – see under Broad bean, p. 209, for details.

The pyrethroids alpha-cypermethrin, deltamethrin and lambda-cyhalothrin are approved for pea and bean weevil control on peas. Spray treatments should be applied as soon as damage is observed. Alpha-cypermethrin has a 1-day harvest interval on vining peas and an 11-day harvest interval on combining peas. No harvest intervals are specified for deltamethrin or lambda-cyhalothrin.

Pea aphid (*Acyrthosiphon pisum*)
Periodically, heavy infestations occur on peas. Pea aphids live throughout the year on leguminous plants, overwintering as eggs or, occasionally, as adults on clovers, lucerne, sainfoin and trefoils. Pea crops can be attacked at any time from early May to autumn, but those growing in June and July are generally most seriously affected. Severe infestations can reduce crop yield directly. In addition, the aphid is a vector of pea leaf-roll virus, pea enation mosaic virus and pea mosaic virus. Crops which are beginning to flower are most susceptible and should be treated when aphids are present on 15% of plants. A predictive model has been developed by CSL and PGRO. This uses weather data and local crop monitoring information to predict the development of pea aphid populations. Details are available from the HDC.

Insecticides approved for aphid control on peas include alpha-cypermethrin, deltamethrin, deltamethrin + pirimicarb, dimethoate, fatty acids, lambda-cyhalothrin, lambda-cyhalothrin + pirimicarb, nicotine and pirimicarb. Alpha-cypermethrin has harvest intervals of 1 day and 11 days on vining and combining peas, respectively. The remaining insecticides have harvest intervals of 0, 3, 14, 0, 0, 3, 2 and 3 days, respectively. Only cypermethrin and nicotine are approved for use on mange-tout, with harvest intervals of 7 and 2 days, respectively. Sprays should be applied when infestations increase, taking guidance from the predictive model.

Pea cyst nematode (*Heterodera goettingiana*)
This nematode causes 'pea sickness' where peas, or broad or field beans have been grown too frequently; the pea plants are stunted and yellow and senesce prematurely, the cysts being readily visible on the roots. The pods turn yellow quickly, and reduce the quality of the crop on the fresh market. Damage is usually patchy in the field. To keep populations below the economic threshold, at least 4 years should be allowed between successive host crops.

Pea midge (*Contarinia pisi*)
This pest is very localized to parts of eastern England. Outbreaks of damage occur infrequently. The small, whitish larvae cause most injury by feeding on the flowers and growing points, but they also infest and malform the pods.

Dimethoate (peas) and lambda-cyhalothrin + pirimicarb (combining peas and vining peas) are approved for pea midge control. They have harvest intervals of 14 and 3 days, respectively. The sprays should be applied at the early green-bud stage when the largest flower buds are about 6 mm long, still enclosed by the leaves of the terminal shoot, and midges can be found by folding back the protecting leaves. A second treatment may be necessary if the attack is severe.

Pea moth (*Cydia nigricana*)
The caterpillars feed on developing peas inside the pods, reducing their value as food or seed. The pest can be found in most areas of England where peas are grown. All cultivars are attacked but dry-harvested peas with a long growing period suffer most. Quick-maturing cultivars sown early for vining, and those sown after the middle of June, usually escape attack. Peas that come into flower during the flight period of the moth, between mid-June and mid-August, normally suffer most damage. Vining crops should be grown well away from sites cropped in the previous year with dry-harvested peas.

The timing of insecticide spray applications is critical. Advice should be sought from PGRO regarding the operation of pheromone traps that enable the moth population to be monitored and, hence, spray thresholds and spray dates to be determined for the dry-harvested pea crops. PGRO run a phone-in service, providing spray dates and treatment thresholds. This uses a model developed by ADAS. The pheromone traps should be used in vining and fresh pea crops, but only to indicate the presence or absence of the pest; if any moths are caught on these crops in the period up until full flower, the crop should be sprayed.

Pea moth can be controlled on peas with foliar sprays of alpha-cypermethrin, *Bacillus thuringiensis*, deltamethrin, lambda-cyhalothrin and lambda-cyhalothrin + pirimicarb. Alpha-cypermethrin has harvest intervals of 1 day and 11 days on vining and combining peas, respectively. Harvest intervals for the other insecticides are 0, 0, 0 and 3 days, respectively. Only cypermethrin (harvest interval 7 days) is approved for use on mange-tout.

Pea thrips (*Kakothrips pisivorus*)
Adults and nymphs of pea thrips feed on the surface tissues of the young pods and foliage of peas and beans, causing silvery, mottled patches and malformation of pods. Heavy attacks may lead to severe stunting. Peas are affected particularly and the main attacks occur in June or July.

Foliar sprays of dimethoate can be applied to control thrips on peas. There is a 14-day harvest interval.

Slugs
Biological details and recommended treatments for all edible vegetable crops are given under Lettuce, p. 228.

Stubby-root nematodes (*Paratrichodorus* spp. and *Trichodorus* spp.)
Pea early browning virus is transmitted by several of these nematode species, which tend to be prevalent in sandy soils. Growers can have soil tested prior to drilling to determine the risk from nematode attack.

Diseases

Chocolate spot and grey mould (*Botryotinia fuckeliana* – anamorph: *Botrytis cinerea* and *Botrytis fabae*)
For a description of disease symptoms see under Broad bean, p. 210. Vinclozolin, applied as a protectant spray at mid-flower or at first pod set, is approved for control of both species of pathogen on peas in the UK. Iprodione has off-label approval (SOLA 1565/98) for use on infected mange-tout.

Damping-off diseases
Germinating seedlings of peas and of many vegetables can be attacked by *Phytophthora* spp. and *Pythium* spp., either before or after emerging above the soil. The damage caused is usually most severe when germination is slow, e.g. early in the season or in cold, wet soils. Soil-borne pathogens can be controlled by treating seeds with metalaxyl + thiabendazole + thiram or with thiram alone. Damping-off in peas can be controlled by applying either carbendazim + cymoxanil + oxadixyl + thiram or metalaxyl + thiabendazole + thiram. Both of these formulated mixtures have on-label approval for use on pea seedlings in the UK. Thiram alone is approved as a seed treatment for control of damping-off in peas. Where seedlings, e.g. cauliflower and celery, are raised in blocks for transplanting later into the field, incorporation of etridiazole into the compost, according to the manufacturer's instructions, will control damping-off fungi. Propamocarb hydrochloride is also approved and active against a range of pathogens, including *Pythium* spp., and should be applied as a compost drench or as a seedling drench following the methods and rates given by the manufacturer. Benomyl (SOLA 0875/97), metalaxyl-M (SOLA 1149/99) and fosetyl-aluminium (SOLA 2390/97) hold specific off-label approvals for use against damping-off diseases on a range of crops. Tolclofos-methyl, applied to disease-free established brassica seedlings, has on-label approval for control of *Thanatephorus cucumeris* (anamorph: *Rhizoctonia solani*), another cause of damping-off.

Downy mildew (*Peronospora viciae*)
Downy mildew affects vining and protein peas in the UK. The disease is seedborne, with mycelia and oospores present on the seed coat. Systemically infected

seedlings, which are stunted and distorted, are randomly scattered through crops in a manner resembling the foci of infection which seed-borne pathogens produce. The fungus sporulates profusely on these stunted, dying seedlings, from which sporangia of *P. viciae* spread to infect other crop plants. Secondary infections appear as purple-brown lesions on the undersides of leaves, with corresponding chlorotic areas on the upper surfaces. The fungus is prevalent under cool, moist conditions and can infect all parts of the pea plant.

Diseased haulm, containing oospores of the fungus, provides a reservoir of soil-borne infection, which may persist for several years. Neither seed and root exudates (from pea roots) nor light affect oospore germination. Oospore germination is optimal at 5 and 10°C. There is some resistance to the pathogen in certain commercial cultivars of vining peas. Seed treatments containing mixtures of either carbendazim, cymoxanil, oxadixyl and thiram or metalaxyl, thiabendazole and thiram have on-label approval for the reduction and prevention of infection on seedlings. Fosetyl-aluminium holds off-label approval (SOLA 2390/97) as a seed treatment on combining and vining peas.

Leaf and pod spot (*Ascochyta pisi*)

A. pisi is the main pathogen causing leaf and pod spot of peas, and is commonly seed-borne. The pathogen is more prevalent in home-raised seeds (e.g. marrowfat cultivars) than in seeds of wrinkled or smooth cultivars, harvested in dry climates for the production of canning and frozen food crops.

The disease causes sunken, brown cankers on the stems of plants, and tan-coloured lesions on the leaves. Lesions eventually become covered with black fungal structures (pycnidia). Ultimately, these appear on the pods, affecting the seeds. Crop losses are due to reduction of stand and yield, and to the production of stained seeds unfit for processing. Spores of the fungus are disseminated through pea crops by rain and wind; thus, the disease is more prevalent in wet years. The disease persists on debris but is also seed-borne. Metalaxyl + thiabendazole + thiram or carbendazim + cymoxanil + oxadixyl + thiram are approved as seed treatments for control of this disease. The related seed-borne fungus *Mycosphaerella pinodes*, which causes a foot rot with leaf spot, has become more prevalent recently in protein peas in the UK. This pathogen differs from *Ascochyta pisi* in that it produces pseudothecia (the fungal sexual stage) which, when mature, give rise to airborne ascospores. Ascospores can be dispersed over considerable distances. This pathogen is restricted but not controlled by seed treatment.

Powdery mildew (*Erysiphe polygoni*)

This is a relatively unimportant disease in the UK. The disease appears on leaves and pods as a whitish, powdery mould. Lesions have abundant conidia. Presence on the pods renders them unmarketable for the fresh market. Infected peas grown in protected environments can be sprayed with fenpropidin (off-label (SOLA 0320/97) or triadimefon (off-label) (SOLA 0158/97).

Viruses and other disease problems
Peas are affected by a number of viruses (bean yellow mosaic, enation mosaic) and other leaf diseases such as rust (*Uromyces pisi* and *U. viciae-fabae*) but these are of minor importance in the UK and there are no approved chemical controls. Rust-infected peas, grown in protected environments, can be sprayed with triadimefon (off-label) (SOLA 0158/97).

Radish

See under Brassica crops, p. 189.

Red beet

Red beet is grown on approximately 2100 ha in the UK.

Pests

Aphids
Black bean aphid (*Aphis fabae*) is common on red beet, causing malformation of the leaves and stunting of the crop. Some loss in yield may result. Peach/potato aphid (*Myzus persicae*) may also infest red beet. Some populations of peach/potato aphid have developed resistance to many of the insecticides applied currently to crops. If this species is present, then a resistance management strategy should be considered (see under Lettuce, p. 226).

Foliar sprays of pirimicarb (off-label) (SOLA 0328/96), dimethoate (excluding peach/potato aphid) or nicotine can be applied to control aphids on red beet. Nicotine and pirimicarb have 2- and 3-day harvest intervals, respectively. Crops should be treated as soon as they become infested (usually in May). The last application of dimethoate must be made by 30 June in the year of harvest.

Beet cyst nematode (*Heterodera schachtii*)
Populations of this nematode are increased by too-frequent growing of host plants – mainly Brassicaceae (Cruciferae) and Chenopodiaceae. The nematode is primarily a pest of sugar beet, but severe infestations can stunt or even kill red beet.

Granules of aldicarb (off-label) (SOLA 0001/96) may be broadcast or applied as a row treatment at drilling. There is a harvest interval of 84 days. The only other economic method of control is crop rotation. Ideally, brassica or beet crops should be grown only once in 5 years.

Capsids (*Calocoris norvegicus* and *Lygus rugulipennis*)
These insects damage the leaves of red beet only very occasionally; when feeding,

they exude toxic saliva, which causes necrotic spots to appear. Capsids are likely to be confined to the headlands of crops. Routine treatment of crops with insecticide is not necessary.

Crops should be monitored regularly. If necessary, foliar sprays of nicotine may be applied to control capsids. There is a 2-day harvest interval.

Caterpillars, including cutworms (*Agrotis segetum* and other species)
Red beet is one of the most susceptible crops to cutworm attack. Cutworms are the caterpillars of certain noctuid moths that typically attack plants at or below soil level; turnip moth (*Agrotis segetum*) is the most common species in the UK. Cutworm attacks are sporadic, and severe infestations do not occur every year. They are favoured particularly by hot, dry weather conditions. Usually, cutworm damage is most severe in light, sandy soils. The adult moths often lay their eggs on weeds, and crops planted immediately following dense weed cover are more likely to be infested with cutworms than those planted in weed-free soil.

Pheromone traps can be used to capture male turnip moths. Trap catches provide a warning of the start of egg laying, but cannot be used to estimate the size of an infestation in any crop. A warning of the risk of cutworm attack is available from ADAS. This is based on a mathematical model using pheromone trap catches and daily records of maximum and minimum temperature and rainfall.

Turnip moth caterpillars can be controlled using irrigation. As an increasing number of crops are being irrigated to improve quality, fewer insecticide treatments are being applied to control cutworms.

In addition to cutworms, red beet may also be infested by foliar-feeding caterpillars, such as those of silver y moth (*Autographa gamma*).

Foliar sprays of cypermethrin or lambda-cyhalothrin (off-label) (SOLAs 1133/97, 1134/97, 0286/2000) should be applied to control cutworms, according to warnings from ADAS or other advisors, or as soon as damage is seen in the crop. Cypermethrin or lambda-cyhalothrin (off-label) (SOLAs 1133/97, 1134/97, 0286/00) can be used also to control caterpillars of silver y moth, which may be sporadic pests of the foliage. The control of large caterpillars is difficult, but young caterpillars can be killed with insecticides or by irrigation (if they are cutworms). No harvest interval is specified for cypermethrin, whereas lambda-cyhalothrin has a harvest interval of 3 days.

Field thrips (*Thrips angusticeps*)
The damage caused by thrips feeding on the leaves of red beet is rarely serious, although leaves may be disfigured in April and May. Routine insecticide treatments are not necessary. If required, foliar sprays of nicotine can be applied to control thrips infestations on red beet. There is a 2-day harvest interval.

Mangold fly (*Pegomya hyoscyami*)
The larvae of mangold fly (commonly known as 'beet leaf miners') feed on the

leaves and cause characteristic blisters, reducing or preventing photosynthesis. The damage is most serious when the plants are small; large plants may be attacked quite heavily without their yield being much affected. Crops can be helped to grow away from an attack by providing a good tilth for seedling emergence and by top dressing with a nitrogenous fertilizer. If an attack develops in the seedling stage, singling should be delayed. In East Anglia, pest management strategies in sugar beet have reduced mangold fly populations and it is now uncommon as a pest on red beet.

If insecticidal control is necessary, nicotine sprays should be applied to the crop as soon as the adult flies appear and a total of three sprays should be applied at weekly intervals. Nicotine has a 2-day harvest interval.

Pygmy mangold beetle (*Atomaria linearis*)
Although adults can often be found in the crop, damage caused by them is rarely serious. No insecticides are approved specifically for their control.

Diseases

Blackleg (*Pleospora bjoerlingii* – anamorph: *Phoma betae*)
Early symptoms are difficult to observe, owing to the colour of the beet. Mature spots are depressed and covered with black fruiting structures (pycnidia). The stems of young seedlings may also become blackened and shrivelled. The pathogen is seed-borne and seed treatment with thiram has on-label approval for control of the disease. This treatment also improves germination. Seeds must be immersed in a thiram solution at 20°C and then dried. Thiram may also give protection against soil-borne damping-off diseases.

Downy mildew (*Peronospora farinosa* f. sp. *betae*)
Downy mildew of red beet is caused by the same species that attacks sugar beet (see Chapter 6, p. 178). The disease is not normally serious in red beet, though infection of seedlings and of the seed production crop can cause damage. Seedling symptoms include chlorosis of the cotyledons and distortion of young leaves. Sporulation produces a grey, felt-like mat which is usually more obvious on the undersurfaces of leaves. Mildew may affect stecklings in the autumn and infections become progressive; in the following spring, they affect the flowers and seed clusters, and cause losses. There are no chemicals with approval for control of this disease in the UK.

Powdery mildew (*Erysiphe polygoni*)
Powdery mildew occurs on both sides of the leaves. Infected leaves are covered superficially in spots, each up to 2 cm in diameter, but the disease soon spreads over the entire leaf surface. Powdery mildew appears on the older leaves first. The disease is of only minor importance in the UK, where there are currently no approved chemicals that can be used for its control.

Rust (*Puccinia aristidae* and *Uromyces betae*)
This disease is more important in the seed-production crop. Orange-brown uredosori are found during the growing season on the undersurfaces of leaves. Fenpropimorph holds specific off-label approval (SOLA 1246/94) for use on crops affected by this disease.

Scab (*Streptomyces scabies*)
This disease is more prevalent on potato, where it is known as common scab. On red beet, scab may take two forms: (a) pitted scab, consisting of deep and scurfy pits, and (b) raised scab, comprising raised corky lumps. Seedlings of red beet are susceptible only from the fourth to the sixth week after sowing. The first form of the disease is caused by *S. scabies*, but it is possible that a second species may be implicated in raised scab. Scab is favoured by high soil pH, low soil moisture and high soil temperatures. Chemical control methods are ineffective against this fungus, and there is no evidence that cultural control measures (e.g. the use of irrigation, which is applied to potatoes, see Chapter 5, p. 142) are effective against scab of red beet.

Silvering (*Curtobacterium flaccumfaciens* pv. *betae*)
This bacterial disease has become less frequent in recent years. It affects mainly seed crops, but can also be found on market crops. The leaves on seed plants become silvery and the plants then wilt and die. The pathogen is seed-borne. At present there is no seed treatment approved for its control in the UK.

Rhubarb

Nearly 670 ha of rhubarb are grown commercially in the UK.

Pests

Rosy rustic moth (*Hydraecia micacea*)
The crowns of rhubarb plants may be damaged by the mining of caterpillars between spring and mid-summer. No insecticide is approved for control of this pest. Infestations tend to be most frequent in weedy sites.

Slugs
Biological details and recommended treatments for all edible vegetable crops are given under Lettuce, p. 228.

Diseases

Bacterial and fungal diseases
There are several minor disease problems on rhubarb. These include bacterial soft

rot (caused by *Erwinia carotovora* ssp. *carotovora* and *Pseudomonas marginalis*), which occurs as a storage problem. *Botryotinia fuckeliana* (anamorph: *Botrytis cinerea*) is also an important pathogen on rhubarb. There are currently no chemical controls that can be applied to crops affected by these diseases. However, chemicals used to control diseases on celery can be applied to rhubarb subject to specific restrictions on the manufacturers' labels.

Virus diseases

Commercial rhubarb crowns may be infected with a number of viruses which reduce the size and weight of the petioles. The causal viruses include arabis mosaic, cherry leaf-roll, cucumber mosaic, strawberry latent ringspot and turnip mosaic, all of which may occur in the same crown. Various combined infections induce leaf mosaic and ringspot symptoms, seen most conspicuously in May–June. There are no chemical controls which can be applied to combat these problems. Practical control has been achieved by producing virus-free meristems from virus-infected plants. The majority of meristems remain healthy and have been grown-on to produce plants which are maintained under virus-free conditions. Virus-free crowns are more vigorous and give heavier petioles than virus-infected ones. Healthy crowns become re-infected very slowly in the field.

Runner bean

See under French bean and runner bean, p. 220.

Savoy

See under Brassica crops, p. 189.

Shallot

Production of shallots in the UK is very limited.

Pests

Onion fly (*Delia antiqua*)

See under Onion, p. 233, for biological information. No insecticides are recommended specifically for the control of onion fly on shallot.

Spinach and spinach beet

Only a small area of spinach (*c.* 350 ha in 1999) is grown annually in the UK. Spinach beet is grown mostly in gardens or allotments but is sometimes a minor overwintered crop for the fresh market.

Pests

Bean seed flies (*Delia florilega* and *D. platura*)
Bean seed flies can cause extensive damage at the seedling stage when a crop is drilled into trash. Spinach and spinach beet seed may be film-coated with tefluthrin (off-label) (SOLA 0234/99) to avoid damage by bean seed flies.

Black bean aphid (*Aphis fabae*)
See under Broad bean, p. 208, for more details. Insecticide sprays should be applied if large colonies develop on the leaves and stems of spinach. Foliar sprays of cypermethrin (off-label) (SOLA 3133/98), nicotine or pirimicarb (off-label) (SOLA1626/95) can be applied to control aphids. No harvest interval is specified for cypermethrin, but treatments should be applied before seven true leaves have developed. Nicotine and pirimicarb have 2- and 3-day harvest intervals, respectively.

Caterpillars
Foliar sprays of cypermethrin can be applied to control caterpillars on spinach (off-label) (SOLA 3133/98). Sprays should be applied as soon as damage is seen. Up to two treatments are allowed per crop and these should be applied before seven true leaves have developed.

Mangold fly (*Pegomya hyoscyami*)
The larvae of this pest ('beet leaf miners') can cause extensive blistering to leaves, which may die subsequently. Crops sown either very early or very late are likely to miss the main period of pest activity. This is a more important pest on spinach than on red beet, because it causes damage to the marketable part of the plant.

If necessary, foliar sprays of nicotine can be used to control leaf-mining maggots on spinach. There is a 2-day harvest interval.

Diseases

Downy mildew (*Peronospora farinosa* f. sp. *spinaciae*)
Downy mildew forms yellow patches on the upper surfaces of leaves, accompanied by grey or violet-grey mould beneath the leaves. Badly affected leaves stop growing and may curl downwards at the edges. The disease can be severe under moist conditions, on badly drained land at low temperatures. Oospores are carried on the seed and in the soil, serving as sources of infection in new crops. Such infections can arise early and the closeness of the affected leaves to the ground makes protective spraying difficult. Mixtures of copper oxychloride + metalaxyl hold specific off-label approval in the UK for use on spinach crops affected by downy mildew (SOLA 1344/98). Alternatively, fosetyl-aluminium (SOLA 1190/97) can be sprayed on infected crops from first signs of disease.

Swede

See under Brassica crops, p. 189.

Sweetcorn

Though seldom grown in the UK on more than a small field scale, sweetcorn has potential. Approximately 1400 ha were grown in the UK in 1999.

Pests

Aphids
Sweetcorn can be attacked by several species of cereal aphid, primarily bird-cherry aphid (*Rhopalosiphum padi*) and grain aphid (*Sitobion avenae*). In general, moderate infestations do not affect yield significantly. However, moderate to severe infestations that occur close to harvest are important and usually require control measures. Foliar sprays of nicotine or pirimicarb (off-label) (SOLAs 1626/95) can be applied to control aphids on sweetcorn; these have harvest intervals of 2 and 3 days, respectively. Up to four sprays of dimethoate (off-label) (SOLA 0778/96) can be used for general insect control. This treatment has a 7-day harvest interval.

Frit fly (*Oscinella frit*)
Severe attacks on sweetcorn cause stunting and distortion of the plants. Where the attack is less severe, there may be little effect on growth, but rows of small holes will be seen across the leaves when they expand.

Generally, routine insecticide treatments are justified, using granular formulations of aldicarb (off-label) (SOLA 2771/96) or phorate incorporated into the soil at sowing. Alternatively, sprays of lambda-cyhalothrin (off-label) (SOLAs 1320/99, 1321/99, 0298/2000) may be applied when seedlings emerge.

Diseases

Maize smut (*Ustilago maydis*)
Maize smut was important in sweetcorn crops in the UK in 1976, but since then has not been such a problem. The disease requires a relatively high temperature for infection and development. It affects actively growing plant tissues and eventually produces galls, mainly on the ears of plants, which decrease yields. The pathogen is soil-borne and also is a contaminant of the seeds. There are no fungicides approved for use either as a seed treatment or as foliar applications to infected plants.

Tomato (outdoor)

At present this crop is grown on a limited area only, but current work with different cultivars could lead to greater commercial production.

Diseases

Blight (*Phytophthora infestans*)
Blight can be severe in cool wet seasons, causing greyish-brown lesions on the leaves and russet-brown, marbled areas on the fruits, which then become unsaleable. The disease can be contracted from infected potatoes and is often known as 'dry phytophthora rot'. Copper oxychloride and cupric ammonium carbonate are approved for control of this disease on outdoor tomatoes. Both fungicides give good protection and should be applied at intervals of 3 weeks from the end of July, or earlier if potato blight is seen in the neighbourhood.

Didymella stem rot (*Didymella lycopersici* – anamorph: *Phoma lycopersici*)
A dark-brown, shrunken canker appears near the base of the stem and, if this girdles it, the plant wilts suddenly. Lesions typically appear at wound sites; later in the season, similar cankers may appear on other parts of plants. There may also be brown spots on the leaves and black, encrusted lesions on the stem-ends of the fruits. Sanitization of affected plants as soon as infection appears, which should be repeated at the end of the season, is an effective control measure. There are currently no chemical treatments that can be applied to outdoor tomatoes to control this disease.

Grey mould (*Botryotinia fuckeliana* – anamorph: *Botrytis cinerea*)
Commonly known as 'ghost spot', the disease is characterized by haloes on the fruit, usually near the point of stem attachment. In humid conditions lesions that bear grey spores develop on the fruit. The pathogen persists as sclerotia in the soil and on infected debris. The appearance of spots on the fruit downgrades their value. The disease can be controlled by applying chlorothalonil, iprodione, pyrimethanil or thiram. Pyrimethanil also holds off-label approval (SOLA 2735/99) for use on tomato crops infected with grey mould.

Turnip

See under Brassica crops, p. 189.

Watercress

This high-value crop occupies about 64 ha and is grown mostly by specialists in localized areas where the demanding environmental requirements can best be met.

Pests

Many watercress crops do not require treatment with insecticide and, on others, only the affected area of crop need be treated. Treatment of watercress beds with insecticides is difficult because of the potential hazard to aquatic organisms and public water supplies. The insecticides are applied at ULV to avoid run-off below the crop canopy.

Aphids

Several species, e.g. buckthorn/potato aphid (*Aphis nasturtii*) and cabbage aphid (*Brevicoryne brassicae*), infest watercress, causing leaf distortion and also transmitting turnip mosaic virus. ULV sprays of dimethoate (off-label) (SOLA 0159/99) can be used to control aphids on watercress. This insecticide should be applied only to crops with 100% leaf cover, and the specific restrictions for use in watercress beds, given in the Notice of Approval, must be followed.

Other insect pests

Flea beetles (*Phyllotreta* spp.) can cause severe damage to watercress and render it unmarketable. Mustard beetles (*Phaedon cochleariae*) often damage leaves during the late spring and summer. Adult midges, and also flies, may act as contaminants in harvested watercress.

ULV sprays of malathion (off-label) (SOLA 2719/97) can be used to control these insects on watercress. This insecticide should be applied only to crops with 100% leaf cover, and the specific restrictions for use in watercress beds, given in the Notice of Approval, must be followed.

Diseases

Crook root (*Spongospora subterranea* f. sp. *nasturtii*)

The roots of watercress are attacked by the swimming spores (zoospores) of the causal fungus which then grows in the cells of the roots. A water temperature of approximately 5°C appears optimal for zoospore survival and infection. Affected roots are stunted, swollen and distorted, and eventually decay. Plants become stunted and their leaves turn yellow. The disease is particularly severe from October to March. Currently, there are currently no approved chemical controls for this disease. *Spongospora subterranea* f. sp. *nasturtii* is considered to be the vector of watercress yellow spot virus.

List of pests cited in the text*

Acrolepiopsis assectella (Lepidoptera: Yponomeutidae)	leek moth
Acyrthosiphon pisum (Hemiptera: Aphididae)	pea aphid
Agrotis segetum (Lepidoptera: Noctuidae)	turnip moth
Aleyrodes proletella (Hemiptera: Aleyrodidae)	cabbage whitefly
Aphis fabae (Hemiptera: Aphididae)	black bean aphid
Aphis nasturtii (Hemiptera: Aphididae)	buckthorn/potato aphid
Atomaria linearis (Coleoptera: Cryptophagidae)	pygmy mangold beetle
Autographa gamma (Lepidoptera: Noctuidae)	silver y moth
Brevicoryne brassicae (Hemiptera: Aphididae)	cabbage aphid
Bruchus rufimanus (Coleoptera: Bruchidae)	bean beetle
Calocoris norvegicus (Hemiptera: Miridae)	potato capsid
Cavariella aegopodii (Hemiptera: Aphididae)	willow/carrot aphid
Cavariella pastinacae (Hemiptera: Aphididae)	parsnip aphid
Cavariella theobaldi (Hemiptera: Aphididae)	willow/parsnip aphid
Ceutorhynchus assimilis (Coleoptera: Curculionidae)	cabbage seed weevil
Ceutorhynchus pallidactylus (Coleoptera: Curculionidae)	cabbage stem weevil
Ceutorhynchus pleurostigma (Coleoptera: Curculionidae)	turnip gall weevil
Cnephasia asseclana (Lepidoptera: Tortricidae)	flax tortrix moth
Contarinia nasturtii (Diptera: Cecidomyiidae)	swede midge
Contarinia pisi (Diptera: Cecidomyiidae)	pea midge
Crioceris asparagi (Coleoptera: Chrysomelidae)	asparagus beetle
Cydia nigricana (Lepidoptera: Tortricidae)	pea moth
Delia antiqua (Diptera: Anthomyiidae)	onion fly
Delia floralis (Diptera: Anthomyiidae)	turnip root fly
Delia florilega (Diptera: Anthomyiidae)	a bean seed fly
Delia platura (Diptera: Anthomyiidae)	a bean seed fly
Delia radicum (Diptera: Anthomyiidae)	cabbage root fly
Deroceras reticulatum (Stylommatophora: Limacidae)	field slug
Ditylenchus dipsaci (Tylenchida: Tylenchidae)	stem nematode
Euleia heraclei (Diptera: Tephritidae)	celery fly
Evergestis forficalis (Lepidoptera: Pyralidae)	garden pebble moth
Heterodera carotae (Tylenchida: Heteroderidae)	carrot cyst nematode
Heterodera cruciferae (Tylenchida: Heteroderidae)	brassica cyst nematode
Heterodera goettingiana (Tylenchida: Heteroderidae)	pea cyst nematode
Heterodera schachtii (Tylenchida: Heteroderidae)	beet cyst nematode
Hydraecia micacea (Lepidoptera: Noctuidae)	rosy rustic moth
Kakothrips pisivorus (Thysanoptera: Thripidae)	pea thrips
Longidorus spp. (Dorylaimida: Longidoridae)	needle nematodes
Lygocoris pabulinus (Hemiptera: Miridae)	common green capsid
Lygus rugulipennis (Hemiptera: Miridae)	tarnished plant bug
Macrosiphum euphorbiae (Hemiptera: Aphididae)	potato aphid
Mamestra brassicae (Lepidoptera: Noctuidae)	cabbage moth
Meligethes aeneus (Coleoptera: Nitidulidae)	pollen beetle
Meloidogyne hapla (Tylenchida: Heteroderidae)	northern root-knot nematode
Myzus persicae (Hemiptera: Aphididae)	peach/potato aphid
Nasonovia ribisnigri (Hemiptera: Aphididae)	currant/lettuce aphid
Oscinella frit (Diptera: Chloropidae)	frit fly
Paratrichodorus spp. (Dorylaimida: Trichodoridae)	stubby-root nematodes
Pegomya hyoscyami (Diptera: Anthomyiidae)	mangold fly
Pemphigus bursarius (Hemiptera: Pemphigidae)	lettuce root aphid

Phaedon cochleariae (Coleoptera: Chrysomelidae)	mustard beetle
Phyllotreta spp. (Coleoptera: Chrysomelidae)	flea beetles
Phytomyza rufipes (Diptera: Agromyzidae)	*larva* = cabbage leaf miner
Pieris brassicae (Lepidoptera: Pieridae)	large white butterfly
Pieris rapae (Lepidoptera: Pieridae)	small white butterfly
Plutella xylostella (Lepidoptera: Yponomeutidae)	diamond-back moth
Psila rosae (Diptera: Psilidae)	carrot fly
Psylliodes chrysocephala (Coleoptera: Chrysomelidae)	cabbage stem flea beetle
Rhopalosiphum padi (Hemiptera: Aphididae)	bird-cherry aphid
Sitobion avenae (Hemiptera: Aphididae)	grain aphid
Sitona lineatus (Coleoptera: Curculionidae)	pea & bean weevil
Tetranychus urticae (Prostigmata: Tetranychidae)	two-spotted spider mite
Thrips angusticeps (Thysanoptera: Thripidae)	field thrips
Thrips tabaci (Thysanoptera: Thripidae)	onion thrips
Trichodorus spp. (Dorylaimida: Trichodoridae)	stubby-root nematodes

* The classification in parentheses refers to order and family.

List of pathogens/diseases (other than viruses) cited in the text*

Albugo candida (Oomycetes)	white blister of brassicas
Alternaria brassica (Hyphomycetes)	dark leaf spot of brassicas
Alternaria brassicicola (Hyphomycetes)	dark leaf spot of brassicas
Alternaria dauci (Hyphomycetes)	leaf blight of carrot
Alternaria radicina (Hyphomycetes)	black rot of carrot
Ascochyta caynarae (Coelomycetes)	leaf spot of globe artichoke
Ascochyta fabae (Coelomycetes)	– anamorph of *Didymella fabae*
Ascochyta pisi (Coelomycetes)	pea leaf and pod spot
Aspergillus fumigatus (Hyphomycetes)	blue/green mould of onion
Aspergillus niger (Hyphomycetes)	black mould of onion
Botryotinia fuckeliana (Ascomycota)	chocolate spot of broad bean, collar rot of onion, grey mould of lettuce, pod rot of French bean
Botrytis allii (Hyphomycetes)	neck rot of onion
Botrytis cinerea (Hyphomycetes)	– anamorph of *Botrytinia fuckeliana*
Botrytis fabae (Hyphomycetes)	chocolate spot of broad bean
Botrytis squamosa (Hyphomycetes)	leaf rot of onion
Bremia lactucae (Oomycetes)	downy mildew of lettuce
Cladosporium allii (Hyphomycetes)	leaf blotch of leek
Cladosporium allii-cepae (Hyphomycetes)	leaf blotch of onion
Cladosporium cucumerinum (Hyphomycetes)	gummosis of cucumber
Colletotrichum lindemuthianum (Coelomycetes)	anthracnose of French bean
Curtobacterium flaccumfaciens pv. *betae* (Firmicutes)†	silvering of red beet and sugar beet
Didymella bryoniae (Ascomycota)	stem and fruit rot of cucumber
Didymella fabae (Ascomycota)	leaf and pod spot of beans
Didymella lycopersici (Ascomycota)	(didymella) stem rot of tomato
Erwinia carotorora ssp. *carotovora* (Gracilicutes: Proteot. bacteria)†	soft rot of brassicas
Erwinia herbicola (Gracilicutes: Proteobacteria)†	bacterial rot of onion
Erysiphe cichoracearum (Ascomycota)	powdery mildew of courgette and cucumber

Erysiphe cichoracearum f.sp. *lactucae* (Ascomycota)	powdery mildew of lettuce
Erysiphe cruciferarum (Ascomycota)	powdery mildew of brassicas
Erysiphe polygoni (Ascomycota)	powdery mildew of pea and red beet
Fusarium moniliforme (Hyphomycetes)	wilt of asparagus
Fusarium oxysporum f. sp. *asparagi* (Hyphomycetes)	wilt of asparagus
Helicobasidium purpureum (Basidiomycetes)	violet root rot
Itersonilia pastinacea (Hyphomycetes)	black canker of parsnip
Lactobacillus-like spp. (affinity uncertain)†	bacterial rot of onion
Leptosphaeria maculans (Ascomycota)	canker of brassicas
Leveillula taurica (Ascomycota)	powdery mildew of globe artichoke
Microdochium panattonianum (Hyphomycetes)	ring spot of lettuce
Mycocentrospora acerina (Hyphomycetes)	black canker of parsnip, crown rot of celery, liquorice rot of carrot
Mycosphaerella brassicicola (Ascomycota)	ringspot of brassicas
Mycosphaerella pinodes (Ascomycota)	pea leaf and pod spot with foot rot
Penicillium spp. (Hyphomycetes)	blue mould of onion
Peronospora destructor (Oomycetes)	downy mildew of onion
Peronospora farinosa f. sp. *betae* (Oomycetes)	downy mildew of beet
Peronospora farinosa f. sp. *spinaciae* (Oomycetes)	downy mildew of spinach
Peronospora parasitica (Oomycetes)	downy mildew of brassicas
Peronospora viciae (Oomycetes)	downy mildew of pea
Phoma apiicola (Coelomycetes)	root rot of celery
Phoma betae (Coelomycetes)	– anamorph of *Pleospora bjoerlingii*
Phoma lingam (Coelomycetes)	– anamorph of *Leptosphaeria maculans*
Phoma lycopersici (Coelomycetes)	– anamorph of *Didymella lycopersici*
Phoma spp. (Coelomycetes)	parsnip black canker
Phytophthora infestans (Oomycetes)	late blight of potatoes, tomatoes
Phytophthora porri (Oomycetes)	white tip of leek
Phytophthora spp. (Oomycetes)	damping-off (e.g. of peas)
Plasmodiophora brassicae (Plasmodiophoromycetes)	clubroot of brassicae
Pleospora bjoerlingii (Ascomycota)	black leg of red beet, sugar beet
Pleospora herbarum f. sp *lactucum* (Ascomycota)	stemphyllium rot of lettuce
Pseudomonas marginalis (Gracilicutes: Proteobacteria)†	soft root of rhubarb
Pseudomonas syringae (Gracilicutes: Proteobacteria)†	spear rot of asparagus
Pseudomonas syringae pv. *maculicola* (Gracilicutes: Proteobacteria)†	bacterial leaf spot of brassicas
Pseudomonas syringae pv. *phaseolicola* (Gracilicutes: Proteobacteria)†	halo blight of French bean
Puccinia allii (Teliomycetes)	rust of leek
Puccinia aristidae (Teliomycetes)	rust of beet
Puccinia asparagi (Teliomycetes)	rust of asparagus
Pyrenopeziza brassicae (Ascomycota)	light leaf spot of brassicas
Pythium spp. (Oomycetes)	damping-off of peas, etc.
Pythium sulcatum (Oomycetes)	cavity spot of carrot
Pythium violae (Oomycetes)	cavity spot of carrot
Rhizoctonia carotae (Hyphomycetes)	crater rot of carrot
Rhizoctonia solani (Hyphomycetes)	– anamorph of *Thanatephorus cucumeris*
Sclerotinia minor (Ascomycota)	watery soft rot of asparagus
Sclerotinia sclerotiorum (Ascomycota)	white mould of artichoke, bean, carrot, sclerotinia rot of bean, carrot

Sclerotium cepivorum (Ascomycota)	white rot of onion
Septoria apiicola (Coelomycetes)	leaf spot of celery
Septoria lactucae (Coelomycetes)	septoria spot of lettuce
Septoria petroselini (Coelomycetes)	leaf spot of parsley
Spongospora subterranea f. sp. *nasturtii* (Plasmodiophoromycetes)	crook root of watercress
Stemphyllium sp. (Ascomycota)	leaf spot of asparagus
Streptomyces scabies (affinity uncertain)†	scab of beet, carrot
Thanatephorus cucumeris (Basidiomycetes)	damping-off and wirestem of brassicas
Thielaviopsis basicola (Hyphomycetes)	root rot of carrot
Uromyces appendiculatus (Teliomycetes)	rust of French bean and runner bean
Uromyces betae (Teliomycetes)	rust of beet
Urocystis cepulae (Ustomycetes)	smut of onion
Uromyces pisi (Teliomycetes)	rust of peas
Uromyces viciae-fabae (Teliomycetes)	rust of beans
Ustilago maydis (Ustomycetes)	maize smut, sweetcorn smut
Xanthomonas campestris (Gracilicutes: Proteobacteria)†	black rot of brassicas

* For fungi, the classification in parentheses refers to class, although this is not possible within the phylum Ascomycota where classes have yet to be satisfactorily defined (see *Mycological Research*, February 2000). Oomycetes are now classified in Chromista with the brown algae, rather than as true fungi. Plasmodiophoromycetes are now classified as Protozoa rather than as true fungi. Some fungi have an asexual (anamorph) and a sexual (teleomorph) state, and the convention is to refer to them by their teleomorph name. However, where anamorph names are still in common use these are listed and cross-referenced to the teleomorph name. Strictly, fungi classified as Coelomycetes and Hyphomycetes should be known as 'hyphomycetous anamorphs' and 'coelomycetous anamorphs' of the relevant teleomorph taxon (e.g. hyphomycetous anamorphic Sclerotiniaceae, for *Botrytis fabae*), respectively. These problems highlight the continual changes in the classification of the fungi.

† Bacteria – the classification in parentheses refers to division and class, or to division only.

Chapter 8
Pests and Diseases of Fruit and Hops

M.G. Solomon
Horticulture Research International, East Malling, Kent

T. Locke
ADAS Rosemaund, Herefordshire

Introduction

The production of high yields of good-quality fruit necessitates a high standard of pest and disease control, and this has been achieved, traditionally, by adopting routine spray programmes. However, in recent years the emphasis has shifted, so that now pest and disease management is built around the aims of minimizing the impact of pesticide use on non-target organisms and the environment, and optimizing the exploitation of natural regulating factors. Whilst routine pesticide applications are still required for some pests and diseases, for others inputs can be minimized by monitoring the occurrence and severity of pests and diseases, and applying treatments only if the threat of damage justifies such action.

Where treatment is required, then the impact on non-target organisms is minimized if the most selective of the available materials is chosen. Natural enemies spared in this way are often able to contribute to the control of pests. The major groups of predatory insects that attack pests of fruit are anthocorid and mirid (capsid) bugs (Anthocoridae and Miridae, respectively), earwigs (Dermaptera), hover fly larvae (Syrphidae), lacewing larvae (e.g. Chrysopidae and Hemerobiidae) and ladybirds (Coccinellidae). Of the materials approved for use in fruit, the alkaloid nicotine, OPs, pyrethroids and tar oils are generally the most damaging to predatory insects. The insecticide/acaricide amitraz, the acaricides dicofol, dicofol + tetradifon, the insect growth regulator insecticides diflubenzuron and fenoxycarb, and the acaricide tebufenpyrad are moderately safe to predatory insects. In the safest category are the acaricides fenazaquin, fenpyroximate and tetradifon, and the insecticide pirimicarb. On crops where predatory phytoseiid mites have a role in the regulation of mite pests, then the impact of pesticides on these mites is an important consideration. For details see under Apple, fruit tree red spider mite, p. 266, and under Strawberry and under Hop, two-spotted spider mite, p. 305 and p. 311, respectively.

Many fruit pests are also attacked by parasitoids. The most important of these are (often small) parasitoid wasps, the commonest being members of the following families: Braconidae, Encyrtidae and Ichneumonidae, parasitizing moth caterpillars (particularly tortricids); Eulophidae, parasitizing leaf miners;

Aphidiidae, parasitizing aphids; and Aphelinidae, parasitizing mussel scale and woolly aphid. Additionally, caterpillars of many moth species are parasitized by members of the Tachinidae, a group of parasitoid flies. In general, these various parasitoid species are more sensitive to pesticides than are the predators mentioned above and, again, OPs and pyrethroids are the most damaging. As the use of broad-spectrum pesticides continues to decline, the currently rather modest contribution of naturally occurring parasitoids towards reducing pest numbers is likely to increase.

The pattern of availability of pesticides is currently in a period of rapid change and, in particular, the anticholinesterase insecticides are under review. Approval has been revoked for several materials used in fruit (the carbamate carbaryl and the OPs fenitrothion, heptenophos, phosalone and trichlorfon), and they cannot be used after April 2001; they are excluded from this chapter. It is essential to consult up-to-date sources of information on pesticide approvals (e.g. *The UK Pesticide Guide*, and the Assured Produce Crop Protocols). Some of the pesticides approved for use in fruit are likely to be used by organic growers, but very little by conventional growers because of the relatively low efficacy and high cost. In this category are the bacterial insecticide *Bacillus thuringiensis*, the natural product rotenone and a soap concentrate containing fatty acids.

Whenever pesticides are used, it is essential that the product label is read, and all instructions followed. As well as giving details of the uses to which the product can legally be put, and dose rates, timings, application methods, etc., the label includes information on crop safety restrictions, protective clothing and handling precautions, latest times of application and harvest intervals, special operator safety precautions and environmental safety requirements. This last category includes buffer-zone restrictions to protect surface water, and details of precautions to be taken to avoid danger to livestock, game, wildlife, bees and fish. Products hazardous to bees should not be used on crops in flower; if weeds are flowering in the orchard or plantation they should be cut down well before spraying.

Apple

Pests

The apple fauna is very extensive, and many species of insects and mites may be damaging. The major pests are apple rust mite, apple sawfly, codling moth, fruit tree red spider mite, fruit tree tortrix moth, rosy apple aphid, summer fruit tortrix moth and winter moth. Treatment thresholds have been established for the major pests of apple, and a programme of monitoring provides the basis for rational decision-making on pesticide use, so that treatments are applied only if the threat of pest damage justifies it. When treatment of a pest is required, it is preferable to choose the pesticide that has the least impact on natural enemy populations. This approach to managing apple pests has made great advances in the past 10 years and, in particular, has led to the widespread implementation of an integrated mite management approach, in which apple rust mite and fruit tree red spider mite are usually controlled by predatory mites.

Aphids

Apart from woolly aphid (*Eriosoma lanigerum*), which is described separately, there are four principal species: apple/grass aphid (*Rhopalosiphum insertum*), green apple aphid (*Aphis pomi*), rosy apple aphid (*Dysaphis plantaginea*) and rosy leaf-curling aphid (*D. devecta*). These begin to emerge from winter eggs on apple trees at about the time of bud break of Cox's Orange Pippin or Bramley's Seedling. Egg hatch is virtually complete by the green-cluster stage of these cultivars, but it may be a little later in the case of green apple aphid and rosy leaf-curling aphid.

Apple/grass aphid causes slight curl of rosette leaves and migrates soon after petal fall to grasses, especially annual meadow grass (*Poa annua*), returning to apple in the autumn. Heavy infestations on apple are more likely to follow summers with sufficient rainfall to maintain continuous growth of grass. This aphid is regarded as non-damaging unless present in very large numbers. An indication of the degree of infestation to be expected on apple in the following spring can be obtained by counting the small, wingless, yellow-green aphids (the egg-laying females) on the undersides of the leaves in late October. An average of one or fewer aphids per leaf, from a sample of 20 leaves from each of about eight trees across the orchard, indicates a light infestation. During the growing season, an appropriate treatment threshold is 50% trusses with five or more aphids in a sample of about 100 trusses taken at late green cluster.

Rosy apple aphid causes severe leaf curl and, more importantly, causes the fruitlets developing close to the infested leaves to remain small and to become distorted. This species disperses to plantains (*Plantago* spp.) in June and July, although some colonies may persist on apple into August; aphids return to apple in autumn. Because of the indirect damage to the fruits themselves, rosy apple aphid is damaging even at low population densities. An appropriate treatment threshold, in a sample of 100 trusses taken at late green cluster, is just one or more clusters infested. Later, as the developing aphid colonies cause more conspicuous leaf curling, a suitable threshold is a single tree infested in a sample of 50 trees inspected.

Rosy leaf-curling aphid is a localized pest and tends to appear (unless controlled) on the same trees year after year, causing severe leaf curl with conspicuous red areas; it lays its winter eggs in June, deep in crevices under the bark. This aphid is restricted to apple and spreads very slowly from tree to tree, so spot-treatment is usually adequate.

Green apple aphid is less common in the spring than apple/grass aphid. In late May, June and July it disperses to other apple trees, and to related hosts such as pear, hawthorn (*Crataegus*) and rowan (*Sorbus aucuparia*). It infests mainly the young extension growth, and is more a pest of young trees which become re-infested in the summer. An appropriate treatment threshold is five shoot tips with curled leaves, or 15 shoot tips infested in a sample of 100 taken from late June onwards.

Anthocorid bugs (Anthocoridae), mirid bugs (Miridae), earwigs (Forficulidae) and ladybirds (Coccinellidae) all attack aphids on apple, and their activity no doubt decreases the frequency with which aphid numbers exceed treatment thresholds. The most widely used approach to the chemical treatment of aphids is the application of sprays in spring. Pesticides available for this use are the OPs chlorpyrifos, dimethoate, malathion and pirimiphos-methyl, the pyrethroids cypermethrin and deltamethrin, the carbamate pirimicarb, and the alkaloid insecticide nicotine. Less used nowadays is tar oil, applied as a dormant-season winter wash against the eggs of aphids. See the Introduction, p. 258, and under Apple, fruit tree red spider mite, p. 266, for information on the impact of pesticides on beneficials and integrated mite management.

Apple blossom weevil (*Anthonomus pomorum*)

Adult weevils winter under loose bark, in leaf litter, etc. They emerge in early spring to feed on young apple foliage. From bud burst onwards the female bores holes into blossom buds, laying one egg in each blossom. The larva feeds on the stamens and base of the flower, and the petals are prevented from expanding and turn brown ('capped blossom'). Medlar, pear and quince may also be attacked. Historically, apple blossom weevil was a serious pest of apple, but the introduction of DDT in the late 1940s virtually eradicated it from commercial orchards. In recent years, however, this pest has begun to reappear in some orchards, particularly those close to woodlands, which provide ample overwintering sites.

The OP chlorpyrifos is available for use against this pest. Incidental control may be achieved if this material or other OPs are applied at green cluster against other pests, but the best timing, if infestations are heavy, is bud burst. See the Introduction, p. 258, and under Apple, fruit tree red spider mite, p. 266, for information on the impact of pesticides on beneficials and integrated mite management.

Apple leaf midge (*Dasineura mali*)

Although generally regarded as a minor pest, this insect is widespread and common, and became damaging in many orchards in the late 1990s. Eggs are laid in the unopened or partly uncurled leaves and, as the leaf expands, the margins become tightly rolled inwards. Affected leaves turn reddish then black, and eventually fall from the tree. Most of the larvae, which are bright pink and feed within the rolled leaves, drop to the ground when fully fed to pupate. There are three overlapping generations in the year, from May to August. There are no specific approvals for this pest. It is possible that OP compounds may be effective from petal fall onwards, but no information is currently available. See the Introduction, p. 258, and under Apple, fruit tree red spider mite, p. 266, for information on the impact of pesticides on beneficials and integrated mite management.

Apple leaf miner (*Lyonetia clerkella*)
This leaf miner is often very common on apple. Infestations also occur on cherry and various rosaceous ornamentals, including cherry laurel (*Prunus laurocerasus*) and snowy mespilus (*Amelanchier laevis*). The larvae form very long mines in the leaves and pupate externally in hammock-like cocoons formed on the bark or under the leaves of host plants. Damage caused, although often noticeable, is rarely important and there are no approved chemical treatments.

Apple rust mite (*Aculus schlechtendali*)
In recent years, heavy infestations of this small, straw-coloured mite have been reported in many orchards. The mites cause severe browning or 'rusting' of leaves, and russeting and cracking of fruits, particularly when the mites are abundant early in the season. This damage is seen particularly in cv. Bramley's Seedling.

The predatory mite *Typhlodromus pyri* feeds on apple rust mite and, in orchards in which the predator is abundant, the pest is usually regulated at non-damaging levels. The preferred food source for *T. pyri* is fruit tree red spider mite (*Panonychus ulmi*), p. 266, and the predator feeds on rust mite when spider mite numbers are low. Rust mite thus constitutes a secondary food source for *T. pyri*, helping to stabilize populations of the predator.

Rust mite can be monitored in its overwintering sites, sheltering behind buds; a suitable treatment threshold is ten mites per bud. From bud burst until petal fall, an appropriate treatment threshold is five mites per leaf but later in the season this threshold can be raised: by mid-summer, 50 per leaf is tolerable, as are even greater numbers later in summer. Available pesticides are the OP compound pirimiphos-methyl, and from other chemical groups amitraz and the insect growth regulator diflubenzuron. Dinocap and sulfur, if used against powdery mildew, have a suppressant effect on rust mite. See the Introduction, p. 258, and under Apple, fruit tree red spider mite, p. 266, for information on the impact of pesticides on beneficials and integrated mite management.

Apple sawfly (*Hoplocampa testudinea*)
In recent years, apple sawfly has become one of the most damaging pests of apple. Adult sawflies emerge about the time when mid-season cultivars are in flower; they are active in warm, sunny weather and are attracted only to trees in blossom. Eggs are laid, usually one per flower, in a slit-like cut just below the calyx. Hatching normally begins 4–5 days after 80% petal fall and is complete within 14–15 days. The larva at first mines under the skin of the fruit and then tunnels to the core. It later leaves to bore straight into another fruitlet, making a large entry hole where sticky frass accumulates. When fully fed, the larva drops to the ground and builds a cocoon in the soil. Most adults emerge the following spring

but some will not emerge until the second spring. Adult emergence can be monitored with white sticky traps.

In cider apple orchards, apple sawfly holes in fruit may provide the means of entry of *Sclerotinia fructigena*, the pathogen causing brown rot (see p. 271).

Available materials to control apple sawfly are the OPs chlorpyrifos and dimethoate, and the pyrethroids cypermethrin and deltamethrin. The appropriate timing is within 7 days after 80% petal fall. There is evidence that the fungicides carbendazim, fenarimol and thiophanate-methyl, when used against diseases at this time or shortly before, give some reduction in sawfly numbers. See the Introduction, p. 258, and under Apple, fruit tree red spider mite, p. 266, for information on the impact of pesticides on beneficials and integrated mite management.

Apple sucker (*Psylla mali*)

This insect overwinters as eggs on the bark. The eggs hatch over a fairly long period, which may extend through April into May. There is only one generation per year. The nymphs feed on the leaves and flowers; as a result, blossom trusses may turn brown, as if killed by frost. Damage as severe as this is seldom seen in dessert and culinary apple orchards, but is more common in cider apple orchards. The insect remains on the trees throughout the summer and eggs are laid in the autumn.

If infestations are sufficiently severe to require treatment, materials available for use in spring are the OPs chlorpyrifos, dimethoate and malathion, and the pyrethroids cypermethrin and deltamethrin. Pre-blossom use of OPs against other pests gives incidental control of apple sucker. Tar oil is available as a dormant-season treatment. See the Introduction, p. 258, and under Apple, fruit tree red spider mite, p. 266, for information on the impact of pesticides on beneficials and integrated mite management.

Bud moth (*Spilonota ocellana*)

See under Tortrix moths, p. 268.

Capsids

Two pest species occur on apple: apple capsid (*Plesiocoris rugicollis*) and common green capsid (*Lygocoris pabulinus*). Apple capsid is now uncommon in commercial orchards, whereas common green capsid has increased in importance and is a serious pest in some orchards in some years.

Common green capsid overwinters as eggs on shoots, the eggs hatching in the spring over a period corresponding to early pink bud to petal fall of cv. Bramley's Seedling. The young nymphs puncture fruitlets and young shoots. When adult the capsids leave the trees and pass through a second generation on herbaceous plants. Adults of this second generation return in autumn to fruit trees, as well as to bush fruit, and lay overwintering eggs. It is difficult to predict the risk of attack in a particular orchard but the pest does seem to recur in certain sites, perhaps because of the presence nearby of suitable host plants for the second (summer) generation.

Apple capsid passes through a single generation per year, remaining on apple. Overwintered eggs hatch at the green-cluster stage of cvs Bramley's Seedling or Cox's Orange Pippin, and the young nymphs puncture leaves, shoots and fruit. Reddish spots form on the leaves, which may become distorted; shoots are scarred and stunted, and rough russeted areas with scattered pits and pimples appear on the fruit. The capsids lay overwintering eggs from about mid-June to mid-July.

Materials available for use against capsids are the OPs chlorpyrifos and dimethoate, the pyrethroids cypermethrin and deltamethrin, and the alkaloid insecticide nicotine. Insecticides applied against other pests before or immediately after blossom may give incidental control of common green capsid, but the timing for applications specifically aimed at this pest is petal fall. In the event of treatment being required against apple capsid, spray at the green-cluster stage. See the Introduction, p. 258, and under Apple, fruit tree red spider mite, p. 266, for information on the impact of pesticides on beneficials and integrated mite management.

Not all capsids are pests; indeed, many are useful predators of pests. Of particular potential importance on apple, as predators of aphids, leafhoppers, mites and psyllids, are *Atractotomus mali*, *Blepharidopterus angulatus*, *Phytocoris tiliae*, *Pilophorus perplexus* and *Psallus ambiguus*.

Clouded drab moth (*Orthosia incerta*)

Damage by caterpillars of this common species occurs locally. Adults appear in the spring, and lay eggs in April and early May. The caterpillars feed on the foliage in the spring and early summer, and also excavate cavities which may penetrate deep into the developing fruitlets. When required, diflubenzuron may be used at petal fall to control this pest.

Codling moth (*Cydia pomonella*)

In an average season, moths emerge from late May or early June until early August, with the main flight period from late June to mid-July. In some years, a small second generation occurs in late August or September. Eggs are generally laid in the evening, on leaves and fruit, and are laid in higher numbers when temperatures exceed about 15.5°C. The caterpillars, which hatch in 10–14 days, bore into the fruit, feeding for the first few days in a cavity just beneath the skin; they then tunnel to the core. The entry hole is small and covered with dry frass, in contrast with the large hole with a mass of wet frass typical of apple sawfly. About half of the early-appearing caterpillars enter by the calyx. When fully fed, in about 4 weeks, the larvae leave the fruit and spin a cocoon under loose bark or other shelter. Second-generation moths arise only from larvae that have formed cocoons by early August. Most caterpillars overwinter in their cocoons, to produce moths the following summer. Since egg-hatch extends over at least 2 months, even with the most persistent available insecticides several sprays are required for complete control. Usually, two sprays are applied to kill all except the latest-emerging caterpillars, and on most commercial farms, where codling moth infestations have for some years been light, this proves adequate. The first spray should be applied just before the earliest eggs hatch, which in southern England is about mid- to late June in average seasons; the second spray should follow 2–3 weeks later.

Forecasting models are available which define the optimum timing for spray application, based on temperature records. Pheromone traps are available for codling moth, providing a means of assessing the size of the population in the orchard, and thus making a rational decision about the need for, and timing of, treatment with pesticide. The usual treatment threshold is five moths per trap in each of two successive weeks. When treatment is necessary, available materials are the OPs chlorpyrifos and malathion, the pyrethroids cypermethrin and deltamethrin, and the insect growth regulator diflubenzuron. The insect growth regulator fenoxycarb, if used against summer fruit tortrix moth (see p. 269), may provide incidental control of codling moth. See the Introduction, p. 258, and under Apple, fruit tree red spider mite, p. 266, for information on the impact of pesticides on beneficials and integrated mite management.

Common earwig (*Forficula auricularia*)

Earwigs lay their eggs in the soil from December to March, and again in May and June. Earwigs shelter by day in dark crevices, ascending plants at night to feed. Numbers appear to be greatest from July to September, and in some orchards damage to fruit occurs in the form of deep rounded cavities with small entry holes. Much of the damage caused by earwigs appears to be secondary in nature, being the enlargement of holes resulting from other causes. Cultivars such as Discovery, with soft tissue, seem to be more susceptible to primary damage by earwigs. Apples are also soiled by frass, where the insects shelter between the fruit and the stalk of an adjacent leaf.

When making decisions about the need to use insecticides against earwigs, it should be borne in mind that they are effective natural enemies of some pests and, in particular, that they consume large numbers of woolly aphids and other aphid species on apple trees. Where treatment is necessary, the insect growth regulator diflubenzuron is effective. See the Introduction, p. 258, and under Apple, fruit tree red spider mite, p. 266, for information on the impact of pesticides on beneficials and integrated mite management.

Dock sawfly (*Ametastegia glabrata*)

There are two and sometimes three generations of this sawfly in the year, and females of the last brood lay eggs in August or September. The light-green caterpillars feed on docks (*Rumex* spp.) and fat-hen (*Chenopodium album*). They hibernate in hollow stems or suitable crevices, and sometimes tunnel into apples in search of overwintering sites. They may also injure young trees by tunnelling into the pith of branches, entering at pruning cuts. Keeping the ground weed-free should prevent this pest from moving into apple trusses and damaging fruit. Where weeds are present under the trees, insecticides used against codling moth (see above) should also control dock sawfly caterpillars.

Fruit tree red spider mite (*Panonychus ulmi*)
This species overwinters as bright red, spherical eggs on the bark. Hatching normally begins at the pink-bud stage of cvs of Bramley's Seedling or Cox's Orange Pippin, in late April or early May; usually, half the eggs have hatched by petal fall, the remainder hatching during the next 3–4 weeks, i.e. up to about mid-June. The mites feed on the undersides of the leaves, causing a minute speckling, and in heavy attacks the leaves become dull green and then bronzed. Summer eggs are laid mainly on the undersides of the leaves, and five generations occur during a warm summer. The development from egg to adult normally takes about 4 weeks. In the first half of September, when day length is about 14 hours, winter eggs are laid. Winter eggs may be laid earlier on severely bronzed trees.

Fruit tree red spider mite is a secondary pest of apple, and was unimportant until the introduction and widespread use of broad-spectrum insecticides in the 1940s to 1950s. These materials killed the predatory mites and insects that had hitherto regulated spider mite numbers; released from this constraint, the mite thrived and became a serious pest. Fruit tree red spider mite has exhibited a well-developed facility for developing resistance to the acaricides developed to combat it, usually within a few years of the introduction of each material. During the 1990s, however, the approach to the management of this pest changed, and most growers now employ an 'integrated mite management' strategy, based on the action of the predatory phytoseiid mite *Typhlodromus pyri*. During the 1980s strains of this predator developed resistance to most OP and carbamate insecticides and were thus able to survive in orchards in which these compounds were used against insect pests. Resistant strains of *T. pyri* are now widespread in apple-growing regions in England, and usually colonize orchards when a suitable pesticide regime is employed. Pyrethroids (e.g. bifenthrin, cypermethrin, deltamethrin and fenpropathrin) are particularly harmful to *T. pyri*; for this reason these materials are little used in commercial apple orchards. The OPs dimethoate and pirimiphos-methyl are harmful, but there is evidence that some populations may be developing resistance to pirimiphos-methyl, so this material may do little damage to those particular populations of *T. pyri*. Other pesticides harmful to *T. pyri* are amitraz and tar oil. Fungicides moderately harmful to *T. pyri* are carbendazim, dinocap, mancozeb, maneb + zinc and, depending on the concentration, sulfur. In most orchards where OP-resistant *T. pyri* are established, and where the use of compounds harmful to the predator is avoided or minimized, fruit tree red spider mite is effectively regulated by the predator in most years, and acaricide use is seldom necessary.

When assessing the requirement for acaricide use, the usual threshold from petal fall onwards is seven leaves with four or more mites per leaf (in a sample of

50 leaves) or, alternatively, an average of two mites per leaf (again, in a sample of 50 leaves). When the predator *T. pyri* is present, these thresholds can be adjusted upwards. Materials available for use against fruit tree red spider mite are the OPs chlorpyrifos, dimethoate and malathion (but note that most populations of fruit tree red spider mite are resistant to OPs, so these materials are unlikely to be effective), the pyrethroids bifenthrin and fenpropathrin (but see note above about the impact of these materials on *T. pyri*), and from other chemical groups amitraz (see above – harmful to *T. pyri*), clofentezine, dicofol, dicofol + tetradifon, fenazaquin, fenpyroximate, tebufenpyrad, tetradifon and a soap concentrate containing fatty acids. Of these materials, clofentezine is for use before winter eggs have hatched (usually between bud burst and pink bud), amitraz at 60–80% hatch of winter eggs and again 3 weeks later, and the other materials during the summer after petal fall. Dinocap, when used against powdery mildew (see p. 274), will give an incidental reduction in fruit tree red spider mite numbers (but is also harmful to *T. pyri*). See the Introduction, p. 258, for information on the impact of pesticides on beneficials.

March moth (*Alsophila aescularia*)
See under Winter moth, p. 269.

Mottled umber moth (*Erannis defoliaria*)
See under Winter moth, p. 269.

Scale insects
The commonest scale insect on apple is mussel scale (*Lepidosaphes ulmi*). The adult scale is about 3 mm long, shaped like a mussel shell, grey in colour, and lies flat on the bark. Eggs, laid beneath the scale in late summer, hatch in the following May, the young nymphs settling in a suitable place and gradually developing the waxy scale. Infestations are seldom damaging, though when they are severe some nymphs settle and develop scales on the fruit.

Occasionally, oystershell scale (*Quadraspidiotus ostreaeformis*) and pear scale (*Q. pyri*) are seen on apple. These produce circular wax scales on the bark. Nut scale (*Eulecanium tiliae*) has become more common in orchards in recent years. The conspicuous female scale is almost spherical, about 6 mm in diameter and brown in colour. Eggs are laid in mid-summer and these hatch in late summer. The insects overwinter in the nymphal stage, becoming adult in the spring.

It is not usually necessary to apply an insecticide against scale insects, and they are probably kept in check by insecticides used in spring and summer against other pests. Where specific treatment is required, the available materials are tar oil, applied as a winter wash during the dormant season, and a soap concentrate containing fatty acids as a summer spray. See the Introduction, p. 258, and under Apple, fruit tree red spider mite, p. 266, for information on the impact of pesticides on beneficials and integrated mite management.

Straw-coloured apple moth (*Blastobasis decolorella*)

This widespread and common moth feeds and breeds on several plant species, including beech (*Fagus sylvatica*), and occasionally causes severe damage in some apple orchards. Adults are active in June and July, with a partial second generation from September to November. Larvae feed from July onwards, usually sheltered under a dead leaf or where two or more apples are held closely in a cluster. The larvae feed on the surface of the fruits, creating quite large shallow wounds.

The OP chlorpyrifos has been found to be effective as a summer spray against this species, but also available are the pyrethroids cypermethrin and fenpropathrin, the insect growth regulator diflubenzuron (although this was ineffective against this pest in recent trials), the alkaloid nicotine and the bacterial insecticide *Bacillus thuringiensis*. See the Introduction, p. 258, and under Apple, fruit tree red spider mite, p. 266, for information on the impact of pesticides on beneficials and integrated mite management.

Tortrix moths

Over 20 tortrix species can be found on apple trees, but the caterpillars of relatively few regularly cause damage to the fruit. In the UK, the main pests are codling moth (*Cydia pomonella*) (considered separately, p. 264), fruit tree tortrix moth (*Archips podana*), fruitlet-mining tortrix moth (*Pammene rhediella*) and summer fruit tortrix moth (*Adoxophyes orana*).

Fruit tree tortrix moth hibernates as young caterpillars in cocoons fixed to twigs or buds. These emerge in spring over a fairly long period, from late March to the green-cluster or pink-bud stages, first boring into fruit buds and then feeding on the young leaves, which are frequently spun together. Pupation occurs in late May to early June, between leaves that have been spun together. Moths begin emerging 1–2 weeks later than codling moth, and they occur until mid-August or September. The greatest numbers appear in late June or July, according to the season. The scale-like eggs are laid on the leaves in flat, green batches. Caterpillars soon emerge and feed first on the leaves and then on the fruit, eating out deep, irregular areas under the protection of a leaf attached to the fruit surface with silk. These caterpillars usually hibernate in the autumn but some, especially in warm summers, mature earlier and produce a partial second generation of adults from late August to October. These lay eggs and the caterpillars may cause further damage to the apples before overwintering. However, the most important damage is usually caused by caterpillars produced by moths of the first generation. Caterpillars taken into apple stores continue to feed on the fruit.

Fruitlet-mining tortrix moth occurs locally in many fruit-growing areas. The moths emerge in May and lay eggs on the undersides of leaves. The caterpillars attack the fruitlets, usually where two are touching, producing groups of round, black-rimmed holes. They are fully fed by early July and hibernate in cocoons under loose bark.

Summer fruit tortrix moth occurs mainly in south-eastern England, where it has become the dominant tortrix species in apple orchards. This species passes through two complete generations a year. The pest overwinters as young caterpillars, under leaf fragments fastened with silk to crevices in the bark, etc.; the caterpillars reappear in late March or early April to feed on the leaves. The caterpillars eventually pupate and produce a first flight of moths, which occurs over a similar period to that of the codling moth. These moths lay eggs and the caterpillars feed on leaves and also remove extensive areas of skin from the fruit. A second generation of moths emerges from late July to September; they lay eggs, and the caterpillars feed a little before hibernating. This second flight is usually much larger than the first.

Pheromone traps are available for all three of the above-mentioned tortrix species, providing a rational basis for decisions about the necessity for, and timing of, pesticide applications. The usual treatment threshold for fruit tree tortrix and summer fruit tortrix is 30 moths per trap per week, and for fruitlet mining tortrix ten moths per trap per week.

Insecticides available for all three tortrix species are the OP chlorpyrifos, the pyrethroids cypermethrin and deltamethrin, and the bacterial insecticide *Bacillus thuringiensis*. The insect growth regulator diflubenzuron is available for use against fruit tree tortrix moth. The insect growth regulator fenoxycarb is available for use against summer fruit tortrix moth, and is also likely to be effective against fruit tree tortrix moth. For fruit tree tortrix moth and summer fruit tortrix moth the usual timing for treatment is late June, followed by a second application 2 or 3 weeks later (but see note below on fenoxycarb). This timing is a little later than that for codling moth, and when this pest also requires treatment, the same application may control both species. If control is required in the spring, one of these materials may be applied at green cluster. A spray at this stage may also control bud moth (*Spilonota ocellana*), the brown caterpillars of which bore into blossom buds in the spring. For fruitlet-mining tortrix moth, the usual timing for treatment is two weeks after petal fall. The exception to the above is the timing of application of fenoxycarb, which is specifically approved for control of summer fruit tortrix moth, and which should be aimed against fifth-instar caterpillars resulting from the first generation of moths. See the Introduction, p. 258, and under Apple, fruit tree red spider mite, p. 266, for information on the impact of pesticides on beneficials and integrated mite management.

Winter moth (*Operophtera brumata*)

The 'looper' caterpillars of this often abundant species commonly attack apple foliage, flowers and fruitlets. Those of related species, e.g. March moth (*Alsophila aescularia*) and mottled umber moth (*Erannis defoliaria*), also occur but are usually far less numerous. The females of winter moth have vestigial wings and cannot fly; females of the other two species are entirely wingless. Adults of winter moth appear from October to December, those of mottled umber moth from October to December, but occasionally in January and February, and those of

March moth in March. Eggs are laid on the bark, those of the March moth in bands around twigs. Caterpillars hatch from about bud break to green cluster and, when fully fed, they drop to the ground and pupate in the soil. Young winter moth larvae may be blown from one tree to another, and from woods or hedgerows into neighbouring orchards.

Treatment thresholds for winter moth are 5–10% of trusses infested at the late green-cluster stage. Insecticides available for spring treatment are the OP chlorpyrifos, the pyrethroid cypermethrin, the insect growth regulator diflubenzuron and the bacterial insecticide *Bacillus thuringiensis*. Tar oil can be used as a dormant-season treatment. See the Introduction, p. 258, and under Apple, fruit tree red spider mite, p. 266, for information on the impact of pesticides on beneficials and integrated mite management.

Woolly aphid (*Eriosoma lanigerum*)

This aphid passes its whole life-cycle on apple. It overwinters as young aphids, devoid of 'wool', sheltering in cracks or under loose bark. These aphids become active in March or April, secreting the typical waxy 'wool'. Some winged forms appear in July, but they are not usually important sources of new infestations; natural spread is by young, wingless nymphs that crawl or are blown from tree to tree. Breeding continues throughout the summer. Eggs are laid in September, but are usually sterile, and the adults die as winter approaches. Aphid infestations cause galling of the wood, which is often not in itself damaging on mature trees, but is more serious on nursery stock. Sticky, woolly secretions may contaminate foliage and fruit, and be troublesome at harvest.

Where treatment is necessary, the available materials are the OPs chlorpyrifos and malathion, and the alkaloid nicotine; the usual timing for application is late June. See the Introduction, p. 258, and under Apple, fruit tree red spider mite, p. 266, for information on the impact of pesticides on beneficials and integrated mite management.

Diseases

Annual routine sprays are necessary for control of scab (bud burst to the fruitlet stage or later) and powdery mildew (pink bud to the cessation of extension growth). In addition, the control of storage rots may require post-harvest fungicide treatment, supplemented by orchard sprays in later summer.

Blossom wilt (*Sclerotinia laxa* f. sp. *mali*)

This disease is prevalent in some seasons and particularly in wetter parts of the country. Spores of the fungus are carried to the blossom where they germinate, causing blossom wilt. The mycelium may continue along the flower stalk into the spur, and even into the spur-bearing branch, where it causes die-back. In moist weather, spore pustules develop on the flowers shortly after infection and these continue the spread. Spurs and killed wood release spores in the following spring, so renewing the disease cycle. The cvs Cox's Orange Pippin, James Grieve and Lord Derby are very susceptible. The removal of infected parts will reduce spread of the disease and is best done in spring and early summer when the disease is easily recognisable. Good control can be obtained by sprays of vinclozolin, applied at first open flower and again 7 days later. A winter wash of tar oil will also reduce infection.

Brown rot (*Sclerotinia fructigena*)

Brown rot is widespread and often causes severe losses, particularly in hot summers. Infection in the orchard occurs initially through wounds in the skin of the fruit caused by insect attack, bird damage or growth cracks and russeting. The fungal spores are easily spread by wind and insects; the disease can also spread by contact from diseased fruit to healthy fruit, both in the orchard and during storage. Characteristic 'mummified' fruits, with numerous buff-coloured fungal pustules, remain on the tree through the winter; the fungus can also form cankers on spurs and branches. The cv. James Grieve is particularly susceptible.

Orchard sprays are not generally effective in controlling brown rot, although captan has some beneficial effect. A post-harvest dip (or drench) treatment with carbendazim gives excellent control of the secondary spread of the disease during storage.

Canker (*Nectria galligena*)

The pathogen *N. galligena* is responsible for most cankers on apple trees and occurs in the majority of orchards. Canker is frequently severe on old trees, especially where root restriction or impeded drainage occurs. The cvs Cox's Orange Pippin, Crispin, James Grieve and Spartan are particularly susceptible. The disease is usually seen as sunken zones of bark around buds, leaf scars or open wounds or around the bases of small, dead, side shoots. Small branches are often encircled but on larger branches the canker is restricted to one side. Fruits can also be infected, especially if weather conditions are wet during August. Two types of spore are produced: (a) summer spores, that ooze from cankers as white pustules; and (b) winter spores, that develop on the canker in red pear-shaped receptacles (perithecia) – the latter are sometimes mistaken for spider mite eggs. The winter spores may be shot out of the perithecia at any time of the year, but particularly in winter. Spores can infect only through breaks in the bark layer, including pruning or other wounds, leaf scars during autumn, woolly aphid galls or wood scab.

All badly infected small shoots should be removed and the cankered areas on large boughs pared away. Cut surfaces should be protected immediately with a wound-covering material; the application of paint containing octhilinone will prevent new infections occurring and, if made to established cankers, will suppress spore production and reduce the spread of infection. Paints should be applied only when the trees are dormant; they should not be applied to maiden trees.

Spray programmes containing carbendazim or dithianon for the control of scab will help to suppress the spread of canker in the orchard. Where disease is severe, spray with a copper-based fungicide just before leaf fall and again at about 50% leaf fall. Dipping or drenching the fruit immediately before storing will reduce the incidence of fruit infection. Such fruit infection may occur as an eye rot, as a stalk-end rot or, occasionally, as a cheek rot.

Collar rot (*Phytophthora cactorum* and *P. syringae*)
These pathogens cause severe losses in some localities and mainly attack mature trees. Almost exclusively, they affect cv. Cox's Orange Pippin, although they have been noted occasionally on cv. James Grieve; collar rot rarely occurs on trees less than 10 years of age. Infection of scion bark usually occurs at or above soil level, as a result of the fungus splashing up from the soil. Infected areas of bark are best seen in spring and autumn, often with cracks at margins. Small, oily or water-soaked patches may occur on the infected areas, as well as an exudate of reddish-brown droplets. If extension of the area ceases, the bark shrinks and appears smooth and shrunken, and cracks away clearly from the surrounding healthy bark. Fosetyl-aluminium can be applied as a paste to affected areas of bark.

The disease is soil-borne, and all trees in the vicinity should be protected as follows: (a) clear trunk bases of debris and weeds, and keep free with herbicides; (b) wherever feasible, remove soil from graft unions; (c) avoid mechanical injury to base of tree; and (d) remove fallen fruit.

All newly planted trees should be worked to resistant rootstocks, at least 30 cm above soil level. The rootstocks M.26, MM.104 and MM.106 are susceptible to infection.

Crown rot (*Phytophthora cactorum* and *P. syringae*)
This disease occurs on the rootstock below soil level. Damage is not usually seen directly, although the rot may extend above soil level or on to the rootstock stem piece. The first indications of crown rot are symptoms in the tree canopy, such as premature autumn coloration of leaves, or failure to 'leaf out' in the year following attack. Most of the commonly used rootstocks are resistant to crown rot; only MM.104 and MM.106 are susceptible. Trees on MM.104 are susceptible throughout their life, whereas those on MM.106 are rarely attacked after the first 3–4 years following planting.

Since most damage is done before foliar symptoms betray the presence of the problem, there is little that can be done by way of a cure. Nevertheless, preventive treatments can be applied in high-risk orchards. Copper oxychloride + metalaxyl can be applied as a drench to the soil around non-fruiting trees in their establishment years. Fosetyl-aluminium can be applied as a foliar spray once the trees have a full canopy, the treatment being repeated after an interval of 6 weeks.

Fireblight (*Erwinia amylovora*)
First recorded in the UK in 1957 on pears, this disease has spread rapidly and by

1969 it had infected a large number of apple trees in Kent and Suffolk. As far as is known, all apple cultivars are susceptible. Infection is usually through the shoots. Initially, the tip of an infected shoot wilts and droops, and at this stage golden droplets of bacterial exudate are often seen on the affected stem. As the bacteria progress down the affected shoot, the leaves and stem become brown. The disease does not appear to spread as rapidly in apple tissue as it does in pear, and scaffold branches or trunks of apple trees are rarely affected. Occasionally, fruit infection occurs, especially on cv. Grenadier. Fruit lesions are visible as slightly sunken, brown areas with a red halo, often with a water-soaked appearance. Under damp conditions, golden droplets of bacterial ooze are visible on the lesion.

The disease became endemic in cider apple orchards by the early 1980s in the south-west (mainly Somerset), and varietal (cultivar) susceptibility appears to depend primarily on time of flowering in relation to optimum infection conditions. However, the cider cvs Brown Snout, Chisel Jersey, Michelin, Somerset Redstreak and Vilberie are particularly prone to attack. Infection is mainly through blossom trusses, the disease often resembling blossom wilt but often with a marked line between infected and apparently healthy tissue. Infection can also occur through young shoots, producing the characteristic 'crook' effect. Movement of infection within the shoot is slow, and much can be done to contain the disease by heavy pruning. Most of the infection observed in apple orchards can be related to spread down-wind from infected hawthorn (*Crataegus*) hedges during storms and spread by insects, including bees. Precautions recommended for pear (p. 282) should be taken where the disease is known to exist.

Gloeosporium rot (*Gloeosporium album* and *G. perennans*)

Prior to the introduction of post-harvest fungicide treatments, losses of fruit in store (particularly long-term storage) were mainly due to *G. perennans*. With improvements in disease control, gloeosporium rot is now less prevalent but still widespread in occurrence. The cv. Cox's Orange Pippin is very susceptible, and fruit with a low calcium content is particularly prone to the disease. The fungi exist on small cankers that may cause die-back of shoots or may be insignificant. Spores are produced from the cankers all the year round but especially in autumn, when pruning cuts and other wounds may be infected. During wet periods, the spores are washed to the fruit surfaces where, after entering the lenticel chambers, they remain dormant until the fruit reaches a stage of maturity in store that permits further penetration by the fungus.

Where the disease has been prevalent, both orchard sprays and a post-harvest treatment are necessary for good control. Sprays of carbendazim or thiophanate-methyl in mid-July, mid-August and 1–2 weeks before harvest have given excellent control of the disease, but the use of these fungicides in orchards may lead to the selection of resistant strains. Captan or thiram at similar timings gives less effective control, but such treatment does not carry the risk of encouraging fungal tolerance. A post-harvest dip or drench treatment with carbendazim or thiophanate-methyl is very effective in reducing the level of fruit infection.

Phytophthora fruit rot (*Phytophthora syringae*)
This disease became prevalent on farms in Kent in 1973; it has subsequently caused serious losses in many parts of the country when wet conditions prevail at harvest. The fungus is soil-borne, and fruit infection occurs when soil or soil water containing the spores comes into contact with the fruit; this may be by direct contact with the ground, by splashing during rain storms or by mud contamination at harvest. Low-hanging fruit is more liable to infection, particularly if a strip of ground is kept free of vegetation by the use of herbicides. Affected fruits show a firm brown rot, often with a 'marbled' appearance. If placed in store, the fungus spreads to adjacent fruits; this results in severe losses in fruit stored long-term.

Orchard sprays of captan may reduce initial infection and post-harvest dip, or drench treatments with captan or a formulated mix of carbendazim + metalaxyl will give some reduction of secondary spread of the disease. Care should be taken to prevent infected fruits being stored. Mancozeb + metalaxyl (off-label) (SOLAs 0195/97, 0197/97), applied to the orchard floor, will also reduce fruit infection. Strains of the pathogen resistant to metalaxyl have been detected. Control is dependent on good orchard hygiene.

Powdery mildew (*Podosphaera leucotricha*)
This pathogen is present in all apple-growing areas and can cause severe reduction in yield and quality.

Routine annual spraying is essential for good control. The fungus overwinters in buds, and when an infected spur bud breaks the emerging growth appears white and mealy owing to the presence of a large number of spores. Diseased blossoms and leaves wither and drop from the tree. A terminal bud in which the fungus has overwintered often produces a 'silvered' shoot with mildewed leaves. Primary outbreaks give rise to secondary infections on young leaves, new shoots and growing points. Infections may also occur on young buds from which the cycle of infection begins again during the following year. Removal or eradication of overwintering sources of infection greatly facilitates control of secondary infection, and can be achieved by cutting out diseased parts in the spring.

Growing-season sprays for control of mildew should be applied, to give protection throughout the period from pink bud until the end of extension growth. It is essential to achieve good fungicide cover of all parts of the trees, especially the growing tips. Fungicides that kill or suppress spore germination are recommended, particularly in the period before petal-fall. Fungicides available for the control of powdery mildew are mainly in the demethylation inhibitor (DMI) group: namely, fenarimol, myclobutanil, penconazole, pyrifenox and triadimefon (with partial control being given by fenbuconazole). Those based on alternative chemistry are bupirimate, kresoxim-methyl and sulfur.

Scab (*Venturia inaequalis*)
The scab fungus overwinters within the tissues of fallen apple leaves. Spore cases

(perithecia) develop in early spring, and during wet periods spores (ascospores) are ejected. The spores germinate on young leaves and fruitlets when the weather is warm and wet, causing scab infections. Diseased parts soon produce summer spores (conidia) which themselves continue the spread. Infected fruits become spotted, distorted and unsaleable. In some cultivars, including Cox's Orange Pippin and Laxton Superb, young extension shoots become infected (wood scab), and during the following spring produce cushions of conidia beneath blister-like swellings. Wood scab is rare on cv. Bramley's Seedling, except where conditions are very favourable for the fungus.

Routine scab control by the application of fungicides is an essential part of apple growing. Two methods of determining the intervals between applications are recognized, according to the type of programme (preventive or curative). With preventive spraying, the aim is to ensure that there is always a sufficient deposit of fungicide on the tree during the vulnerable period. This can be achieved by spraying at 10-day intervals from bud burst until late June, although the time intervals between sprays may be modified according to the manufacturer's instructions. With a curative programme, the aim is to apply a suitable fungicide immediately after weather conditions have been favourable for infection; that is, as soon as possible after an infection period assessed from a Mills table giving periods of leaf wetness needed at various temperatures (Table 8.1). From experience, it seems best to follow a protective programme, applying an extra spray (containing a curative fungicide) after a long infection period. The extra spray is particularly useful when new growth is developing rapidly.

Table 8.1 Hours of wetness needed for ascospores to infect leaves on the tree*

Average temperature (°C) during wet period	Time that wetness must persist for scab infection
0.6–5.0	48 h or more
5.6	30 h
7.2	20 h
10.0	14 h
12.8	11 h
14.4	10 h
16.7	9 h

* Data from Western New York State, US (Mills, L.D. & Laplante, A.A. (1951). *Cornell Extension Bulletin*, No. 711, pp. 21–27).

The fungicides available for scab control are numerous, and some are available in formulated mixtures. Those in the DMI group are fenarimol, fenbuconazole, myclobutanil, penconazole and pyrifenox; other types are Bordeaux mixture, captan, carbendazim, dithianon, dodine, kresoxim-methyl, mancozeb, pyrimethanil, thiophanate-methyl and thiram. Strains of scab resistant to carbendazim and thiophanate-methyl, and to dodine, are known; nowadays, decreased sensitivity to DMI fungicides also occurs. Growers should make use of the wide range of products with different modes of action in their programmes, to minimize any risk of resistance.

Post-harvest sprays of thiophanate-methyl or urea are effective in preventing perithecial development; however, if applied just before bud burst they are ineffective in preventing the release of ascospores from mature perithecia. Urea is far less effective in reducing sporulation from wood scab.

Specific apple replant disease (SARD)

Recently, this disease has been shown to be caused by *Pythium* spp., mainly *Pythium sylvaticum*. The symptoms are characteristic and may occur when land previously cropped with apples is replanted with apples. Affected trees have a much reduced root system, which results in poor growth and cropping. The economic effects of the disease vary with the severity of SARD present in the soil and the rootstock/scion combination planted; cvs Cox's Orange Pippin and Golden Delicious on M.9 rootstock are particularly susceptible, whereas cv. Bramley's Seedling is more resistant.

The effects of SARD can be reduced by a pre-planting fumigation of the orchard soil with chloropicrin. This is usually done by a contractor. Where replanting a single grubbed tree, or for other small areas, a hand-operated injector may be used.

Chloropicrin is a noxious substance and is included in the Agriculture (Poisonous Substances) Regulations as a Part I (Schedule 2) substance. It is essential that fumigators are aware of these regulations, which must be observed for the protection of employees who carry out scheduled operations with chloropicrin. A code of practice for the fumigation of soil with chloropicrin has been devised by the MAFF for the guidance of contractors and fumigators, and is obtainable free of charge from the Ministry's Divisional Offices.

Use of peat compost in the planting hole and/or trickle irrigation offer cheaper alternatives to soil fumigation. However, in trials with these methods results have been inconsistent.

Pear

Pests

Pear sucker is the pest that normally dominates pest management decision-making in pear. Because of its actual and potential resistance to insecticides, most growers base their pest management strategy on the need to avoid damaging the predatory anthocorids that contribute to the control of pear sucker. Other pests, including pear/bedstraw aphid and tortrix moths, threaten damage in some years, and may require pesticide treatment.

Aphids

Pear/bedstraw aphid (*Dysaphis pyri*) is the most important aphid pest on pear. It overwinters as eggs on the tree, and egg hatch is complete by the white-bud stage. The pinkish aphids cause severe leaf curling and may spread over the tree. They persist into July and then depart for bedstraws (*Galium* spp.), returning to pear in the autumn. Pear/coltsfoot aphid (*Anuraphis farfarae*) causes the leaves to fold upwards and turn red; the adults are brown and the nymphs yellow-green. This species overwinters on pear and disperses to coltsfoot (*Tussilago farfara*) in late May, returning to pear in the autumn. Other aphids occurring on pear include apple/grass aphid (*Rhopalosiphum insertum*), green apple aphid (see under Apple, aphids, p. 260), and pear/grass aphid (*Longiunguis pyrarius*).

Pear/bedstraw aphid is potentially very damaging, and it is usual to treat it if it is detected in the orchard before flowering. A suitable threshold from petal fall onwards is 1% of trees infested. The other aphid species can be tolerated in greater numbers. Insecticides may be applied at green cluster or at petal fall. Materials available for spring use are the OPs chlorpyrifos and dimethoate, the pyrethroid cypermethrin, the alkaloid insecticide nicotine, a soap concentrate containing fatty acids, and a natural product rotenone. It is possible to use a tar oil winter wash during the dormant season against the overwintering eggs. See the Introduction, p. 258, and under Pear sucker, p. 279, for information on the impact of pesticides on beneficials and psyllid (pear sucker) management in pear orchards.

Apple blossom weevil (*Anthonomus pomorum*)

See under Apple, p. 261.

Codling moth (*Cydia pomonella*)

See under Apple, p. 264, for details of the life history. The treatment strategy is as for apple, but the insecticides available against codling moth on pear are the OPs chlorpyrifos and malathion, and the insect growth regulator diflubenzuron. See the Introduction, p. 258, and under Pear sucker, p. 279, for information on the impact of pesticides on beneficials and psyllid (pear sucker) management in pear orchards.

Common green capsid (*Lygocoris pabulinus*)

See under Apple, capsids, p. 263, for details of the life history. If treatment is needed against common green capsid on pear, available insecticides are the OPs chlorpyrifos and dimethoate, the pyrethroid cypermethrin, and the alkaloid nicotine. See the Introduction, p. 258, and under Pear sucker, p. 279, for information on the impact of pesticides on beneficials and psyllid (pear sucker) management in pear orchards.

Fruit tree red spider mite (*Panonychus ulmi*)

This is not a common pest on pear. See under Apple, p. 266, for an outline of the life history. The predatory mite *Typhlodromus pyri* is uncommon on pear, so does not contribute effectively to the biocontrol of spider mite on this crop. Where chemical treatment is required, the materials available for use on pear are the OPs chlorpyrifos, dimethoate and malathion (but note that most populations of fruit tree red spider mite are resistant to OPs, so these materials are unlikely to be effective), the pyrethroid bifenthrin, and from other chemical groups amitraz, clofentezine, dicofol + tetradifon, a soap concentrate containing fatty acids and tetradifon. See the Introduction, p. 258, and under Pear sucker, p. 279, for information on the impact of pesticides on beneficials and psyllid (pear sucker) management in pear orchards.

Pear leaf blister mite (*Eriophyes pyri*)
This small, narrow, pale-coloured mite overwinters behind bud scales, emerging in spring and feeding on the underside of leaves. The feeding sites develop into blister galls, which become hollow. The mites gain access to these galls and continue to breed within them. The galls become yellow, then red, and finally black. Mites may also feed on fruitlets, on which the feeding damage develops as red or black pustules. Once inside the galls the mites are protected from pesticides. There is no material specifically approved for use against this pest, but pirimiphos-methyl, when used against pear rust mite, may provide some control.

Pear leaf midge (*Dasineura pyri*)
Midges lay eggs in folds of young leaves in May; larvae feed on leaves, the edges of which become rolled upwards. There are several overlapping generations a year. The midge is sometimes a pest of nursery trees, but is seldom damaging on mature trees. There is no material specifically approved for this pest, but see note under Apple leaf midge, p. 261.

Pear rust mite (*Epitrimerus piri*)
This mite is similar in appearance and habits to apple rust mite (p. 262). On pear, however, the predatory mite *Typhlodromus pyri* is uncommon, and so does not contribute to the biocontrol of this pest. Materials available for use against pear rust mite are the OPs pirimiphos-methyl and the insect growth regulator diflubenzuron. The appropriate time for treatment is petal fall. When used against powdery mildew or scab, sulfur may have a suppressant effect on pear rust mite. See the Introduction, p. 258, and under Pear sucker, p. 279, for information on the impact of pesticides on beneficials and psyllid (pear sucker) management in pear orchards.

Pear sawfly (*Hoplocampa brevis*)
This is a local pest, similar to the apple sawfly in habits and damage caused (p. 262). If treatment is required, the materials available are the OPs chlorpyrifos and dimethoate, and the alkaloid nicotine. See the Introduction, p. 258, and

under Pear sucker, p. 279, for information on the impact of pesticides on beneficials and psyllid (pear sucker) management in pear orchards.

Pear slug sawfly (*Caliroa cerasi*)
The adults appear in May and June; eggs are laid in slits cut in the leaf. The larvae are at first yellowish-white but soon appear dark greenish or black, as they become coated in slime, and are then slug-like in appearance. They feed on the upper leaf surfaces and leaves may become skeletonized. There are two to three generations per year. If treatment is necessary, the materials available are those listed above for pear sawfly, and also the natural product rotenone. See the Introduction, p. 258, and under Pear sucker, below, for information on the impact of pesticides on beneficials and psyllid (pear sucker) management in pear orchards.

Pear sucker (*Psylla pyricola*)
Adults hibernate on the bark, or among dead leaves and other shelter in pear orchards, and in nearby hedgerows, woodland and other orchards. Eggs are laid on the shoots and spurs from March to petal fall, and the nymphs feed on the buds and blossom trusses; later, they feed on the leaves. Two further generations follow, in which eggs are laid on the leaves. Nymphs feeding on the leaves produce copious quantities of watery honeydew. If the insects are numerous this honeydew trickles over the leaf surface on to the fruit, and sooty moulds grow in the residue.

The predatory bug *Anthocoris nemoralis* is an effective natural enemy of pear sucker. The most sustainable approach to the management of the pest is to optimize the biocontrol potential of anthocorids by making decisions on pesticide use such that as little damage as possible is done to populations of these important predators. Earwigs also probably contribute to the biocontrol of pear sucker in most orchards and, in particular, they are capable of consuming large numbers of pear sucker eggs. Anthocorids do not provide regular and reliable control of the pest, however, and pesticides should be used if infestations become severe. A possible treatment threshold from petal fall onwards is 30% of leaves infested with pear sucker nymphs.

Materials available for use against pear sucker are the OPs chlorpyrifos, dimethoate and malathion, the pyrethroids cypermethrin, deltamethrin and lambda-cyhalothrin, and from other groups amitraz and the insect growth regulator diflubenzuron. There is evidence from trials and grower experience that the insect growth regulator fenoxycarb, if applied against summer fruit tortrix moth at petal fall, also reduces numbers of nymphs of pear sucker. Some fungicides, particularly mancozeb, when used against scab on pear, may also reduce pear sucker numbers.

There are some important points to consider when choosing an insecticide for pear sucker control. Resistance to OPs has been widespread in pear sucker populations since the late 1970s, so these materials are unlikely to be effective against most populations of the pest; additionally, these compounds are damaging to anthocorids. Pyrethroids are generally very effective against adults and nymphs of pear sucker, although resistance appears to be developing in some populations; pyrethroids are very damaging to anthocorids. Anthocorids generally colonize pear orchards in spring, and relatively few of them appear to overwinter in the orchard. If overwintering populations of pear sucker are high, it is possible to apply a pyrethroid in late February or March, just before the beginning of egg laying, without jeopardizing the biocontrol potential of summer populations of anthocorids. Amitraz and diflubenzuron are effective against nymphs of pear sucker and non-damaging to anthocorids; thus, they can be used against the pest in summer without disrupting the biocontrol potential of the predators. Fenoxycarb also affects only the nymph stage of pear sucker, and is non-damaging to anthocorids. Because these materials are not effective against adult pear suckers, for maximum effect they should be applied when most of the population is in the nymphal stage. This can be determined by monitoring; a forecasting model is also available (from HortiTech, HRI-Wellesbourne) for predicting this timing. See the Introduction, p. 258, for information on the impact of pesticides on beneficials.

Scale insects

See under Apple, p. 267.

Tortrix moths

In addition to codling moth (*Cydia pomonella*) (see p. 264), several other species of tortrix moth occur on pear, including fruit tree tortrix moth (*Archips podana*) and summer fruit tortrix moth (*Adoxophyes orana*). See under Apple, p. 268, for details of the life history and treatment strategies. Fruit tree tortrix moth is rarely important but the summer fruit tortrix moth may be damaging in some orchards in south-eastern England. Insecticides available for use against tortrix moth caterpillars on pear are the OP chlorpyrifos, the pyrethroid cypermethrin and the bacterial insecticide *Bacillus thuringiensis*. The insect growth regulator diflubenzuron is available against fruit tree tortrix moth. The insect growth regulator fenoxycarb is available for use against summer fruit tortrix moth, and is also likely to be effective against fruit tree tortrix. See the Introduction, p. 258, and under Pear sucker, p. 279, for information on the impact of pesticides on beneficials and psyllid (pear sucker) management in pear orchards.

Winter moth (*Operophtera brumata*)

See under Apple, p. 269, for details of the life history and other information. Insecticides available for spring treatment of winter moth (and related species) on pear are the OP chlorpyrifos, the pyrethroid cypermethrin, the insect growth regulator diflubenzuron and the bacterial insecticide *Bacillus thuringiensis*. Tar oil can be used as a dormant-season treatment. See the Introduction, p. 258, and under Pear sucker, p. 279, for information on the impact of pesticides on beneficials and psyllid (pear sucker) management in pear orchards.

Diseases

Routine annual sprays are necessary for scab, and a post-harvest fungicide treatment is advised for stored fruit.

Botrytis fruit rot (*Botryotinia fuckeliana* – anamorph: *Botrytis cinerea*)
This disease is the main cause of losses in stored pears. The fungus invades the fruit through small wounds, and particularly through stalks damaged at harvest. The disease develops during storage and can spread to adjacent fruits, resulting in severe losses.

A substantial reduction in occurrence of the disease can be obtained by extreme care in removal of fruit from the tree and subsequent handling operations prior to storage. A post-harvest dip or drench treatment with iprodione (off-label) (SOLA 0693/98) gives excellent control of both initial infection and subsequent spread. Carbendazim (in mixture with metalaxyl) may also be used, but strains of *Botrytis* resistant to carbendazim are relatively common. Although, occasionally, strains resistant to iprodione may also be found, this fungicide gives good control of the fruit rot.

Brown rot (*Sclerotinia fructigena*)
Brown rot can cause losses in fruit in the orchard and during storage, particularly during hot summers. For biology and control, see under Apple, p. 271.

Canker (*Nectria galligena*)
Common canker of pear is caused by the same fungus as that causing apple canker. Biology and control measures are the same as for apple, p. 271.

Fireblight (*Erwinia amylovora*)
This disease has spread rapidly since first having been reported in the UK on pear in 1957. In addition to pear, a number of apple cultivars and also other members of the sub-family Pomoideae are also attacked; these include hawthorn (*Crataegus*), *Cotoneaster*, firethorn (*Pyracantha*), rowan (*Sorbus aucuparia*), *Stranvaesia* and whitebeam (*Sorbus* spp.). In pears, infection by the causal bacterium generally occurs through late summer (secondary) blossom, killing this and progressing via the stalk into the twig, branch and, finally, the trunk. Secondary blossom is usually produced in warm conditions, and is particularly prone to attack. For this reason, cv. Laxton's Superb has now largely been grubbed or top worked. During the late 1970s/early 1980s, fireblight became endemic in south-west England, and much of the area of perry pear was either killed or had to be grubbed because the orchards were no longer economically viable. Infection normally occurs at flowering time and, under ideal conditions for the pathogen,

the tree can be killed within 12 months. The most susceptible cultivars of pear appear to be Barnett, Blakeney Red, Judge Amphlett and Moorecroft.

In summer and autumn, parts of the bark containing active bacteria usually show a red discoloration on cutting; however, during winter months the discoloration may be dark brown. In a mild winter the bacteria may continue to advance along the affected part but in certain circumstances a limited canker is formed and this may crack at the margin, thus becoming isolated from healthy tissue adjacent to it. The bacteria in some of these 'holdover cankers' may survive until the following spring when they can initiate new outbreaks. Although infection is usually through the blossom, shoots and leaves may also be attacked.

The grubbing of badly infected trees may be required, but where cutting-out of branches is permissible, this should be done at not less than 60 cm below visible signs of the disease within the bark. To reduce the chance of re-infection, the cut surface should be painted immediately with a wound-sealing material. The cutting parts of all pruning tools should be immersed in a strong disinfectant between each cut on both affected and healthy trees.

Powdery mildew (*Podosphaera leucotricha*)
Infection occurs sporadically and can be severe on cv. Comice. Symptoms appear as a russeting of the fruit. Bupirimate, myclobutanil, pyrifenox and sulfur are recommended for the control of pear mildew. Partial control is given by fenbuconazole when applied for scab control.

Scab (*Venturia pirina*)
Pear scab is caused by a different fungus from that causing apple scab. Nevertheless, the life histories of the two fungi are almost identical and the reader is referred to the description of apple scab, p. 274.

Protectant sprays of captan, carbendazim (off-label) (SOLA 2256/99), dithianon, dodine, fenbuconazole, mancozeb, myclobutanil, pyrifenox, thiophanate-methyl or thiram are effective to varying degrees. Sprays of copper or sulfur cause damage on some cultivars (e.g. Doyenné du Comice) and dodine may cause fruit russeting under some conditions.

Cherry

Pests

Cherry blackfly is the major pest of cherries, and most growers use a routine insecticide application against it.

Apple leaf miner (*Lyonetia clerkella*)
See under Apple, p. 262.

Cherry-bark tortrix moth (*Enarmonia formosana*)
The pinkish-white caterpillars tunnel under the bark of the trunk, often just below the crotch; successive generations use the same galleries, which may become extensive. Heavy infestations sometimes occur in old orchards, and there is some evidence that extensive bark injury can kill large branches or even whole trees. Apple, cherry, pear and plum may be attacked. The moths appear from mid-May until early September.

There is no specific approval for this pest on cherry. However, tar oil (when used against aphids, scale insects or winter moth, while the trees are still dormant) may be effective, particularly if loose bark is removed first. See the Introduction, p. 258, for information on the impact of pesticides on beneficials.

Cherry blackfly (*Myzus cerasi*)
This is the only aphid species found on cherry. It overwinters as eggs on the bark and these hatch in March and April, hatching being complete by the white-bud stage. Successive generations of black aphids are produced, then winged forms which disperse in June and July to bedstraws (*Galium* spp.). Infested leaves are severely curled and new growth is checked.

Materials available for use in spring against this pest are the OPs dimethoate and malathion, the pyrethroid cypermethrin, the carbamate pirimicarb, the alkaloid nicotine, the natural insecticide rotenone, and a soap concentrate containing fatty acids. Tar oil may be used as a dormant season winter wash. See the Introduction, p. 258, for information on the impact of pesticides on beneficials.

Cherry fruit moth (*Argyresthia pruniella*)
Moths appear in late June and July and lay eggs under the bud scales, in crevices in the bark etc., especially towards the tips of branches. Most eggs hatch in the autumn, the caterpillars hibernating in silk cocoons in bark crevices, but some do not hatch until the spring. The caterpillars bore into flower buds in the spring, feed on the flowers and, later, attack the young fruitlets. When fully fed they drop to the ground and pupate.

The pyrethroid cypermethrin and the bacterial insecticide *Bacillus thuringiensis* are available for use against caterpillars on cherry. The timing for application is late March, when buds are breaking, and again at the white-bud stage if infestations are heavy. A winter wash of tar oil, against scale insects and winter moths, may kill overwintering eggs of cherry fruit moth, but they will not reach hibernating caterpillars. See the Introduction, p. 258, for information on the impact of pesticides on beneficials.

Fruit tree red spider mite (*Panonychus ulmi*)
See under Apple, p. 266, for details of the life history. This species is rarely a problem on cherry. Where treatment is required, clofentezine is available for use between early white-bud and the onset of flowering, and the OPs dimethoate and malathion (but note that most populations of fruit tree red spider mite are

resistant to OPs, so these are unlikely to be effective), the compound tetradifon and a soap concentrate containing fatty acids, timed as for apple. See the Introduction, p. 258, for information on the impact of pesticides on beneficials.

Pear slug sawfly (*Caliroa cerasi*)
See under Pear, p. 279, for details of the life history. The only material with specific approval against this pest on cherry is the alkaloid nicotine, but when used against other pests, the OP dimethoate, and the natural product rotenone, may provide some control.

Winter moth (*Operophtera brumata*)
See under Apple, p. 269, for details of the life history and other information. If treatment is required against winter moth (or related species), the materials available are the pyrethroid cypermethrin and the bacterial insecticide *Bacillus thuringiensis*, for use at the white-bud stage, and tar oil as a dormant-season wash. See the Introduction, p. 258, for information on the impact of pesticides on beneficials.

Diseases

Bacterial canker (*Pseudomonas syringae* pv. *mors-prunorum*)
This disease occurs in all cherry orchards and is frequently severe, causing the death of branches and whole trees, and annual treatment is necessary. There is a well-defined seasonal cycle to the disease, with a winter canker phase alternating with a summer leaf-spot phase on the foliage. During autumn the bacteria enter bark wounds and fresh leaf scars to form small cankers but making little progress until spring when they extend rapidly, killing large areas of green bark. Soon after petal fall the growth of the cankers ceases and the bacteria die in the diseased tissues. During the summer, wood infection does not occur but the bacterium invades the foliage, causing leaf spotting and providing a plentiful supply of new bacteria to repeat the cycle in the autumn. Cultivars differ in their susceptibility but none is immune. Infected trees should be pruned or cut back between May and the end of August.

In order to reduce the numbers of bacteria during the most vulnerable period, a drenching spray of Bordeaux mixture should be applied three times at intervals of 3–4 weeks from mid-August. An extra application at petal fall may be advisable on young trees of susceptible cultivars, in order to build a canker-free framework. The petal-fall application may cause damage. The addition of cotton-seed oil to the first two sprays is recommended. Alternatively, copper oxychloride could be used.

Blossom wilt (*Sclerotinia laxa*)
See under Plum and damson, p. 288.

Specific cherry replant disease

When cherries are replanted on land that has previously grown cherries or plums, growth and cropping are frequently poor. This replant disease has been shown to be associated with colonization of the roots of the replanted trees with the fungus *Thielaviopsis basicola*. Affected trees are stunted as a result of a much reduced root system.

Soil fumigation with chloropicrin (see comments on p. 276), as for specific apple replant disease, will control cherry replant disease.

Plum and damson

Pests

Three species of aphid attack plum, of which damson/hop aphid is particularly difficult to control chemically because of its facility for developing resistance to insecticides. Because of the off-label approval of pirimicarb, an IPM approach to aphids is now possible. Fruit tree red spider mite and plum rust mite are usually controlled by naturally occurring predatory mites, provided that pesticides applied against other pests are not damaging to these predators.

Aphids

The three common species are damson/hop aphid (*Phorodon humuli*), leaf-curling plum aphid (*Brachycaudus helichrysi*) and mealy plum aphid (*Hyalopterus pruni*). Damson/hop aphid and leaf-curling plum aphid are both vectors of plum pox virus (Sharka disease), see p. 288. All species overwinter as eggs laid in autumn on the twigs. Eggs of leaf-curling plum aphid hatch early, and aphids feed on the dormant buds; in the spring, successive generations feed on the foliage and produce severe leaf curl; from May to July, winged forms disperse to various summer host plants. Eggs of mealy plum aphid hatch later, but have done so by the white-bud stage; from the end of June, winged forms disperse to grasses or reeds and to other plum trees. Eggs of damson/hop aphid hatch in early spring; winged forms disperse to hops from mid-May until late July.

Materials available for spring use on plum are the OPs chlorpyrifos, dimethoate and malathion, the pyrethroids cypermethrin and deltamethrin, the alkaloid nicotine, the natural insecticide rotenone, and a soap concentrate containing fatty acids. Additionally, the carbamate pirimicarb has off-label approval for this use (SOLA 2178/96). Tar oil is available as a winter wash, applied when buds are dormant (in December to early January for early cultivars, and to the end of January for maincrop cultivars). The buds of some plum cultivars are very susceptible to tar oil damage. Tar oil should not be used on myrobalan, and not after about mid-January on cvs Belle de Louvain, Victoria and Yellow Egg or on gages. Infestations of mealy plum aphid are often overlooked and, as they can persist until August, may require additional sprays in May or June if not effectively controlled in spring, or if re-infestation occurs in summer. Resistance to OPs is widespread in populations of damson/hop aphid in hop-growing areas, so these compounds are unlikely to be effective against this pest. See the Introduction, p. 258, and under Apple, fruit tree red spider mite, p. 266, for information on the impact of pesticides on beneficials and integrated mite management.

Many natural enemies of aphids occur on plum, particularly predatory anthocorids (Anthocoridae) and mirids (Miridae), and they are potentially effective biocontrol agents against damson/hop aphid and mealy plum aphid, but not plum leaf-curling aphid, because this species causes rapid leaf damage early in the spring before the predators become active. The use against leaf-curling plum aphid of the carbamate pirimicarb, which is non-damaging to these predators, is likely to optimize the biocontrol potential of these predators against damson/hop aphid and mealy plum aphid.

Fruit tree red spider mite (*Panonychus ulmi*)

See under Apple, p. 266, for details of the life history. Suitable treatment thresholds are as for apple. Materials available for use against this pest on plum are the OPs chlorpyrifos, dimethoate and malathion (but note that most populations of fruit tree red spider mite are resistant to OPs, so these materials are unlikely to be effective), and from other chemical groups clofentezine, a soap concentrate containing fatty acids and tetradifon. The timings for these compounds are as detailed under apple. The predatory mite *Typhlodromus pyri* occurs on plum trees, and provided that pesticides applied do not damage it, the predator contributes to the control of fruit tree red spider mite. See the Introduction, p. 258, and under Apple, p. 266, for information on the impact of pesticides on beneficials and integrated mite management.

Plum fruit moth (*Cydia funebrana*)

Moths are on the wing at about the same time as codling moth. Eggs are laid from mid-June to August, and the whitish caterpillar bores into the fruit towards the stone; in its final growth stage it is reddish in colour (hence the name 'red plum maggot'). When fully fed, in late August and September, it leaves the fruit and builds a cocoon in which to overwinter in bark crevices, etc. Although usually single-brooded, in favourable seasons there may be a partial second generation.

Pheromone traps are available for the plum fruit moth, providing a rational basis for decisions on the necessity for, and timing of, pesticide applications. The usual treatment threshold is five moths per trap per week. The materials available for this pest are the pyrethroid deltamethrin, and the insect growth regulator diflubenzuron. Deltamethrin should be applied in late June and again if necessary 2–3 weeks later. The timing for diflubenzuron is a week earlier. Annual sprays should not be necessary, once good control in the orchard is achieved. See the Introduction, p. 258, and under Apple, fruit tree red spider mite, p. 266, for information on the impact of pesticides on beneficials and integrated mite management.

Plum rust mite (*Aculus fockeui*)
This small, brownish mite has a life-cycle similar to that of apple rust mite. The feeding damage causes leaves to become speckled, with a browning of the lower surface and development of a silvery appearance on the upper surface. With severe infestations, shoot growth may become stunted and terminal buds killed. Infestations as severe as this are seen most frequently on nursery trees but they do occur, occasionally, on mature trees. The treatment thresholds suggested for apple rust mite, p. 262, are appropriate also for plum rust mite. The insect growth regulator diflubenzuron is the only material specifically approved for use against this pest on plum. The predatory mite *Typhlodromus pyri* occurs on plum trees and, provided that the pesticides applied do not damage it, usually prevents outbreaks of plum rust mite.

Plum sawfly (*Hoplocampa flava*)
Adults appear at blossom time, when the female lays eggs in the flowers. The creamy-white larvae bore into the fruitlets; one larva may attack as many as four fruitlets before it is fully fed and finally drops to the ground to build a cocoon in the soil where it will overwinter. Marked preference for cultivars is shown, Czar being particularly susceptible. Materials available for plum sawfly are the OP dimethoate, the pyrethroid deltamethrin, and the alkaloid nicotine. The timing of application is the cot-split stage (7–10 days after petal fall). See the Introduction, p. 258, and under Apple, fruit tree red spider mite, p. 266, for information on the impact of pesticides on beneficials and integrated mite management.

Scale insects
Brown scale (*Parthenolecanium corni*), mussel scale (*Lepidosaphes ulmi*), nut scale (*Eulecanium tiliae*) and oystershell scale (*Quadraspidiotus ostreaeformis*) may occur on plum. For further details see under Apple, p. 267, and black currant, p. 291. Materials available to control scale insects on plum are a soap concentrate containing fatty acids (for use in summer) and tar oil (for dormant-season use). The buds of some plum cultivars are very susceptible to tar oil damage. Tar oil should not be used on myrobalan, and not after about mid-January on cvs Belle de Louvain, Victoria and Yellow Egg or on gages. See the Introduction, p. 250, and under Apple, fruit tree red spider mite, p. 266, for information on the impact of pesticides on beneficials and integrated mite management.

Tortrix moths
Caterpillars of plum tortrix moth (*Hedya pruniana*) feed on foliage and tunnel in the shoots from April to June. Several of the tortrix moth caterpillars found on apple also occur on plum (see p. 268). Materials for use on plum are the OP chlorpyrifos, the insect growth regulator diflubenzuron, and the bacterial insecticide *Bacillus thuringiensis*. The caterpillars are also kept in check by spring sprays applied against winter moth caterpillars. See the Introduction, p. 258, and

under Apple, fruit tree red spider mite, p. 266, for information on the impact of pesticides on beneficials and integrated mite management.

Winter moth (*Operophtera brumata*)
See under Apple, p. 269, for details of the life history and other information. Materials available for control of winter moth (and related species) on plum at the white-bud stage are the OPs chlorpyrifos, the pyrethroid cypermethrin, the insect growth regulator diflubenzuron and the bacterial insecticide *Bacillus thuringiensis*. Tar oil can be used in the dormant season. The buds of some plum cultivars are very susceptible to tar oil damage. Tar oil should not be used on myrobalan, and not after about mid-January on cvs Belle de Louvain, Victoria or Yellow Egg or on gages. See the Introduction, p. 258, and under Apple, fruit tree red spider mite, p. 266, for information on the impact of pesticides on beneficials and integrated mite management.

Diseases

Blossom wilt and brown rot (*Sclerotinia fructigena* and *S. laxa*)
Both of these species of *Sclerotinia* cause brown rot of plum fruits; the latter species also causes blossom wilt, sometimes followed by spur blight and canker and wither tip of shoots. Cherry is also affected by blossom wilt. Both species overwinter in mummified fruits and cankers; these produce spores in spring, so continuing the cycle of infection. *S. laxa* can enter shoots and spurs via damaged leaves. Plums are not attacked by the same race or strain of *S. laxa* that attacks apples.

Diseased trusses, shoots and cankered spurs and branches should be cut out and burnt, preferably in the spring or summer when their presence can more easily be recognized. Mummified fruits should also be collected and burnt. Spraying with tar oil, as for aphids, late in the following dormant period gives partial control by destroying the cushions of spores that appear on any diseased parts overlooked during the earlier cutting out. Spraying at early flowering with myclobutanil (off-label) (SOLA 1825/98) will reduce infection levels of blossom wilt; myclobutanil (off-label) (SOLA 1535/99) is also available for use on cherry. Brown rot can be controlled partially by spraying with carbendazim at the first sign of fruit colouring.

Plum pox virus (Sharka disease)
In addition to plum, this virus disease affects apricot, damson, greengage and peach, and a range of ornamental and hedging trees and shrubs; these include blackthorn (*Prunus spinosa*) and the ornamental flowering and purple-leaved species of *Prunus*. Affected trees often show indistinct pale spots or blotches on the leaves but affected fruit may be severely damaged, with uneven ripening and the appearance of dark-coloured rings, lines or bands in the flesh. The cv.

Victoria is particularly prone to fruit symptoms. The disease is normally introduced into an orchard by infected planting stock, and the subsequent spread by aphid vectors is fairly slow. There is no cure for this disease and control is based on the use of certified planting stock, eradication of sources of infection and control of the aphid vectors (see p. 285).

Rust (*Tranzschelia pruni-spinosae* var. *discolor*)
This disease occurs in all the main plum-growing areas, and appears as orange-yellow pustules on the leaves. Severe infection can result in premature defoliation and, rarely, disfigurement of fruit by the presence of orange-brown spores on the surface. The cv. Victoria is particularly prone to infection and the disease is severe in cool, moist summers.

The full economic effects of the disease are not known, but the early leaf fall will lead to general tree debilitation. Myclobutanil (off-label) (SOLA 1536/99), applied at the first sign of infection or as a routine treatment in mid-August, will give some control.

Silver leaf (*Chondrostereum purpureum*)
Silver leaf is widespread and causes severe losses in plums, and also in cherries and some other fruit crops. The fungus attacks the wood of the tree after gaining entry through pruning wounds, cracked branches and damaged trunks. Silvering of the foliage occurs as a result of a toxin produced by the fungus in the wood and carried to the leaves in the sap stream. On wood killed by the fungus, bracket-shaped fruit bodies are formed; in wet weather, these produce spores that are capable of establishing new infections through wounded tissue. The cv. Victoria is very susceptible to infection. Control is based on prevention of infection and removal of sources of disease. It is important to remove dead or dying branches and to prevent injury to the bark, particularly during autumn and winter when the fungus is most active. At this time all wounds and pruning cut surfaces should be covered immediately with a paint containing octhilinone.

Currant and gooseberry

Pests

Aphids and black currant gall mite are the major pests on currant and gooseberry, with pesticide applications being required in some years against other pests such as sawflies and capsids.

Aphids
Several species of aphid overwinter as eggs on currant or gooseberry. The eggs hatch in spring; aphids then feed and reproduce, and in the summer disperse to various host plants, returning to currant and gooseberry in the autumn. Currant/

sowthistle aphid (*Hyperomyzus lactucae*) causes the leaves of red currant and black currant to curl downwards and also stunts young growth; it disperses to sow-thistles (*Sonchus* spp.) in summer. Lettuce aphid (*Nasonovia ribisnigri*), a darker-green species, normally infests gooseberry and disperses to lettuce. Red currant blister aphid (*Cryptomyzus ribis*) is a pale, yellowish-green, delicate-looking species; it causes red leaf blisters on red currant and white currant, and yellowish-green ones on black currant. It disperses to hedge woundwort (*Stachys sylvatica*). Gooseberry aphid (*Aphis grossulariae*) is dark green in colour, and causes severe curling and distortion of young leaves of both currant and gooseberry; some aphids are present on the bushes all summer. The permanent currant aphid (*Aphis schneideri*) is blue-green in colour, and causes similar damage on red currant and black currant. Other species occur occasionally, but cause little or no damage.

Materials available for spring use against aphids on currant and gooseberry are the OPs chlorpyrifos, dimethoate and malathion, and the carbamate pirimicarb. Tar oil can be used as a winter wash in December and January, but should not be used later. The red currant cv. Raby Castle is prone to tar-oil injury and should not be treated. If the OC endosulfan is used just before flowering against black currant gall mite, this will also give some control of aphids. See the Introduction, p. 258, for information on the impact of pesticides on beneficials.

Black currant gall mite (*Cecidophyopsis ribis*)

This small, whitish mite feeds and multiplies inside the buds of black currant, which become swollen 'big buds' by the autumn. In the following spring many fail to produce flower trusses or may fail to open. It is also the vector of the virus causing reversion disease; this is often the limiting factor in the life of a plantation. Mites disperse from the swollen buds from March onwards, the main emergence period being from early April to the end of June, with a peak usually in May. Emergence is accelerated by rising temperatures; mites may be dispersed to fresh sites by air currents or transported by rain or insects.

The difficulty in chemical control is to protect the young buds against mite infestation for a sufficiently long period, without leaving residues on the fruit. Materials available are the OC endosulfan, and the inorganic material sulfur. The pyrethroid fenpropathrin is specifically approved for use against caterpillars and two-spotted spider mite in black currant, and this has also been used effectively against black currant gall mite. The usual practice is to apply a spray of endosulfan or fenpropathrin at the beginning of flowering, a second at the end of flowering, and a third at first fruit set, usually about 14 days after the second application. In addition, some growers apply sulfur shortly after bud-burst. On nursery stock and non-fruiting bushes, endosulfan should be applied when growth starts, followed by up to four further applications at intervals of 10–14 days (fenpropathrin is also effective; see above). See the Introduction, p. 258, for information on the impact of pesticides on beneficials.

Black currant leaf midge (*Dasineura tetensi*)
There are three, sometimes four, generations a year; the first appears in April to June, the second in late June and July, and the third in late July and August, but actual timing is variable. Eggs are laid between folds of young leaves, and feeding by the white or orange-coloured maggots causes the leaves to become tightly twisted and folded. Shoot growth may be checked and lateral branches may develop, but more important is the masking of symptoms of reversion in nursery stock. Pupation occurs in the soil. Endosulfan sprays against gall mite, or dimethoate against aphids, may give some control of leaf midge, and fenpropathrin used against black currant gall mite gives effective incidental control. See the Introduction, p. 258, for information on the impact of pesticides on beneficials.

Black currant sawfly (*Nematus olfaciens*)
There are two or more overlapping generations a year. Adults emerge from overwintering cocoons in the soil in May and June, and those of the later broods from mid-June to mid-September. Eggs are laid on the underside of leaves, especially those near the middle of the bushes. The green, black-spotted caterpillars at first feed gregariously, but later spread through the bushes and may occasionally cause considerable defoliation. Bushes should be examined in May and early June; if caterpillars are present, a pesticide should be applied in early June. The only material approved against this pest is the alkaloid nicotine, but black currant sawfly is often effectively controlled by OP or pyrethroid insecticides applied against other pests. See the Introduction, p. 258, for information on the impact of pesticides on beneficials.

Brown scale (*Parthenolecanium corni*)
The full-grown female scale of this common pest is about 3–6 mm long, hemispherical, and chestnut brown in colour. Eggs are laid beneath the scales in summer, after which the female dies. Nymphs emerge in the autumn and overwinter on the branches, often under loose bark. After a short period of activity in the spring they settle, becoming adult in June. If infestation is severe, tar oil can be applied against this pest as a winter wash, when the buds are dormant in December and January but not later. The red currant cv. Raby Castle is prone to tar-oil injury and should not be treated with a winter wash. See the Introduction, p. 258, for information on the impact of pesticides on beneficials.

Common earwig (*Forficula auricularia*)
These insects sometimes roost in black currant bushes and contaminate fruit when mechanical harvesting techniques dislodge them into the collecting trays. If treatment is required, diflubenzuron (applied in May or June) may provide a reduction in numbers of earwigs in the plantation.

Common green capsid (*Lygocoris pabulinus*)
See under Apple, p. 263, for details of the life history. Materials available for use

against this pest on currants and gooseberry are the OP chlorpyrifos and the alkaloid nicotine. Also available for use on gooseberry are the OP dimethoate and the pyrethroid lambda-cyhalothrin (off-label) (SOLA 2340/97). The timing for treatment on currant is when damage is first seen (usually when the first flowers are about to open) and again, if necessary, 3 weeks later. The timing for treatment on gooseberry is at the end of flowering. It is probable that incidental control of capsids is provided by OPs and pyrethroids used against other pests on black currant and gooseberry, and endosulfan when used against black currant gall mite. See the Introduction, p. 258, for information on the impact of pesticides on beneficials.

Gooseberry bryobia (*Bryobia ribis*)

At one time this mite was a serious pest but it is now uncommon. It overwinters as eggs under loose bark; hatching begins near the beginning of March and continues well into April. There is just one generation per year, the laying of winter eggs beginning in May. The mites feed on the leaves in warm, sunny conditions, retiring to the wood and under bud scales in cold weather or when moulting. On severely infested bushes the foliage becomes yellowed, and leaves may wither and drop. The pyrethroid lambda-cyhalothrin is available off-label (SOLA 2340/97) for use against this pest. OPs applied against aphids probably provide incidental control. See the Introduction, p. 258, for information on the impact of pesticides on beneficials.

Gooseberry sawfly (*Nematus ribesii*)

The adult sawflies first appear in April and May. Eggs are laid on the undersides of the leaves, especially near the centre of the bushes. The black-spotted, green caterpillars feed together for a few days, and later spread through the bush and may cause defoliation. When fully grown, they moult to an active, non-feeding, prepupal stage, which is light green with an orange patch behind the head and another near the tail and lacks the black spots. There are three overlapping generations in the year. The need for treatment can be assessed by examining bushes in May and early June; if appreciable numbers of caterpillars are seen, then apply a treatment. Materials available for use against this pest are the OP malathion, the alkaloid nicotine, and the natural insecticide rotenone. Other OPs, and pyrethroids, may provide incidental control of gooseberry sawfly when used against other pests. See the Introduction, p. 258, for information on the impact of pesticides on beneficials.

Snails

Several species (*Cepaea hortensis*, *C. nemoralis*, *Helix aspersa* and *Hygromia striolata*) may crawl from ground vegetation into trays of picked fruit. Also, individuals 'roosting' in the bushes may fall into picking containers. An important source of contamination is the indiscriminate placing of fruit trays on the ground, which allows snails to move into them for shelter; this can be avoided by placing the trays on a hard standing or on plastic sheeting to separate them from the soil or vegetation. There is also an increased risk of contamination by snails when harvesters are used in wet conditions or early in the morning. Control of weeds, particularly on headlands, leads to a depletion of snail populations in black currant plantations, but this may also remove the shelter that enhances populations of predators of vine weevil. Metaldehyde and the carbamate methiocarb are available for use against snails. See the Introduction, p. 258, for information on the impact of pesticides on beneficials.

Two-spotted spider mite (*Tetranychus urticae*)
This is the same species that is common on many greenhouse and outdoor plants, including black currant and strawberry. See under Strawberry, p. 305, for details of the life history. Materials available for use against this pest on black currant and gooseberry are the OPs chlorpyrifos, dimethoate and malathion (but note that resistance to OPs is widespread in two-spotted spider mite populations, so these compounds are unlikely to be effective), and tetradifon. Additionally the pyrethroids bifenthrin and fenpropathrin, and the acaricides clofentezine (off-label) (SOLA 1250/95) and dicofol + tetradifon can be used on black currant, and the pyrethroid lambda-cyhalothrin (off-label) (SOLA 2340/97) on gooseberry.

Vine weevil (*Otiorhynchus sulcatus*)
See under Strawberry, wingless weevils, p. 306, for details of the life history and natural enemies/biocontrol.

Winter moth (*Operophtera brumata*)
See under Apple, p. 269, for details of the life history and other information. Treatment against winter moth (and related species) is rarely necessary on currant or gooseberry but, if required, the insect growth regulator diflubenzuron is available for use on black currant, applied when the first flowers are about to open and repeated if necessary. Tar oil can be used when the buds are dormant in currants and gooseberry. The red currant cv. Raby Castle is prone to tar-oil injury and should not be treated with a winter wash.

Diseases

Annual treatment is necessary for control of leaf spot and American gooseberry mildew and, in some localities, *Botryotinia* (*Botrytis*).

American gooseberry mildew (*Sphaerotheca mors-uvae*)
This disease is prevalent on both black currant and gooseberry, and can result in premature defoliation and disfigurement of fruit. The white, powdery (sporing) fungal growth occurs on young leaves, fruits and shoots, and is favoured by warm weather and soft, luxurious growth. In late summer and autumn the fungal

growth becomes a brown, felt-like layer that contains black spore cases (perithecia) which are involved in the overwintering of the disease. Most of the commonly grown cultivars of black currant and gooseberry are susceptible.

Various fungicides are available for control of the disease, and some of the chemicals also control leaf spot and *Botrytis*. Bupirimate, fenarimol, fenpropimorph (off-label) (SOLA 0787/95), myclobutanil, penconazole, pyrifenox, sulfur and triadimefon (off-label) (SOLA 0024/95) are recommended. Sprays for mildew control should commence at or before first open flower on gooseberry and at the grape stage on black currant. Sprays of sulfur may cause damage to some cultivars of black currant and gooseberry. It is important to consult the chemical manufacturers' labels for precise spray timings, harvesting intervals and information on acceptability of sprays on fruit for processing. Additional post-harvest spray applications may be necessary if the disease is severe. Infected shoots should be pruned out after the wood is ripe.

Black currant rust (*Cronartium ribicola*)

This rust spends part of its life-cycle on black currant and part on five-needled pines, particularly Weymouth pine (*Pinus strobus*). On black currant, the disease is rarely seen in well-sprayed plantations, and control measures are needed only if the disease becomes prevalent. Spores (aeciospores) from pine trees infect nearby currant bushes, the fungus then appearing in early summer as yellow outgrowths on the underside of the leaves. From these outgrowths, the spores (uredospores) are produced and these spread the disease within the currant plantation. Later, yet other kinds of spores (teleutospores and basidiospores) are produced on the leaves, and these cause re-infection of pine trees. Where rust is troublesome, the bushes should be sprayed with copper oxychloride or with thiram.

Botrytis grey mould (*Botryotinia fuckeliana* – anamorph: *Botrytis cinerea*)

This disease is widespread, and can cause severe losses of fruit on black currant and gooseberry. Fruit damage is particularly severe when infection is associated with low-temperature injury in the spring. The fungus also attacks shoots, leading to die-back, and is favoured by wet weather. Spores are produced throughout the year and these can infect plant parts via damaged or senescent tissue.

Control can be obtained by applying sprays of chlorothalonil, dichlofluanid, fenhexamid or pyrimethanil (off-label) (SOLA 1939/99). Sprays for botrytis control should be applied at the late-grape stage on black currant and at first early flower on gooseberry. Treatment should be repeated according to the manufacturer's label recommendation.

European gooseberry mildew (*Microsphaera grossulariae*)

This disease is far less serious than American gooseberry mildew and rarely occurs in commercial plantations. It is seen as a delicate sporing mould, mainly on the upper side of the leaf; it rarely occurs on the berries. Overwintering spore cases fall to the ground with leaves, and these restart the cycle of infection by ejecting spores (ascospores) in the following spring. If required, spray with one of the fungicides listed for American gooseberry mildew (p. 294).

Leaf spot (*Drapenopeziza ribis*)
Leaf spot is widespread, and can cause severe defoliation and crop loss of black currant, particularly during wet summers. Leaf spot is generally not so severe on gooseberry. The fungus overwinters on dead leaves, and spores (ascospores) from these start the infection in the spring. Symptoms are visible on both black currant and gooseberry from about May, and appear as brown spots or patches on the leaves. On these spots, spores (conidia) are formed, which spread infection through the plantation. The leaf spots gradually coalesce, until a large part of the leaf area is affected and the leaves take on a scorched appearance and fall. Premature defoliation results in weakened growth and yield reduction the following season.

On black currant, routine sprays should commence at the early-grape stage, using Bordeaux mixture, chlorothalonil, copper ammonium carbonate, dodine, mancozeb, pyrifenox or zineb. Treatments should be repeated according to the recommendation on the manufacturer's label, on which details of harvest interval and acceptability of spray materials on fruit for processing are also given.

Blackberry, loganberry and raspberry

Pests

Raspberry beetle is the major pest of cane fruit grown in the open. During the past ten years, the production season has been extended by growing plants in heated or cold glasshouses or in polythene tunnels. In these situations two-spotted spider mite has become more important, but these conditions are also more favourable for naturally occurring and introduced predators, so there are biocontrol possibilities.

Aphids
Several species of aphid are found on these crops, overwintering as eggs on the canes. On raspberry, eggs of large raspberry aphid (*Amphorophora idaei*) begin hatching in March; this is a large, pale-green aphid that causes slight leaf curl. It disperses between canes in the summer. Raspberry aphid (*Aphis idaei*) is smaller and greyish-green in colour. It causes pronounced curling of young leaves, and infests fruiting laterals, dispersing to raspberry and raspberry hybrids in June and July. Both are important as virus vectors. Eggs of a third species, blackberry/cereal aphid (*Sitobion fragariae*), hatch in February and March, and the nymphs feed on the tips of the buds; heavy infestations cause severe leaf curling. Dispersal to grasses occurs in May and June. Only blackberry is severely attacked by this species.

Where treatment is required, the materials available for spring use against aphids are the OPs chlorpyrifos and dimethoate, and the carbamate pirimicarb. The appropriate timing for application is late April. Tar oil may also be used as a winter wash when the buds are dormant. See the Introduction, p. 258, for information on the impact of pesticides on beneficials.

Blackberry mite (*Acalitis essigi*)
Small numbers of mites overwinter under bud scales, becoming active in the spring. During the summer they feed on the basal drupelets of the fruits, causing uneven ripening and hardening of the berries. In the typical 'redberry' condition associated with this mite it is the drupelets near the calyx that remain red; uneven ripening is also caused by other factors. Both wild and cultivated blackberries are attacked. Treatment is not normally necessary, but when required the OC endosulfan is available. The appropriate timing is before flowering (in late April or early May), and again 2 and 4 weeks later.

Bramble shoot moth (*Epiblema uddmanniana*)
Moths occur from late June to late July, laying eggs on the leaves and shoots of blackberry and loganberry. The caterpillars hibernate when about 3 weeks old, in a cocoon on the lower parts of the plant. Activity is resumed in March or April, when the caterpillars web together young leaves on the shoots, and burrow into flower buds. There are no specific approvals for insecticides for use against this pest, but the OP chlorpyrifos, if used at bud burst against aphids, is likely to provide some control.

Common green capsid (*Lygocoris pabulinus*)
See under Apple, capsids, p. 263, for details of the life history. This pest is occasionally damaging on cane fruit. The OP dimethoate and the alkaloid nicotine are available for use against this pest on cane fruit. See the Introduction, p. 258, for information on the impact of pesticides on beneficials.

Raspberry beetle (*Byturus tomentosus*)
This is the most important pest of cane fruits. The beetles hibernate in the soil, emerging in April and May. They are active in sunny weather, frequenting flowers of apple, hawthorn (*Crataegus*), raspberry, etc. Eggs are laid in the flowers of raspberry and other *Rubus* spp., hatching in 10–12 days. The larvae feed on the surface of the fruit and, as it begins to ripen, tunnel into the plug. They may leave one fruit and attack a neighbouring one before becoming fully grown and eventually pupating in the soil. The adult beetles will also feed on the flower buds and tips of young canes, particularly on raspberry.

The materials available against this pest are the OP chlorpyrifos, the pyrethroid deltamethrin, and the natural insecticide rotenone. On raspberry, a single application at the first pink fruit is often adequate, but to avoid slight damage to the basal drupelets of the earliest berries, growers of high-quality dessert fruit sometimes apply an additional earlier spray, when about 80% of the blossom is over. On loganberry, two applications are usual, when 80% of the blossom is over and again when the first fruit is colouring, usually about 2 weeks later. On blackberry, an application just before first open flower is usually sufficient. Chlorpyrifos can be used a maximum of two times per year. See the Introduction, p. 258, for information on the impact of pesticides on beneficials.

Raspberry cane midge (*Resseliella theobaldi*)
Adults of this localized pest normally emerge from the soil in early May, though not until 2–3 weeks later in cold springs. In Scotland, emergence is about a month later than in southern England. Eggs are laid on the young spawn of raspberry in breaks in the rind, such as growth splits. The pink larvae feed under the rind, and the damaged tissues are susceptible to fungal attack, which may lead eventually to the death of the cane ('cane blight') (see p. 298). Two further generations of midges appear (in July/August and in September) but these overlap considerably. The later generations of larvae are often very large, and considerable damage may result. The winter is passed in cocoons in the soil, the full-grown larvae pupating in the spring. The raspberry cv. Glen Prosen and the hybrid berries loganberry and tayberry, in which the rind does not split readily in the spring, are only lightly attacked by raspberry cane midge.

The OP chlorpyrifos is available for control of this pest. The usual timing for a first application is early May, when most spawn growth on cv. Malling Promise is 25–30 cm high, followed by a second application 2 weeks later. In cold springs, sprays should be delayed by a week or two. A temperature-based forecasting model is used as the basis for advice on the timing of pesticide applications against this pest. See the Introduction, p. 258, for information on the impact of pesticides on beneficials.

Two-spotted spider mite (*Tetranychus urticae*)
See under Strawberry, p. 305, for details of the life history. This pest can be particularly damaging on raspberries grown under protected cultivation. It is likely that naturally occurring populations of the predatory mites *Amblyseius* spp. and *Typhlodromus pyri* contribute to its control, except where pyrethroids, damaging to these predators, are used against other pests.

Materials available for use against this pest on raspberry are the OPs chlorpyrifos and dimethoate (but note that resistance to OPs is widespread in two-spotted spider mite populations, so these compounds are unlikely to be effective), and from other chemical groups clofentezine (off-label) (SOLA 1250/95) (SOLA 1645/98 for protected raspberry and blackberry) and tetradifon. See the Introduction, p. 258, and under Apple, fruit tree red spider mite, p. 266, for information on the impact of pesticides on beneficials and integrated mite management.

The non-native predatory mite *Phytoseiulus persimilis* is available from commercial biocontrol suppliers, and can be introduced as a biological control agent for two-spotted spider mite. In outdoor, summer-fruiting raspberries, this predator is usually applied a week after the last insecticide application before harvest. Several introductions may be needed in autumn-fruiting raspberries. In raspberries grown under protection, *Phytoseiulus persimilis* can be introduced as soon as two-spotted spider mite becomes active in early spring.

Diseases

Annual treatment is necessary for grey mould of raspberry and for cane diseases. If fruit is to be used for processing, the processors should be consulted before any spray is applied.

Blackberry purple blotch (*Septocyta ramealis*)
This disease frequently occurs in blackberry plantations and can cause severe crop loss. New infections are seen in early spring as small, light-green blotches, usually near the base of the canes. These enlarge, coalesce and quickly turn purple. In severe attacks, infections occur along the length of the canes, which may be killed. Spores of the fungus are released from the purple blotches, spreading the disease during the growing season. For control, copper oxychloride should be applied (a) immediately before blossom, (b) at fruit set, (c) immediately after harvest, and (d) again, 14 days later. It is important to direct the spray at the young growing canes and to ensure good cover.

Cane blight (*Leptosphaeria coniothyrium*)
Although blackberry can be infected, the disease is rare on this crop; however, it is widespread and prevalent on raspberry, often causing severe losses. The fungus invades the stems of developing spawn through wounds in the bark (including mechanical injury, natural splits or damage caused by feeding of raspberry cane midge (*Resseliella theobaldi*) larvae (see p. 297). If the fungus invades the vascular tissue, the cane is killed, and spores are produced from affected tissue and spread the disease.

Control is based on prevention of mechanical injury to young canes and an effective spray programme for raspberry cane midge (see p. 297). Spread of the disease can be reduced by regular sprays of dichlofluanid, directed at the base of the canes. Post-harvest sprays are also required, and all dead canes should be cut out and removed.

Cane spot (*Elsinoë veneta*)
Although present in many raspberry and loganberry plantations, this disease is easily controlled by routine spray programmes and is rarely serious. New infections are seen as small, purple spots on the young canes from early June onwards. The spots enlarge, become elliptical and up to 6 mm long, and have a light-grey centre with a purple border; the centres of the spots split, leaving cavities which give the fruiting canes a rough and cracked appearance.

Where spots have coalesced, the tips of canes may be killed. Leaves and fruits are sometimes attacked, the latter becoming distorted. Spores (conidia) of the cane spot fungus are released from the spots, so spreading the disease during the growing season. The fungus overwinters in the canes, and produces a second type of spore (ascospore) in the spring to restart the cycle of infection. Control is based on sprays of chlorothalonil, thiram or a copper-based fungicide. On loganberry, sprays should be applied immediately before flowering and again after fruit set. Sprays should be applied at HV and directed at both the fruiting canes and the developing spawn.

Grey mould (*Botryotinia fuckeliana* – anamorph: *Botrytis cinerea*)
This disease occurs wherever raspberry crops are grown, and routine sprays are needed to prevent severe fruit infection. Blackberry may also be attacked but the disease is less prevalent on loganberry. The fungus invades the floral parts and not infrequently attacks the canes. Under moist conditions the fruits become infected, the fungus producing a grey, furry mould. Large, black bodies (sclerotia) develop in the bark of the canes and eventually fall on to the soil. Under suitable conditions the sclerotia produce spores which spread the disease.

Sprays should commence as soon as the flowers begin to open, and should be repeated according to the manufacturer's label; ensure good cover of all floral parts. Fungicides for the control of grey mould include chlorothalonil, dichlofluanid, fenhexamid pyrimethanil (off-label) (SOLA 2182/99) and thiram.

Raspberry mildew (*Sphaerotheca macularis*)
This raspberry disease, which attacks the leaves and fruit, has become prevalent in recent years, particularly on cvs Glen Clova, Joan Squire and, under protection, Glen Ample. White mildew growth occurs on the upper surface of the leaves, and affected fruit is often disfigured severely. The disease is favoured by an absence of rainfall. Fungicide sprays should be applied during the flowering period and repeated as necessary, allowing a 7-day interval before harvest. Bupirimate, fenarimol, fenpropimorph (off-label) (SOLA 0787/95) and triadimefon (off-label) (SOLA 0024/95) are recommended; some mildew suppression may also be obtained with dichlofluanid. Bupirimate, fenarimol, myclobutanil (off-label) (SOLA 1881/99) and pyrimethanil (off-label) (SOLA 1938/99) are available for use on protected crops.

Raspberry root rot and die-back (*Phytophthora fragariae* var. *rubi* and other *Phytophthora* spp.)
Over recent years, this disease has become an increasingly important problem in raspberry plantations. The soil-borne fungus, which thrives in wet conditions, infects the roots, causing a dark-brown staining. The above-ground symptoms are wilting of the primocane, with a characteristic crooking of the tip, often accompanied by a dark-purple staining at the base of the cane. Fruiting canes are

also affected; fruiting laterals may be produced, but these suddenly wilt and fail to develop. The symptoms are most likely to be seen when a period of wet weather is followed by much drier conditions. The disease is spread along a row, or along several rows, by the movement of spores in soil water, so that a distinct patch of cane death occurs. The species of *Phytophthora* that affect raspberry also affect deciduous trees, and the disease is frequently found on plantations on or adjacent to old woodland sites. The cvs Glen Ample, Glen Clova, Glen Garry, Glen Moy, Glen Prosen, Julia, Malling Leo and Tulameen are all very susceptible; although none of the cultivars commonly grown in the UK is resistant, cvs Gaia and Glen Magna are least susceptible. Of the autumn-fruiting cultivars, Autumn Bliss appears more tolerant of infection but Joan Squire is very susceptible.

Improvements in soil drainage will reduce the risk of disease spread, and drenches of a formulated mix of metalaxyl + mancozeb (off-label) (SOLA 1189/96) or oxadixyl + mancozeb (off-label) (SOLA 0750/95), applied as a band spray either side of the row, are recommended for protection against raspberry root rot. The most effective timing of fungicide treatment is between mid-September and mid-October, and again in March. Where there is an advanced attack of root rot, treatment will be less effective.

Raspberry spur blight (*Didymella applanata*)

This disease is probably the main cause of reduced yields in raspberry plantations and is widespread in occurrence, particularly where high rates of nitrogen are applied. The characteristic symptoms of the disease are dark-purple blotches arising at the nodes around the bases of the leaf petioles. These extend longitudinally and can coalesce to form long, discoloured lengths of cane. Buds arising at infected nodes are weakened or killed. Occasionally, the leaves are also attacked. Spores (conidia) are released from spore sacs on the discoloured areas, so spreading the disease during the season. The fungus overwinters within the canes, and produces another type of spore (ascospore) in the spring to restart the cycle of infection. Sprays of Bordeaux mixture or thiram should be applied when the buds are not more than 1 cm long and repeated at 14-day intervals until the end of blossom. The sprays should be directed at both fruiting canes and young spawn growth.

Raspberry yellow rust (*Phragmidium rubi-idaei*)

This disease has recently become more common in the UK, particularly in some Scottish plantations. Symptoms of the disease are first seen in the spring or early summer, when yellow pustules (aecia) appear on the upper surface of leaves of the young or the fruiting canes. During the summer, more yellow pustules (uredinia) are produced on the underside of leaves. Under favourable conditions, spores from these pustules (urediniospores) can spread the disease, giving rise to further uredinia. In the early autumn, a third type of spore, the overwintering teliospore, is produced in black telia that are easily seen on the underside of leaves. These teliospores survive the winter, attached to the bark of fruiting canes (or to supporting posts and wires, or to nearby weeds) from where they germinate to produce further spore types; these eventually result in the visible aecia appearing. Severe infections can lead to premature defoliation but there is little information on the effect of the disease on yield. The cvs Glen Ample, Glen Clova, Glen Lyon, Glen Moy and Tulameen are among the most susceptible, whereas cvs Glen Prosen, Julia and Malling Leo show some resistance. Triadimefon, applied for mildew control, may give some control of the disease.

Strawberry

Pests

Cultural methods for strawberry production are changing, in that a proportion of the production is now on everbearer cultivars, in which flowering and fruiting are continuous over a period of several months. This has led to the appearance of tarnished plant bug as a new pest. The growing of strawberries under protection, particularly polythene tunnels, is increasing. This is creating a situation that favours two-spotted spider mite, but also improves opportunities for exploiting introduced predatory mites as biocontrol agents.

Aphids

Several species are found on strawberry. Strawberry aphid (*Chaetosiphon fragaefolii*) is the most important as it is the main vector of damaging virus diseases. This is a creamy-white aphid, with reddish eyes and knobbed hairs on its body. It occurs on the plants all year, with peak numbers in early summer on established fruiting plants, and in September on first-year plants. Winged forms appear in May and June, dispersing to other strawberry plants, and small numbers also occur from October to December. Shallot aphid (*Myzus ascalonicus*), a greenish-brown species that sometimes colonizes strawberries in autumn, may cause severe damage in the following spring, distorting leaves and blossom and destroying the crop, especially after a mild winter. This species is also a virus vector. Melon & cotton aphid (*Aphis gossypii*) is a pest of protected crops, and is becoming more common on strawberries grown under glass or polythene protection.

Because shallot aphid is damaging even when present in small numbers, it is usual to treat if its presence is detected. The direct damage caused by strawberry aphid is serious only at high population densities; a possible treatment threshold is one aphid per leaf. If virus is present in the immediate vicinity, however, this aphid requires treatment if its presence is detected.

Materials available for use against aphids in strawberry are the OPs chlorpyrifos, demeton-S-methyl, dimethoate, disulfoton and malathion, the alkaloid nicotine, and the carbamate pirimicarb. Nicotine is less effective than the other materials against strawberry aphid and shallot aphid. Melon & cotton aphid is resistant to many insecticides, however, and nicotine is likely to be the only available material that is effective against it. See the Introduction, p. 258, under Apple, fruit tree red spider mite, p. 266, and under Strawberry, two-spotted spider mite, p. 305, for information on the impact of pesticides on beneficials and integrated mite management.

Capsids

Several species of capsid are found on strawberry, but tarnished plant bug (*Lygus rugulipennis*) has become the most damaging on late-season strawberries. This species overwinters as adults, laying eggs on various weeds in the spring. Adults from this generation fly to strawberry from June to August and lay eggs. Feeding by nymphs and adults causes damage to the flowers and developing fruits.

Everbearer strawberries flower and fruit throughout the season, so if insecticides are used they need to be of short persistence. There are no specific approvals for insecticides for use against capsids on strawberry, but growers are achieving control with the application of the OP malathion. See the Introduction, p. 258, for information on the impact of pesticides on beneficials.

Chafer grubs

The large, white grubs of various species of chafer, e.g. cockchafer (*Melolontha melolontha*), garden chafer (*Phyllopertha horticola*) and summer chafer (*Amphimallon solstitialis*), attack the roots, causing the plants to wilt and die. Damage usually occurs only where crops are planted after pasture. If chafer grubs are seen during cultivation, gamma-HCH can be applied and worked into the soil pre-planting.

Cutworms

In some seasons, the plump, greenish-brown caterpillars of turnip moth (*Agrotis segetum*) feed on the roots and crowns, and may eat away the growing point. Other, related species of cutworm also occur. Cutworms feed at night and are most likely to be troublesome in hot, dry summers from late June or July onwards. It is not usually necessary, or practicable, to apply a treatment aimed specifically against cutworms, but the OP chlorpyrifos, if applied as a drench against vine weevil, is likely to reduce cutworm numbers.

Nematodes

Leaf nematode (*Aphelenchoides fragariae*) and chrysanthemum nematode (*A. ritzemabosi*) feed in the crowns and in the folds of young unopened leaflets. Damaged leaves may be puckered, and show a pale-grey or silver patch near the base of the midrib when they expand. The main crown may become blind, with secondary crowns developing. Stem nematode (*Ditylenchus dipsaci*) causes a marked corrugation of leaves, and a shortening and thickening of the stalks of leaves and blossom trusses. Strawberries are attacked by the stem nematode races that affect onion, oats, red clover, *Narcissus*, parsnip and other vegetables. The purchase of certified plants means that they should be virtually free of nematodes. Nematode-free runners should not be planted in infested soil. This can often be avoided by crop rotation.

Treatment against stem nematode in the soil before planting should be contemplated only if soil sample assessment by a specialist indicates there is a need. Treatment with the materials below will also check the spread of soil-borne virus diseases, by controlling migratory nematodes such as *Xiphinema diversicaudatum*, a vector of arabis mosaic virus. Materials approved for this purpose are the soil nematicide 1,3-dichloropropene, and the highly toxic soil fumigants chloropicrin and methyl bromide with chloropicrin. All three treatments require special equipment for application. See the manufacturers' labels for details and for required safety procedures. Chloropicrin and methyl bromide with chloropicrin are subject to Poison Rules and the Poisons Act. Methyl bromide with chloropicrin may be used only by professional operators trained in its use and familiar with the precautionary measures to be observed.

Strawberry blossom weevil (*Anthonomus rubi*)
Adult weevils emerge from hibernation in April and May. The female lays eggs in the unopened flower buds, then partially severs the flower stalk below. The larvae develop inside the flower buds, young weevils appearing in July. The damage is often less serious than it looks; slight thinning of the blossom may in fact result in larger fruit. Severe infestations can, however, greatly reduce yield. If treatment is required, the only material available for this use is the OP chlorpyrifos; it should be applied as soon as damage is seen. See the Introduction, p. 258, under Apple, fruit tree red spider mite, p. 266, and under Strawberry, two-spotted spider mite, p. 305, for information on the impact of pesticides on beneficials and integrated mite management.

Strawberry mite (*Phytonemus pallidus* ssp. *fragariae*)
These minute mites feed amongst the young folded leaflets, which may remain undersized, and become wrinkled and eventually turn brown; heavy infestations stunt the plants. The pest occurs sporadically, and is usually more in evidence in hot, dry summers and on older strawberry beds. With the exception of crops under protection, attacks do not normally become severe on June bearers until after cropping, and are more likely to occur in the second and subsequent years in fruiting beds. On everbearers, mite infestations can be severe throughout the summer. Certified planting material should be free of this mite.

The predatory mite *Amblyseius cucumeris* is available from commercial biocontrol suppliers for introduction as a biocontrol agent against this pest. Releases would typically start in May. When pesticide treatment is required, the materials available for this use are dicofol, dicofol + tetradifon and endosulfan. The timing for application on June bearers is after picking, when new growth appears after mowing or burning off. Chemical control on everbearers is impractical because fruiting continues into the autumn, by which time the mites are sheltered in their overwintering sites. See the Introduction, p. 258, under Apple, fruit tree red spider mite, p. 266, and under Strawberry, two-spotted spider mite, p. 305, for information on the impact of pesticides on beneficials and integrated mite management.

Strawberry seed beetle (*Harpalus rufipes*)
The adult beetles overwinter under rough vegetation, and may enter strawberry fields when the fruit is forming. They bite the seeds from the fruit, spoiling its appearance and market value. Strawberry seed beetle is a local and sporadic pest, and much of the damage attributed to it is in fact caused by linnets (*Carduelis cannabina*). Linnets pick the seeds out cleanly, usually from only the upper surface of exposed fruits, and cause little damage to the flesh. The beetles, however, usually damage the surrounding flesh and attack the lower surface of fruits next to the ground.

Strawberry seed beetle is also a predator, and contributes to the natural control of vine weevil. When pesticide treatment is required, the carbamate methiocarb is available. The pelleted formulation can be applied before strawing-down in plantations where the beetle is known to be abundant, or early in the fruiting period as soon as damage is observed. Methiocarb is also likely to be toxic to other predatory ground beetles (Carabidae) likely to be contributing to the biocontrol of vine weevil. See the Introduction, p. 258, for information on the impact of pesticides on beneficials.

Strawberry tortrix moth (*Acleris comariana*)
The moth has two flights, in June/July and August/September. It overwinters as eggs on the plants, which hatch in April to early May. The caterpillars feed on leaves and, to a lesser extent, the flowers, and damage is usually visible before flowering. Caterpillars of several other species may sometimes occur on strawberry, e.g. carnation tortrix moth (*Cacoecimorpha pronubana*), dark strawberry tortrix moth (*Olethreutes lacunana*), flax tortrix moth (*Cnephasia asseclana*) and straw-coloured tortrix moth (*Clepsis spectrana*). These species overwinter as young caterpillars and can cause damage early in the season, especially on protected plants; blossoms are particularly liable to be attacked.

When chemical treatment is required the materials available are the OP chlorpyrifos, the bacterial insecticide *Bacillus thuringiensis*, and the alkaloid nicotine. The timing of application is before flowering, as soon as the damage is seen; a second application may be needed in late summer after picking, against second-generation caterpillars. Populations of strawberry tortrix moth are often kept in check by naturally occurring parasitoids, especially the chalcid wasp *Litomastix aretas* (Encyrtidae). See the Introduction, p. 258, under Apple, fruit tree red spider mite, p. 266, and under Strawberry, two-spotted spider mite, p. 305, for information on the impact of pesticides on beneficials and integrated mite management.

Thrips

Several species of thrips (*Thrips atratus*, *T. major* and *T. tabaci*) are found on strawberry. These slender insects are 1–2 mm long, are yellow to dark brown in colour, and have narrow, feathery wings in the adult stage. The adults and wingless nymphs feed in the strawberry flowers and on young fruitlets, and are sometimes damaging in everbearer cultivars; damaged fruits have a bronzed appearance. Recently, western flower thrips (*Frankliniella occidentalis*) has become damaging on strawberries grown under protection.

The predatory mite *Amblyseius cucumeris* is available from commercial biocontrol suppliers as a control agent for western flower thrips grown under protection.

There are no specific approvals for insecticides for use against thrips on strawberry, but where infestations threaten damage, growers are achieving control on field- and tunnel-grown strawberries with the application of the OP malathion during flowering. This insecticide is of short persistence, with a short harvest interval; this is necessary because everbearer strawberries are picked at intervals throughout summer and autumn. See the Introduction, p. 258, for the impact of insecticides on beneficials.

Two-spotted spider mite (*Tetranychus urticae*)

This is the spider mite species that occurs on a range of soft fruits, and on hops, as well as on protected crops. These mites overwinter as adult females on the underside of old strawberry leaves, in the soil and in other shelter. They become active in April, feeding on the foliage. Eggs are laid on the lower leaf surface, and up to seven overlapping generations may follow during the summer. Damaging infestations chiefly occur in warm summers but tend to be more frequent in some intensive strawberry-growing areas. They are more likely to occur in the second and third years of fruiting beds, and in outdoor crops do not usually become severe until the crop has been picked. In crops grown under glass or in plastic tunnels, severe infestations are more common, and may occur during flowering and fruiting. To reduce the risk of infestations arising, it is important to plant runners as free as possible from spider mites.

Natural populations of the predatory mites *Amblyseius* spp. and *Typhlodromus pyri* often occur in strawberry, and these will contribute to the control of spider mite. The non-native predatory mite *Phytoseiulus persimilis* is available from commercial biocontrol suppliers, and can be introduced as an effective control agent for spider mite; this biocontrol system is now in widespread use on strawberries grown in the open as well as under protection.

When chemical treatment is required, the materials available for this use are the OPs chlorpyrifos, demeton-S-methyl and dimethoate (but note that resistance to OPs is widespread in populations of two-spotted spider mite, so these compounds are unlikely to be effective), the pyrethroids bifenthrin and fenpropathrin, and from other chemical groups clofentezine (off-label) (SOLA 1250/95) (SOLA 1646/98 for protected cropping), dicofol, dicofol + tetradifon, fenbutatin oxide (under protection only), tebufenpyrad and tetradifon. The pyrethroids are harmful to the naturally occurring predatory mites (*Amblyseius* spp. *Typhlodromus pyri*) and to the introduced *Phytoseiulus persimilis*. See the Introduction, p. 258, and under Apple, fruit tree red spider mite, p. 266, for information on the impact of pesticides on beneficials and integrated mite management.

Vine weevil (*Otiorhynchus sulcatus*)
See under Wingless weevils, below.

Wingless weevils (*Otiorhynchus* spp.)
Several species occur as pests, the larvae killing or weakening plants by feeding on the roots. Two common species are strawberry root weevil (*Otiorhynchus rugosostriatus*) and vine weevil (*O. sulcatus*). The latter is also an important pest on black currant and various other soft fruits. Eggs are found mostly in late July, August and September, in the surface soil beneath the canopy of leaves. The larvae hatch in the late summer and autumn, and feed on the roots during the autumn and the following spring; they then pupate in the soil. Most adult weevils emerge in June or July. As a direct result of larval feeding, plants often collapse during the fruiting period. The adults also feed on the foliage but the damage they cause is not serious. A few adults overwinter in the soil under the plants, or under black polythene sheeting, and these become active in the following spring.

In parts of the south-west, red-legged weevil (*Otiorhynchus clavipes*) is also a pest. Adults appear in two waves, the minority emerging in the spring from pupae formed in the autumn, and then in a succession from mid-June to the end of August from pupae formed in late spring and summer. Eggs are laid from late May to the end of August.

Several species of predatory ground beetles (Carabidae) and rove beetles (Staphylinidae) have been shown to prey on all stages of vine weevil, and to contribute to the control of the pest. Populations of these predators are usually greatest in plantations with grass and other low plants as cover. Some species of entomopathogenic nematodes (*Heterorhabditis* and *Steinernema* spp.) are available commercially for control of vine weevil larvae, but currently available strains are not active at low soil temperatures, which imposes a constraint on their effectiveness.

Where pesticide treatment is required, the available materials are the carbamate carbofuran and the OP chlorpyrifos. Carbofuran granules are broadcast, or applied as a band treatment; chlorpyrifos is applied as a soil drench after fruiting. Both of these materials are harmful to the predatory ground beetles and rove beetles that prey on vine weevil. See the Introduction, p. 258, under Apple, fruit tree red spider mite, p. 266, and under Strawberry, two-spotted spider mite, p. 305, for information on the impact of pesticides on beneficials and integrated mite management.

Wireworms (*Agriotes* spp.)
Strawberries are susceptible to damage by these pests, which may occur where crops are grown in broken-up grassland. Roots are bitten through and holes drilled into the crowns. When treatment is necessary, gamma-HCH is available for use against this pest as a pre-planting soil application.

Diseases

Annual treatment is necessary for grey mould and for fields known to be infested with red core. If fruit is to be used for processing, the processors should be consulted before any spray is applied.

Crown rot (*Phytophthora cactorum*)
The pathogen is soil-borne and infection results in a reddish-brown discoloration of the crown, followed by a collapse of the foliage and rapid death of the plant. Some alleviation of symptoms can be achieved by treatment with fosetyl-aluminium (off-label) (SOLA 0564/99).

Grey mould (*Botryotinia fuckeliana* – anamorph: *Botrytis cinerea*)
This disease is widespread and causes severe loss of crop in wet seasons. The fungus is ubiquitous and, under moist conditions, large quantities of spores are produced which can invade the floral parts of the strawberry. This infection subsequently appears as fruit rot; secondary spread to ripening berries is rapid in wet conditions.

A fungicide should be applied very early in the flowering period (white-bud stage) and repeated according to the manufacturer's instructions. Additional late sprays will not compensate for the omission of early sprays. Suitable fungicides include captan, chlorothalonil, dichlofluanid, fenhexamid, iprodione, pyrimethanil and thiram. It is imperative to cover all floral parts with the fungicide and to apply the fungicide in at least 1000 litres of water (preferably 2000 litres)/ha. Disease control can be improved by the application of an additional HV spray of dichlofluanid immediately before cloching or, on outdoor crops, during late March or April at the second-expanded-leaf stage. Strains of *B. cinerea* that are resistant to iprodione are known to occur in some strawberry fields, and spray programmes based on the exclusive or extensive use of this fungicide are not recommended.

Mildew (*Sphaerotheca macularis*)
This disease is more severe on protected crops and may cause disfigurement of berries. The cvs Elsanta, Elvira, Hapil, Honeoye, Ostara, Sophie and Symphony are particularly susceptible; some resistance is present in cv. Eros and, particularly, in cv. Florence. The disease is seen during spring as dark patches on the upper side of the leaf. These patches correspond to a whitish-grey sporing growth on the underside. Affected leaves may curl upwards as if with drought, and the

mildew spreads to other leaves, the blossoms and the berries. The latter may become shrivelled or otherwise unmarketable. The fungus overwinters on old green leaves.

Mildew can be controlled by applications of bupirimate, fenarimol, fenpropimorph (off-label) (SOLA 0787/95), myclobutanil, pyrifenox or triadimefon (off-label) (SOLA 0024/95). A full programme of dichlofluanid sprays for botrytis control also gives a useful early-season control of mildew. Sulfur, applied just before flowering and at 10- to 14-day intervals, will also give some control of mildew. Post-harvest applications may be necessary if the disease is severe.

Red core (*Phytophthora fragariae* var. *fragariae*)
This disease can cause extensive losses in strawberry plantations, particularly following wet soil conditions in autumn and spring. The fungus infects the root tips and develops in the central root core, producing the characteristic reddening associated with the disease. Severely affected plants are stunted and will eventually die. The disease is usually spread by planting diseased runners, or by infected soil adhering to machinery, workers' boots, etc., or by the movement of spores in soil water. Some cultivars, e.g. Eros and Symphony, show immunity to some strains of the fungus, but none is totally immune.

Control of red core is based on preventing the introduction of the disease to clean land by the use of disease-free planting material. Once infection is present, disease incidence is reduced by improving soil drainage, which could be achieved by planting into ridges. Drench treatments with copper oxychloride + metalaxyl, or sprays of fosetyl-aluminium can give effective control of the disease.

Wilt (*Verticillium dahliae*)
This fungus is widespread in soils, but severe symptoms of wilt in strawberry occur only in some localities, particularly on light land and where very susceptible cultivars (such as Elsanta and Hapil) are grown. The majority of commonly grown cultivars are susceptible to wilt to some degree, with only cv. Florence exhibiting good resistance. The disease is soil-borne and affects the vascular tissue of the plant, resulting in wilting and death. Adverse soil conditions greatly favour the development of wilt. Soil fumigation with chloropicrin (or with methyl bromide, while available) reduces wilt infection levels; such treatment should be undertaken by a contractor (see p. 303).

Grapevine

Diseases

Downy mildew (*Plasmopara viticola*)
This disease is not generally widespread but can cause severe damage. Lightish-green patches occur on the upper surfaces of leaves and these correspond to

growth of the sporulating fungus on the undersides; diseased areas later become dry and brittle. Berries can also be affected and may then shrivel. Overwintering spores are produced in affected leaves, and these can renew the disease the following spring.

Routine sprays should be applied in the spring, commencing when the shoots are 5 cm long and repeated according to the manufacturer's recommendation. The most effective fungicide is metalaxyl + copper oxychloride (off-label) (SOLA 1362/97), but copper oxychloride alone or chlorothalonil give some control. Mancozeb (off-label) (SOLA 1666/99) is also available. Where practicable, affected leaves and plant debris should not be allowed to remain on the soil during the winter.

Grey mould (*Botryotinia fuckeliana* – anamorph: *Botrytis cinerea*)
This disease is the major limiting factor for outdoor grape production in the UK. The fungus is ubiquitous and can infect leaves and stems, but most damage occurs through fruit infection. It is thought that fruit infection occurs via the floral parts, immediately after fruit set. The fungus attacks leaves and tendrils after they have been damaged by wind or rain; it then works back into the stem, resulting in loss of the following year's fruit buds. Severe infection occurs in wet summers and autumns, and overcrowded vineyards are particularly prone to the disease. Most of the commonly grown cultivars are susceptible.

The disease is less prevalent in well-spaced vineyards with good air movement. Protectant HV fungicide sprays should be applied at the following times: (a) before flowers open, (b) at 70% caps-off stage, and (c) thereafter at fortnightly intervals, ensuring one spray is applied just before closure of the bunches. The sprays should be repeated according to the manufacturer's instructions, allowing a sufficient interval between the last spray and harvest so as not to interfere with fermentation. Suitable fungicides are chlorothalonil, dichlofluanid, iprodione (off-label) (SOLA 0478/93) and pyrimethanil (off-label) (SOLA 1203/98). Overuse of iprodione can lead to the selection of resistant strains of the pathogen in the vineyard.

Powdery mildew (*Uncinula necator*)
This disease, also known as oidium, occurs widely and often causes loss in developing fruit. The mildew forms white sporulating patches on young leaves and shoots, but its development may be so sparse that a grey or purplish discoloration of the diseased parts is the most obvious symptom. Flowers and berries also become infected; these either drop off or develop as distorted cracked fruitlets.

Sprays of dinocap (off-label) (SOLA 1543/99), sulfur or triadimefon (off-label) (SOLA 0024/95) should be applied at the first sign of the disease and repeated every 10–14 days. A winter wash of tar oil should give a useful reduction in overwintering inoculum.

Hop

Pests

The damson/hop aphid is the dominating pest on hop and routine pesticide application is required, at least at the beginning of the aphid season. Dwarf cultivars of hop are now beginning to be grown commercially; this different growing system has led to an increase in the importance of two-spotted spider mite, but also increases the opportunities for exploiting naturally occurring, and introduced, predatory mites as biocontrol agents.

Caterpillars
Caterpillars of several species of moth occur from time to time in hop gardens, but infestations are not usually sufficiently damaging to warrant treatment. Species in this category include cabbage moth (*Mamestra brassicae*), knotgrass moth (*Acronycta rumicis*) and rosy rustic moth (*Hydraecia micacea*). Additionally, currant pug moth (*Eupithecia assimilata*) seems to be on the increase in hop gardens. This species has two generations each year, with larvae causing sometimes extensive leaf damage in June and July, and again in August and September. Should this, or any other moth pest, require treatment, the only material available for use against caterpillars on hop is the pyrethroid fenpropathrin; this material is damaging to predatory phytoseiid mites and so is incompatible with biocontrol of two-spotted spider mite. See the Introduction, p. 258, for information on the impact of pesticides on beneficials.

Dagger nematode (*Xiphinema diversicaudatum*)
This nematode is the vector of arabis mosaic virus, the hop strain of which appears to be an essential component of nettlehead and split leaf blotch diseases. However, these diseases are much less common than they were in the past, probably because very few growers are planting new hop gardens. There is no treatment for the diseases. Therefore, where possible, growers planning new hop plantings should choose sites that appear to be free from the virus vector. To this end, soil samples should be examined for *Xiphinema diversicaudatum* by a specialist. If the use of a *Xiphinema*-infested site is unavoidable, and particularly if virus-infested hops have been grubbed, the site should be fallowed for 2 years. As an alternative, the soil nematicide 1,3-dichloropropene is approved for this use. This material is applied by soil injection. See the manufacturer's label for application techniques and safety procedures.

Damson/hop aphid (*Phorodon humuli*)
This aphid is one of the main limiting factors to hop production. It overwinters as eggs on the twigs of *Prunus* spp., especially blackthorn (*Prunus spinosa*), bullace, damson and plum, and the eggs begin hatching in February or March. After one or two generations of wingless aphids, winged forms begin appearing in the latter

half of May; these then disperse to hops. Individual aphids may visit several hop plants and most eventually settle at the tips of the bines or laterals. The migration usually begins in earnest in early June, reaching a maximum in the second half of the month. It then declines, to end in late July or early August (or sometimes later). A return flight to the winter hosts occurs in September and October. There is no evidence that winged aphids from hop spread infestations to other hops.

Crops need protection by insecticides from the time the first adult wingless aphids mature, usually in the second week of June. Several species of natural enemies of damson/hop aphid occur on hop, in particular the anthocorid bug *Anthocoris nemoralis*, earwigs and ladybirds (Coccinellidae). All of these contribute to the biocontrol of the aphid. Unless predator populations have built up sufficiently to provide continuing control, pesticide protection needs to continue until the infestation is completely controlled after immigration ends (see paragraph above). The time at which migration to hops is completed is critical for gaining control of the aphid on conventional, tall hops. From mid-July the canopy of foliage near the top wires becomes very dense, and on some cultivars the mature leaves on the lower part curl downwards, making thorough spraying at this stage extremely difficult. At the same time, growth of the bines slows down and movement of systemic insecticides appears to be restricted. Thus, if migration continues into August it is considerably more difficult to control the infestation than if migration is completed in July. If hot, dry weather follows during August, surviving aphids are able to multiply at a very rapid rate and there may be severe infestation of the cones when they are harvested in September.

The insecticides available for use against damson/hop aphid are the pyrethroids bifenthrin, cypermethrin, deltamethrin, fenpropathrin and lambda-cyhalothrin, and from other chemical groups imidacloprid and tebufenpyrad. Resistance to the pyrethroids is now widespread in damson/hop aphid, and these materials are unlikely to be effective; the pyrethroids are also damaging to the natural enemies of the pest. Imidacloprid is applied as a direct stem-base spray, when the soil is moist and the hops are growing well; the material is then translocated in the plant. The usual strategy is to make this application in May, before the aphids begin to colonize hop. The combination of this early treatment and the subsequent action of natural enemies may be sufficient to control the aphid for the whole season. If further treatment is required, then tebufenpyrad may be applied as a foliar spray. It is challenging to achieve good coverage on tall hops after mid-July when growth has become dense, but it is important to do so. On dwarf hops, good coverage is much easier to achieve. Up to three full doses of tebufenpyrad are permitted on hops per year. See the Introduction, p. 258, for further information on the impact of pesticides on beneficials.

Two-spotted spider mite (*Tetranychus urticae*)

The mite overwinters as adult females in soil, crevices in poles and wirework and, in dwarf hops, on the dead hop material that remains in the plantation over the winter. It emerges in late April and feeds on the leaves, where eggs are laid. Up to seven generations may follow during the summer. Young hops in their first season should be watched closely for spider mite infestations. Mite infestations tend to be more severe on dwarf hops than on traditional tall hops. The predatory mites *Typhlodromus pyri* and *Amblyseius* spp. are natural enemies of two-spotted spider mite, and while their numbers are usually negligible on tall hops, they are being found in greater numbers on dwarf hops, where opportunities for successful overwintering are much greater. The non-native predatory mite *Phytoseiulus persimilis* is available from commercial biocontrol suppliers, and on some crops it is an effective biocontrol agent against two-spotted spider mite. Whilst it appears to have only limited potential in tall hops, it has been very successful in trials on dwarf hops; the hedge-like structure of the plants in the rows provides greater opportunity for movement of the predator within the plantation.

If pesticide treatment is required against two-spotted spider mite in hops, the materials available are the pyrethroids bifenthrin, fenpropathrin and lambda-cyhalothrin, and from other chemical groups dicofol, dicofol + tetradifon, tebufenpyrad and tetradifon. The pyrethroids are damaging to predatory phytoseiid mites, and are thus incompatible with biocontrol of two-spotted spider mite. See the Introduction, p. 258, for information on the impact of pesticides on beneficials.

Diseases

Routine treatment is necessary for downy mildew and hop mould.

Downy mildew (*Pseudoperonospora humuli*)

This is a common disease and can be severe on susceptible cultivars. It begins each spring from systematically infected shoots (basal spikes) arising from the crown of the rootstock. Spores from these result in more basal spikes. In suitable weather, the disease infects new leaves and cones at any stage of development. Infection produces dark-brown, velvety pustules on the underside of leaves, and severe infection causes distorted and dwarfed shoots. If routine preventive methods are not adopted, the whole crop may become worthless. Downy mildew is probably the most common cause of death of hop rootstocks, being introduced through short shoots (secondary basal spikes) and through the base of the bine; some cultivars are very prone to this form of the disease.

Early-season control consists of treatments to the base of the plant. Fungicides should be applied to the foliage within 10 days of the last basal treatment and then at intervals of 10–14 days until immediately after burr. More frequent and later applications may be given on susceptible cultivars if the weather is warm or humid.

Various copper-based fungicides are available, including copper oxychloride + metalaxyl. Other suitable materials are chlorothalonil, fosetyl-aluminium and zineb. Copper-containing dusts may be used for foliar application instead of a spray. Intervals between dust applications should be shortened to 7 days in rainy weather. To control downy mildew on susceptible cultivars a full protective spray programme is needed in most years. Coverage of all growth, particularly the highest, is of utmost importance. Late applications of some chemicals produce unacceptable residues in the cones. The grower is advised to consult his hop factor if in doubt about any specific spray after burr.

Hop mould or powdery mildew (*Sphaerotheca humuli*)

This disease is widespread and in some seasons causes severe losses, particularly on susceptible cultivars, e.g. Northern Brewer.

Hop mould is a single fungus that appears in two forms. From May onwards the white, powdery stage develops on the leaves and sometimes on the young shoots. Likewise, the burr and cones may be attacked and become distorted and useless. If effective treatment has not been given, and if the weather is humid and warm, the later-summer stage develops. This is seen as foxy-red spots or patches on leaves and cones. On these red patches, minute, black spore cases (perithecia) develop. When cones shatter, the spore cases fall to the ground, where they overwinter. In spring, the spores within the cases mature and are ejected on to the shoots or lower leaves of the plant, where they germinate and so again begin the cycle of infection. The fungus may also overwinter in the form of mycelium within the bud scales and, in April, a small number of shoots smothered with white powdery fungal growth appear.

Bupirimate, fenpropimorph (off-label) (SOLA 2078/96), myclobutanil (off-label) (SOLA 1560/99), penconazole, sulfur or triadimefon should be applied in early May and then at intervals of 10–14 days according to the proneness of the garden to mould. A wetting agent may be needed when using some proprietary preparations at HV. If the full programme has been carried out, and if no mould can be detected, applications may cease at burr. Otherwise, they may be continued for a few more weeks.

When a crop has not been picked because of excessive mould, the bines should be cut and burnt before the cones shatter.

Nettlehead

See under Hop, dagger nematode, p. 310.

Split leaf blotch

See under Hop, dagger nematode, p. 310.

List of pests cited in the text*

Acalitis essigi (Prostigmata: Eriophyidae)	blackberry mite
Acleris comariana (Lepidoptera: Tortricidae)	strawberry tortrix moth
Acronycta rumicis (Lepidoptera: Noctuidae)	knotgrass moth
Aculus fockeui (Prostigmata: Eriophyidae)	plum rust mite
Aculus schlechtendali (Prostigmata: Eriophyidae)	apple rust mite

Adoxophyes orana (Lepidoptera: Tortricidae)	summer fruit tortrix moth
Agriotes spp. (Coleoptera: Elateridae)	*larvae* = wireworms
Agrotis segetum (Lepidoptera: Noctuidae)	turnip moth
Alsophila aescularia (Lepidoptera: Geometridae)	March moth
Ametastegia glabrata (Hymenoptera: Tenthredinidae)	dock sawfly
Amphimallon solstitialis (Coleoptera: Scarabaeidae)	summer chafer
Amphorophora idaei (Hemiptera: Aphididae)	large raspberry aphid
Anthonomus pomorum (Coleoptera: Curculionidae)	apple blossom weevil
Anthonomus rubi (Coleoptera: Curculconidae)	strawberry blossom weevil
Anuraphis farfarae (Hemiptera: Aphididae)	pear/coltsfoot aphid
Aphelenchoides fragariae (Tylenchida: Aphelenchoididae)	leaf nematode
Aphelenchoides ritzemabosi (Tylenchida: Aphelenchoididae)	chrysanthemum nematode
Aphis gossypii (Hemiptera: Aphididae)	melon and cotton aphid
Aphis grossulariae (Hemiptera: Aphididae)	gooseberry aphid
Aphis idaei (Hemiptera: Aphididae)	raspberry aphid
Aphis pomi (Hemiptera: Aphididae)	green apple aphid
Aphis schneideri (Hemiptera: Aphididae)	permanent currant aphid
Archips podana (Lepidoptera: Tortricidae)	fruit tree tortrix moth
Argyresthia pruniella (Lepidoptera: Yponomeutidae)	cherry fruit moth
Blastobasis decolorella (Lepidoptera: Blastobasidae)	straw-coloured apple moth
Brachycaudus helichrysi (Hemiptera: Aphididae)	leaf-curling plum aphid
Bryobia ribis (Prostigmata: Tetranychidae)	gooseberry bryobia
Byturus tomentosus (Coleoptera: Byturidae)	raspberry beetle
Cacoecimorpha pronubana (Lepidoptera: Tortricidae)	carnation tortrix moth
Caliroa cerasi (Hymenoptera: Tenthredinidae)	pear slug sawfly
Carduelis cannabina (Passeriformes: Fringillidae)	linnet
Cecidophyopsis ribis (Prostigmata: Eriophyidae)	black currant gall mite
Cepaea hortensis (Stylommatophora: Helicidae)	white-lipped banded snail
Cepaea nemoralis (Stylommatophora: Helicidae)	dark-lipped banded snail
Chaetosiphon fragaefolii (Hemiptera: Aphididae)	strawberry aphid
Clepsis spectrana (Lepidoptera: Tortricidae)	straw-coloured tortrix moth
Cnephasia assesclana (Lepidoptera: Tortricidae)	flax tortrix moth
Cryptomyzus ribis (Hemiptera: Aphididae)	red currant blister aphid
Cydia funebrana (Lepidoptera: Tortricidae)	plum fruit moth
Cydia pomonella (Lepidoptera: Tortricidae)	codling moth
Dasineura mali (Diptera: Cecidomyiidae)	apple leaf midge
Dasineura pyri (Diptera: Cecidomyiidae)	pear leaf midge
Dasineura tetensi (Diptera: Cecidomyiidae)	black currant leaf midge
Ditylenchus dipsaci (Tylenchida: Tylenchidae)	stem nematode
Dysaphis devecta (Hemiptera: Aphididae)	rosy leaf-curling aphid
Dysaphis plantaginea (Hemiptera: Aphididae)	rosy apple aphid
Dysaphis pyri (Hemiptera: Aphididae)	pear/bedstraw aphid
Enarmonia formosana (Lepidoptera: Tortricidae)	cherry-bark tortrix moth
Epiblema uddmanniana (Lepidoptera: Tortricidae)	bramble shoot moth
Epitrimerus piri (Prostigmata: Eriophyidae)	pear rust mite
Erannis defoliaria (Lepidoptera: Geometridae)	mottled umber moth
Eriophyes pyri (Prostigmata: Eriophyidae)	pear leaf blister mite
Eriosoma lanigerum (Hemiptera: Pemphigidae)	woolly aphid
Eulecanium tiliae (Hemiptera: Coccidae)	nut scale
Eupithecia assimilata (Lepidoptera: Geometridae)	currant pug moth
Forficula auricularia (Dermaptera: Forficulidae)	common earwig

Frankliniella occidentalis (Thysanoptera: Thripidae)	western flower thrips
Harpalus rufipes (Coleoptera: Carabidae)	strawberry seed beetle
Hedya pruniana (Lepidoptera: Tortricidae)	plum tortrix moth
Helix aspersa (Stylommatophora: Helicidae)	garden snail
Hoplocampa brevis (Hymenoptera: Tenthredinidae)	pear sawfly
Hoplocampa flava (Hymenoptera: Tenthredinidae)	plum sawfly
Hoplocampa testudinea (Hymenoptera: Tenthredinidae)	apple sawfly
Hyalopterus pruni ((Hemiptera: Aphididae)	mealy plum aphid
Hydraecia micacea (Lepidoptera: Noctuidae)	rosy rustic moth
Hygromia striolata (Stylommatophora: Helicidae)	strawberry snail
Hyperomyzus lactucae (Hemiptera: Aphididae)	currant/sowthistle aphid
Lepidosaphes ulmi (Hemiptera: Diaspidae)	mussel scale
Longiunguis pyrarius (Hemiptera: Aphididae)	pear/grass aphid
Lygocoris pabulinus (Hemiptera: Miridae)	common green capsid
Lygus rugulipennis (Hemiptera: Miridae)	tarnished plant bug
Lyonetia clerkella (Lepidoptera: Lyonetiidae)	*larva* = apple leaf miner
Mamestra brassicae (Lepidoptera: Noctuidae)	cabbage moth
Melolontha melolontha (Coleoptera: Scarabaeidae)	cockchafer
Myzus ascalonicus (Hemiptera: Aphididae)	shallot aphid
Myzus cerasi (Hemiptera: Aphididae)	cherry blackfly
Nasonovia ribisnigri (Hemiptera: Aphididae)	currant lettuce aphid
Nematus olfaciens (Hymenoptera: Tenthrididae)	black currant sawfly
Nematus ribesii (Hymenoptera: Tenthrididae)	gooseberry sawfly
Olethreutes lacunana (Lepidoptera: Tortricidae)	dark strawberry tortrix
Operophtera brumata (Lepidoptera: Geometridae)	winter moth
Orthosia incerta (Lepidoptera: Noctuidae)	clouded drab moth
Otiorhynchus sulcatus (Coleoptera: Curculionidae)	vine weevil
Otiorhynchus clavipes (Coleoptera: Curculionidae)	red-legged weevil
Otiorhynchus rugosostriatus (Coleoptera: Curculionidae)	strawberry root weevil
Pammene rhediella (Lepidoptera: Tortricidae)	fruitlet-mining tortrix
Panonychus ulmi (Prostigmata: Tetranychidae)	fruit tree red spider mite
Parthenolecanium corni (Hemiptera: Coccidae)	brown scale
Phorodon humuli (Hemiptera: Aphididae)	damson/hop aphid
Phyllopertha horticola (Coleoptera: Scarabaeidae)	garden chafer
Plesiocoris rugicollis (Heteroptera: Miridae)	apple capsid
Phytonemus pallidus ssp. *fragariae* (Prostigmata: Tarsonemidae)	strawberry mite
Psylla mali (Hemiptera: Psyllidae)	apple sucker
Psylla pyricola (Hemiptera: Psyllidae)	pear sucker
Quadraspidiotus ostreaeformis (Hemiptera: Diaspididae)	oystershell scale
Quadraspidiotus pyri (Hemiptera: Diaspididae)	pear scale
Resseliella theobaldi (Diptera: Cecidomyiidae)	raspberry cane midge
Rhopalosiphum insertum (Hemiptera: Aphididae)	apple/grass aphid
Sitobion fragariae (Hemiptera: Aphididae)	blackberry cereal aphid
Spilonota ocellana (Lepidoptera: Tortricidae)	bud moth
Tetranychus urticae (Prostigmata: Tetranychidae)	two-spotted spider mite
Thrips atratus (Thysanoptera: Thripidae)	carnation thrips
Thrips major (Thysanoptera: Thripidae)	rubus thrips
Thrips tabaci (Thysanoptera: Thripidae)	onion thrips
Xiphinema diversicaudatum (Dorylaimida: Longidoridae)	a dagger nematode

* The classification in parentheses represents order and family.

List of pathogens/diseases (other than viruses) cited in the text*

Botryotinia fuckeliana (Ascomycota)	botrytis fruit rot, (common) grey mould
Botrytis cinerea (Hyphomycetes)	– anamorph of *Botryotinia fuckeliana*
Chondrostereum purpureum (Basidiomycetes)	silver leaf
Cronartium ribicola (Teliomycetes)	black currant rust
Didymella applanata (Ascomycota)	raspberry spur blight
Drapenopeziza ribis (Ascomycota)	leaf spot
Elsinoë veneta (Ascomycota)	cane spot
Erwinia amylovora (Gracilicutes: Proteobacteria)†	fireblight
Gloeosporium album (Coelomycetes)	gloeosporium rot
Gloeosporium perennans (Coelomycetes)	gloeosporium rot
Leptosphaeria coniothyrium (Ascomycota)	cane blight
Microsphaera grossulariae (Ascomycota)	European gooseberry mildew
Nectria galligena (Ascomycota)	canker
Phragmidium rubi-idaei (Teliomycetes)	raspberry yellow rust
Phytophthora cactorum (Oomycetes)	collar rot, crown rot
Phytophthora fragariae var. *fragariae* (Oomycetes)	red core
Phytophthora fragariae var. *rubi* (Oomycetes)	raspberry root rot
Phytophthora syringae (Oomycetes)	collar rot, crown rot, phytophthora fruit rot
Plasmopara viticola (Oomycetes)	downy mildew
Podosphaera leucotricha (Ascomycota)	powdery mildew
Pseudomonas syringae pv. *mors- prunorum* (Gracilicutes: Proteobacteria)†	bacterial canker
Pseudoperonospora humuli (Oomycetes)	downy mildew
Pythium sylvaticum (Oomycetes)	a causal agent of specific apple replant disease (SARD)
Sclerotinia fructigena (Ascomycota)	brown rot
Sclerotinia laxa (Asomycota)	blossom wilt
Sclerotinia laxa f. sp. *mali* (Ascomycota)	blossom wilt of apple
Septocyta ramealis (Coelomycetes)	blackberry purple blotch
Sphaerotheca humuli (Ascomycota)	hop mould, powdery mildew
Sphaerotheca macularis (Ascomycota)	mildew
Sphaerotheca mors-uvae (Ascomycota)	American gooseberry mildew
Thielaviopsis basicola (Ascomycota)	a causal agent of specific cherry replant disease
Tranzschelia pruni-spinosae var. *discolor* (Teliomycetes)	rust
Uncinula necator (Ascomycota)	powdery mildew
Venturia inaequalis (Ascomycota)	apple scab
Venturia pirina (Ascomycota)	pear scab
Verticillium dahliae (Hyphomycetes)	wilt

* For fungi, the classification in parentheses refers to class, although this is not possible within the phylum Ascomycota where classes have yet to be satisfactorily defined (see *Mycological Research*, February 2000). Oomycetes are now classified in Chromista with the brown algae, rather than as true fungi. Some fungi have an asexual (anamorph) and a sexual (teleomorph) state, and the convention is to refer to them by their teleomorph name. However, where anamorph names are still in common use these are listed and cross-referenced to the teleomorph name. Strictly, fungi classified as Coelomycetes and Hyphomycetes should be known as 'hyphomycetous anamorphs' and 'coelomycetous anamorphs' of the relevant teleomorph taxon (e.g. hyphomycetous anamorphic Sclerotiniaceae, for *Botrytis fabae*), respectively. These problems highlight the continual changes in the classification of the fungi.

† Bacteria – the classification in parentheses refers to division and class.

Chapter 9
Pests and Diseases of Protected Vegetables and Mushrooms

T.M. O'Neill
ADAS Arthur Rickwood, Cambridgeshire

J.A. Bennison
ADAS Boxworth, Cambridgeshire

R.H. Gaze
Horticulture Research International, Wellesbourne, Warwickshire

Introduction

The protected environment

Management of pests and diseases on vegetable crops in glasshouses and polythene tunnels is influenced profoundly by the fact that the crops are enclosed. Control systems have been developed to maintain the protected environment at an economic and physiological optimum for the crop. However, often, the conditions that promote crop growth also promote the growth of pathogens and pests. Many UK glasshouse crops are either subtropical species, which will not survive or do not grow well outdoors, or they are native plants grown at steady and elevated temperatures for increased productivity or uniformity. The elevated temperature of a glasshouse favours development of many pests and diseases. Once they are established in a crop, they may be more prolific on, and cause more damage to, their hosts under protection than in their natural (outdoor) habitat.

Greenhouses also give shelter from rain and wind, elements that commonly influence pest and disease development. Many fungal pathogens need moisture, particularly at certain stages of their life-cycles, e.g. spore germination, and if this is unavailable they will not multiply. Watering within the glasshouse needs to be managed so that moisture does not persist on plant surfaces for prolonged periods. Ventilation and heating are used to dispel pockets of humid air within the crop, although care needs to be taken to ensure that the environment does not become so dry that the crop is moisture-stressed (as this will check growth and may make the crop more susceptible to disease). The well-controlled protected environment consequently reduces the risk of some diseases. The enclosed nature and relative stability of the glasshouse environment are advantageous in that they allow natural enemies of pests to be used as effective means of control.

Pest and disease carryover

Because successful protected crop production requires specialist skills, facilities and equipment, the same crop species is generally grown every year. Thus, there is a considerable danger that pests and diseases may be carried over from one crop to the next. This is particularly true of the pests and pathogenic fungi that persist in the soil in which crops are directly grown. Many protected vegetable crops are now grown in inert substrates, partly to avoid this risk. Where crops are still grown in the soil, sterilization between crops usually becomes necessary at some stage and may be undertaken on a regular basis. Some pests and pathogenic fungi may also persist on weeds, the glasshouse structure and concrete pathways. Thorough cleaning of the glasshouse, coupled with hygiene measures as outlined below, reduces the risk of carryover.

Greenhouse hygiene

Strict attention to hygiene is most important in reducing the incidence of pests and diseases in greenhouses. When diseases have occurred, the internal structure of greenhouses and concrete pathways should be treated with a disinfectant. Although formaldehyde was once widely used for this purpose and is a very effective glasshouse disinfectant, it is now used only rarely because of its harmful nature. Many alternative disinfectants are used, including products based on glutaraldehyde, hydrogen peroxide + peracetic acid, organic acids, phenols, quaternary ammonium compounds, sodium hypochlorite, and various combinations of these.

After removing a crop grown hydroponically, the whole irrigation system and the tanks should be flushed through with a disinfectant (e.g. sodium hypochlorite) and rinsed with water. Once the greenhouse structure has also been cleaned and disinfected, new polyethylene sheets should be placed over the soil floor. Fresh plants introduced into the house must never be stood on bare soil or on a dirty surface; otherwise, disease may be introduced into the clean system.

Changing cropping practices

Methods of growing plants under protection have continued to change over the last decade, generally towards more intensive production methods, a longer growing season and ways of minimizing labour requirement. Hydroponic growing systems now predominate, where plants are no longer grown in soil. These include (a) nutrient film technique (NFT), in which plants grow with their roots directly immersed in a thin film of flowing nutrient solution; (b) systems where roots grow in an inert inorganic substrate (e.g. a slab of foam or rockwool, or a container of perlite or pumice); or (c) systems where roots grow in an organic substrate (e.g. on mats of coir or in bags of peat). Efforts to improve crop uniformity and reduce labour requirement have led to automatic drip irrigation systems. The waste nutrient solution may be collected and recycled to save on costs and to minimize any potential soil and water pollution from residual fertilizer (especially nitrate and phosphate), and from pesticides applied in the irrigation water.

Supplementary carbon dioxide, increasingly the by-product of an on-site combined heat and power unit, is now commonly distributed through houses to increase yields of cucumber and tomato crops. Greater plant populations are grown, to allow greater yields. In tomato crops, side shoots are trained as extra 'heads' in the spring, to increase the plant population as light levels increase. The resultant crop canopy can be very dense. Bumblebees (*Bombus* spp.) are now used widely to pollinate tomato fruit, and this has encouraged increased uptake of biological pest control methods, in order to avoid harmful effects of pesticides on these non-target organisms.

These changes in cropping practice have modified the spectrum of pest and disease risk. Now that most cucumber, tomato and peppers crops are no longer grown directly in the soil, they are no longer vulnerable to damage by soil-dwelling pests such as millepedes, nematodes, symphilids and woodlice. Soil-less systems, while generally free of typical soil-borne diseases such as brown and corky root rot, fusarium root rots and *Thanatephorus* (*Rhizoctonia*), tend to be more prone to widespread attack by *Phytophthora* spp. and *Pythium* spp. Significant losses have been encountered in NFT-grown tomatoes where roots have become colonized with *Phytophthora* spp. and with *Thielaviopsis basicola* (synanamorph: *Chalara elegans*), probably arising from contamination of the nutrient solution with soil. Every effort needs to be made to prevent soil entering water-distribution, -collection and -recirculation pipework. Irrigation lines needs to be cleaned and disinfected thoroughly between crops, especially after an outbreak of a root disease.

A number of methods are used to aid continuity of cropping and the continued production of high-quality produce. Cucumber crops may be replanted once or twice a year to maintain fruit quality. Some growers replant the whole crop at once, whereas others practice inter-planting to maintain fruit production. Commonly, tomatoes are grown for 11 months of the year by 'layering' the crop (the lower stem is laid horizontal at the same speed as the plant head grows upwards, so that the productive part of the plant continues to receive maximum light). The final length of a layered tomato plant can exceed 12 m. Lettuces are often grown without rotation, sometimes six crops a year in the same soil. Replanting, inter-planting, layering and repeated cropping all have profound influences on pest and disease management strategies. Some examples of problems commonly encountered as a result of these production practices are given in the relevant crop sections of this chapter.

Specialist plant propagators

Most young vegetable plants are now raised by specialist plant propagation nurseries, where scrupulous attention to hygiene is imperative because of the potential consequences of distributing a pest or disease to many nurseries. The propagation nurseries usually adhere to detailed and comprehensive protocols to minimize pest and disease risk. These specify procedures on, for example: (a) tray cleaning and washing; (b) disinfection and/or new polythene floor covering between each crop; (c) specified disinfectants for different crops (e.g. iodine to control *Olpidium*, the vector of lettuce big-vein virus); (d) restriction of visitor entry; (e) foot dips; and (f) regular crop inspections. Specifications on pesticide treatments to be applied during propagation are often agreed formally between the propagator and the customer.

Integrated pest and disease management (IPM)

IPM is now widely practised in UK greenhouse crops. The technique combines biological, cultural and genetic control, with minimal use of pesticides; pesticides that are used are selected for their safety to biological control agents and other beneficial (non-target) organisms. Biological control methods form the basis of IPM strategies against pests in most UK aubergine, cucumber, herb, pepper and tomato crops, and similar systems are being developed for other protected vegetables (such as leafy salads and lettuce). The main stimulus for the development and commercial uptake of biological control is still pesticide resistance. Additional factors encouraging the expansion of biological control methods have been the withdrawal of certain pesticides, use of bumblebees for tomato pollination, and increasing market pressures to reduce the use of pesticides and to use environmentally responsible production systems.

Biological control agents include parasitoids, predators, and insect-pathogenic fungi and bacteria which act as biological pesticides. Once natural enemies have been released into the crop, it is important to monitor their progress and to manage the biological control programme effectively within IPM. Other aspects of crop husbandry need to be harmonized to ensure that the biological control systems are not affected adversely; careful choice of compatible pesticides is very important. Pesticides still play an important part in IPM, (a) to ensure the crop starts off as pest-free as possible, (b) to control pests for which there are no available biological control agents, or (c) to restore balance where the natural enemy is not giving sufficient control of the pest. An IPM-compatible pesticide is not always available for all pests or diseases occurring on the crop, and these 'gaps' are constantly being addressed with research to find solutions which can be integrated, whether these are new biological control agents for new or 'minor' pests, or cultural control methods for both pests and diseases.

Biological control of diseases is still in the early stages of development, and the basis of integrated disease management is the use of disease-resistant cultivars whenever possible. Where suitable disease-resistant cultivars are unavailable, and where pathogens have overcome cultivar resistance, other components in IPM strategies, such as greenhouse hygiene, weed control (to avoid alternative hosts for pests and diseases), environmental control practices to reduce disease development and the use of selective fungicides, assume even greater importance. IPM is constantly evolving, to offer sustainable strategies for pest and disease control, by developing robust biological and cultural control systems, which reduce selection pressure for cultivar and pesticide resistance and prolong the effectiveness of compatible selective pesticides.

Organic production

With the recent expansion in organic cropping, and a predicted further rapid increase over the next few years, management of pests and diseases in crops grown to organic standards will receive increased attention. As in IPM, wherever possible, pest control relies on the use of natural enemies; otherwise, organic production relies on a limited range of permitted pesticides. Disease management in organic crops is a particular challenge, because of the lack of any biological control options; cultural control is central to effective IPM strategies. This includes: (a) greater use of resistant cultivars; (b) grafting plants on to disease-resistant rootstocks (e.g. cucumbers and tomatoes); (c) careful selection of seeds from disease-free crops; (d) crop rotation (where possible); (e) less-intensive crop production; and (f) the use of heat and ventilation (e.g. for control of cucumber downy mildew). Recent improvements in the understanding of interactions between microorganisms, plants, pathogens and the environment offer the potential to manage specific disease problems, should they occur. Examples are the use of soil amendments or specific propagation composts that suppress root diseases, and the use of plant extracts applied for foliar disease control. The use of novel crop covers with altered UV transmission, which reduce fungal sporulation (e.g. in the case of downy mildew and grey mould), is an additional strategy suitable for organic cropping.

The pesticide use recommendations given in this chapter refer to non-organic crop-production systems. A limited number of chemical pesticides (e.g. some copper and sulfur compounds) are permitted in crops grown to organic standards; refer to the relevant organic standards authority for details.

Pesticides

Because the value of crops grown in greenhouses is usually high, effective pest and disease control methods are necessary to ensure the desired quality of produce. Where pesticides are necessary on greenhouse crops, great care is needed in their selection and in the choice of application method, to ensure maximum efficacy and cost-effectiveness and minimal side effects on natural enemies, and to avoid unsightly deposits on (or damage to) the harvested produce.

Pesticide availability

In addition to on-label approvals, this chapter includes mention of various

Specific Off-label Approvals (SOLAs). Although approved, off-label uses are not endorsed by manufacturers and such treatments are made entirely at the risk of the user. Also, as mentioned elsewhere, products can be used under the provisions of the *Revised Long Term Arrangements for Extension of Use (2000)*. Specifically, extrapolation from some major protected edible crops to some minor edible crops (e.g. from tomato to aubergine) is permitted, subject to certain restrictions (again, entirely at the risk of the user). Mention in the text of use of a pesticide under the provisions of these arrangements is marked with an asterisk (*).

Methods of pesticide application

HV spraying is still the most effective application method for certain pests and diseases. However, crops grown within structures that can be more or less sealed may be treated with pesticides, using methods of application that take advantage of the closed environment. Chemicals may be applied in mists, fogs and smokes and as vapours, often with considerable savings in labour. Fogs, mists and smokes are used almost exclusively in the greenhouse environment (see below).

Fogs

Fogging machines are still used by some growers. These produce a stream of hot gas from a petrol-driven, pulse-jet engine. Metered by a valve, pesticide is pumped into the jet stream and broken up into airborne particles (5–100 μm in diameter). The distribution of the pesticide may be limited but, nevertheless, satisfactory disease control has been obtained in practice by appropriate positioning of the machine within the cropping structure. Most of the deposit arrives at leaf surfaces by sedimentation from the air and, therefore, the deposit on the undersurface will be small. Chemicals with a strong systemic or vapour action will help to minimize irregular distribution of the fog. In trials it has been found that the limit of 'throw' of a fogging machine may be as little as 10–15 m. Thus, it is necessary to carry the machine through the crop and to fog strips no more than 20 m in width. Horizontal distribution is improved when the fog is directed over the top of the crop.

Mists

An increasingly common method of application is as a mist generated by a low-volume mister (LVM). These machines are automated and usually set to apply the chemical during the night when the greenhouse is closed and unoccupied. Strategically placed fans distribute the chemical through the crop. Their ease of use and low labour requirement has made them very popular with growers, where regular pesticide treatment is common (e.g. for applying fungicides against cucumber powdery mildew; for applying insecticides against western flower thrips). Certain pesticides must not be applied in low volumes, and label recommendations regarding application methods and dilution rates must always be followed.

Smokes

The distribution of a chemical applied as a smoke formulation will probably be similar to that of a fog, but there is no means of directing the flow of smoke. The dose is regulated by the size of canister that is recommended by the manufacturer for a given volume of glasshouse. Nicotine can be applied as a fumigant pesticide, but is supplied as 'shreds' that are lit in the greenhouse to generate a smoke.

Fogs, mists and smokes should be applied only during calm conditions and, preferably, in the evening. Many smokes are most effective if glasshouse temperatures are above 16°C and this is specified on the label.

Application of pesticides to soil, growing media and nutrient solution

Pesticides are sometimes applied to the soil before planting out, or as drenches to the soil or growing medium during the life of the plant. Soil drenching is labour-intensive and expensive and, thus, increasingly unpopular, although this method may still be cost-effective for the control of certain diseases. Some pesticides can be applied through the irrigation system, though care needs to be taken as automatic irrigation systems may become blocked if some pesticide formulations, such as wettable powders, are added to the nutrient solution. Systemic pesticides applied as soil drenches generally use more chemical, but will protect the plant for longer than the same compound applied as a spray to the foliage. Soil drenches with systemic pesticides are more effective if the drench reaches actively growing roots. This is more difficult to achieve with mature plants growing in soil. Drench application to peat bags or rockwool blocks can be most effective although, currently, this method is restricted to certain fungicides.

Some pesticides can be applied to the seed or incorporated into blocking compost before use. This type of application, which allows uniform treatment of compost in bulk, is generally recommended in preference to compost drenching after sowing or potting, but may mean unnecessary use if, subsequently, there is no challenge from the pest or the disease.

Imperfect mixing of pesticides in the nutrient solution will result in locally high concentrations that may be damaging to crop growth. Wettable powder formulations are particularly prone to this problem and, therefore, may have to be dissolved in warm water (see manufacturer's instructions) before pouring into the stock tank. It should be remembered that the pesticide will be active within the solution for only a limited period, depending on the chemical used.

Pesticide resistance

Pesticide resistance is much more of a problem in greenhouse crops than in outdoor crops, owing to the favourable conditions allowing more rapid pest and disease development and, thus, increased selection for resistance. As discussed on p. 320, the use of IPM is the most sustainable method to reduce the need for pesticides and to avoid continued pesticide resistance problems in pest and disease control. If pesticides are needed, it is important to use the recommended rate and not to exceed the maximum number of applications per crop and to follow any guidelines given on the label (including the product literature) to reduce the development of resistance. As a general rule, it is advisable not to rely exclusively on one pesticide but to adopt a programme that, sequentially, uses pesticides from different chemical groups. If full control is not given, advice should be sought on the possible presence of resistant pests or diseases, and on the best course of action.

Restrictions on pesticide use

Restrictions on pesticide use over and above those specified by pesticide legislation may be agreed by the propagator, grower and retailer. Some of the major supermarkets, in particular, have developed codes of practice which, among many other specifications, may limit the range of pesticides that can be used or the number of treatments that can be applied. The overriding aim is to grow the product with the minimum necessary pesticide use and, where their use is required, to utilize the least hazardous chemicals for the task in hand. Nurseries are audited by independent inspectors to confirm that they comply with the agreed codes of practice. Pesticide spray records are examined and crops may be sampled and tested for pesticide residues. Where there is deviation from the agreed protocol it is likely that the product will not be accepted by the retailer.

Sterilization of soil and other growing media

The use of soil sterilization in greenhouse vegetable production has declined greatly, as growers have adopted hydroponic growing systems. Lettuce remains the only major crop where soil sterilization is commonly used, although it is still used occasionally to assist continued cropping of some minor soil-grown crops (e.g. celery). With the proposed ban on use of methyl bromide, alternative methods of soil sterilization are being re-evaluated and new treatments sought. The relative effectiveness of current soil sterilization treatments against major diseases and nematodes is given in Table 9.1.

Steaming

Steaming is an effective way to eliminate many pests and diseases from well prepared greenhouse soils. Although it is physically hard and demanding work, a slow operation to treat large areas and one with inherent safety hazards (steam under pressure), it does allow replanting as soon as the soil has cooled. Also, it can be done while there are still crops in other areas of the greenhouse and it matches methyl bromide in its broad spectrum of activity and consistency of results. Care is needed to ensure the greenhouse soil is not re-contaminated by cultivating too deeply (below the treated depth), or by bringing in a pathogen on

Table 9.1 Effectiveness of soil treatments against soil-borne pests and diseases of protected vegetable crops

Treatment	Fusarium and verticillium wilt diseases	Tomato brown root rot (*Pyrenochaeta lycopersici*)	*Phytophthora*	Rhizoctonia (*Thanatephorus*)	Insects	Nematodes*
dazomet	+	++	++	++	++	++
1, 3-dichloropropene	–	+	–	–	–	++
formaldehyde	++	+	++	++	–	+
metham-sodium	+	++	++	++	++	++
methyl bromide	++	++	+++	+	++	+++
steam	++	+++	+++	++	+++	+++

– No control.
\+ Some control.
++ Moderate control.
+++ Good control.
* Effectiveness can vary according to type of nematode (cyst, migratory or root-knot).
Note: The success achieved with soil disinfection depends on the original population levels of pests and diseases present, soil type and preparation, and efficiency of the operation.

shoes or tools as certain fungal pathogens (e.g. *Fusarium oxysporum* f. sp. *radicis-lycopersici*, the cause of tomato crown and root rot) rapidly colonize sterilized soil. Steaming has declined in popularity as a method of soil sterilization, because of the lack of steam boilers on many nurseries and the poor availability of mobile steam boilers. Steaming of used rockwool slabs and peat is a highly effective treatment.

A common method is 'sheet steaming', whereby steam is fed under a heavy-duty plastic sheet (25–30 m long), anchored at the edges by a heavy chain or by bags of sand. To obtain rapid steam penetration, the soil should be moist and cultivated to a coarse tilth. Depth of cultivation varies from 200 to 300 mm, depending on the crop. Steam is applied until the soil temperature is raised to 70°C at the required depth for at least 30 minutes. The steaming period or 'cook' usually takes 2–4 hours to complete.

Other techniques include grid steaming, using moveable hand-buried pipes or semi-automated grid steaming, using a winch-drawn steam plough. A few greenhouses have the facility for steaming from below soil level, using permanently buried pipes or tile drains.

Chemical soil sterilants

These can provide effective broad-spectrum control of soil-borne pathogens and weeds but there are potential disadvantages to their use. A relatively long period is required between treatment and planting, to ensure the release of phytotoxic vapours from treated soils; there is a risk of phytotoxicity to crops in adjoining greenhouses; soils with a high organic content or heavy soils which cannot be broken down to a fine tilth may be unsuitable for treatment by chemicals, owing to excessive retention in the former and poor penetration in the latter. Methyl bromide has proved to be the best chemical sterilant under most conditions and has a short turn-around time, but use of this ozone-depleting chemical is being subjected to increasingly strict control. With all chemical sterilants, the condition of the soil during treatment, soil temperature, the application technique and state of the pathogenic organisms, all have a significant effect on the success of these treatments.

Soils must be moist and must be worked thoroughly before treatment. After treatment, the soil surface is sealed by compacting and wetting, or is covered with polythene sheeting where the active ingredient is volatile. Because the speed of sterilization and release of the chemical is faster at higher soil temperatures, chemical sterilization should not be used during the winter and only when soil temperatures are high during the late autumn. In unheated houses, release of fumes after treatment will be hastened by forking or rotary cultivation to a depth of 250–300 mm.

When using the following chemicals it is essential, strictly, to obey the manufacturer's instructions with respect to the interval between treatment and planting.

Dazomet

This should be used only on light to medium soils with less than 5% organic matter. The soil temperature should be above 7°C for effective treatment, ideally during the spring or early summer. Evenly incorporate the prill to a depth of 180–200 mm. The surface of the soil should be sealed with an anchored polythene sheet for 7–28 days, according to soil temperature. With warm soils, the surface needs to be sealed soon after incorporation to prevent loss of the active chemical and reduced efficacy. After treatment, cultivate and ventilate the soil. Test for residual chemical, using the cress germination method: half fill a jar with a representative sample of moistened soil, scatter cress seeds on the surface and seal the jar. If the soil is safe for use, the seed germinates normally.

Formaldehyde

When used as a pesticide, this chemical is classified as a Commodity Substance under the Control of Pesticides Regulations and use must be within the terms of approval specified. It is subject to the Poisons Rules (1992) and the Poison Act (1972) and operators must observe the Occupational Exposure Standards set out by the HSE. Spray or drench the soil surface and keep the greenhouse closed for 3 days. Allow vapour to disperse before planting or laying polythene sheets for soil-less culture.

Metham-sodium

Water the soil with metham-sodium. In summer, keep the soil surface sealed by spraying with water, cultivate after 7 days and repeat a number of times at 4-day intervals. Test for residues with cress seed 4 weeks after application. In autumn, allow from 7 to 10 weeks for treatment, which must be completed by the end of October. After application, close the greenhouse for 14 days, then ventilate and fork to 300 mm. Leave the soil for 2–3 weeks and fork again. Test with cress seed 7 weeks after application; if the seed germinates normally, the first crop (lettuce or tomato) may be planted.

Methyl bromide

This has been widely used for the control of pathogenic fungi and nematodes in greenhouse soils, when the soil temperature is 10°C or above. It has a high mammalian toxicity and, for its use, special equipment is required. Therefore, it can be applied only by licensed operators and according to the Code of Practice. Methyl bromide has the advantage of rapid diffusion into and from soils and, consequently, the interval between treatment and planting is short (6–8 days). In addition to the use for soil sterilization, methyl bromide is also used to sterilize peat in troughs, growing bags, or loose piles. In the case of bags or piles of loose peat, they are usually placed on a sheet of polythene and a gas-tight envelope is made by bringing up the sides of the sheet and sealing the two edges together. Used rockwool slabs and capillary matting may also be treated in this way.

Control of virus diseases in greenhouse crops

Virus diseases can cause severe losses in many protected vegetable crops and cannot be controlled directly with chemicals. Spread of mechanically transmitted viruses – e.g. cucumber green mottle virus, pepino mosaic virus and tomato mosaic virus – can be rapid in greenhouse crops because the crop is handled regularly. Large populations of pests, which may occur under glass, can also lead to increased risk from insect-vectored viruses, e.g. cucumber mosaic virus, associated with large numbers of aphids. The use of hydroponic growing systems generally avoids the danger of soil-borne virus infection, whether transmitted by nematodes or chytrid fungi or through contact with infected root debris in the soil, but a widespread problem can occur if the system is contaminated by the virus or virus vector – for example big-vein virus transmitted by *Olpidium* in NFT lettuce. Resistant cultivars have largely overcome the problem of several once-common virus diseases.

Control of weeds around greenhouses

Weeds around the outside of greenhouses can harbour pests and some diseases. Aphids, leafhoppers, thrips and whiteflies, in particular, are found commonly on weed hosts and these can be the source of crop infestation and reservoirs for re-infestation after control measures have been applied to the crop. With biological control, the immigration of flying pests from outside the greenhouse can upset the balance between pest and natural enemy in the crop. An area of mown grass or weed-free soil at least 3 m wide should surround each greenhouse. The risk of crop damage must be considered carefully before the application of any weed-killer close to greenhouses.

Management of individual pests and diseases

There are many publications giving details of the life histories and classifications of the organisms responsible for crop loss in the greenhouse. A brief description of the important pests and diseases, their damage symptoms and the most effective management strategies follows for each of the crops grown widely in protected cultivation. IPM strategies are described if this is the most effective and commonly used method. Selective IPM-compatible pesticides are specified, if available. Full details of the compatibility of pesticides in IPM and their side effects on beneficial species are available from biological control suppliers or consultants. Where IPM strategies are not yet developed for a crop, pest or disease, the most effective or appropriate pesticide or pesticides are cited. For a comprehensive listing of available pesticides, consult the UK Pesticide Guide.

Notifiable pests and diseases

Several non-indigenous, notifiable pests and diseases can occur on protected vegetable crops, usually originating from imported plant material, although in some instances the pests have flown in from adjacent nurseries or overwintered on an infested nursery. Any suspected alien pests should be reported immediately to MAFF Plant Health and Seeds Inspectorate (PHSI). PHSI can provide further information on recognition of notifiable pests and diseases, and will stipulate eradication or containment measures as appropriate. The most commonly found notifiable pests and diseases are described below.

Leaf miners

Non-indigenous leaf miner species include South American leaf miner (*Liriomyza huidobrensis*) and American serpentine leaf miner (*L. trifolii*). Both of these species have a very wide host range and have been confirmed on various UK protected crops, including celery, Chinese leafy salads, cucumber, lettuce, spinach and tomato. Adults of *L. huidobrensis* resemble those of tomato leaf miner (*L. bryoniae*) (p. 357), being small, black flies, approximately 2 mm long, with a yellow spot on the back between the wings. *L. trifolii* adults are more yellow in appearance. Larvae of both species are cream-coloured, and mine between the upper and lower surfaces of the leaf. *Liriomyza* species pupate outside the leaf, although sometimes the puparium will hang off the leaf, whereas puparia of the indigenous pest chrysanthemum leaf miner (*Chromatomyia syngenesiae*) (commonly found on lettuce and leafy salads, see under Lettuce, leaf miners, p. 348) occur in the undersides of leaves. Identification of leaf miners can be done only by specialist examination, and this can be arranged by PHSI.

Tobacco whitefly (*Bemisia tabaci*)

This pest is very similar to glasshouse whitefly (*Trialeurodes vaporariorum*) (see under Cucumber, p. 337), but the adults are slightly smaller and tend to hold their wings slightly apart when at rest, exposing the yellow abdomen. The mature scales of tobacco whitefly, found on the undersurfaces of leaves, tend to be yellow rather than white, as found in glasshouse whitefly. *B. tabaci* has a wide host range and, potentially, can transmit several very damaging viruses. The pest has been confirmed on several UK crops, including cucumber, although to date no problems with viruses have occurred. As with notifiable leaf miner species, identification of whitefly species can be done only by specialist examination, and this can be arranged by PHSI.

Bacterial wilt (*Ralstonia solanacearum*)

This non-indigenous disease affected a few rockwool-grown tomato crops at two locations in England in 1997 and 1998, where irrigation water was abstracted from a river contaminated with the bacterium. The disease causes a pale-brown discoloration in the stem base and also wilting. The bacterium persists in roots of bittersweet (*Solanum dulcamara*) growing as a riparian plant. Statutory action is being taken to remove infected weed hosts in the UK.

Pepino mosaic virus

See p. 361.

Aubergine

Pests

Aphids

The main aphids damaging aubergine are glasshouse & potato aphid (*Aulacorthum solani*), peach/potato aphid (*Myzus persicae*) and potato aphid (*Macrosiphum euphorbiae*). These aphids can be controlled effectively with the parasitoid wasps *Aphidius colemani* (against peach/potato aphid) and *A. ervi* (against glasshouse & potato aphid, and potato aphid). If necessary, biological control can be supplemented with use of the predatory midge *Aphidoletes aphidimyza*. Pirimicarb* (use as for tomato) or nicotine may be integrated if necessary, although very resistant strains of peach/potato aphid may occur.

Broad mite (*Polyphagotarsonemus latus*)

Broad mite can cause leaf and flower distortion, leaf bronzing and fruit russeting. The tiny, white mites tend to hide in growing points and are difficult to detect. This pest is now uncommon, probably owing to incidental control by the predatory mites *Amblyseius cucumeris* introduced against thrips. It is unlikely that an acaricide will be required if using IPM, and the most effective acaricide, dicofol + tetradifon* (use as for tomato) is harmful to biological control agents.

Capsids

The capsid *Liocoris tripustulatus* has recently become a pest on some nurseries. It is similar in general appearance to the species found on cucumber. For description, damage and control, see under Cucumber, p. 337.

Glasshouse whitefly (*Trialeurodes vaporariorum*)

Aubergine is very susceptible to this pest. For description and symptoms see under Cucumber, p. 337. The parasitoid *Encarsia formosa* can give effective control, as on cucumber (see p. 337). Buprofezin* (use as on tomato) can be integrated safely with biological control agents but whiteflies (already resistant to many other pesticides) are now becoming increasingly resistant to this insect growth regulator. Nicotine can give some control of adult whiteflies and can be integrated with IPM if timed carefully.

Thrips

Both onion thrips (*Thrips tabaci*) and western flower thrips (*Frankliniella occidentalis*) can damage aubergine. For description, damage symptoms and biological control with *Amblyseius cucumeris* see under Cucumber, thrips, p. 338.

Two-spotted spider mite (*Tetranychus urticae*)

This pest is common on aubergine; infestations of carmine spider mite (*Tetranychus cinnabarinus*) can also occur. Both species can be controlled biologically using the predatory mite *Phytoseiulus persimilis*, as for cucumber and tomato, pp. 339 and 357. If necessary, the predatory midge *Feltiella acarisuga* can be used to supplement control, as for tomato, p. 358. Fenbutatin oxide* (use as on tomato) can be integrated safely with biological control agents if required.

Diseases

Grey mould (*Botryotinia fuckeliana* – anamorph: *Botrytis cinerea*)
This disease can be particularly troublesome on fruit, and probably results from infected flowers adhering to the developing fruit. It often occurs at the end of the season, during cold, humid weather. Regular protective fungicide applications, e.g. dichlofluanid (off-label) (SOLA 0167/93) or pyrimethanil (off-label) (SOLA 0509/99), are necessary, especially for cold-house crops. Pesticides approved for control of grey mould on protected tomato may be used on protected aubergine, subject to the *Revised Long Term Arrangements for Extension of Use (2000)* regulations.

Powdery mildew (*Erysiphe orontii*)
This fungus also affects tomato. A white, powdery mould develops on the upper leaf surface and causes small, yellow spots. HV sprays of sulfur (off-label) (SOLA 2080/98) provide effective control.

Sclerotinia rot (*Sclerotinia sclerotiorum*)
This very damaging disease usually affects mature plants during the summer. All parts of the plant can be affected, but the light-brown lesions that are typical of the disease are usually found on leaf scars or in the axils of leaves. The lesions extend rapidly under suitable conditions, often causing the death of the affected shoot. Dense, white, fluffy, mycelial growth develops on these lesions under humid conditions, and the large black sclerotia of the fungus develop under this or inside the stem.

Careful removal and destruction of affected plants or plant parts, to avoid sclerotia falling on the floor, will reduce the risk of subsequent crop infection. For soil-grown crops, thorough soil sterilization with steam or methyl bromide will reduce the risk of carryover in the soil.

Wilt (*Verticillium albo-atrum* and *V. dahliae*)
Aubergines are particularly susceptible to wilt, especially in the first 6–8 weeks of planting. Initially, affected plants show yellow lower leaves, and plants may be stunted. As the disease progresses, staining of the vascular system, one-sided wilting and leaf death will usually precede death of the plant. Once the pathogen is established in soil, it is difficult to eradicate. Sterilize the soil with steam or methyl bromide. For chemical control measures, see under Tomato, p. 364.

Bean (French and climbing French)

Pests

Glasshouse whitefly (*Trialeurodes vaporariorum*)
For a description and biological control see under Cucumber, p. 337. Fatty acids, applied as a HV spray, can give some control if necessary and can be integrated with biological control if timed carefully.

Two-spotted spider mite (*Tetranychus urticae*)
For description and biological control with *Phytoseiulus persimilis* see under Cucumber, p. 339. If necessary, tetradifon can be safely integrated with biological control, although spider mite resistance to this acaricide is common and only eggs and the pre-adult motile stages are killed in susceptible populations.

Western flower thrips (*Frankliniella occidentalis*)
For description and biological control see under Cucumber thrips, p. 338. If necessary, dichlorvos (off-label) may be used as a HV spray (SOLA 0625/99) or as a fog (SOLA 0626/99).

Diseases

Grey mould (*Botryotinia fuckeliana* – anamorph: *Botrytis cinerea*)
Beans grown under protection are very susceptible to this disease and should be grown in well ventilated conditions.

Celery

Pests

Aphids
Peach/potato aphid (*Myzus persicae*) is the most common species to infest celery. A spray of pirimicarb or cypermethrin (at HV) should give control, although very resistant strains may not respond to treatment. These strains, and the more occasional aphid pest melon & cotton aphid (*Aphis gossypii*), should be controlled by an HV spray of nicotine.

Celery fly (*Euleia heraclei*)
The larvae mine between the upper and lower surfaces of the leaves causing large blisters. Cypermethrin applied against aphids usually gives incidental control of the adult flies.

Leaf miners
See under Notifiable pests and diseases, p. 329.

Slugs
Slugs can cause grazing damage at the base of the stalks. Control by using methiocarb (off-label) (SOLA 1599/98) or metaldehyde pellets. Often, application around the edge of the glasshouse or polythene tunnel is sufficient.

Diseases

Crater spot (*Thanatephorus cucumeris* – anamorph: *Rhizoctonia solani*)
This sporadic disease causes sunken, brown lesions near the stem base. Often, symptoms are not apparent until close to harvesting. Sometimes, on close examination, fungal webbing can be seen between adjacent petioles. The disease is favoured by high humidity and can be reduced by increased ventilation. Soil sterilization, or a pre-planting soil surface spray of tolclofos-methyl (off-label) (SOLA 2009/99), reduce the risk of a severe attack.

Grey mould (*Botryotinia fuckeliana* – anamorph: *Botrytis cinerea*)
The fungus colonizes damaged petioles, and produces clusters of grey spores. Infected tissue will rot and, eventually, the leaf stalk collapses. Infected but apparently healthy sticks may rot in cold store. High humidity and over-maturity encourage the disease. For control, harvest as soon as the crop is mature. Sprays of carbendazim (off-label) (SOLA 2078/99), used against *Sclerotinia*, may give some control, although carbendazim-resistant strains of *B. fuckeliana* are common and will not be controlled.

Leaf spot (*Septoria apiicola*)
This disease initially causes small, brown spots on leaves and stems. The disease can progress rapidly, causing extensive necrotic lesions on the petiole, and may render the crop unmarketable. While infected seeds are usually the primary source of inoculum, the fungus can also carry over for at least 9 months on affected celery debris in the soil. Overhead watering spreads the disease rapidly, once infection is established. Spores are splashed from pycnidia in leaf lesions, and leaf wetness favours infection.

The most effective method of control is to use seeds that have been soaked for 24 hours in an agitated suspension of thiram maintained at 30°C. Additionally, a 2-year break should be left between celery crops on the same ground. Prompt removal of affected leaves or seedlings can delay epidemic development. Avoid irrigation at times when the crop is likely to remain wet for more than 12 hours. HV sprays of carbendazim (off-label) (SOLA 2078/99), copper oxychloride and copper ammonium carbonate provide protection (Table 9.2). Treatment at short intervals is necessary when conditions favour the disease.

Table 9.2 Chemical control of various diseases on protected celery

Compound	Diseases controlled (or partially controlled)				Max. number of treatments	Minimum harvest interval (days)	Remarks
	Crater spot	Leaf spot	Pythium root rot	Sclerotinia rot			
Sprays							
carbendazim (off-label)	(✓)	✓*	–	✓	4	14	SOLA 2078/99
copper oxychloride	–	✓	–	–	–	0	–
cupric ammonium carbonate	–	✓	(✓)	–	5	0	–
Soil treatment							
tolclofos-methyl (off-label)	✓	–	–	–	1	Pre-planting	SOLA 2009/99
Block incorporation/drench							
etridiazole	–	–	✓	–	1	–	–

* Resistant strains will not be controlled.
(✓) Partial control.

Pythium root rot (*Pythium hydnosporum* and other *Pythium* spp.)
This is the main disease of seedlings and, occasionally, it affects plants raised from seed in peat blocks. Patches of seedlings collapse and are easily pulled out to reveal shrivelled and sometimes reddish-brown roots. Root damage caused by transplanting seedlings increases the risk of pythium root rot. Although affected plants may grow away from the disease, crop growth may be irregular. Propagate in sterile compost and avoid overwatering and root damage. Ensure peat blocks are not stood on dirty surfaces during propagation. Etridiazole may be incorporated into seed compost, or seedlings may be drenched with this compound or with copper ammonium carbonate.

Root and crown rot (*Phoma apiicola*)
Occasionally, this root rot of seedlings may cause losses of the same order as *Pythium*. The disease typically causes a wilting of outer leaves and sometimes progresses to kill the plant. Brown or black lesions are found at the base of the stem. The disease can be seed- or debris-borne and is spread by water splash. Use thiram-soaked seed. Good hygiene, particularly in the disposal of old crop debris, and sterilization of the soil are the most effective means of avoiding the disease. There is no approved chemical control measure for this disease.

Sclerotinia rot or pink rot (*Sclerotinia sclerotiorum*)
The first symptoms of the disease are usually the presence of pink lesions at the base or tip of the leaf stalks. The lesions become covered with a white, fluffy growth within which large, black sclerotia (resting bodies) are formed. These sclerotia can remain viable in the soil for several years. Where successional celery crops are grown it is important to remove and destroy all infected plants and debris, and to consider sterilizing the soil by steam or methyl bromide after an infected crop. HV sprays of carbendazim (off-label) (SOLA 2078/99) will give useful protection.

Chinese cabbage

Pests

Peach/potato aphid (*Myzus persicae*)
This pest should be controlled with a HV spray of deltamethrin (off-label) (SOLA 0125/99) or pirimicarb, although resistant strains may occur.

Slugs
For damage and control, see under Celery, p. 333.

Diseases

Bacterial soft rot (*Erwinia carotovora* ssp. *atroseptica*)
This disease usually results from water stress. Outer leaves that wilt because of water shortage can become infected and develop water-soaked lesions followed by a light-brown, slimy rot. Whole plants may become affected, leading to wilting and collapse. Heart leaves may develop glassiness as a result of excess or insufficient water. Bacterial invasion of this tissue can lead to necrosis, which often progresses to a complete heart rot. Attention to watering and avoidance of high temperatures and humidity will help to reduce the risk of infection, whilst careful disposal of debris will reduce the source of infection.

Courgette and marrow

Pests

Aphids (*Aphis gossypii* and *Myzus persicae*)
For damage symptoms and transmission of viruses, see under Cucumber, p. 337. Weed control will help to reduce sources of aphids. Biological control with the parasitoid wasp *Aphidius colemani* is possible. Pesticide resistance is common in both aphid species. If an aphicide is required, a HV spray or fumigation with nicotine should give control.

Diseases

Grey mould (*Botryotinia fuckeliana* – anamorph: *Botrytis cinerea*)
Extensive losses can be caused by this fungus under suitable conditions of high humidity and cool weather. The disease usually attacks the flowers and developing fruit, causing flower abortion and fruit rotting. It may also attack stems and leaves, and may cause the collapse of whole plants if lesions are low down on the stem. Avoid high humidity, if possible by heat and ventilation, and reduce damage to the plants during picking and trimming. Although there are no fungicides with recommendations for control of the disease on these crops, on courgette imazalil (off-label) (SOLA 1491/99), as used for control of powdery mildew, will also give some control of grey mould.

Powdery mildew (*Erysiphe orontii* and *Sphaerotheca fusca*)
In the warm, humid conditions of late summer and autumn the disease can spread rapidly, causing loss of photosynthetic leaf area and premature death of plants. Spots of white, powdery mould are seen first and these soon coalesce to cover the leaf surface completely. Affected leaves desiccate and hang dead on the plants. On courgette HV sprays of bupirimate or imazalil (off-label) (SOLA 1491/99) should be used at intervals of 10–14 days as soon as the disease is seen. If possible, lower the humidity. Also, destroy infected debris at the end of the crop.

Cucumber

Pests

Aphids

Melon & cotton aphid (*Aphis gossypii*) can be a serious pest, although it usually infests only summer-replanted crops. This aphid varies in colour from yellowish-green to dark green or black, and is found on the undersides of leaves, in growing points and (in severe infestations) on the fruits. Damage symptoms include severe leaf distortion and sooty moulds which grow on honeydew excreted by the aphids. Both *A. gossypii* and peach/potato aphid (*Myzus persicae*) are vectors of cucumber mosaic virus (see p. 340) but *M. persicae* is not common on cucumber. Biological control with the parasitoid wasp *Aphidius colemani* is possible but not widely used. *A. gossypii* is resistant to many aphicides, but spraying with pymetrozine at the first sign of the pest should be effective and can be integrated with biological control programmes for other pests. Fumigation with nicotine is also effective and is not too harmful to biological control programmes if timed carefully.

Capsids

Tarnished plant bug (*Lygus rugulipennis*) has recently become a pest of summer-planted crops on some nurseries, although the bugs are difficult to find on the crop. The adults are 5–7 mm long, varying from green to brown, with large, prominent eyes. The nymphs vary from pale-green to brown. Other species of capsid (including *Liocoris tripustulatus*) also occur on protected crops. Capsid damage symptoms include irregular leaf holes, distorted leaves and fruit, and death of growing points. Nicotine, as recommended for aphids, will give some control; other broad-spectrum pesticides are too harmful to biological control agents used against other pests. Research is being done on the efficacy of the entomopathogenic fungus *Beauvaria bassiana* against capsids, and this pathogen may play a role in future integrated control strategies.

Glasshouse whitefly (*Trialeurodes vaporariorum*)

This is a common pest of cucumber. The adults are small, white, moth-like insects, found on the undersides of leaves and in growing points. Their yellowish, conical eggs are laid on the undersides of leaves, and these turn black just before hatching. The nymphal stages or 'scales' are immobile and greenish-white when young and white when fully grown. Whitefly damage is caused by the excretion of honeydew on to leaves and fruit, which leads to the growth of black sooty moulds. Biological control using the parasitoid wasp *Encarsia formosa* should be effective. The parasitoids are introduced weekly from planting, as black parasitized whitefly 'scales' stuck on to cards. Occasionally, a pesticide may be required for use in whitefly 'hotspots' or for 'clean-up' towards the end of the season. Buprofezin can be used safely within IPM programmes but, as with many other pesticides, whitefly populations are now showing resistance to this insect growth regulator.

Sciarid flies

The most common glasshouse sciarid fly is *Bradysia paupera*. The larvae of these flies are mostly saprophagous, feeding on algae on the surface of the rockwool cube. However, occasionally, they can cause root damage if present in sufficient numbers; the adults have also been recorded as transmitting root pathogens, e.g. *Pythium* spp. The adults are small, fragile flies and the larvae are white and translucent, with shiny, black head capsules. Treatment is rarely required but, if necessary, larvae can be controlled by introductions of predatory mites (*Hypoaspis* spp.). Adult flies in damage 'hotspots' can be trapped on yellow sticky traps placed near the base of the plants or by using a localized spray of a sticky polybutene plus deltamethrin mixture on the polythene covering the rockwool slabs. Use of pesticides against sciarid fly adults is inadvisable, as this would interrupt biological control programmes for other pests.

Thrips

Western flower thrips (WFT) (*Frankliniella occidentalis*) is the main pest of cucumbers, although onion thrips (*Thrips tabaci*) can also occur. WFT adults are small, slender, brown or yellow insects, approximately 2 mm long, with fringed wings; the nymphs are yellow and wingless. Both species feed on the leaves, causing small, white flecks or patches, within which tiny black faecal specks can be seen. WFT also feed in the flowers and on young developing fruits, causing deformed 'pigtail' fruit. Nymphs of onion thrips, and a proportion of those of WFT, drop from the leaves to pupate on the polythene floor or rockwool slab coverings. A mixture of sticky polybutenes and deltamethrin is available for spraying on to the floor covering, which controls thrips pupating on the floor. Although this method is compatible with IPM, it is now unpopular with growers as it is sticky to walk on and gives poor control of WFT. The majority of growers use *Amblyseius cucumeris* (a predatory mite) for control of thrips within IPM programmes. These small, white predators feed on the thrips nymphs and are introduced at intervals of approximately 6 weeks, starting from planting. 'Controlled release' paper sachets containing a culture of the predators are hung on the plants. This system now gives generally reliable control of WFT, as long as introductions are managed correctly. However, use of an aerosol formulation of dichlorvos is often necessary to reduce numbers of thrips at the start of each crop, for 'cleanup' between successive plantings and towards the end of the season. Dichlorvos must not be used at the flowering stage, as it can cause flower or fruit abortion.

Two-spotted spider mite (*Tetranychus urticae*)

This is a major pest, which feeds on the undersides of leaves, causing patches of fine speckling and chlorosis. In severe attacks, leaf or whole-plant senescence can occur, and webbing produced by the mites can be extensive. The translucent, round eggs are laid on the undersides of the leaves and these hatch into the pale-green, oval, immature stages. Older nymphs and adults are green, with two black patches on their backs. In autumn, in response to shortening day-length and plant senescence, female mites tend to develop a brick-red colour and migrate towards the greenhouse structure where they remain in a state of diapause over the winter. Any diapausing females emerge in the following spring, as day-length and temperatures increase, and migrate to the new crop.

Biological control with the predatory mite *Phytoseiulus persimilis* should be effective, if introduced at the first sign of damage and repeated as required until established in the crop. For successful management, regular monitoring of predator establishment is essential. The use of an acaricide may be needed to reduce numbers of spider mites in 'hotspots', and restore the balance between predators and pests. Fenbutatin oxide is a specific acaricide, which is safe to both *P. persimilis* and other biological control agents used in IPM. Abamectin can also be used as a spot treatment, without long-term adverse effects on biological control agents, and as a 'cleanup' treatment towards the end of the crop. This 'cleanup' procedure is a very important component in spider mite control strategies, to prevent diapausing females entering the structure of the house towards the end of the season.

Diseases

Basal stem rot (*Erwinia carotovora* ssp. *carotovora*)
This is a slimy, soft-rot of the stem base, found most commonly in soil-grown crops. It occurs following stem damage as a result of pest attack or through natural growth cracks. The stem base should be kept dry as far as practicable.

Black root rot (*Phomopsis sclerotioides*)
This very common disease of soil-grown crops causes rotting of smaller roots, followed by the tap root and hypocotyl; plants subsequently wilt. *Phomopsis* attack is characterized by black spotting on small roots and black lesions on larger roots and hypocotyls below ground. Typically, the stem base thickens and, occasionally, greyish lesions extend above ground level. In cold soil, where root growth is slow, rotting may be severe and black lesions may not develop. Regular steaming of the soil is the most effective control; methyl bromide fumigation is not recommended. This disease is uncommon in rockwool-grown crops; where it occurs, drenches of carbendazim (off-label) (SOLA 1476/95) can slow disease spread.

Cucumber green mottle mosaic virus (CGMMV)
This highly infectious virus disease can spread rapidly throughout a crop if appropriate management action is not taken. The virus causes a light-green to

dark-green mottle in young leaves, with the darker areas bubbled. Older leaves are symptomless and the disease is generally less conspicuous than cucumber mosaic virus (CMV). Fruit are symptomless but yield can be reduced by up to 25% if plants become infected at an early stage. Spread occurs by handling the crop, and via knives and clothes. Seed should be heat-treated (3 days at 70°C). If a small number of affected plants are seen, these should be removed promptly and carefully, together with at least six plants either side. Rockwool slabs should be replaced with new slabs after the whole area has been treated with a suitable disinfectant. Some growers use a UHT milk suspension to limit spread of the virus, as a hand and knife dip and as a spray to plants. Restrictions on staff entry to an affected house, and the use of separate coveralls, reduce the risk of further spread. For soil-grown crops, steaming reduces the risk of carryover in the soil. Rockwool slabs that are to be re-used should be steamed thoroughly.

Cucumber mosaic virus (CMV)

Although CMV is not an uncommon disease in cucumber crops, it causes significant yield loss only rarely. It is most damaging when root diseases are also present (e.g. *Pythium*), as plants then wilt and die within 7–10 days. A variety of symptoms occur according to virus strain, plant age and growing conditions. Most commonly, leaves and shoots show a yellow-green mosaic. The virus is spread by aphids, both melon & cotton aphid (*Aphis gossypii*) and peach/potato aphid (*Myzus persicae*), and to a much lesser extent by handling the crop. Numerous weeds are a potential source of the virus, including annual nettle (*Urtica urens*) and chickweeds (e.g. *Stellaria* spp.). Control of CMV can be achieved by prompt removal of affected plants, combined with good control of aphids (see under Cucumber, aphids, p. 337).

Downy mildew (*Pseudoperonospora cubensis*)

Severe outbreaks of this disease occurred in the UK in 1986, and until 1991 it was a notifiable disease. Recently, it has been found in occasional crops most years, usually towards the end of the season, and can be kept at a low level by good management. Symptoms show on the upper surface of leaves as a mosaic of angular-shaped, yellowish areas that develop a purplish or grey-black felt of the fungus on the lower surface. These areas later become necrotic. The disease can rapidly reach epidemic proportions in warm, humid conditions, spores being dispersed in air currents. Reducing prolonged leaf wetness and glasshouse humidity, by increased heat and ventilation, provides very effective control. Severe attacks are most likely in polythene tunnels and in unheated crops. HV sprays of metalaxyl + copper oxychloride (off-label) (SOLA 1564/98) give some control, although it should be noted that metalaxyl-resistant strains of the fungus have been reported in some countries and these will not be controlled. This obligate parasite is restricted to the Cucurbitaceae, so effective disposal of crop debris helps prevent carryover between crops.

Fusarium wilt (*Fusarium oxysporum* f. sp. *cucumerinum*)
In recent years, this soil-borne disease has become more troublesome in crops grown in rockwool slabs. Usually, it is first noticed 3–4 weeks after planting out, with wilting of one or more of the lower leaves. Wilting increases until the whole plant is affected and dies. The vascular strands are discoloured brown and an orange or pink sporulation occurs at the nodes and develops along the stem. All affected plants should be removed carefully as soon as they occur, preferably before the sporing stage is present. On crops grown on inert media, drenching with carbendazim (off-label) (SOLA 1476/95) will give some control, providing it is done in the early stages of infection. The glasshouse needs to be cleaned and disinfected thoroughly at the end of cropping. For soil-grown crops the soil should be sterilized. Some cultivars show resistance to the disease. Grafting on to a resistant rootstock (e.g. *Cucurbita ficifolia*) is an effective method of control.

Grey mould (*Botryotinia fuckeliana* – anamorph: *Botrytis cinerea*)
This disease is most troublesome in unheated or partially heated crops and in long-season crops. Yield loss can occur from direct fruit infection but more commonly from stem infections, which lead to shoot or plant death. It is encouraged by high humidity and by failure to remove damaged tissue and yellowing foliage, which are readily colonized by the fungus. Weekly leaf trimming is more effective at preventing grey mould than occasional leaf removal. Thinning lateral shoots reduces the risk of grey mould in the upper canopy. HV fungicide sprays used to protect against the disease include chlorothalonil and iprodione (Table 9.3).

Gummosis (*Cladosporium cucumerinum*)
Gummosis is now rare because of the resistance that has been bred into modern cultivars. It causes sunken, scab-like depressions on fruit, from which the sap exudes to form an amber gum. Fruits are often distorted, especially if infected when young. Remove all diseased fruits and reduce humidity, if possible, by heating. Spray with chlorothalonil while the disease persists.

Penicillium stem rot (*Penicillium oxalicum*)
Occasionally, this disease is very damaging, spreading rapidly at high humidity. It causes stem lesions at the plant base or nodes and, occasionally, fruit rotting from the tip. It is recognized by the typical blue-green fungal growth, which releases a cloud of spores on touching. It is partially controlled by prompt removal of affected plants and by HV sprays of iprodione.

Powdery mildew (*Erysiphe orontii* and *Sphaerotheca fusca*)
A white, powdery mould forms on both surfaces of the leaves, causing chlorotic spots which spread and desiccate the leaf tissue. Spores produced on cucumber are viable for only a short period but overlapping crops provide a continuous supply of inoculum. The disease is most common in the summer and autumn. If

Table 9.3 Chemical control of various diseases on protected cucumber

Compound	Diseases controlled (or partially controlled)					Max. number of treatments	Minimum harvest interval (days)	Remarks
	Downy mildew	Grey mould	Powdery mildew	Pythium root rot	Stem rot (*Mycosphaerella*)			
Sprays								
bupirimate	–	–	✓*	–	–	6	2	[Note 1]
chlorothalonil	(✓)	✓	✓	–	(✓)	2	2	–
copper oxychloride + metalaxyl (off-label)	✓	–	–	–	–	8	2	SOLA 1534/95
imazalil	–	(✓)	✓*	–	(✓)	–	1	–
iprodione	–	✓	–	–	(✓)	4	2	–
Root drenches								
carbendazim (off-label)	–	✓*	(✓)*	–	(✓)	8	2	SOLA 1476/95
etridiazole	–	–	–	✓	–	–	3	–
propamocarb hydrochloride (off-label)	(✓)	–	–	✓	–	4	2	SOLA 2032/99**

* Resistant strains will not be controlled.
** For crops grown on inert media/substrates or by NFT.

Note 1: May cause leaf damage when light levels are low.

mildew is at a low level, prompt removal of affected leaves delays the need for chemical treatment. Fungicides are still used widely for mildew control and are usually applied preventatively by LVM application. At the first sign of the disease, apply one of the treatments listed in Table 9.3, p. 342. Careful selection of fungicide treatments is required as resistance has occurred to several fungicide groups, including the sterol biosynthesis inhibitors (SBIs); resistant strains of mildew can rapidly come to predominate if the same fungicide group is used for several successive sprays. The development of mildew-tolerant cultivars has reduced the need for frequent spraying in summer planted crops. However, such cultivars tend to show chlorosis under low light conditions and, consequently, are not used for early plantings; there is also concern over their yield potential compared with susceptible cultivars. Often, they are planted just at the row ends or around the perimeter of a susceptible crop. Increased silicon nutrition has been demonstrated to reduce the severity of powdery mildew on cucumber and is practised by some growers, the silicon being applied to the feed solution as potassium metasilicate. Some common weeds, e.g. sow-thistles (*Sonchus* spp.), may also become infected by powdery mildew and should be removed from the glasshouse and its neighbourhood.

Pythium root and stem base rot (*Pythium aphanidermatum* and other *Pythium* spp.)

Pythium root and stem base rot is a major disease of cucumber. Infection causes a water-soaked rot of roots at the base of propagation cubes and, in some instances, an orange-brown rot at the stem base. Affected plants wilt progressively in sunny weather and may die. Crops replanted in mid-summer on to once-used rockwool slabs, when root temperatures may be high because of the lack of crop shading, can be particularly badly affected. Species of *Pythium* are also responsible for damping-off of seedlings.

For soil-grown crops, steam-sterilize the soil and, if the disease is troublesome, drench roots with etridiazole or propamocarb hydrochloride (off-label) (SOLA 2032/99). In rockwool-grown crops, a preventive treatment may be given as a drench to the rockwool cube, using propamocarb hydrochloride (off-label) (SOLA 2032/99). Rockwool slabs can be treated by drenching, or by applying propamocarb hydrochloride in the nutrient solution. Occasional gaps left between adjacent rockwool slabs minimize the risk of extensive spread along a row in water films on the floor. Fungus flies (Mycetophilidae) and shore flies (Ephydridae) need to be kept at low levels as these are recognized vectors of *Pythium*.

Root mat (*Agrobacterium* sp.)

This disease is caused by rhizogenic strains of *Agrobacterium* bv. 1, and both soil-grown and hydroponic crops are affected. The bacterium is soil-borne and is also found commonly in the run-off solution from affected hydroponic crops. Infection results in abnormal plant growth, typically upward growth of roots from the surface of the propagation block (or soil), swelling of the propagation block and slab (owing to excessive root production), and swelling of the stem base. Affected plants may grow slowly, and produce an increased proportion of bent fruit or excessive vegetative growth, to the detriment of fruit production. Often, only occasional plants in a crop are affected and these may appear to grow normally.

Good hygiene during plant propagation reduces the risk of a damaging attack. Disinfectants shown to be effective against *Agrobacterium* include (a) glutaraldehyde plus a quaternary ammonium compound, (b) hydrogen peroxide plus peracetic acid, and (c) sodium hypochlorite. Thorough steaming of rockwool slabs which are to be re-used prevents carryover of the disease.

Sclerotinia stem rot (*Sclerotinia sclerotiorum*)

This disease is most commonly found on shoots in the upper canopy, although it can occur on fruits and stems. It is characterized by a fluffy, white fungal growth and subsequent development of hard, black sclerotia. Control by HV sprays of chlorothalonil and iprodione. Remove affected shoots to prevent sclerotia falling to the floor.

Stem and fruit rot (*Didymella bryoniae*)

This disease, sometimes known as 'black stem rot', is one of the most common problems of cucumber, under whatever system the crop is grown. Lesions develop on the stem, leaves and fruit and, occasionally, on roots. Stem infections usually originate at pruning wounds or from damaged tissue; fruits are infected from the blossom-end or from contact with an infected stem or leaf; on leaves, spreading, light-brown patches develop, usually from the leaf margin, which later collapse and rot. The growing point of the plant can also be infected. Rotting of the stem base can be very damaging in replanted crops. Black pycnidia develop on the lesions and exude pinkish spore masses.

HV sprays of chlorothalonil or iprodione help protect against the disease; a key spray is to the stem base after replanting. Use heat and ventilation at least one hour before sunrise to prevent condensation on fruit. Partial resistance to *Didymella* stem rot has been introduced recently in some new cultivars.

Verticillium wilt (*Verticillium albo-atrum*)

Both soil-grown and rockwool-grown crops may be affected. The plants wilt and the leaves become yellow and desiccated, from the base upwards. For crops grown on inert media, drenching with carbendazim (off-label) (SOLA 1476/95) will give some control, providing it is done before symptoms are obvious.

Herbs

Pests

Aphids

Aphids are the main pest of herbs grown under protection and include glasshouse & potato aphid (*Aulacorthum solani*), mint aphid (*Ovatus crataegarius*), peach/

potato aphid (*Myzus persicae*) and violet aphid (*Myzus ornatus*). Damage symptoms include distorted leaves with chlorosis, and the presence of white, cast-off aphid skins on the leaves. Retailers have a zero tolerance of pests or damage on herbs, so it is important to maintain high levels of control. Biological control using the parasitoid wasps *Aphidius colemani* and *A. ervi* is successful, as long as temperatures are adequate and parasitoid introductions are managed correctly. It is essential to identify aphids to species, in order to introduce the appropriate parasitoid species. *O. crataegarius* does not seem to be attacked by either parasitoid species. The predatory midge *Aphidoletes aphidimyza* can be used to complement control by parasitoids. Pirimicarb* (use as on protected lettuce) can be used within IPM programmes and will control most aphid species occurring on herbs, although resistant strains of *M. persicae* may occur. Nicotine may also be used, either as a spray or as a fumigant, as it has only short persistence against biological control agents and controls resistant aphids. If IPM is not being used, or if 'cleanup' is needed before sale, cypermethrin* (use as on protected lettuce) should be effective against all aphids, except for resistant strains of *M. persicae*.

Brown soft scale (*Coccus hesperidum*)
This is a common pest of bay. The flat, oval, brown scales, with dark stripes, are found on the leaves and stems, and sooty moulds often colonize the sticky honeydew that they produce in abundance. If using IPM, the scales are often parasitized by naturally occurring parasitoids. If necessary, a HV spray of malathion* (use as for protected lettuce) will give some control but this is not compatible with IPM.

Caterpillars
Various species occur on herbs, including *Pyrausta* spp. on mint. The caterpillars cause 'windowing' of the upper surface of leaves and web leaves together. If using IPM, a HV spray of *Bacillus thuringiensis** (use as for protected lettuce) is safe to biological control agents. HV sprays of cypermethrin* (use as for protected lettuce) will give control but this insecticide is not compatible with IPM.

Leafhoppers
Chrysanthemum leafhopper (*Eupteryx melissae*) is the most common species on a range of herbs (both outdoors and under protection), although glasshouse leafhopper (*Hauptidia maroccana*) can also occur in glasshouses. Adults of *E. melissae* are green with dark spots. For a description of glasshouse leafhopper, and damage symptoms, see under Tomato, p. 356. Although the parasitoid wasp *Anagrus atomus* will attack the eggs of *H. maroccana* it does not parasitize those of *E. melissae*. A HV spray of nicotine* (use as for protected lettuce) will give some control within IPM. A HV spray of cypermethrin* (use as for protected lettuce) or deltamethrin (off-label) (SOLA 0945/99) will give control, but these pyrethroids are not compatible with IPM.

Leaf miners
Chrysanthemum leaf miner (*Chromatomyia syngenesiae*) occasionally causes damage to herbs, particularly mints – see also under Notifiable pests and diseases, p. 329. Introductions of the parasitoid wasps *Dacnusa sibirica* and *Diglyphus isaea* can give effective control. A HV spray of abamectin* (use as for protected lettuce) will control larvae within the leaf, but has some harmful effects on biological control agents.

Western flower thrips (*Frankliniella occidentalis*)
This pest can damage various herb species and is a particular problem if herbs are allowed to flower. For description of the pest and leaf damage symptoms, see under Cucumber, thrips, p. 338. Because of pesticide resistance, biological control is the best option. The predatory mite *Amblyseius cucumeris* is usually effective, if managed correctly. If a pesticide is needed, a HV spray of abamectin* or malathion* (use both pesticides as for protected lettuce) or dichlorvos (off-label) (SOLA 0625/99) may be used.

Whiteflies
Glasshouse whitefly (*Trialeurodes vaporariorum*) is a common pest of herbs, particularly rue, sages, mints, oregano and thymes. See also under Notifiable pests and diseases, tobacco whitefly (*Bemisia tabaci*), p. 329. Biological control of whiteflies on herbs with the parasitoid wasp *Encarsia formosa* is not always reliable. Also, whiteflies are resistant to many pesticides approved for use on herbs, although a HV spray of nicotine* (use as for protected lettuce) or fatty acids, or fumigation with nicotine, can give some control of adults.

Diseases

Grey mould (*Botryotinia fuckeliana* – anamorph: *Botrytis cinerea*)
Grey mould can attack many herbs but is particularly common on basil. It causes a leaf and stem rot on basil seedlings in pots and a stem die-back of soil-grown plants cropped for cut basil. The characteristic grey mass of spores is produced on affected tissue. Increased ventilation to reduce humidity provides some control. Where the harvest interval allows, and providing treatment is in compliance with maximum residue level (MRL) requirements, a HV spray of iprodione* (use as for protected lettuce) can be used.

Leaf spot of parsley (*Septoria petroselini*)
This common disease causes a pale leaf spot with a brown border; pycnidia are clearly visible within spots. The quality of the crop is markedly reduced. The disease is very similar to that caused by *Septoria apiicola* on celery, but *S. petroselini* does not attack celery. It can be controlled by the use of clean or treated seed.

Powdery mildew (*Erysiphe heraclei*)
Powdery mildew affects several herbs grown under protection, including fennel and parsley. For control, sulfur (off-label) (SOLA 1715/97) can be used on parsley and some other leafy herbs.

Pythium root rot (*Pythium* spp.)
This is sometimes a problem in pots of herb seedlings, especially cress and parsley. Uneven growth and wilting of occasional seedlings can make pots unmarketable. Disease is controlled by attention to cleanliness of benches and irrigation water and by providing optimum growing conditions (both compost and aerial environment) with no checks to plant growth.

Rust (*Puccinia menthae*)
This pathogen is responsible for the most important disease of cultivated mint, and symptoms are readily seen as obvious orange pustules on the lower leaf surfaces. Overwintering rhizomes are infected by teliospores in the soil, resulting in systemic infection, and pale, swollen, distorted shoots emerge in the spring. Control by heat treatment of infected rhizomes (44°C per 10 minutes) pre-planting, by minimizing prolonged leaf wetness. Some cultivars are resistant.

Lettuce

Pests

Aphids
Several species attack protected lettuce, of which peach/potato aphid (*Myzus persicae*) is the most common. Other species which may occur include currant/lettuce aphid (*Nasonovia ribisnigri*), glasshouse & potato aphid (*Aulacorthum solani*), potato aphid (*Macrosiphum euphorbiae*) and, occasionally, melon & cotton aphid (*Aphis gossypii*). HV sprays of cypermethrin or pirimicarb should give control of most species, although *A. gossypii* and resistant strains of *M. persicae* will not respond to treatment. A HV spray of nicotine should control resistant aphids. Research is being carried out on integrated control of aphids on protected lettuce and this may be a viable strategy in the future.

Caterpillars
Caterpillars of various pests feed on the leaves, including angle-shades moth (*Phlogophora meticulosa*), silver y moth (*Autographa gamma*), tomato moth (*Lacanobia oleracea*) and, occasionally, carnation tortrix moth (*Cacoecimorpha pronubana*) and related species. Damage symptoms are holes in the leaves; in the case of tortrix moth caterpillars, the leaves are rolled and webbed together. HV sprays of cypermethrin applied against aphids usually give control.

Leaf miners

The most common leaf miner attacking lettuce is chrysanthemum leaf miner (*Chromatomyia syngenesiae*). See also under Notifiable pests and diseases, p. 329. The adults are small, grey-brown flies, approximately 2 mm long, and cause small, white, feeding spots on the leaves. The creamy-white larvae mine between the upper and lower surfaces of the leaf. HV sprays of cypermethrin, applied against aphids, usually give incidental control of the adult flies. If leaf mines are seen, abamectin will give control but this may be used on protected lettuce between early March and late October only. Leaf mines are often most numerous on the lower leaves, which can be trimmed off at harvest. Careful disposal of discarded leaves is important, to reduce infestation of following crops.

Shore flies (*Scatella* spp.)

Two species of shore fly (*Scatella stagnalis* and *S. tenuicornis*), also known as 'glasshouse wing-spot flies', occur commonly in greenhouses. The adults, which are small and black-bodied, do not cause direct damage to lettuce. However, they can be present in large numbers, as both adults and larvae feed on algae growing on the soil surface. The presence of flies on the lettuce at harvest can cause crop rejection, owing to retail intolerance of insect contaminants. There are no insecticides approved for use against shore flies on lettuce. Current research on integrated control methods has indicated that use of algicides can reduce the problem, and several biological control agents have potential for future control strategies.

Slugs

Slugs are common pests of lettuce, causing ragged feeding damage to the edges of leaves and shallow cavities in the petioles. For control with molluscicide pellets, see under Celery, p. 333. Research is being carried out on the use of entomopathogenic nematodes for biological control of slugs on horticultural crops, and this may be a cost-effective method in the future.

Diseases

Big-vein

This disease is caused by a virus and is spread by the common soil fungus *Olpidium brassicae*. The disease is most damaging when plants become infected pre-planting, and can cause serious losses in winter-grown crops. Symptoms do not become visible until 4–5 weeks after infection. Puckering of the leaves, clearing of areas along the veins, reduced growth rate and, in extreme cases, stunting and the production of unmarketable plants, may result. If the disease becomes established in NFT systems it is difficult to eradicate and it spreads rapidly, causing serious losses to slow-maturing winter crops.

Regular sterilization of soil with steam or methyl bromide is necessary to maintain the vector at a low level. The main way of avoiding the disease is to plant only healthy plants. Make sure plants are raised in systems that are isolated from the soil. Carbendazim, incorporated into the blocking compost, will protect young plants. The use of non-ionic wetting agents in the circulating solution restricts the spread of the organism in NFT systems. Agral, added to the NFT solution twice a week, is recommended. Grower reports suggest that fosetyl-aluminium, applied against downy mildew, provides some control of big-vein.

Bottom rots

Five diseases cause rotting of the stem and leaf bases: grey mould; phoma basal rot; pythium basal rot; rhizoctonia rot and sclerotinia rot (see below). Infected plants may be killed or may require excessive trimming at cutting.

Downy mildew (*Bremia lactucae*)

This is the most important fungal disease, and it can attack the plant at any stage of growth. It is first noticed on lower leaves as pale, angular areas bounded by the veins. Masses of white spores form on the lower leaf surface. Free (unbound) water on the leaf surface is essential for infection by spores; checks to growth also predispose the plant to infection. Thus, good management of watering and conditions in the glasshouse will aid control. Exclusion of the disease during propagation and immediately after planting by regular spraying is also essential. Because there are restrictions on the levels of residues permitted on marketed lettuce, sprays must be applied well before harvest. Resistant lettuce cultivars are available but these should still be subject to routine chemical measures because pathotypes of the fungus capable of overcoming host resistance may occur.

An integrated management strategy is commonly used successfully to control the disease in protected crops. This strategy combines the use of appropriate resistant cultivars to combat metalaxyl-resistant pathotypes (known in the UK since 1983), continued use of metalaxyl to control metalaxyl-sensitive pathotypes, and use of other fungicides, with modes of action different from that of metalaxyl, to minimize the risk of selecting new metalaxyl-resistant pathotypes.

The most effective means of chemical control is to spray HV during propagation, then at intervals during crop growth. Compounds available are fosetyl-aluminium (off-label) (SOLA 0057/99), mancozeb, mancozeb + thiram, propamocarb hydrochloride (off-label) (SOLAs 1971/99, 1972/99), thiram and zineb. Fosetyl-aluminium may also be incorporated into the blocks for crops grown between September and April. It is essential to follow instructions on spray number and harvest interval to avoid residue problems. Most compounds can be applied only in the first 14–21 days after planting, although propamocarb hydrochloride has a 14-day harvest interval (Table 9.4, p. 350).

Grey mould (*Botryotinia fuckeliana* – anamorph: *Botrytis cinerea*)

This fungus can cause severe losses, especially in crops planted in autumn and winter. *B. fuckeliana* is a weak pathogen and plants are predisposed to attack

Table 9.4 Chemical control of various diseases on protected lettuce

Compound	Diseases controlled (or partially controlled)			Max. number of treatments	Minimum harvest interval (days)	Remarks
	Downy mildew	Grey mould	Rhizoctonia (*Thanatephorus*)			
Sprays						
fosetyl-aluminium (off-label)	✓	–	–	2	14†	SOLA 0057/99
iprodione	–	✓*	(✓)	3	28 winter	October–February
				7	7 summer	March–September
mancozeb	✓	(✓)	–	2‡	21	DTC restrictions [Note 1] apply
metalaxyl + thiram	✓*	–	–	2 to 3‡	21	DTC restrictions [Note 1] apply
prochloraz (off-label)	–	(✓)	–	4	21	SOLA 2002/99§
propamocarb hydrochloride (off-label)	✓	–	–	–	14	SOLAs 1971/99, 1972/99¶
pyrimethanil (off-label)	–	✓	–	–	14	SOLA 1590/00
thiram	✓	–	–	2 to 3‡	21	DTC restrictions [Note 1] apply
Soil treatment (pre-planting only)						
quintozene	–	(✓)	✓	1	–	–
tolclofos-methyl	–	–	✓#	1	–	Variable results reported
Block incorporation						
fosetyl-aluminium	✓	–	–	1	–	Alternative to spray application

* Resistant strains will not be controlled.
† Latest application permitted is 21 days after planting out.
‡ Sprays must not be applied within 14 days of planting (April–October) or within 21 days (November–March).
§ Not to be used during propagation, or on hydroponic crops. Main use is for ring spot control.
¶ For use on soil-growth crops only.
# Do not disturb the soil surface after application.

Note 1: DTC restrictions – the maximum number of treatments of mancozeb, zineb or other EBDC fungicide, or thiram, is two per crop post-planting up to 2 weeks later, and none thereafter. If thiram-based products are used post-planting on crops that will mature from November to March, three treatments are permitted within 3 weeks of planting out.

following damage due to handling at planting, slugs, other fungi (e.g. downy mildew) and poor establishment or checks to growth after planting. In plants approaching maturity, red-brown lesions are often seen on the outer leaves close to the soil surface and at the stem base. Sudden wilting and characteristic masses of grey spores accompany extensive rotting. The crop should be well ventilated to reduce the RH of the air. Remove debris from the soil surface. Plant into a moist soil and irrigate well, soon after planting. Chemical control measures include quintozene applied to the soil surface, and HV sprays of iprodione, pyrimethanil (off-label) (SOLA 1590/00) or thiram applied post-planting. Great care needs to be taken to allow the recommended harvest intervals, in order to avoid residues in the crop at harvest (Table 9.4, p. 350). Prochloraz (off-label) (SOLA 2002/99) applied for ring spot also provides some control of grey mould.

Phoma basal rot (*Phoma exigua*)
Occasionally, this disease causes significant losses, particularly in winter crops in poorly drained soil and under gutters. A dry, brown rot occurs at soil level. Large (10–20 mm), grey-black spots may also occur on lower leaves. Pycnidia of the fungus develop in affected tissue. Control by improving soil drainage and minimizing water drip. HV sprays of prochloraz (off-label) (SOLA 2002/99), applied against ring spot, reduce losses due to *Phoma*.

Pythium basal rot (*Pythium* spp.)
This disease is uncommon, but can cause widespread damage in late autumn and winter. A slimy, black, wet rot affects the lower leaves and stem. Resting spores (oospores) of *Pythium* occur abundantly in affected tissue. Avoid overwatering. Fungicides applied to control downy mildew may give some control. Soil sterilization may be necessary if the problem persists in successive crops.

Rhizoctonia rot (*Thanatephorus cucumeris* – anamorph: *Rhizoctonia solani*)
This disease is common but severe only occasionally. The fungus causes a superficial, slimy rot of the stem and petiole bases, beneath an apparently healthy head. Red flecks on petioles, and fine, mycelial strands may be associated with the rot; small, brown sclerotia are sometimes present. Consider sterilizing the soil after a severe outbreak. Spray or dust quintozene or spray tolclofos-methyl on to the surface of the soil before planting out. However, do not apply either chemical to the growing crop. Where attacks occur, remove debris. Iprodione, applied to control grey mould, may suppress *Rhizoctonia*.

Ring spot (*Microdochium panattonianum*)
Occasionally, this disease causes losses in glasshouse crops, usually under vents where water splash occurs. Symptoms are small (4–5 mm diameter), brown, circular spots on the older leaves, and sunken, brown lesions on the veins; the latter can be mistaken for symptoms of slug damage. The fungus survives on crop debris and in the soil. It is spread readily by water splash and the most effective

method of control is to reduce water splash. Should it be necessary to apply fungicides, HV sprays of prochloraz (off-label) (SOLA 2002/99) or thiram can be used.

Sclerotinia rot (*Sclerotinia minor* and *S. sclerotiorum*)
Outbreaks of the disease are erratic and probably less common than the other bottom rots but they may cause severe losses, especially in warm weather. Lower leaves and stem bases develop a soft rot and become covered with a dense, white mycelium, which may contain large, black sclerotia. The plant wilts progressively. Remove infected debris and the surrounding soil because sclerotia can infect the next crop. Sterilize soil or flame the surface of affected areas. Do not dump infected debris from lettuce or other crops close to glasshouses because wind-blown ascospores can cause infections. HV sprays for grey mould control (Table 9.4, p. 350) are likely to reduce the incidence of this rot.

Pepper

Pests

Aphids
Glasshouse and potato aphid (*Aulocorthum solani*), melon & cotton aphid (*Aphis gossypii*) and peach/potato aphid (*Myzus persicae*) are all common pests of peppers. The aphids are found on the undersides of leaves, and damage symptoms include sooty moulds on leaves and fruit, and (with *A. solani*), bright yellow patches on the leaves. Biological control with the parasitoid wasps *Aphidius colemani* (for *A. gossypii* and *M. persicae*) and *Aphidius ervi* (for *A. solani*) is usually successful, as long as the appropriate parasitoid species is used and managed effectively within IPM. The predatory midge *Aphidoletes aphidimyza* can be used to supplement control, although it does not always establish in rockwool-grown crops, owing to poor survival of pupae on the polythene flooring. Pirimicarb can be used within IPM but *A. gossypii* and the most resistant strains of *M. persicae* are not controlled by this aphicide. Nicotine, either as a spray or as a fumigant, should control all species and strains of aphids.

Capsids
The capsid *Liocoris tripustulatus* has recently become a pest in some nurseries. It is similar in general appearance to the species found on cucumber. For description, damage and control, see under Cucumber, p. 337.

Caterpillars
The most common species attacking peppers is tomato moth (*Lacanobia oleracea*). Damage symptoms are large holes in the leaves and, in heavy infestations, holes in the fruit. HVsprays of *Bacillus thuringiensis* are compatible with IPM.

Glasshouse whitefly (*Trialeurodes vaporariorum*)
This is an occasional pest on peppers. For description and biological control with the parasitoid *Encarsia formosa*, see under Cucumber, p. 337.

Sciarid flies (e.g. *Bradysia paupera*)
See under Cucumber, p. 338. As on cucumber, this is an occasional and localized pest.

Thrips (*Frankliniella occidentalis* and *Thrips tabaci*)
For description, see under Cucumber, p. 338. Biological control with the predatory mite *Amblyseius cucumeris* is very successful on pepper, as the predators can establish in advance of thrips infestation, by feeding on pollen. The predators are introduced in slow-release packs, at the first flowers. The application of pesticides against thrips on peppers is not usually necessary, but dichlorvos (off-label) as a HV spray (SOLA 0625/99) or as a fog (SOLA 0626/99) may be used.

Two-spotted spider mite (*Tetranychus urticae*)
This is an occasional pest on peppers. For a description see under Cucumber, p. 338. The predatory mite *Phytoseiulus persimilis*, as used on cucumber, should give good control. If an acaricide is needed, a HV spray of fenbutatin oxide (off-label) (SOLA 0857/97) is compatible with IPM.

Diseases

Damping-off and foot rot (*Phytophthora* spp., *Pythium* spp. and *Thanatephorus cucumeris* – anamorph: *Rhizoctonia solani*)
These soil-borne fungi cause damping-off of seedlings, foot rot of young plants or root rot of mature plants. Peppers are very slow to root and are predisposed to attack if sown or planted too deeply, in compost below 15°C or if over-watered. Crop production in sterilized soil, or a hydroponic system, reduces the risk of these diseases. If *Pythium* or *Phytophthora* infections do occur, remove diseased plants and drench the remainder with propamocarb hydrochloride (off-label, for crops on inert substrates or NFT) (SOLA 2032/99).

Fusarium stem and fruit rot (*Fusarium solani*)
This is an uncommon disease, which causes stem rotting and, occasionally, a fruit end rot. It is characterized by pale-brown lesions on the stem which develop an orange spore mass. Disease development is primarily on wounded or weakened tissue and is believed to be favoured by high humidity. Remove affected plants, especially where infection occurs early in the crop life; minimize stem damage and high humidity.

Grey mould (*Botryotinia fuckeliana* – anamorph: *Botrytis cinerea*)
For a description see under Tomato, p. 360. To prevent attack regulate heating,

ventilation and watering to avoid high humidity in the crop. If necessary, apply HV sprays of chlorothalonil or dichlofluanid (off-label) (SOLA 0167/93).

Pepper mild mosaic virus

This tobamovirus causes a mild mosaic of leaves and fruit; affected fruits may be distorted and bear occasional necrotic patches. This debris- and seed-borne virus is spread rapidly by handling the crop. Isolation of affected plants, and working towards them, reduces the rate of spread, as do sprays of dried milk powder. The latter is also used as a hand dip between working on adjacent plants. See under Cucumber, cucumber green mottle mosaic virus, p. 339.

Powdery mildew (*Leveillula taurica*)

This is an important disease of peppers, and causes a severe premature leaf fall. Initially showing as scattered yellow spots on the upper leaf surface, the fungus produces a brown, felt-like mat of conidia on the lower surface. The pathogen is wind-borne, and can occur at low humidities. HV sprays of fenarimol (off-label) (SOLA 0645/94) or sulfur (off-label) (SOLA 1714/97) give control.

Sclerotinia rot (*Sclerotinia sclerotiorum*)

This fungus can cause damping-off of seedlings; if airborne spores infect petioles, drooping leaves may be the first sign of attack. A fluffy, white growth develops on infected tissue and black sclerotia form in the pith cavity. Remove affected tissue to prevent sclerotia falling to the floor.

Radish

Diseases

Damping-off and crater rot (*Thanatephorus cucumeris* – anamorph: *Rhizoctonia solani*)

These diseases can occur at any stage of growth, particularly when growing conditions are unfavourable and the plants are not growing normally. Seedlings may damp-off or develop a constricted stem at soil level ('wirestem'). Black, sunken lesions may develop on the hypocotyl. Sometimes the brown mycelium of the fungus can be seen on the plant and soil surface; occasionally, sclerotia are also seen. Old crop debris should be destroyed and the soil sterilized with steam or methyl bromide. The seedbed should be well prepared and kept moist to encourage good growth. Tolclofos-methyl (off-label) (SOLA 2007/99), applied to the soil surface pre-sowing, will give reasonable control.

Downy mildew (*Peronospora parasitica*)

This fungus causes chlorotic areas on leaves, frequently limited by the veins. On the undersurface of leaves a profuse greenish-white growth of spores develops.

Lesions may become necrotic and dry out. In seedlings, it can be very damaging as systemic invasion of cotyledons, hypocotyl and roots can occur, leading to seedling death. The disease is most prevalent under cool (15°C), moist conditions. It is controlled by good ventilation, to reduce humidity and persistent water films on leaves, and by use of mancozeb + metalaxyl (off-label) (SOLA 0936/99) or the systemic fungicide propamocarb hydrochloride (off-label) (SOLA 2030/99), applied as a drenching spray to prevent the disease.

Spinach

Pests

Aphids

The most common species found on spinach is peach/potato aphid (*Myzus persicae*). A HV spray of pirimicarb (off-label) (SOLA 1626/95) will control susceptible strains, although resistant strains may occur. A HV spray of cypermethrin (off-label) (SOLA 3133/98) can be used up to the seven-leaf stage.

Caterpillars

These pests form holes in the leaves. A HV spray of cypermethrin (off-label) (SOLA 3133/98) can be used up to the seven-leaf stage.

Leaf miners

See under Lettuce, p. 348.

Slugs

See under Lettuce, p. 348.

Diseases

Downy mildew (*Peronospora farinosa* f. sp. *spinaciae*)

This disease causes yellow lesions on older leaves and can be severe in cool, humid conditions. Control is by the use of resistant cultivars, good ventilation and HV sprays of copper oxychloride + metalaxyl (off-label) (SOLA 1344/98).

Tomato

Pests

Aphids

Aphids, e.g. glasshouse & potato aphid (*Aulacorthum solani*), peach/potato aphid (*Myzus persicae*) and potato aphid (*Macrosiphum euphorbiae*), are only occasional pests of commercially grown tomatoes. They are found on the undersides of the leaves, although the white, cast skins can be seen on both leaf surfaces. Damage symptoms include distorted leaves, yellow patches on the leaves (*A. solani*) and sooty moulds colonizing aphid honeydew. Parasitoids can be used for biological control – see under Pepper, p. 352. Alternatively, a spot spray with nicotine or pirimicarb can be used within IPM. Pirimicarb may not control *M. persicae* if resistant strains are present.

Glasshouse leafhopper (*Hauptidia maroccana*)
This insect has become a more common pest with the reduced use of broad-spectrum pesticides in IPM programmes. The adults are 3–4 mm long, and pale yellow with two, dark, chevron-shaped marks on the wings. The nymphs are pale yellow. Both stages are found on the undersides of leaves, together with white, cast skins left when the nymphs moult during their development. Feeding damage to the leaves appears as white, indistinct spotting and bleaching. Weeds such as chickweeds (e.g. *Stellaria* spp.) are alternative hosts, so weed control both in and around the glasshouse plays an important role in preventing infestations. The parasitoid wasp *Anagrus atomus* can give effective control if introductions start at the first sign of damage. This tiny, delicate parasitoid lays its eggs in the leaf-hopper eggs. Parasitized eggs turn from green to brick red and can be seen in the leaf veins on the undersides of leaves. If necessary, buprofezin, as used against whiteflies (see under Tomato, glasshouse whitefly, below), will give control of leafhopper nymphs, and can be integrated safely with biological control agents.

Glasshouse whitefly (*Trialeurodes vaporariorum*)
For description, damage and biological control with the parasitoid wasp *Encarsia formosa*, see under Cucumber, p. 337. If a pesticide is needed to control patches of whitefly infestation, buprofezin may be used but there is increasing whitefly resistance to this pesticide. Alternatives are spot sprays of fatty acids or nicotine. The predatory bug *Macrolophus caliginosus* can also give effective control of whiteflies and other pests on tomatoes, but large numbers of the predator (which is also phytophagous) can build up during the season and, when whiteflies and other prey are scarce, crop damage can occur, particularly on cherry tomatoes. Current research is evaluating improved management strategies for *M. caliginosus* to avoid crop damage.

Mealybugs
Mealybugs, e.g. glasshouse mealybug (*Pseudococcus viburni*), are becoming an increasing pest on some nurseries. Cultural control by preventing carryover between crops on irrigation lines, glasshouse structure, packing trays, etc. can reduce sources of infestation. If necessary, buprofezin as recommended for control of whiteflies should give control.

Sciarid flies (e.g. *Bradysia paupera*)
These are only occasional, localized pests. For description and control see under Cucumber, p. 338.

Thrips

Thrips seldom cause serious direct damage to tomatoes but western flower thrips (WFT) (*Frankliniella occidentalis*) can transmit tomato spotted wilt virus (TSWV), particularly if infected ornamentals are grown on the same or adjacent nurseries. TSWV-infected plants should be rogued to reduce the source of virus. WFT can be controlled, if necessary, by dichlorvos (off-label) applied as a HV spray (SOLA 0625/99) or as a fog (SOLA 0626/99). Malathion will also give some control but has persistent harmful effects against biological control agents. Abamectin will give some control of thrips nymphs but is less effective against the adults, which are usually more of a problem in tomatoes. See under Tomato, tomato leaf miner for restrictions on using abamectin.

Tomato leaf miner (*Liriomyza bryoniae*)

This is a common pest of glasshouse-grown tomato. Adults are small, black flies with a yellow spot on their backs. The first signs of damage are small, round, feeding punctures made by the adults on the leaves; these are followed by leaf mines, caused by the creamy-white larvae feeding between the upper and lower leaf surfaces. Damage starts in the spring when any overwintered puparia in the soil give rise to adults, and serious infestations can develop unless managed carefully. Parasitoid wasps are used to control leaf miners within IPM programmes. Either a mixture of *Dacnusa sibirica* and *Diglyphus isaea*, or *D. isaea* alone, can be used. *D. isaea* is more effective at high pest densities, so the current strategy is to wait until leaf mines reach a certain density before starting introductions of this species. Regular monitoring of parasitism levels is essential for successful control. Use of pesticides for leaf miner control should be avoided if possible, as they can seriously interrupt the biological control system. Nicotine can give useful control of adult leaf miners, if required, and abamectin can give control of larvae, but these must be timed carefully to avoid disruption of biological control programmes. Abamectin must not be used between 1 November and the end of February, nor on cherry tomatoes at any time. It is important to avoid the overwintering of puparia in the soil, so end of season 'cleanup' with either of these pesticides may be appropriate, together with careful disposal of the trimmed leaves and crop debris. See also under Notifiable pests and diseases, leaf miners, p. 329.

Tomato moth (*Lacanobia oleracea*)

The green or brown caterpillars cause holes in the leaves and can also feed on fruits and stems. HV sprays of *Bacillus thuringiensis* can be used safely within IPM.

Two-spotted spider mite (*Tetranychus urticae*)

This is a serious pest of tomatoes, together with the carmine spider mite (*T. cinnabarinus*), which is reddish-brown in colour. For a description of *T. urticae* see under Cucumber, p. 338. Both species can cause fine yellow speckling on the leaves and also bright-yellow patches, known as 'hypernecrotic' damage. Biological control with the predatory mite *Phytoseiulus persimilis*, as described for

Cucumber (p. 339), can be successful, although control, particularly of *T. cinnabarinus*, is not always reliable. The predatory midge *Feltiella acarisuga* can give useful supplementary biological control. When used for whitefly control (see above), the predatory bug, *Macrolophus caliginosus* can also give useful control of spider mites. If necessary, fenbutatin oxide can be used safely to control spider mites in IPM, although strains of mites suspected to be resistant have now been recorded. Abamectin can be used if necessary, but as this pesticide has some harmful effects on biological control agents, it is best used as a spot treatment or for 'cleanup' at the end of the season – see under Tomato, tomato leaf miner, p. 357, for restrictions on use of abamectin. As on cucumbers, this cleanup procedure at the end of the season is important to prevent the overwintering of diapausing spider mites – see under Cucumber, two-spotted spider mite, p. 339.

Diseases

Bacterial canker (*Clavibacter michiganensis* ssp. *michiganensis*)
This is a serious though uncommon disease, causing loss of vigour, unmarketable spotted and disfigured fruit and premature plant death. The pathogen is a listed quarantine organism, which is notifiable if infection occurs at a registered nursery where plants are being propagated. Symptoms are very variable. Early in the season, bird's-eye spotting on fruit, mealiness on the stems and white spotting on leaves may be found in crops where spraying has been done regularly. Later, symptoms associated with blocking of the vascular tissues are typical and include one-sided wilting of leaves, fruit marbling and dropping, and yellow, straw-coloured staining of tissue surrounding the vascular bundles, particularly in the petioles. Plant death usually follows the onset of high summer temperatures.

The disease can be contained by strict isolation of infected plants and preventing spread by any form of water splash. Copper sprays (e.g. copper oxychloride) may give some reduction in spread during the early part of the season. A high standard of hygiene, both within the glasshouse and in the disposal of crop debris, is necessary to prevent re-infection of the next tomato crop.

Bacterial wilt (*Ralstonia solanacearum*)
See under Notifiable pests and diseases, p. 329.

Brown root rot
See under Root rots, p. 363.

Buck-eye rot (*Phytophthora nicotianae* var. *parasitica*)
This disease is now rarely seen, except where watering is by overhead sprinklers. The lower fruits show grey and red-brown patches with concentric dark rings. Infection is caused by water-splash from, or by contact with contaminated soil. Keep trusses out of contact with soil and avoid splashing when hand-watering.

Soil should be sterilized with steam to reduce contamination. Where the disease occurs, apply HV sprays of chlorothalonil or copper oxychloride to the lower part of the plant and the surface of the soil. Symptoms of buck-eye rot may be confused with those of blossom-end rot and other ripening disorders.

***Calyptella* root rot** (*Calyptella campanula*)
This disease has been found only in soil-grown crops, where it can cause significant losses. The first signs of the disease are wilting on bright days, which usually coincides with the commencement of picking. Plants initially recover turgor at night but, eventually, as the disease progresses, plants do not recover and they die. Rusty-brown lesions are present on all sizes of root, similar to the lesions caused by brown and corky root rots (see p. 363) but without the corkiness. Fruiting bodies of the fungus are sometimes produced on the surface of the soil, in clusters near the base of the plant. They are lemon-yellow to white in colour, saucer-shaped, 2–4 mm in diameter, and borne on a stalk 2–4 mm long. There is a close association between high soil-moisture content and the incidence and severity of the disease. Where the soil is sufficiently moist to support the growth of mosses, conditions are usually suitable for the disease. The disease does not appear to spread easily from house to house but rather to spread gradually within a house.

Conventional soil sterilization techniques do not appear to give effective control. Avoiding prolonged periods of excessively wet soil offers the best solution. Reduce the quantity and frequency of watering and move the drip outlet from the plant base to about 20 cm away.

***Colletotrichum* root rot**
See under Root rots, p. 363.

Corky root
See under Root rots, p. 363.

Foot rot (*Phytophthora cryptogea*)
These fungi cause foot rot of plants soon after planting out. Infection arises from the soil, or from contaminated drip lines or water. Attack by *Phytophthora* spp. is associated with cold, wet soils and is less likely in heated glasshouses. Good hygiene during plant propagation is essential. After planting, etridiazole can be applied as a drench to soil-grown or hydroponic crops. In nutrient film culture, species of *Phytophthora* can cause extensive wilting, with associated suppressed root regeneration.

Protect plants by the addition of etridiazole. Propamocarb hydrochloride may also be used in the nutrient solution. Heating the solution to 20–22°C and avoiding checks to growth will reduce the risk of root rotting owing to infection with *Phytophthora*. Where plants are propagated in rockwool cubes, it is very important to avoid placing the base of the cubes on an unsterilized surface. A precautionary drench of propamocarb hydrochloride, applied to the cubes, can be followed by adding propamocarb hydrochloride (off-label) (SOLA 2032/99) to the feed solution. A maximum of four applications should be made.

***Fusarium* crown and root rot** (*Fusarium oxysporum* f. sp. *radicis-lycopersici*)
Although this disease caused considerable losses in some UK crops over the last decade, the introduction of resistant cultivars has now restricted outbreaks to crops of susceptible, specialist ones (e.g. some cherry and plum tomatoes). It is characterized by a sudden wilt just before the first fruits are ready for picking. After repeated wilting, severely affected plants may die. It is distinguished from fusarium wilt by an extensive chocolate-brown root rot and a strong, reddish-brown vascular staining in the lower 25 cm of the stem base. The optimum temperature for disease expression is lower (15–18°C) than that for fusarium wilt (28°C). The disease can be readily mistaken for phytophthora root rot unless, as sometimes occurs at an advanced stage, near-dead plants affected by *Fusarium* bear a pinkish-white mass of mycelium and spores at the stem base.

Use of resistant cultivars is the most effective method of control. For soil-grown crops, soil sterilization can worsen the problem if there is re-infestation of the sterile soil by *F. oxysporum* f. sp. *radicis-lycopersici*. Mounding peat around the stem base of affected plants encourages growth of adventitious roots, which remain relatively free of the disease. Reducing the fruit load early in the season can reduce the risk of rapid wilting and plant death.

Grey mould (*Botryotinia fuckeliana* – anamorph: *Botrytis cinerea*)
The disease is common, especially when the weather is wet and cool. The fungus can invade all parts of the plant above soil level and is characterized by the masses of grey spores produced on infected tissues. Lesions on the stem are caused by spread of infection from damaged or senescent leaves, petiole stumps, de-leafing wounds, side shoots trapped in the supporting string or old fruit trusses; in layered crops, lesions may spread by stem contact. Stem lesions are particularly damaging because, if they become aggressive, they destroy the structural and conducting tissue and kill the plant. Airborne *Botrytis* spores can cause a reaction in matt-surfaced immature fruit that persists to disfigure the ripe fruit with 'ghost-spots'. Fruits are lost when flowers become infected before setting, or the calyx or corolla may be colonized after the fruit has set and cause premature drop.

To control the disease, reduce humidity. Remove all plant debris from the rows, and also remove decaying leaves and fruits from plants; remove dead plants promptly, including the stem base. Allow circulation of air through the lower parts of the plants by removing leaves and shoots cleanly and by supporting layered stems off the floor. Lesions of grey mould on stems can be painted with food-grade vinegar to reduce *Botrytis* sporing and the rate of lesion spread. Removal of old fruit trusses can help in crops where these decay to cause stem lesions. Maintain a minimum pipe temperature throughout the life of the crop.

To protect foliage and fruit from infection, use HV sprays of chlorothalonil, dichlofluanid, iprodione or pyrimethanil (off-label) (SOLA 0509/99) (Table 9.5). Control ghost spot by the use of heat and ventilation to prevent condensation on fruit. Dichlofluanid is effective if heat is unavailable. Fungal strains resistant to certain fungicides (e.g. iprodione) are often found in tomato crops. Use several different fungicides during a season, subject to the constraints of avoiding adverse effects on natural enemies of pests and subject to following manufacturer's guidelines.

Late blight (*Phytophthora infestans*)
Crops in unheated or well-sealed or insulated glasshouses are at risk during wet weather in August, when the disease may be very common on potato crops, and considerable losses can occur owing to attack on leaves and fruit. On the leaves, large, brown spots with pale margins and a downy, white, fungal growth on the underside spread rapidly in humid conditions. Stems may show dark lesions, and green fruit are disfigured with large, hard, brown patches. High temperatures or low humidity will control the disease but, in unheated crops near potato fields, a routine spray application may be worth while. Apply chlorothalonil or dichlofluanid. Young plants raised in late autumn may also be affected, but the disease rarely spreads after plants are spaced-out in the cropping houses and no action other than removing affected tissue is usually necessary.

Leaf mould (*Fulvia fulva*)
Following the introduction of resistant cultivars, this once-common disease is now found only on susceptible 'heritage' cultivars (e.g. Gardener's Delight). Leaf mould is favoured by warm weather and high overnight humidities, and generally occurs in mid-summer. Yield loss occurs only some weeks after the disease has become established. Older leaves show pale-yellow patches that, on the undersurface, become covered with a pale-grey mould which changes to brownish-violet. The disease is controlled by growing resistant cultivars, by increased ventilation and, if necessary, by HV sprays of carbendazim (off-label, for use on crops grown in soil or peat bags) (SOLA 2079/99), chlorothalonil or dichlofluanid.

Pepino mosaic virus (PepMV)
Outbreaks of this non-indigenous disease occurred in a few tomato crops in the UK in 1999 and 2000. It is mechanically transmissible and can spread rapidly through a house. Symptoms observed were yellowing and distortion of young leaves, and an altered plant growth habit, resulting in a nettle-like head to the plant. Leaf mosaic, chlorosis and bubbling can also occur on older leaves fruit marbling may also occur. Prompt removal of affected plants, the use of disposable clothing, restriction of staff working in the affected house and observation of strict hygiene measures are all important to minimize the rate of disease spread. PepMV is now a notifiable disease, and eradication is required where it is confirmed on a nursery.

Table 9.5 Chemical control of various diseases on protected tomato

Compound	Diseases controlled (or partially controlled)				Max. number of treatments	Minimum harvest interval (days)	Remarks
	Grey mould	Powdery mildew	Phytophthora root rot	Stem rot (*Didymella*)			
Sprays							
carbendazim (off-label)	✓*	✓	–	✓*	1	1	SOLA 2079/99†
chlorothalonil	✓	(✓)	–	–	2	2	–
dichlofluanid	✓	(✓)	–	–	8	3	–
iprodione	✓*	–	–	✓*	–	14	–
maneb (off-label)	–	–	–	✓	8	14	SOLA (2257/99)
pyrimethanil (off-label)	✓	–	–	–	4	3	SOLA (0509/99)
sulfur (off-label)	–	✓	–	–	–	0	SOLA (0909/96)
Root drench treatments							
carbendazim (off-label)	(✓)	(✓)	–	(✓)	1	4	SOLA (0965/00)†
etridiazole (off-label)	–	–	✓	–	–	1	SOLA (0600/99)‡
propamocarb hydrochloride (off-label)	–	–	✓	–	–	2	SOLA (2032/99)‡

* Resistant strains will not be controlled.
† For crops grown in soil/peat bag (spray or root treatment) or on inert substrate/NFT (root treatment only).
‡ For crops grown on inert media/substrate or NFT.

Powdery mildew (*Erysiphe orontii*)
Although, in the UK, this disease was first recorded as recently as 1987, it is now a common problem on commercial tomato crops. White, powdery spots (*c.* 5–10 mm in diameter) develop on the upper leaf surface, and these may expand and coalesce so that the whole surface is covered with a thin, white growth. Occasionally, the fungus develops on the lower leaf surface, stems or fruit calyxes. Infection can occur at low humidities but generally the disease is more damaging at high humidities. HV sprays of sulfur (off-label) (SOLA 1717/97), applied promptly after infection first occurs, are very effective; sprays of chlorothalonil or dichlofluanid, applied for grey mould, will give protection against powdery mildew.

Root rots
Tomato roots are commonly attacked and rotted by several fungi. The most important are: (a) *Pythium* spp., which can be troublesome in hydroponic crops; (b) *Pyrenochaeta lycopersici*, which is responsible for brown root rot and for corky root, especially in soil grown crops; (c) *Colletotrichum coccodes* (black dot), which destroys the cortex of the roots and stem below soil level or towards the end of the season in hydroponic crops; and (d) *Thielaviopsis basicola* (synanamorph: *Chalara elegans*), which causes black root rot and is particularly damaging in NFT crops. The effect of these fungi is to reduce plant vigour and crop production, and plants tend to wilt in bright, hot weather.

For soil-grown crops, control is best achieved by soil sterilization. Drench treatment with carbendazim is effective against brown and corky root rot. Tomatoes grown in NFT or rockwool systems should remain free of these diseases but, should they succumb, carbendazim (off-label) (SOLA 0095/00) can be used in the nutrient solution to control black root rot, and etridiazole (off-label) (SOLA 0600/94) or propamocarb hydrochloride (off-label) (SOLA 2032/99) used against *Pythium*.

Stem rot (*Didymella lycopersici*)
This disease has potential to cause considerable crop loss and, once established, it is difficult to eradicate. Dark-brown, sunken, girdling lesions are usually found first at the base of stems. In hydroponic crops the initial lesion is often at the top of the rockwool propagation cube, and may be mistaken for nutrient scorch. Pinhead-sized pycnidia, which exude spores when wet, are often found embedded in the soft and rotting tissues of the lesion. The lower leaves become yellow, and plants wilt and die. Later, stem rot lesions may occur on the upper part of the stem and on the fruit. They may be distinguished from grey mould lesions by the absence of grey spores and mycelium and by the presence of pycnidia. Young and wounded tissues are most susceptible to infection, especially in wet conditions and when the temperature is 15–20°C. Inoculum survives in debris, in soil and on the glasshouse structure and wires. It spreads in the crop by splash dispersal and on implements, hands and clothes.

Scrupulous hygiene is vital for the exclusion and control of the disease. When infection occurs, completely enclose affected tissues in an air-tight bag before prompt and clean removal from the crop. Do not cut out stem rot lesions. Treat other plants with HV sprays, as shown in Table 9.5, p. 362, at weekly intervals if the disease is severe. Carbendazim (off-label) (SOLA 0965/00) can be used as a drench but is likely to be effective only on young plants. At the end of the crop, dispose of all debris safely, and thoroughly sterilize the glasshouse, equipment and used growing substrates (if they are to be re-used), so that there is no risk of spread or reinfection. Wash down all structures and implements with 2% formaldehyde or with another disinfectant effective against *Didymella*. On nurseries with a history of stem rot, soil should be sterilized with steam or methyl bromide, even if soil-less cultivation is to be used. Protect with HV sprays, drenches or addition to the nutrient solution of carbendazim (off-label; restriction on application method according to the production system) (SOLA 0965/00) or HV sprays of iprodione or maneb (off-label) (SOLA 2257/99).

Tomato mosaic virus (ToMV)
This serious and once-common disease is now generally well controlled by the use of resistant cultivars. The $Tm2^2$ resistance gene has proved very effective and durable. Occasional outbreaks occur, however, in crops of susceptible speciality cultivars. Symptoms are very variable, being influenced by several factors – including strain of the virus, crop cultivar, age of plant at infection and growing conditions. The most common symptom is a leaf mosaic, varying from a pale mottle to bright yellow and green demarcated areas. Other common symptoms are the development of narrow leaves ('fern leaf'), fruit bronzing and reduced fruit set; marketable yield can be reduced by 15 to 90%. The disease is transmitted readily by handling the crop. If a susceptible cultivar is grown, it is essential to make efforts to prevent the disease by seed treatment (70°C for 4 days), by steam sterilization of the soil or hydroponic production out of the soil, and by inoculation of young plants with a mild strain of ToMV (e.g. MII-16) to provide cross-protection.

Tomato spotted wilt virus (TSWV)
This virus disease is transmitted by thrips and has become more common since the establishment of western flower thrips (*Frankliniella occidentalis*) in the UK. It is an occasional problem in tomato crops, most frequently found on mixed tomato/ornamental nurseries, or where ornamentals (e.g. chrysanthemum plants) are grown in close proximity to a tomato crop. Symptoms are pale rings or brown spots on leaves, which later become bronze in colour, and a pale mottle on fruit, sometimes with distinct brown rings or line patterns. For control measures, see under Tomato, thrips, p. 357.

Wilt diseases (*Verticillium albo-atrum*, *V. dahliae* and *Fusarium oxysporum* f. sp. *lycopersici*)
Isolation and sterile culture are required to identify the organism responsible for

wilting, because the diseases are not easily distinguished by symptoms. Fusarium wilt is more serious at higher temperatures (around 28°C). The first symptoms of wilt are usually yellowing and temporary wilting of lower leaves, often on one side of the plant. The woody stem tissues become brown and the discoloration can be extensive or limited to the lower stem region. The plants grow poorly with thin stems in the plant head. Yield can be significantly reduced. Most modern cultivars are resistant to both pathogens but it should be noted that races of *Fusarium* occur, and probably of *Verticillium* also, that can overcome the resistance; recently, there have been an increasing number of outbreaks of verticillium wilt in resistant cultivars.

Where verticillium wilt occurs, symptoms may be suppressed by raising the glasshouse temperature to 25°C and shading lightly. Root drenches of carbendazim (off-label) (SOLA 0965/00) may give some control of both diseases. Dissemination occurs via soil, via infected plants at planting, via infected crop debris or from contamination persisting in the glasshouse or on equipment (e.g. drip pegs) from a previous infected crop. Thorough cleaning and disinfection of the glasshouse and equipment should be done after removal of an affected crop. Check that there is no contamination of hydroponic systems with soil. Consider soil sterilization where there is a severe or persistent problem.

Mushrooms

Introduction

The advent of bulk peat-heating of compost (phase II) and bulk spawn-running (phase III) has presented both new opportunities for pest and disease infestation and, in some instances (particularly pest infestation), some additional protection. The implications of these changes are dealt with, where relevant, in the sections dealing with individual pests and diseases.

The current diversity of growing systems – trays, shelves, blocks and bags – has had less effect than one might imagine. There are some exceptions to this generalization, which will be referred to in the relevant items. It is, however, the increasing use of bulk compost handling techniques that has brought about changes from a situation that has existed for several decades.

With all pesticide treatments to mushrooms it is essential to consult manufacturers' product labels for minimum harvest intervals between final application and harvest, and other details.

Disinfection of mushroom houses

To destroy spores of pathogenic fungi on walls, floors, woodwork etc., the cropping houses should be treated with a disinfectant after the diseased crop has been removed, even though the crop may have been 'cooked out' by steaming to 65–70°C. Fogs or HV sprays of formaldehyde (see p. 327) are suitable for this purpose. There are a number of non-persistent disinfectants which are suitable and which are currently supplied to the industry. Timber cropping trays should be dipped in azaconazole, dichlorophen or sodium orthophenyl phenate tetrahydrate. The disinfectant treatments of the houses have the added advantage that they will also discourage the carryover of pests from one crop to the next.

Pests

Flies

Flies from three families are common pests of mushrooms: Cecidomyiidae (midges), Phoridae (phorid flies = scuttle flies) and Sciaridae (sciarid flies). The life history and behaviour of each group differ and, therefore, the pests are considered separately below.

Cecid midges (*Heteropeza pygmaea, Mycophila barnesi, M. speyeri*)
Cecid midge larvae are white or orange in colour, and distinguishable from other mushroom maggots by a pair of dark 'eye spots' on the body just behind the minute head. They are paedogenetic, giving birth to 'daughter' larvae. This process may continue for a long time before this kind of development changes and the small, delicate, adult midges (that are rarely seen) are produced. With heavy infestations, larvae climb the mushroom stems, rendering the mushrooms unsaleable. Modern pasteurization methods normally preclude compost as the primary source of these pests. Once they are introduced on to a farm, however, possibly in small quantities in casing peat, their numbers can quickly increase and they are readily spread from crop to crop. In recent years some of the more notable infestations have arisen from stored, infested tray timber. Cecid midge larvae may be partially eliminated by efficient cook-out treatments but effective hygiene is essential. The larvae become a contaminant, being transferred easily to new crops by farm staff and, most effectively, back to casing materials where they can begin the infestation cycle once more. There are currently no chemical controls since the withdrawal of diazinon.

Phorid flies (*Megaselia halterata* and *M. nigra*)
These are small, dark, humped, stout-bodied flies; the maggot-like larvae are white and lack the obvious black head of sciarid larvae (see below). Larvae of *M. halterata* feed on mycelium in the compost. Attacks are most frequent in summer and autumn, and come mainly from flies entering spawn-running rooms, attracted by the smell of the mushroom mycelium. The increasing use of recycled cooled air in traditional spawn runs and the excellent sealing and filtration of bulk phase III rooms have reduced the importance of this pest; however, for smaller farms (particularly those employing phase II blocks and bags and which are without cooling) it remains a significant problem. This situation has been worsened by the loss of diazinon as a compost treatment.

Aerial treatments employing permethrin and pyrethrins + resmethrins are currently the only treatments available. For those farms without good protection during phase III (spawn-running) *M. halterata* remains an important pest.

M. nigra is a very occasional pest. Its presence is usually first detected by finding heavily burrowed mushrooms, both caps and stipes. This pest lays its eggs only in the light. Its presence therefore indicates light leakage into the cropping house. If this is rectified the fly is controlled.

Sciarid flies (*Lycoriella auripila* and others)
The larvae of sciarid flies are elongate, white and shiny, with conspicuous black heads. The adults are small, dark-bodied, 'leggy' insects with conspicuous antennae; sciarid flies have a tendency to 'run' rather than fly. The damage caused by sciarid larvae is most commonly the severance and internal burrowing of pins and small buttons. In extreme cases, stipes of more mature mushrooms may be tunnelled; this, however, is extremely uncommon. Sciarid flies are often implicated in the spread of diseases, particularly *Verticillium*; they are also vectors of nematodes. A more insidious problem now is product contamination. Even one fly in a pre-pack is unacceptable. Pasteurization will kill the sciarids but the flies are attracted to the compost as it cools.

Owing to a combination of pesticide withdrawal, pest resistance and pesticide unacceptability on the part of retailers, few pesticides remain for the control of sciarids. A three-tier control strategy, however, still remains. Compost protection from pasteurization cool-down to completed spawn-run now relies entirely on the physical exclusion of flies from the compost. Emptying bulk phase II rooms is perhaps the most vulnerable entry point for egg-laying females. Protection can be achieved only by exclusion of the flies from the compost.

Casing may be protected by drenches of diflubenzuron, methoprene or the biological agent *Steinernema feltiae* (an entomopathogenic nematode). Some degree of aerial control may be achieved by means of permethrin and pyrethrins + resmethrins.

Mushroom mite (*Tarsonemus myceliophagus*)
This scarcely visible mite (sometimes called 'tarsonemid mite') causes a chestnut-brown discoloration and rounding of the base of mushroom stalks. Since there is no effective chemical control, strict attention to hygiene is essential, and buildings and equipment should be cleaned thoroughly as described under Disinfection of mushroom houses, p. 365. This pest is now rare and its occasional presence indicates a lapse of hygiene measures.

Mycophagous nematodes (*Aphelenchoides composticola* and *Ditylenchus myceliophagus*)
These and other species of nematode feed on mycelium, causing it to disappear in patches. Where the pests are numerous the compost may become dark, sunken and wet. Nematodes can survive for at least 2 years in dry compost and wood-

work, and can be carried from place to place on tools, boxes or by adult flies (notably, sciarid flies, see above).

There are no acceptable methods of controlling nematodes in spawned or cropping compost. Pasteurization is effective, and structures and spent compost may be disinfected by cooking-out. Nowadays, these pests are extremely rare. Outbreaks can arise from small infestations in casing material or compost, which are then allowed to increase owing to lapses in hygiene measures.

Diseases

Mushroom sporophores may be affected severely by several fungal and bacterial pathogens. Crop loss may be either direct or by a reduction in quality, e.g. marked or spotted mushrooms.

Crop yield may be affected adversely by a number of compost moulds. These are seldom now indicators of poor composting and are more likely to be the result of hygiene failures in bulk compost handling systems, either phase II or phase III. Mushrooms are also affected by several viruses that can, in extreme cases, result in almost total loss of yield.

Bacterial blotch (*Pseudomonas tolaasi*)

This is an increasingly uncommon disease. The symptoms are brown staining of the mushroom cap. The staining may be in the form of small spots, blotches or even extensive areas of the cap. The colour of the staining ranges from light to dark chocolate-brown. Efficient air movement, avoiding areas of poor evaporation, and improved control of temperature and humidity appear incidentally to have largely eradicated this disease. Routine watering with sodium hypochlorite is an effective additional control.

Bacterial pit

This disease is now very infrequent. Its cause is unknown but bacteria appear to be associated with the symptoms. It starts as small cavities in the cap beneath the skin, which eventually collapses to leave open pits. Should the problem occur, poor environmental control is invariably the cause.

Cobweb (*Cladobotryum* sp.)

The disease has become common and can cause severe losses. The fungus envelops entire mushrooms with a white, loose, cobweb-like mycelium, occasionally with pink tinges, that spreads on the surrounding casing. A secondary symptom is brown spotting on the mushroom caps. It is this symptom which causes major economic loss. Partial chemical control can be achieved by either carbendazim or prochloraz but benzimadazole resistance is widespread. Individual pathogen outbreaks should be tested for resistance before pesticide choice is made. A SOLA (1665/99) for pyrifenox has recently been granted, but commercial control has yet to be thoroughly assessed.

Cladobotryum spores are readily airborne. This has been shown to play an important role in disease spread. Patches of cobweb disease must be covered with wet tissue (to prevent spore dispersal) and then salt. Salting alone (i.e. without pre-covering), together with crop watering, have been shown to be the most efficient disease-spreading mechanisms.

Dry bubble (*Verticillium fungicola*)
This is currently the most important fungal disease of mushrooms. Bubble symptoms range from small, grey, undifferentiated lumps to grotesque, phallic distortions, dependent on the time and concentration of infection. Spores splashed from the 'bubble' on to adjacent mushrooms rapidly result in the development of more 'bubbles' or spotted caps, depending on the stage of development of the mushrooms infected. Spores can also be easily transferred by pickers or flies or, when washed on to the floor, in dust. Control by prochloraz, applied as casing drenches, is significant but not total, owing to partial resistance and pesticide breakdown. Control depends heavily, therefore, on stringent hygiene, an essential part of which is immediate identification and isolation of any infected mushrooms. In many cases, it is not possible visually to distinguish this disease from wet bubble (see below).

Green mould (*Trichoderma* spp.)
The causal pathogens are ubiquitous in modern mushroom-production systems, and several species are found commonly in compost; others occur on tray bedboards, casing or the chogs remaining after mushrooms are harvested.

The single most significant pathogen is *Trichoderma harzianum* type Th. 2. In the past, this has caused significant crop loss when present in compost, particularly in bags and blocks. *T. harzianum* Th. 2 is usually visually obvious, when present at problematic levels. Strict hygiene during spawning, together with carbendazim (off-label) (SOLA 1144/95) treatment of spawn, have proved to be effective control measures. Other *Trichoderma* spp. can be the cause of significant crop loss without any visible symptoms being apparent, if introduced into compost at spawning particularly prior to bulk phase III.

At times, *Trichoderma* on trays, casing or chogs can cause significant economic loss, as a result of cap spotting. Spot treatment of the patches of sporulating mould are the only, if often impractical, control method. Bed hygiene and tray treatments, however, undoubtedly help to control this problem.

Truffle (*Diehliomyces microsporus*)
Although now very rare, this competitor can cause very serious crop losses. The cream-coloured mycelium grows rapidly in compost. Mushroom mycelium disappears from affected areas and does not recolonize. Fruiting bodies of truffle are cream-coloured, with a convoluted surface ('calves' brains') up to 25 mm in diameter, and are produced in the compost and casing. Spores are resistant to disinfectants and can survive pasteurization and 'cook-out' but are killed if maintained at 70°C for 3 hours. The initial source of this disease is likely to be

contaminated compost. Maintaining spawn-running temperatures of 20°C, as opposed to 25°C +, will help control the problem. The only other recourse is rigorous post-crop hygiene.

Wet bubble (*Mycogone perniciosa*)
This disease is less common than *Verticillium* but is often confused with it. The mould on distorted mushrooms is often whiter and may be associated with amber-coloured droplets but the visual diagnosis is unreliable. One symptom alone is diagnostic – large, even fist-sized, grotesque lumps are invariably caused by *Mycogone*. Disease spread is not dissimilar to that of dry bubble, although wet bubble seems more commonly to arise from casing contamination, possible in wind-blown soil dust.

Hygiene control for dry bubble and wet bubble is similar. However, owing to the lack of resistance, carbendazim (in addition to prochloraz) can be used for chemical control, although it is unlikely, on account of pesticide degradation, to be very effective against late infections.

Virus disease
Virus disease has many symptoms, ranging from symptomless crop loss, through off-coloured mushrooms, distorted mushrooms and bare patches to almost complete absence of mushroom sporophores.

Until recently the disease has been associated predominantly with 35 nm particles mixed with small amounts of 25 nm and bacilliform particles. Recently, a putative novel virus has been detected, which causes similar symptoms at a significantly lower level of presence. This virus has yet to be identified. Virus disease is transmitted via mushroom spores and mycelium. Techniques to prevent the transfer of virus from old crops to new have proved effective. The most vulnerable areas are compost spawning and retention of live mycelium in the growing containers (trays and shelves).

Bulk composting techniques (phases II and III) are particularly vulnerable to virus infection. The modern practice of growing open mushrooms and, thus, dramatically increasing the mushroom spore load on farms, has undoubtedly also increased the likelihood of mushroom virus outbreaks. These two factors may well prove to have been the cause of the appearance of the novel virus currently being experienced.

List of pests cited in the text*

Aphelenchoides composticola (Tylenchida: Aphelenchoididae)	a mushroom spawn nematode
Aphis gossypii (Hemiptera: Aphididae)	melon & cotton aphid
Aulacorthum solani (Hemiptera: Aphididae)	glasshouse & potato aphid
Autographa gamma (Lepidoptera: Noctuidae)	silver y moth
Bemisia tabaci (Hemiptera: Aleyrodidae)	tobacco whitefly
Bradysia paupera (Diptera: Sciaridae)	a sciarid fly

Cacoecimorpha pronubana (Lepidoptera: Tortricidae)	carnation tortrix moth
Chromatomyia syngenesiae (Diptera: Agromyzidae)	chrysanthemum leaf miner
Coccus hesperidum (Hemiptera: Coccidae)	brown soft scale
Ditylenchus myceliophagus (Tylenchida: Tylenchidae)	a mushroom spawn nematode
Euleia heraclei (Diptera: Tephritidae)	celery fly
Eupteryx melissae (Hemiptera: Cicadellidae)	chrysanthemum leafhopper
Frankliniella occidentalis (Thysanoptera: Thripidae)	western flower thrips
Hauptidia maroccana (Hemiptera: Cicadellidae)	glasshouse leafhopper
Heteropeza pygmaea (Diptera: Cecidomyiidae)	mushroom cecid
Lacanobia oleracea (Lepidoptera: Noctuidae)	tomato moth
Liocoris tripustulatus (Hemiptera: Miridae)	a capsid
Liriomyza bryoniae (Diptera: Agromyzidae)	*larva* = tomato leaf miner
Liriomyza huidobrensis (Diptera: Agromyzidae)	*larva* = South American leaf miner
Liriomyza trifolii (Diptera: Agromyzidae)	*larva* = American serpentine leaf miner
Lycoriella auripila (Diptera: Sciaridae)	a mushroom sciarid fly
Lygus rugulipennis (Hemiptera: Miridae)	tarnished plant bug
Macrosiphum euphorbiae (Hemiptera: Aphididae)	potato aphid
Megaselia halterata (Diptera: Phoridae)	a mushroom scuttle fly
Megaselia nigra (Diptera: Phoridae)	a mushroom scuttle fly
Mycophila barnesi (Diptera: Cecidomyiidae)	a mushroom midge
Mycophila speyeri (Diptera: Cecidomyiidae)	a mushroom midge
Myzus ornatus (Hemiptera: Aphididae)	violet aphid
Myzus persicae (Hemiptera: Aphididae)	peach/potato aphid
Nasonovia ribisnigri (Hemiptera: Aphididae)	currant/lettuce aphid
Ovatus crataegarius (Hemiptera: Aphididae)	mint aphid
Phlogophora meticulosa (Lepidoptera: Noctuidae)	angle-shades moth
Polyphagotarsonemus latus (Prostigmata: Tarsonemidae)	broad mite
Pseudococcus viburni (Hemiptera: Pseudococcidae)	glasshouse mealybug
Pyrausta spp. (Lepidoptera: Pyralidae)	pyraustid moths
Scatella stagnalis (Diptera: Ephydridae)	a shore fly
Scatella tenuicornis (Diptera: Ephydridae)	a shore fly
Tarsonemus myceliophagus (Prostigmata: Tarsonemidae)	mushroom mite
Tetranychus cinnabarinus (Prostigmata: Tetranychidae)	carmine spider mite
Tetranychus urticae (Prostigmata: Tetranychidae)	two-spotted spider mite
Thrips tabaci (Thysanoptera: Thripidae)	onion thrips
Trialeurodes vaporariorum (Hemiptera: Aleyrodidae)	glasshouse whitefly

* The classification in parentheses represents order and family.

List of pathogens/diseases (other than viruses) cited in the text*

Agrobacterium bv. 1 (rhizogenic strain) (Gracilicutes: Proteobacteria)†	root mat of cucumber
Botryotinia fuckeliana (Ascomycota)	(common) grey mould
Botrytis cinerea (Hyphomycetes)	– anamorph of *Botryotinia fuckeliana*
Bremia lactucae (Oomycetes)	downy mildew of lettuce
Calyptella campanula (Basidiomycetes)	calyptella root rot of tomato
Chalara elegans (Hyphomycetes)	– synanamorph of *Thielaviopsis basicola*

Cladobotryum sp. (Hyphomycetes)	cobweb of mushroom
Cladosporium cucumerinum (Hyphomycetes)	gummosis of cucumber
Clavibacter michiganensis ssp. *michiganensis* (Firmicutes)†	bacterial canker of tomato
Colletotrichum coccodes (Coelomycetes)	black dot of tomato
Didymella bryoniae (Ascomycota)	stem and fruit rot of cucumber
Didymella lycopersici (Ascomycota)	stem and fruit rot of tomato
Diehliomyces microsporus (Ascomycota)	truffle
Erwinia carotovora ssp. *atroseptica* (Gracilicules: Proteobacteria)†	bacterial rot of Chinese cabbage
Erwinia carotovora ssp. *carotovora* (Gracilicules: Proteobacteria)†	basal stem rot of cucumber
Erysiphe heraclei (Ascomycota)	powdery mildew of fennel and parsley
Erysiphe orontii (Ascomycota)	powdery mildew of courgette and cucumber
Fulvia fulva (Hyphomycetes)	leaf mould of tomato
Fusarium oxysporum f. sp. *cucumerinum* (Hyphomycetes)	fusarium wilt of cucumber
Fusarium oxysporum f.sp. *lycopersici* (Hyphomycetes)	fusarium wilt of tomato
Fusarium oxysporum f. sp. *radicis-lycopersici* (Hyphomycetes)	fusarium crown and root rot of tomato
Fusarium solani (Hyphomycetes)	fruit and stem rot of pepper
Leveillula taurica (Ascomycota)	powdery mildew of pepper
Microdochium panattonianum (Hyphomycetes)	ring spot of lettuce
Mycogone perniciosa (Hyphomycetes)	wet bubble of mushroom
Olpidium brassicae (Chytridiomycetes)	fungal vector of lettuce big-vein virus
Penicillium oxalicum (Hyphomycetes)	stem rot of cucumber
Peronospora farinosa f. sp. *spinaciae*	downy mildew of spinach
Peronospora parasitica (Oomycetes)	downy mildew of radish
Phoma apiicola (Coelomycetes)	root and crown rot of celery
Phoma exigua (Coelomycetes)	basal rot of lettuce
Phomopsis sclerotioides (Coelomycetes)	black root rot of cucumber
Phytophthora cryptogea (Oomycetes)	foot rot of tomato
Phytophthora infestans (Oomycetes)	late blight of tomato
Phytophthora nicotianae var. *parasitica* (Oomycetes)	buck-eye rot of tomato
Phytophthora spp. (Oomycetes)	damping-off of pepper
Pseudomonas tolaasi (Pseudomonadales)†	bacterial blotch
Pseudoperonospora cubensis (Oomycetes)	downy mildew of cucumber
Puccinia menthae (Teliomycetes)	rust of mint
Pyrenochaeta lycopersici (Coelomycetes)	brown root rot of tomato
Pythium aphanidermatum (Oomycetes)	root and stem base rot of cucumber
Pythium hydnosporum (Oomycetes)	root rot of celery
Pythium spp. (Oomycetes)	root rot of cucumber, tomato and pepper
Pythium spp. (Oomycetes)	basal rot of lettuce
Ralstonia solanacearum (Gracilicutes: Proteobacteria)†	bacterial wilt of tomato
Rhizoctonia solani (Hyphomycetes)	– anamorph of *Thanatephorus cucumeris*
Sclerotinia minor (Ascomycetes)	sclerotinia rot of lettuce
Sclerotinia sclerotiorum (Ascomycota)	sclerotinia rot of aubergine, celery, cucumber, lettuce and pepper
Septoria apiicola (Coelomycetes)	leaf spot ('late blight') of celery
Septoria petroselini (Coelomycetes)	leaf spot of parsley

Sphaerotheca fusca (Ascomycota)	powdery mildew of cucumber
Thanatephorus cucumeris (Basidiomycetes)	crater spot of celery; basal rot of lettuce; foot rot of tomato
Thielaviopsis basicola (Hyphomycetes)	black root rot of tomato
Trichoderma harzianum (Hyphomycetes)	green mould of mushroom
Trichoderma spp. (Hyphomycetes)	green mould of mushroom
Verticillium albo-atrum (Hyphomycetes)	verticillium wilt of aubergine, cucumber and tomato
Verticillium dahliae (Hyphomycetes)	verticillium wilt of aubergine and tomato
Verticillium fungicola (Hyphomycetes)	dry bubble of mushroom

* For fungi, the classification in parentheses refers to class, although this is not possible within the phylum Ascomycota where classes have yet to be satisfactorily defined (see *Mycological Research*, February 2000). Oomycetes are now classified in Chromista with the brown algae, rather than as true fungi. Some fungi have an asexual (anamorph) and a sexual (teleomorph) state, and the convention is to refer to them by their teleomorph name. However, where anamorph (or synanamorph) names are still in common use, these are listed and cross-referenced to the teleomorph name. Strictly, fungi classified as Coelomycetes and Hyphomycetes should be known as 'hyphomycetous anamorphs' and 'coelomycetous anamorphs' of the relevant teleomorph taxon (e.g. hyphomycetous anamorphic Sclerotiniaceae, for *Botrytis fabae*), respectively. These problems highlight the continual changes in the classification of the fungi.

† Bacteria – the classification in parentheses refers to division and class, or to division only.

Chapter 10
Pests and Diseases of Protected Ornamental Flower Crops

T.M. O'Neill
ADAS Arthur Rickwood, Cambridgeshire

J.A. Bennison
ADAS Boxworth, Cambridgeshire

J.M. Scrace
ADAS Wolverhampton, West Midlands

Introduction

This chapter discusses pests and diseases of bedding plants, pot plants, forced bulbs and cut flowers. Hardy nursery stock and herbaceous plants are considered in Chapter 11. A general introduction to the control of pests and diseases on crops grown in glasshouses and polythene tunnels is given in Chapter 9, but aspects that specifically influence pest and disease management of protected ornamental crops are discussed below.

Production of plug plants

Many ornamental species are raised by specialist plant propagators as 'plug plants' and delivered to the production nursery ready for potting into their final pot (e.g. cyclamen and primula), or for planting into the ground (e.g. chrysanthemum and lisianthus). Bedding plants are usually sown by machine, directly into plug-trays or multi-packs. The plug plant ('plug') is, in essence, a small, rooted plant capable of growing away rapidly when planted out. This system of plant production has distinct advantages compared with raising and transplanting seedlings: increased opportunity for automation; opportunity to gap-up trays before dispatch; more rapid establishment of plants after planting out; increased crop uniformity; and reduced disease risk. Damping-off is generally less of a problem, with the reduced opportunity for spread between adjacent seedlings and the use of new (plastic or polystyrene) plug trays for each batch. Nevertheless, occasionally, *Pythium* can affect a whole tray of young plants, for example when the plants are past their optimum stage for potting on or are in transit for an extended period. Black root rot can also be a severe problem, especially on *Viola* (pansy), if plug trays are re-used.

Seeds and cuttings

The health of plant propagation material can have a profound influence on subsequent need for pest and disease control treatments. Important seed- or corm-borne diseases of ornamentals include: *Alternaria alternata* on lobelia; *Fusarium oxysporum* f. sp. *cyclaminis* on cyclamen; *Colletotrichum* on anemone; and some bacterial diseases (e.g. *Xanthomonas campestris* pv. *campestris* on wallflower). Latent infection by fungal diseases (e.g. *Botryotinia fuckeliana* – anamorph: *Botrytis cinerea* on fuchsia cuttings; *Puccinia horiana* on chrysanthemum) can also pose a risk. Cuttings can also be infested with pests. Examples are: leaf miners, leaf nematodes or western flower thrips on chrysanthemum; and sciarid flies or whiteflies on poinsettia. Often, seeds and cuttings are produced overseas, sometimes in areas with less-sophisticated crop inspection and disease management facilities than in the UK, resulting in variable plant health and the risk of importing non-indigenous pests and diseases (see under Notifiable pests and diseases, Chapter 9, p. 328). Control of pathogens and pests associated with plant propagation material is achieved by a variety of measures, including statutory inspections at the seed/cutting production site; additional testing demanded by the customer (e.g. for specific viruses); fungicide treatment on receipt (e.g. unrooted chrysanthemum cuttings may be dipped in propiconazole for control of latent white rust); and seed treatment (e.g. iprodione for control of *Alternaria* on lobelia).

Sub-irrigation

Sub-irrigation is increasingly used in the production of ornamental plants, occasionally in the propagation stage and more commonly for production of pot plants. Typical systems include NFT for growing mother plants for cutting production (e.g. fuchsia and poinsettia), flood-drain sand-beds, and capillary matting over which irrigation trickle tape is laid. Phytophthora, pythium and thielaviopsis root rots can all be major problems in ornamental plants grown in such systems; for example, widespread development of *Phytophthora* in kalanchöe and dieffenbachia can occur in plants grown on flood-drain benches and floors.

Water disinfection (e.g. by heat or UV light) appears to be of minimal benefit in controlling spread of pathogens in flood-drain water; most commonly, the irrigation water is dumped and replenished at regular intervals. Accurate control of depth and duration of flooding, combined with disinfection of the empty floor/bench between crops (especially after a root disease problem), is important in reducing the risk of root diseases. Care needs to be taken to ensure that there is no residual disinfectant when the new crop is placed on the treated surfaces; hypochlorite can be taken up by roots and cause leaf bleaching and plant death; phenolic compounds may inhibit root growth; formaldehyde fumes are phytotoxic.

Disinfection of standing areas

Pot and bedding plant nurseries may have plants standing on the glasshouse floor or benches covered with sand or other water-retentive materials such as capillary matting. These floor and bench coverings can become contaminated by root pathogens such as *Pythium* spp., *Phytophthora* spp. or *Thielaviopsis basicola*, or by plant and compost debris falling on to them. They also accumulate salts from the compost. Most growers cannot afford to replace floor or bench coverings annually, and therefore these should be decontaminated either by drying followed by brushing and steaming, or by treating with a suitable disinfectant. Floor and bench coverings can also become a source of pests such as sciarid flies or shore flies, or thrips pupae, and this should be considered when planning control strategies.

Growing medium amendments

Amendments are sometimes made to the growing media of ornamental crops to suppress root disease and this aspect of disease control has recently received increased attention. Mature or composted bark may be used to suppress fusarium wilt of cyclamen; young, Finnish peats have been shown to suppress pythium root rot and there is experimental work in using these for production of the peat blocks in which chrysanthemum cuttings are rooted. Several biocontrol products for use against diseases on ornamentals are available in Continental Europe but at present none is approved for use in the UK, although cyclamen plants produced in compost containing a *Fusarium* sp. antagonistic to the fusarium wilt pathogen *F. oxysporum* f. sp. *cyclaminis* are produced abroad and grown on in the UK. Products containing microorganisms (e.g. *Gliocladium catenulatum*) which enhance root growth are marketed in the UK and may provide an associated reduction in root disease.

Quality specifications

Slight alteration in appearance or deviation from supermarket specification can result in the product being rejected by the retailer. This may be caused, for example, by pesticide phytotoxicity. Equally, slight disfigurement caused by a pest or disease may also cause rejection, as can the 'zero tolerance' of any pests or even biological control agents. Problems have occurred where systemic pesticides have caused leaf-margin chlorosis and necrosis; where triazole fungicides have caused a reduction in flower stem length; and where drench treatment has resulted in root growth inhibition and leaf distortion. The large number of cultivars of some species, with different sensitivities to pesticides (e.g. in chrysanthemum), different cropping situations (e.g. whether heated or unheated), and in many cases the lack of label recommendations, means that a limited number of plants should always be test-treated before a product is used widely on a crop for the first time.

Pesticide availability

In addition to on-label approvals, this chapter includes mention of various specific off-label approvals (SOLAs). Although approved, off-label uses are not endorsed by manufacturers and such treatments are made entirely at the risk of the user. Also, as mentioned elsewhere, many products can be used on ornamentals under the provisions of the *Revised Long Term Arrangements for Extension of Use (2000)*. Specifically, these arrangements allow the use of pesticides approved for use on any protected growing crop to be used on commercial holdings on protected ornamental crops, including bulbs, subject to certain restrictions (again, entirely at the risk of the user). Mention in the text of the use of a pesticide under the provisions of these arrangements is marked with an asterisk (*).

Integrated pest management (IPM)

See Chapter 12, p. 320, for the general principles and components of IPM.

The main stimulus for the increasing uptake of biological pest control methods on protected ornamental crops has been pesticide resistance. Severe problems in controlling western flower thrips with pesticides (when this pest became established in the UK in the late 1980s) led to increased use of biological agents for its control, particularly on pot and bedding plants. Since then, comprehensive IPM programmes have been developed for control of most pests on many ornamental crops, including pot and bedding plants, hardy nursery stock under protection and certain flower crops. Overall on protected ornamentals, pesticides are still used on a larger area than biological control agents, but this situation is rapidly changing as the uptake of IPM increases every year. Growers using IPM on ornamentals profit from many other benefits in addition to improved pest control, such as better plant quality, savings in labour time and a better working environment for staff. Supermarkets are now beginning to demand reduced pesticide inputs on ornamental crops, as well as on edible produce, and this is further stimulating the uptake and use of IPM. Reliable biological control methods for certain pests on some crops are still in development, including western flower thrips on chrysanthemum. Pesticides are still needed for these crops, and for integrating with biological control agents on crops using IPM. When selecting pesticides (including fungicides) for using in IPM, compatibility with biological control agents should be checked with the biological control supplier or a consultant.

Anemone

Pests

Aphids
See under *Anemone*, Chapter 12, p. 542.

Caterpillars
For damage caused, see under *Anemone*, Chapter 12, p. 543. If necessary, spray with a pyrethroid such as cypermethrin or deltamethrin.

Diseases

Downy mildew (*Peronospora ficariae*)
This soil-borne disease is uncommon in protected crops, unless anemones have been grown frequently in the same soil without sterilization. Affected plants have dull, grey-yellow leaves which are usually down-curled. Plants are often stunted, with affected patches of plants developing in the most humid part of the house first. Control by soil sterilization, and by ventilation to reduce glasshouse humidity.

Grey mould (*Botryotinia fuckeliana* – anamorph: *Botrytis cinerea*)
Grey mould often colonizes old senescing leaves, especially in the autumn. Plants growing rapidly under shortening days, sometimes because of excessive nitrogen, seem particularly susceptible. Losses caused by plant death, infection of young growth and flower spotting may have a significant effect on the production of marketable blooms. Avoid producing too soft a plant by restricting the amount of nitrogen given and avoid excessive damage during picking. Trimming plants to remove senescing leaves may be necessary. Routine applications of a programme of appropriate fungicides, selected from Table 10.1, should be made, starting in early autumn or at the first signs of the disease.

Leaf curl (*Colletotrichum acutatum*)
This is a serious disease, which has become widespread following dissemination in infected corm stocks. The most obvious symptoms are twisting, curling and distortion of leaves and flower stalks. Salmon-pink sporing areas are sometimes seen, initially on leaf and flower primordia, at the centre of affected plants. The disease is favoured by temperatures over 20°C and is spread by water splash and probably by wind and on pickers' hands and clothes. Affected plants should be removed and destroyed as soon as they become evident. A corm dip in carbendazim (off-label) (SOLA 0009/99) may be given immediately before planting. The disease is soil-borne and anemones should not follow an affected crop unless the soil has been sterilized.

Table 10.1 Chemical control of grey mould and powdery mildew on protected ornamentals

Compound	Grey mould	Powdery mildew	Protected crop/crop type on which there are on-label or off-label recommendations
HV sprays			
bupirimate	–	✓	begonia, chrysanthemum, rose
captan (off-label)	(✓)	–	non-edible ornamental bulbs (SOLA 1229/95)
carbendazim	✓*	✓*	protected pot plants
carbendazim (off-label)	✓*	✓*	protected ornamental plants, soil-grown and container-grown; ornamental bulbs and corms; chrysanthemum, ornamental pot and bedding plants (SOLA 0009/99)
carbendazim + prochloraz (off-label)	✓	–	protected ornamentals (SOLA 2217/99)
chlorothalonil	(✓)	–	protected ornamentals
dichlofluanid (off-label)	✓	–	non-edible ornamentals (SOLA 0167/93)
dinocap	–	✓	chrysanthemum, roses
fenarimol†	✓	–	none; use at grower's own risk
imazalil	–	✓	protected ornamentals; protected roses
iprodione	✓*	–	pot plants
prochloraz	✓	–	ornamentals
pyrifenox	–	✓	ornamentals
pyrimethanil†	✓	–	none; use at grower's own risk
sulfur†	–	✓	none; use at grower's own risk
thiram	✓	–	chrysanthemum, freesia, ornamentals (except hydrangea)
quintozene	✓	–	bedding plants, chrysanthemum, dahlia, fuchsia, pelargonium, pot plants, protected carnation
zineb	✓	–	anemone, tulip, carnation
Smokes			
imazalil	–	✓	protected ornamentals; protected roses

(✓) Partial control.
* Resistant isolates will not be controlled.
† Off-label, under Extended Use Arrangements.

Plum rust (*Tranzschelia pruni-spinosae*)
This is an unusual disease but where it does occur on protected crops the source is usually infected corms rather than plum trees. Affected plants have a stiff, upright habit and may be taller than healthy ones. They may also cease to produce further flowers. Leaf margins are down-curled and yellow pustules may be evident on the undersides of infected leaves. Affected plants should be destroyed. Protective sprays of zineb and oxycarboxin can be used.

Powdery mildew (*Erysiphe aquilegiae* var. *ranunculi*)
This disease occurs occasionally on greenhouse crops but is rarely a problem. Typical white, powdery patches appear on leaves and bracts during the summer and autumn, and disappear with the onset of cool weather. At the first sign of symptoms a fungicide should be applied. Imazalil or pyrifenox should give satisfactory control.

Bedding plants

Pests

Aphids
Peach/potato aphid (*Myzus persicae*) can infest various bedding plant species including *Fuchsia, Impatiens, Nicotiana, Petunia, Primula* and *Viola*. Glasshouse & potato aphid (*Aulacorthum solani*) is commonly found on *Fuchsia, Nicotiana, Pelargonium* and *Verbena*. Melon & cotton aphid (*Aphis gossypii*) can infest *Begonia* and *Fuchsia*. If using IPM, the parasitoid wasp *Aphidius colemani* will control *M. persicae* and *A. gossypii*; similarly, *Aphidius ervi* will control *A. solani*. The predatory midge *Aphidoletes aphidimyza* can be used to supplement control of all aphid species, if necessary. If an aphicide is required, pirimicarb or the aphid anti-feedant pymetrozine can be used in IPM, although *A. gossypii* and some strains of *M. persicae* are resistant to pirimicarb. A HV spray or fumigation with nicotine can also be integrated with biological control agents, and should control all aphid species. If using a pesticide programme, imidacloprid can either be used as a drench or incorporated in the compost, and can be useful for persistent control in hanging baskets hung high in the glasshouse roof where regular monitoring for pests is difficult.

Chrysanthemum leaf miner (*Chromatomyia syngenesiae*)
Bellis, Calendula, Primula and *Viola* are particularly susceptible to attack. See under *Dendranthema* (chrysanthemum), p. 401. See also under Notifiable pests and diseases, leaf miners, Chapter 9, p. 328.

Glasshouse leafhopper (*Hauptidia maroccana*)
Fuchsia, Nicotiana, Pelargonium, Primula, Salvia and *Verbena* are particularly susceptible to damage. For further information, see under *Pelargonium*, p. 417.

Glasshouse whitefly (*Trialeurodes vaporariorum*)
This pest can occur on *Begonia, Fuchsia, Impatiens, Nicotiana, Pelargonium, Primula* and *Salvia*. See under *Fuchsia*, p. 412. See also, under Notifiable pests and diseases, tobacco whitefly, Chapter 9, p. 328.

Sciarid flies
Sciarid fly larvae (e.g. those of *Bradysia paupera*) can cause damage, particularly to young seedlings or cuttings, although these pests can often be secondary to disease. Good nursery hygiene and avoiding overwatering will reduce the likelihood of infestations developing. Preventive treatment against larvae may be justified during propagation when there are large surface areas of damp compost which might attract the adult flies and upon which their eggs could be laid. For further information, see under *Euphorbia pulcherrima* (*Poinsettia*), p. 409.

Shore flies (*Scatella stagnalis* and *S. tenuicornis*)
These small, black flies are commonly found on bedding plants but they are not plant pests. The larvae are saprophytic, feeding on algae on the surface of the compost, bench or floor covering. If numerous, the flies can be a nuisance to nursery staff. No pesticides are recommended for control of shore flies. However, good nursery hygiene, avoiding overwatering and ensuring adequate ventilation can reduce the numbers of flies. The predatory mite *Hypoaspis* spp. can give effective control in IPM.

Thrips
Western flower thrips (WFT) (*Frankliniella occidentalis*) is a major pest of bedding plants, although onion thrips (*Thrips tabaci*) can also occur. *Bacopa, Brachycome, Fuchsia, Impatiens*, ivy-leaf *Pelargonium, Primula* and *Verbena* are particularly susceptible. WFT can transmit tomato spotted wilt virus (TSWV), which can cause various symptoms (including chlorosis, necrosis and ring spotting). Biological control with the predatory mite *Amblyseius cucumeris* is effective on many plant species. For biological and chemical control, see under *Pelargonium*, p. 417. Malathion should not be used on *Antirrhinum, Crassula, Fuchsia, Gerbera, Lathyrus* (sweet pea), *Petunia, Pilea, Zinnia* or ferns.

Two-spotted spider mite (*Tetranychus urticae*)
This is an occasional pest of bedding plants, such as *Fuchsia, Primula* and *Verbena*. See under *Dendranthema* (chrysanthemum), p. 402.

Diseases

Many of the diseases that affect bedding plants are found on a wide range of species. In this section, therefore, the information is presented in tabular form by disease, with the symptoms, favourable environmental conditions, commonly affected bedding plant subjects and chemical control options listed in each table. Three tables are presented, dealing with soil-borne diseases (Table 10.2, p. 382), airborne diseases (Table 10.3, p. 384) and seed-borne diseases (Table 10.4, p. 386).

Table 10.2 Soil-borne diseases of bedding plants

Disease	Symptoms	Most susceptible subjects*	Favourable conditions for disease	Chemical control**	Additional information
pythium	Damping-off. Root and stem-base rot. Most severe on young seedlings	*Antirrhinum*, *Alyssum*, *Callistephus*, *Lobelia*, *Nemesia*, *Tagetes*	High soil moisture levels; under or (particularly) overwatering; overcrowding of seedlings; high conductivity; low temperatures	Copper ammonium carbonate, etridiazole, fosetyl-aluminium, furalaxyl (†), propamocarb hydrochloride (†)	See Table 11.2 for additional information on fungicide application techniques, crop safety etc.
phytophthora	Rot of root and stem-bases of mature plants. Most damaging on outdoor beds	*Callistephus*, *Cineraria*, *Petunia*, *Viola*	As above	As above	As above
rhizoctonia (*Thanatephorus*)	*Seedlings:* damping-off; reddish-brown constricted stem-base lesion (wirestem)	*Alyssum*, *Aubrietia*, *Cheiranthus*, *Impatiens*, *Matthiola*, *Salvia*	High humidity; high nutrient levels; slow germination; deep planting; high temperatures	Tolclofos-methyl (compost incorporation or drench); iprodione (HV spray); quintozene (†) (compost or soil incorporation)	
	Mature plants: foot rot and/or root rot. Light brown lesions on stem at soil level (stem canker)				

black root rot (*Thielaviopsis*)	Root rot with blackening of the root system. Rotting and discoloration of stem base. Plants stunted with chlorotic foliage	*Alyssum*, *Antirrhinum*, *Cheiranthus*, *Matthiola*, *Nicotiana*, *Petunia*, *Primula*, *Viola*	High soil moisture; poor soil aeration; high pH (6.5+); temperatures between 17 and 23°C; plant 'stress'	Carbendazim (†) compost drench (off-label) (SOLA 0009/99)	Particularly damaging on winter-flowering pansies
fusarium wilt (*Fusarium oxysporum*)	Wilt follows yellowing of foliage. May be restricted to one side of plant	*Callistephus*, *Dianthus*	High temperatures	Carbendazim (†) drench as above; prochloraz (drench)	Different strains of the fungus affect *Callistephus* and *Dianthus*; cross-infection does not occur
fusarium basal rot (*Fusarium culmorum*)	Basal stem rot with pink discoloration of affected tissues	*Dianthus*	High temperatures; overwatering	As above	The same fungus can cause a die-back of branches (known as stub rot)
sclerotina stem rot	Sunken water-soaked lesions on stem causing stem collapse. Black sclerotia 10 cm in length found inside stem or in rotting tissue	*Antirrhinum*, *Campanula*, *Cineraria*, *Matthiola*, *Petunia*, *Primula*, *Tagetes*, *Zinnia*	Humid, shady conditions; poor nursery hygiene	Quintozene (†) (compost or soil incorporation); prochloraz, iprodione, carbendazim (†) (spray)	Dispose of affected plants carefully, so that sclerotia do not contaminate standing areas

* All bedding plant subjects are susceptible to black root rot, phytophthora, pythium and rhizoctonia. Those listed are very susceptible.

** Compounds marked (†) have label recommendations or specific off-label approval (SOLA) for use on 'bedding plants'. Other compounds listed (unless stated otherwise) have approval for use on 'ornamentals' or a specific ornamental crop. Those in the former category sometimes list bedding plant species in their crop safety lists. However, because of the large number of bedding plant species and cultivars grown, it is always recommended that the compound be tried on a limited number of plants first to assess crop safety. Some of the compounds in the table are available as more than one proprietary product, the label recommendations for which may differ. It is very important to check the label of any product intended for use.

Table 10.3 Airborne diseases of bedding plants

Disease	Symptoms	Susceptible subjects	Favourable conditions for disease	Chemical control*	Additional information
downy mildew	Yellow, sometimes angular, patches on upper leaf surface. Felty covering on undersurface, ranging from white to dark-grey/mauve	*Alyssum*, *Aubrietia*, *Cineraria*, *Cheiranthus*, *Matthiola*, *Myosotis*, *Nicotiana*, *Viola*	High humidity; free water on leaf usually required for infection	Chlorothalonil, dichlofluanid, mancozeb, thiram, zineb	Dichlofluanid (SOLA 0167/93) Metalaxyl + thiram and fosetyl-aluminium are frequently used – these have no label recommendations for ornamentals and are used entirely at the grower's risk
grey mould	Grey, powdery mass of spores covers affected tissue. Flattened, black sclerotial bodies 2-8 mm in length may form on decaying plant debris	All bedding plants	High humidities; still air. Soft, wilted, ageing or damaged plant parts and fallen flowers are most susceptible	Carbendazim (off-label) (†), chlorothalonil, dichlofluanid (off-label), iprodione, prochloraz, quintozene (†), thiram	Carbendazim (SOLA 0009/99), dichlofluanid (SOLA 0167/93). All products applied as sprays except quintozene, which is incorporated into soil or compost
powdery mildew	Foliage, especially young leaves and shoot tips, are covered with a white, powdery deposit. New growth is stunted and distorted	*Alyssum*, *Begonia*, *Calendula*, *Cineraria*, *Myosotis*, *Petunia*, *Viola*	High temperatures; high humidity	Bupirimate, bupirimate + triforine, carbendazim (†), chlorothalonil, fenarimol, imazalil, pyrifenox, thiram	Prolonged leaf wetness may reduce levels of powdery mildew, but will encourage other diseases such as downy mildew and botrytis

ramularia	Brown, papery leaf spot surrounded by yellow halo	*Primula*, *Viola*	High humidity; water splash	Carbendazim (†), chlorothalonil, prochloraz	Carbendazim (SOLA 0009/99)
rusts	Buff, orange or dark brown pustules on leaf surfaces	*Antirrhinum*, *Bellis*, *Calendula*, *Cineraria*, *Dianthus*, *Viola*	High temperatures; free water on leaf usually required for infection	Azoxystrobin (off-label), bupirimate + triforine, difenoconazole (off-label), mancozeb, oxycarboxin, propiconazole (off-label), thiram, zineb	Azoxystrobin (SOLA 1536/00) (chrysanthemum), difenoconazole (SOLAs 1728/97, 1729/97) (sweet william and pinks), propiconazole (SOLAs 0019/92, 1121/96, 1388/97) (chrysanthemum)

* Compounds marked (†) have label recommendations or specific off-label approval (SOLA) for use on 'bedding plants'. Other compounds listed (unless stated otherwise) have approval for use on 'ornamentals' or a specific ornamental crop. Those in the former category sometimes list bedding plant species in their crop safety lists. However, because of the large number of bedding plant species and cultivars grown, it is always recommended that the compound be tried on a limited number of plants first to assess crop safety. Some of the compounds in the table are available as more than one proprietary product, the label recommendations for which may differ. It is very important to check the label of any product intended for use.

Table 10.4 Seed-borne diseases of bedding plants

Disease	Symptoms	Susceptible subjects	Favourable conditions for disease	Chemical control*	Additional information
alternaria	Damping-off, leaf spot, stem-base lesion	*Cheiranthus*, *Cineraria*, *Dianthus*, *Lobelia*, *Zinnia*	High humidity; water splash	Iprodione	Applied as seed treatment or foliar spray. Resistance of alternaria to iprodione is documented on *Lobelia*
bacterial leaf spot (*Pseudomonas*)	Leaf spot with well-defined dark-brown margin and water-soaked dark-green zone around lesions. Most serious on seedlings	*Antirrhinum*, *Impatiens*, *Lobelia*, *Primula*, *Tagetes*, *Verbena*, *Viola*	High temperatures; high humidity; water splash	Copper ammonium carbonate	Chemical control is often ineffective. Dispose of affected plants and avoid leaf wetness
bacterial wilt (*Xanthomonas campestris*)	Variable. Leaf yellowing, with numerous small, brown spots. Staining of vascular tissue in stem. Stunted growth and leaf loss	*Cheiranthus*, *Lobelia*, *Matthiola*	High humidity; water splash	Copper ammonium carbonate	Chemical control is often ineffective. Dispose of affected plants and avoid leaf wetness
septoria	Damping-off, leaf spot	*Antirrhinum*, *Phlox*	High humidity; water splash	Carbendazim (off-label) (†), chlorothalonil, mancozeb, prochloraz, zineb	carbendazim (SOLA 0009/99)

* Compounds marked (†) have label recommendations or specific off-label approval (SOLA) for use on 'bedding plants'. Other compounds listed (unless stated otherwise) have approval for use on 'ornamentals' or a specific ornamental crop. Those in the former category sometimes list bedding plant species in their crop safety lists. However, because of the large number of bedding plant species and cultivars grown, it is always recommended that the compound be tried on a limited number of plants first to assess crop safety. Some of the compounds in the table are available as more than one proprietary product, the label recommendations for which may differ. It is very important to check the label of any product intended for use.

Some plant subjects traditionally grown as pot plants are also produced for bedding, e.g. carnation, chrysanthemum, fuchsia and pelargonium. The diseases of these plants are dealt with elsewhere in this chapter. Diseases of sweet pea (*Lathyrus*) and wallflower (*Cheiranthus*) are described in detail in Chapter 11, although the fungicide tables in this chapter should be consulted if these plants are being grown under protection.

Cultural control of bedding plant diseases

The same basic principles for cultural control used for all protected ornamental crops are applicable to bedding plants.

- Use fresh compost, straight from the bag if possible, or from a covered heap. Do not store compost in uncovered heaps which can become contaminated with plant debris and the resting spores of fungal pathogens.
- Clean benches and standing areas between crops, removing any plant debris. Consider disinfection of these areas, especially if there have been problems from root diseases in the previous crop.
- Clean and disinfect or replace capillary matting periodically.
- Use new seed trays and pots, especially for propagation.
- Stand seed trays off the ground, to avoid contamination by root pathogens.
- Prick-out from healthy trays of seedlings; discard any trays with damped-off patches.
- Ensure that water is free from fungal pathogens such as *Pythium* and *Phytophthora* by using a mains supply, or by installing a suitable disinfection or filtration system for non-mains or recirculated water. Keep water tanks covered.
- Avoid overwatering (which encourages root pathogens) and overcrowding (which encourages foliar diseases). Use of low-level irrigation can prevent many foliar diseases by reducing leaf wetness and water splash.
- Avoid high humidities, as these favour foliar diseases such as mildews, rusts, *Botrytis* and leaf spots.
- Avoid damaging plants during transplanting or routine maintenance, and remove senescing flowers. Damaged or senescent tissue can be colonized rapidly by *Botrytis*.

Soil-borne diseases

Contaminated compost and debris from previous crops are the principal sources of infection. Standing areas and benches, pots, trays, capillary matting and Danish trolleys can all become contaminated. Irrigation water may also carry *Pythium* and *Phytophthora*, particularly if rainwater or water drawn from uncovered tanks is used.

Airborne diseases

Large numbers of spores are usually produced on the infected tissue, and these spores are easily moved in air currents. Spread can occur over long distances, but frequently infection comes from areas close to the nursery where unsold and poor-quality plants from previous crops have been discarded. Air-dispersed diseases usually affect the foliage and can occur at any stage of growth.

Seed-borne diseases

Infection is carried as spores, fungal growth or bacteria either on or inside the seed. Check with suppliers of seed of susceptible subjects to see whether they offer pathogen-tested seed.

Virus diseases

A large number of different viruses are found on bedding plants (which is unsurprising, given the wide range of plant genera and species encompassed within this term). The symptoms of virus infection are numerous, but those most commonly encountered are chlorotic mottles, mosaics and ring spots, leaf distortion and general stunting of plants. Many viruses have insect vectors (such as aphids or thrips), and others can be transmitted mechanically from plant to plant. Prevention of widespread virus infection depends on the control of vectors and the destruction of affected material. In the case of vegetatively propagated plants it is particularly important that mother plants are virus-free.

The most common viruses found in bedding plants are probably tomato spotted wilt virus (TSWV) and impatiens necrotic spot virus (INSV), both of which are transmitted by thrips, predominantly western flower thrips (*Frankliniella occidentalis*), and the aphid-transmitted and mechanically transmitted cucumber mosaic virus (CMV). Confirmation of the presence of virus in suspect material, by laboratory testing, is always recommended.

Begonia

Begonia flowers are particularly susceptible to injury by many pesticides, especially when applied as sprays. Therefore, do not spray plants in bright sunshine or in humid conditions.

Pests

Aphids

Melon & cotton aphid (*Aphis gossypii*) is the main species infesting *Begonia*. The pest can be controlled effectively within IPM by the parasitoid wasp *Aphidius colemani*. See under *Cyclamen*, p. 398.

Glasshouse whitefly (*Trialeurodes vaporariorum*)

Biological control with the parasitoid *Encarsia formosa* is effective within IPM and is the best option for control on *Begonia*.

Thrips

Western flower thrips (*Frankliniella occidentalis*) is the most common species. Biological control with the predatory mite *Amblyseius cucumeris* is effective within IPM and is the best option for control on *Begonia*.

Diseases

Bacterial blight (*Xanthomonas campestris* pv. *begoniae*)

This is a major disease of elatior, fibrous-rooted and tuberous begonias. It is usually first seen as small water-soaked spots towards the margins of older leaves; the spots enlarge and coalesce, leading to a soft rot and premature leaf fall. Infection arises from cuttings carrying latent systemic infection or dried, infested debris. The disease is readily spread by water-splash, by handling plants and on knives. Control by propagating from disease-free stock plants, by preventing water-splash and by watering using a sub-irrigation system. There are no known resistant cultivars or effective means of chemical control.

Grey mould (*Botryotinia fuckeliana* – anamorph: *Botrytis cinerea*)

Foliage, stems and flowers may be attacked by this fungus if plants are grown in humid conditions. Spray at the first sign of attack with a suitable fungicide selected from Table 10.1, p. 379.

Powdery mildew (*Microsphaera begoniae* – anamorph: *Oidium begoniae*)

A typical white, powdery mould forms on young leaves, stems and flower buds; when severely attacked, tissue may become desiccated. The disease is usually associated with large fluctuations in humidity and temperature. Use HV sprays of a suitable fungicide, such as bupirimate (Table 10.1, p. 379).

Root and stem rots

Several fungi, including *Pythium* spp., *Thanatephorus cucumeris* (anamorph: *Rhizoctonia solani*) and *Thielaviopsis basicola*, cause the roots, tubers or stems to rot, particularly when plants are frequently overwatered. Prevent infection by using healthy planting material, but if the disease is suspected use one of the drench or incorporation treatments shown in Table 10.5. Etridiazole, fosetyl-aluminium and propamocarb hydrochloride should give some control of pythium root rot, while carbendazim (off-label) (SOLA 0009/99) will control *Thielaviopsis*; *Rhizoctonia* can be controlled by a drench of tolclofos-methyl.

Bulbs (forced)

Many pest and disease problems occur as a result of 'housing' infested bulbs. It is very important to select only sound, healthy bulbs for forcing and to apply preventive treatments before planting into sterilized soil or new compost.

Table 10.5 Chemical control of root diseases on protected ornamentals

Compound	Diseases controlled	Protected crop/crop type on which there is a label recommendation
Dip		
captan (off-label)	fusarium	non-edible ornamental bulbs (SOLA 1229/95)
carbendazim (off-label)	botrytis, fusarium, penicillium, sclerotinia	ornamental bulbs and corms (SOLA 0009/99)
propamocarb hydrochloride	pythium root rot	tulip bulbs
Drench		
carbendazim (off-label)	fusarium, thielaviopsis	ornamental plant production, soil and container grown crops; freesia (SOLA 0009/99)
etridiazole	phytophthora	container-grown stock
fosetyl-aluminium	phytophthora	protected pot plants, capillary mats
furalaxyl	pythium, phytophthora, damping-off	bedding plants, pot plants
prochloraz (off-label)	fusarium, penicillium	non-edible ornamental bulbs (SOLA 2222/99)
propamocarb hydrochloride	phytophthora, pythium	ornamentals, flower bulbs, pot plants, bedding plants
tolclofos-methyl	damping-off, foot rot, root rot	ornamentals, seedlings of ornamentals
In nutrient solution		
carbendazim (off-label)	thielaviopsis	ornamental crops grown in NFT (SOLA 0009/99)
propamocarb hydrochloride	phytophthora, pythium	ornamentals, flower bulbs, pot plants, bedding plants
Compost incorporation		
etridiazole	phytophthora	container-grown stock
furalaxyl	pythium, phytophthora, damping-off	bedding plants, pot plants
propamocarb hydrochloride	phytophthora, pythium	ornamentals, flower bulbs, pot plants, bedding plants
quintozene	rhizoctonia	anemone, bedding plants, chrysanthemum, dahlia, flower bulbs, fuchsia, hyacinth, iris, *Narcissus*, pelargonium, pot plants, protected carnation, tulip
tolclofos-methyl	damping-off, foot rot	ornamentals, seedlings of ornamentals

Bulbs (forced) – *Hyacinthus* (hyacinth)

Pests

Stem nematode (*Ditylenchus dipsaci*)
See under *Hyacinthus* (hyacinth), Chapter 12, p. 548.

Diseases

Penicillium bulb rot (*Penicillium corymbiferum*, *P. cyclopium* and *P. hirsutum*)
These fungi cause rotting of bulbs stored at high humidity. Plant bulbs as soon as possible on receipt.

Soft rot (*Erwinia carotovora* ssp. *carotovora*)
This bacterial disease results in the complete breakdown of the tissue of the bulb. The first signs of rotting are a stunting of growth and yellowing of leaf tips. Flower buds often fall off. Rotting of the flower stem at soil level may result in the collapse of the whole plant. It is a bulb- or soil-borne disease. Remove and destroy the affected bulbs. Drenches of copper oxychloride* may reduce the spread of the disease if infection appears early in the life of the crop.

Yellow disease (*Xanthomonas campestris* pv. *hyacinthi*)
This is a bacterial disease, which totally rots the bulbs before or soon after planting. Slightly diseased bulbs, when cut across, show small, yellow spots arranged in concentric rings. Diseased bulbs should be removed very carefully and great care taken with watering to avoid splashing the bacteria about. There is no effective chemical treatment.

Bulbs (forced) – *Iris*

Pests

Aphids
See under *Iris*, Chapter 12, p. 549.

Thrips
See under *Iris*, Chapter 12, p. 549.

Diseases

Botrytis rot (*Botryotinia fuckeliana* – anamorph: *Botrytis cinerea*)
This disease causes plants to topple with a rot at the base of the stem and top of the bulb. The fungus sometimes causes flower spotting and often results in die-

back of leaves owing to tip infection. It is a problem in unheated crops only where conditions can sometimes favour the development of the disease. HV sprays of chlorothalonil or iprodione should be used if necessary.

Fusarium basal rot (*Fusarium oxysporum* f. sp. *gladioli*)
This is usually considered to be mainly a bulb-borne disease, although it can result from contaminated soil. Affected plants are stunted and have a characteristic one-sided growth which leads to the shoot growing at an angle of 45° to the ground. These plants have a dark-brown rot of the base plate and disintegrated roots. The bulb rots away completely.

Pre-planting bulb dips with captan (off-label) (SOLA 1229/95), carbendazim (off-label) (SOLA 0009/99) or prochloraz (off-label) (SOLA 2222/99) should give some control. Post-planting soil drenches of carbendazim (off-label) (SOLA 0009/99) are sometimes used. Soil should be sterilized thoroughly.

Grey bulb rot (*Rhizoctonia tuliparum*)
See under *Tulipa* (tulip), p. 555.

Leaf spot (*Mycosphaerella macrospora*)
This disease affects leaves and spathes; although not usually a problem in protected crops, under some circumstances it can cause serious losses of marketable blooms. The first signs of the disease are small, brown spots in the centre of a water-soaked area. The lesion turns pale grey, then the tissue dries to a light grey-brown. Such lesions are typically elliptical in shape, being restricted laterally by the veins, up to 10 mm in length and about 5 mm in width. They may coalesce to form larger lesions. Under suitable conditions of high humidity, particularly when plants are alternately wet and then dry, the centres of the spots turn dark grey and the fungus produces tufts on which spores are produced. Spores are dispersed by air movement and water splashing. The disease perennates on debris. Control can be achieved by keeping the foliage dry, especially during the autumn and winter, and (where possible) by ventilating to reduce humidity. Destroy infected debris from previous crops. Protective sprays of chlorothalonil or mancozeb* have been found to be effective.

Penicillium rot or bulb rot (*Penicillium corymbiferum*)
The disease usually causes a problem only during storage or while bulbs are being transported. Under suitable conditions the pathogen can penetrate through wounds caused by handling and through natural wounds, such as those caused by emergence of root initials. The disease produces a rot which is typically blue-green, often noticeable only by peeling the outer scales from the bulbs. Slightly affected bulbs grow normally. Adverse growing conditions favour the development of the disease, or invasion and rotting by secondary organisms which may lead to poor growth or death of the bulb. Careful handling and storage will reduce the subsequent development of the disease. If storage is needed, bulbs

must be stored in shallow layers, ideally in wire-bottomed trays, to allow good air circulation and ventilation. Pre-planting dips with carbendazim (off-label) (SOLA 0009/99) or prochloraz (off-label) (SOLA 2222/99) may give some control. Carbendazim will not control resistant strains of the fungus.

***Pythium* root rot** (*Pythium irregulare*)
Pythium root rot is a serious root disease of iris. Affected plants stop growing and the root tips are found to be yellow and rotted. The disease develops in clearly defined areas of the crop, in which plants are stunted and fail to flower. Cold, wet conditions encourage development of the disease. It can be serious, especially where irises are monocropped and where soil sterilization is infrequent. A bulb dip of propamocarb hydrochloride* for 20 minutes, combined with pre-planting soil treatment with etridiazole or propamocarb hydrochloride, is effective but soil incorporation must be thorough. Post-planting soil drenches with etridiazole are used but are less effective than pre-planting treatments. Sterilize the soil between crops.

Rhizoctonia (*Thanatephorus cucumeris* – anamorph: *Rhizoctonia solani*)
The disease can be bulb- or soil-borne. It causes a root or neck rot of the bulbs. Affected plants are stunted and the foliage yellows. Symptoms on the bulbs are of a soft, brown rot. Spread can be rapid under conditions of high soil temperature, and can result in significant losses. Sterilize the soil or treat soil with quintozene or tolclofos-methyl before planting.

Bulbs (forced) – *Lilium* (lily)

Pests

Aphids
For details, see under *Lilium* (lily), Chapter 12, p. 550. Melon & cotton aphid (*Aphis gossypii*) is a major pest under glass and is resistant to pirimicarb and certain other aphicides. For chemical control, see under *Dendranthema* (chrysanthemum), p. 401.

Lily beetle (*Lilioceris lilii*)
This pest has not yet been a problem on commercially produced lilies under glass, but is becoming more widespread outdoors so is a potential problem for the future. See under *Lilium* (lily), Chapter 12, p. 550.

Slugs
Lily bulbs are very susceptible to damage caused by slugs, including field slug (*Deroceras reticulatum*) and garden slug (*Arion hortensis*). For control, apply pellet baits of metaldehyde or methiocarb (off-label) (SOLA 1599/98) when slugs are active on the soil surface.

Diseases

Foot rot (*Phytophthora* spp.)
Affected plants may topple over and die. The bulbs of such plants will be found to be in an advanced state of rotting. The disease is worse in poorly drained soils under wet conditions. The effects of the disease can be reduced by using pre-planting incorporation of etridiazole or propamocarb hydrochloride, or drenches of etridiazole after planting.

Fusarium scale rot (*Fusarium oxysporum* f. sp. *lilii*)
Plants wilt slightly and, although bulb scales may appear healthy, roots turn red. Eventually, the scales become infected and rot away. Sterilize the soil. A soil drench with carbendazim (off-label) (SOLA 0009/99) may give some control.

Leaf spot (*Botrytis elliptica*)
This fungus causes leaf spots that enlarge rapidly in humid conditions. Keep the leaves as dry as possible when watering. In spring, spray new growth with a programme of suitable fungicides selected from Table 10.1, p. 379.

Pythium root rot (*Pythium* spp.)
Tips of roots show the first signs of the disease, which gradually spreads to affect whole roots. All roots normally become infected and rot. This often leads to the disintegration of the bulb. Above ground, the plant becomes stunted and then the foliage becomes chlorotic and dies. It is a soil-borne disease but may be spread in non-mains water. Some control may be effected by pre-planting incorporation or by post-planting drenches of etridiazole or by drenches of propamocarb hydrochloride.

Stump rot (*Phytophthora nicotianae* vars *nicotianae* and *parasitica*)
A water-soaked area appears in the centre of the rosette, or at the base of the leaves, soon after emergence of the shoot through the soil. This may lead to the entire shoot being killed, in which case unaffected laterals are often produced; these remain healthy. Where the disease appears later in the season and under dry conditions, only the tips of shoots and the flower buds may be affected. The disease is encouraged by wet soil conditions and the use of a contaminated water supply. Pre-planting soil incorporation or drenches of etridiazole at the first signs of the disease will give some control.

Bulbs (forced) – *Narcissus* (daffodil)

Pests

Bulb mites (*Rhizoglyphus callae* and *R. robini*)
See under *Narcissus* (daffodil), Chapter 12, p. 551.

Bulb scale mite (*Steneotarsonemus laticeps*)
See under *Narcissus* (daffodil), Chapter 12, p. 551.

Stem nematode (*Ditylenchus dipsaci*)
See under *Narcissus* (daffodil), Chapter 12, p. 553.

Diseases

Basal rot (*Fusarium oxysporum* f. sp. *narcissi*)
Diseased bulbs are soft and frequently have a pink mass of spores on the root plate. Infected bulbs should be removed and contaminated soil or compost must be sterilized if it is used again. It is important to distinguish between basal rot (a uniform chocolate-brown rot) and symptoms of stem nematode (*Ditylenchus dipsaci*) (brown concentric rings from the base upwards), a pest that can also cause serious damage. Forced bulbs should not be recovered for forcing if there are appreciable amounts of basal rot, bulb scale mite (see Chapter 12, p. 551) or stem nematode (see Chapter 12, p. 553).

Bulbs (forced) – *Tulipa* (tulip)

Pests

Aphids
The most important aphid species under glass are tulip bulb aphid (*Dysaphis tulipae*), peach/potato aphid (*Myzus persicae*) and potato aphid (*Macrosiphum euphorbiae*). For details and control of *D. tulipae*, see *Tulipa* (tulip), Chapter 12, p. 555. For details and control of *M. persicae* and *M. euphorbiae*, see under *Anemone*, aphids, Chapter 12, p. 543.

Stem nematode (*Ditylenchus dipsaci*)
See under *Tulipa* (tulip), Chapter 12, p. 555.

Diseases

Fire (*Botrytis tulipae*)
This fungus can be seen on outer and fleshy bulb scales as small, black sclerotia (similar in size to onion seed), which may be associated with brown-margined lesions. Before planting, the fungus may be controlled by dipping the bulbs in carbendazim (off-label) (SOLA 0009/99). Primary infections occur on emerging shoots, which produce masses of olive-grey spores. These can cause secondary infection or spotting on other leaves and flowers. Cut off all infected shoots below the soil surface, avoid splashing foliage when watering and use sprays of chlorothalonil, dichlofluanid (off-label) (SOLA 0167/93) or iprodione.

Fusarium bulb rot (*Fusarium oxysporum* f. sp. *tulipae*)
Infection usually occurs in the summer before forcing, and diseased bulbs may become soft and a pinkish-white fungal mat may develop on the shrivelling bulb. 'Gumming' on the bulb surface is a characteristic indicator of infection. If infected bulbs are planted, ethylene gas produced by rotting tissues can cause damage to the embryonic flowers of nearby bulbs. As a preventive measure, dip the bulbs in carbendazim (off-label) (SOLA 0009/99) before planting.

Grey bulb rot (*Rhizoctonia tuliparum*)
The fungus attacks the bulb, causing it to rot (usually from the neck downwards), and often entirely preventing the growth of the shoot. The disease is soil-borne, and if plants become affected after emergence the new foliage is often distorted and poorly developed; affected plants frequently fail to flower. Inspection of affected bulbs shows a dry, grey to purplish rot. Scales stay firm unless secondarily attacked by soft-rotting organisms. Soil often adheres to the neck of bulbs following the development of a profuse white mycelium. In this mycelium, large (5–7 mm), white, later almost black, sclerotia may be seen. Use clean growing media and forcing boxes. Dip bulbs in carbendazim (off-label) (SOLA 0009/99), or dust them with quintozene. Treat soil with quintozene or tolclofos-methyl preplanting.

Root rot (*Pythium* spp.)
This disease has become more prevalent since bulbs for forcing have been planted direct into greenhouse border soil instead of boxes. Root rot causes patches of weak, yellowish plants with poor root systems. As a preventive treatment, dip bulbs in propamocarb hydrochloride for 20 minutes, or use a soil incorporation or soil drench treatment of etridiazole. Bulbs should be planted in sterilized soil.

Shanking (*Phytophthora cryptogea* and *P. erythroseptica*)
Affected plants usually show the first symptoms when flower buds are well developed. Buds often shrivel and leaf tips turn yellow. The roots of these plants will be poorly developed and they often rot. The rot may extend into the base plate. It is a soil-borne disease. Use clean substrate and forcing boxes. Treat the soil or compost with etridiazole.

Cineraria

Pests

Aphids
Cinerarias are very susceptible to several aphid species, including glasshouse & potato aphid (*Aulacorthum solani*), melon & cotton aphid (*Aphis gossypii*), mottled arum aphid (*Aulacorthum circumflexum*), peach/potato aphid (*Myzus*

persicae) and potato aphid (*Macrosiphum euphorbiae*). Biological control with parasitoids is possible, but the cool temperatures at which the crop is grown may reduce their activity. If using a pesticide programme, resistant strains of *A. gossypii* and *M. persicae* will not respond to pirimicarb or to pyrethroid sprays such as deltamethrin. A HV spray of pymetrozine, or a drench or compost incorporation with imidacloprid, should control all species.

Chrysanthemum leaf miner (*Chromatomyia syngenesiae*)
See under *Dendranthema* (chrysanthemum), p. 401.

Glasshouse whitefly (*Trialeurodes vaporariorum*)
Cinerarias are very susceptible to whiteflies, and temperatures at which the crop is grown are usually too cool for effective biological control by the parasitoid *Encarsia formosa*. For chemical control, see under *Dendranthemum* (chrysanthemum), p. 402.

Thrips
For species, see under *Dendranthemum* (chrysanthemum), p. 402. Cinerarias are susceptible to tomato spotted wilt virus (TSWV), which can be transmitted by western flower thrips (*Frankliniella occidentalis*) (see under *Cineraria*, p. 398). The cool temperatures at which the crop is grown should reduce thrips multiplication, whilst allowing some activity of the predatory mite *Amblyseius cucumeris* if using biological control. Pesticides should be used only if absolutely necessary as cinerarias are prone to spray damage. If using a pesticide programme, dichlorvos (off-label) (SOLA 0625/99) should be effective but should be tried on a few plants first to check for any potential plant damage. A HV spray of abamectin should also give some control.

Diseases

Downy mildew (*Bremia lactucae*)
Under cool, moist conditions a dense, felt-like growth of the fungus develops on the lower surface of the leaf. In severe cases, the leaves become curled and deformed. Ventilate well, to avoid high humidity and leaf wetness. A drench of furalaxyl, applied to the compost, will give some protection.

Grey mould (*Botryotinia fuckeliana* – anamorph: *Botrytis cinerea*)
This fungus attacks flowers, leaves and stems; for control measures, see under Cyclamen, p. 400.

Impatiens necrotic spot virus (INSV)
This virus disease, closely related to TSWV, causes necrotic lesions, vein clearing and stunting on cinerarias. It is controlled by removal of affected plants and

maintaining control of western flower thrips (*Frankliniella occidentalis*), see p. 397.

Powdery mildew (*Sphaerotheca fusca*)
For fungicide treatments see Table 10.1, avoiding applications during hot, bright weather.

Rust (*Coleosporium tussilaginis* and *Puccinia lagenophorae*)
Rust develops rapidly under moist conditions, when orange-coloured pustules appear on the leaves. Ventilation should be increased and HV sprays of oxycarboxin can be used as soon as the disease is first seen.

Tomato spotted wilt virus (TSWV)
TSWV can affect cinerarias severely, causing a leaf mosaic and vein blackening. For control of the principal vector, western flower thrips (*Frankliniella occidentalis*), see p. 397.

Cyclamen

Pests

Aphids
Melon & cotton aphid (*Aphis gossypii*), a yellowish-green, olive-green or black aphid, is the most common species on cyclamen. In addition, glasshouse & potato aphid (*Aulacorthum solani*), a shiny-green aphid, and mottled arum aphid (*Aulacorthum circumflexum*), a green aphid with a black horseshoe-shaped mark on its back, can also occur. If using IPM, the parasitoid wasp *Aphidius colemani* will control *A. gossypii* and *Aphidius ervi* will control *A. solani*. The predatory midge *Aphidoletes aphidimyza* can be used to supplement control if necessary. If using a pesticide programme, *A. gossypii* is resistant to pirimicarb and can also be resistant to other aphicides. A HV spray of pymetrozine, a drench or compost incorporation using imidacloprid, or (if temperatures exceed 16°C), a spray or fumigation with nicotine should give control of all species.

Broad mite (*Polyphagotarsonemus latus*)
See under Tarsonemid mites, p. 399.

Caterpillars
Caterpillar damage, such as that not infrequently caused by tomato moth (*Lacanobia oleracea*), appears as holes and the presence of faecal pellets on the leaves. Cyclamen is also attacked by caterpillars of straw-coloured tortrix moth (*Clepsis spectrana*), also known as 'cyclamen tortrix moth'. For control, see under *Dendranthema* (chrysanthemum), p. 401.

Cyclamen mite (*Phytonemus pallidus*)
See under Tarsonemid mites.

Tarsonemid mites
Cyclamen can be attacked by both broad mite (*Polyphagotarsonemus latus*) and cyclamen mite (*Phytonemus pallidus*). The mites are tiny and difficult to find in the growing points. Broad mite damage causes stunted, bronzed and distorted leaves and flower buds, and cyclamen mite feeding causes distorted leaves which often curl inwards, and distorted flowers which can fail to open. Incidence of both pests has become rare since the widespread commercial uptake of biological control methods for thrips on cyclamen, thought to be due to incidental control by the predatory mite *Amblyseius cucumeris*. If using a pesticide programme, a HV spray of dicofol + tetradifon should be effective.

Thrips
Western flower thrips (WFT) (*Frankliniella occidentalis*) is the most common species, although onion thrips (*Thrips tabaci*) can also occur. Thrips damage results in white flecking on the petals and distorted flowers. Cyclamen are also susceptible to tomato spotted wilt virus, which WFT can transmit. Virus symptoms on cyclamen include brown necrotic spots around the edges of the leaves. Biological control of WFT with the predatory mite *Amblyseius cucumeris* is very successful on cyclamen and is a better option than using pesticides.

Vine weevil (*Otiorhynchus sulcatus*)
Adult weevils cause U-shaped notches at the edges of leaves. The creamy-white larvae have brown head capsules and feed on the corms and roots, causing stunted growth and wilting. If using IPM, the larvae can be controlled with a drench of entomopathogenic nematodes (*Heterorhabditis* spp. or *Steinernema* spp.). Alternatively, a drench or compost incorporation using imidacloprid will give control.

Diseases

Bacterial soft rot (*Erwinia carotovora* ssp. *carotovora*)
This bacterium causes a soft rot of the corm, often resulting in wilting and death of the whole plant. Infection occurs through growth cracks or following corm damage (e.g. by fusarium wilt). Soft-rotting of the corm is rapid at temperatures above 20°C. Control is achieved by avoiding pest or pathogen damage to the corm and by regular watering to reduce the risk of growth cracks. Avoid warm, moist conditions and do not plant corms too deeply.

Fusarium wilt (*Fusarium oxysporum* f. sp. *cyclaminis*)
Leaves wilt after one or two older ones have turned bright yellow. The corm is

firm, but the vascular strands become orange-brown. Eventually, the whole plant collapses and white or pink fungal sporulation develops on the petiole bases. This disease can cause widespread losses in a warm autumn. Bench hygiene is very important when this crop is grown year after year, so the capillary matting or other standing surface should be disinfected at the end of each season. When the first symptoms are detected in a batch of plants, the affected ones should be removed before sporulation occurs, and the rest of the pots drenched with carbendazim (off-label) (SOLA 0009/99) to thoroughly wet the compost; repeat after 1 and 2 months. Addition of matured or composted bark to the growing medium can help control the disease. Avoid overwatering.

Grey mould (*Botryotinia fuckeliana* – anamorph: *Botrytis cinerea*)
This disease attacks leaves, petioles and flowers, and can result in considerable crop damage. It particularly affects leaves in the centre of plants. On flowers it causes a spotting symptom or flower rot. To prevent the disease, avoid prolonged high humidity and remove senescent leaves and flowers. Apply HV sprays of chlorothalonil, dichlofluanid (off-label) (SOLA 0167/93), iprodione or pyrimethanil*, directing sprays into the centre of plants. Sprays applied early in crop production appear to be more effective than treatments applied to mature plants.

Root rots (*Cylindrocarpon destructans*, *Pythium* sp. and *Thielaviopsis basicola*)
Drench soil or potting compost with carbendazim (off-label) (SOLA 0009/99) to control *Cylindrocarpon* and *Thielaviopsis*, or with furalaxyl or propamocarb hydrochloride to control *Pythium*.

Dendranthema (chrysanthemum)

There are marked differences in the susceptibility of chrysanthemum cultivars to pests and diseases, and in their tolerance to pesticides. Where there is any uncertainty, e.g. with a new cultivar or with a new pesticide, always check for phytotoxicity on a small batch of plants before extensive use.

Pests

Aphids
Several aphid species can attack chrysanthemum, the most common being melon & cotton aphid (*Aphis gossypii*) and peach/potato aphid (*Myzus persicae*). *A. gossypii* is yellowish-green, olive-green or black, and often forms large colonies underneath the leaves and in the growing points; *M. persicae* is green or pink, and is found underneath the leaves or in the buds and flowers. Other, less common species include leaf-curling plum aphid (*Brachycaudus helichrysi*), a small, yellowish-green aphid which is found in growing points and causes leaf and

flower distortion, and chrysanthemum aphid (*Macrosiphoniella sanborni*), a large, shiny, dark-brown aphid which tends to form colonies on the stems rather than on the leaves.

It is important to control all aphid species before the flower buds open. Only a few commercial growers currently use biological pest control methods on chrysanthemum, owing to the problems of controlling western flower thrips (WFT) (*Frankliniella occidentalis*) within IPM. However, where IPM is used, *A. gossypii* and *M. persicae* can be controlled biologically with the parasitoid wasp *Aphidius colemani,* the entomopathogenic fungus *Verticillium lecanii* and, if necessary, the predatory midge *Aphidoletes aphidimyza.* Where pesticide programmes are used, some aphids will be controlled by the regular pesticides applied for control of WFT; see under *Dendranthemum* (chrysanthemum), thrips, p. 402). However, pesticide-resistant strains of both *A. gossypii* and *M. persicae* are common. If necessary, a HV spray of nicotine or pymetrozine should control resistant aphids. Nicotine should not be used as a fumigant on chrysanthemum, owing to the risk of leaf-edge chlorosis.

Capsids

Capsids are rarely found on chrysanthemum where chemical control is used but they can become a problem in IPM, tarnished plant bug (*Lygus rugulipennis*) being the most frequently encountered species. Damage symptoms are distorted and punctured leaves, malformed flowers and blindness. For further details and for control within IPM, see under Cucumber, Chapter 9, p. 337. If using a pesticide programme on chrysanthemum, a HV spray of cypermethrin or deltamethrin will give some control of capsids.

Caterpillars

Several species have been recorded on chrysanthemum, of which angle-shades moth (*Phlogophora meticulosa*) is the most common. The caterpillars feed on both foliage and flowers, often causing noticeable damage. If using IPM, a HV spray of *Bacillus thuringiensis*, diflubenzuron or teflubenzuron will be compatible with biological control agents. All these products are more effective against young caterpillars, so should be applied at the first sign of damage. If using a pesticide programme, a HV spray of a pyrethroid, such as deltamethrin, should be effective but is incompatible with IPM.

Chrysanthemum leaf miner (*Chromatomyia syngenesiae*)

This is an often common and troublesome pest. The larvae disfigure plants by forming mines in the leaves. For further details, see under Lettuce, leaf miners, Chapter 9, p. 348. If using biological control methods within IPM, the parasitoid wasps *Dacnusa sibirica* and *Diglyphus isaea* are effective – see under Tomato, tomato leaf miner, Chapter 9, p. 357. Where pesticide programmes are used, abamectin will control larvae within the leaves. See also under Notifiable pests and diseases, leaf miners, Chapter 9, p. 329.

Chrysanthemum nematode (*Aphelenchoides ritzemabosi*)
This pest is now rare on chrysanthemum, owing to the widespread use of uninfested stock plants. The first sign of attack is a yellowish-green or purple blotching on the lower leaves, often delineated by the veins. Damage can spread up the plant and, eventually, the infested leaves blacken and become desiccated. Plants should be propagated from clean stock plants. Aldicarb (off-label) (SOLA 1325/95) granules should give effective control.

Glasshouse whitefly (*Trialeurodes vaporariorum*)
For description and damage, see under Cucumber, Chapter 9, p. 337. Chrysanthemum is not a good host for this pest, so serious attacks are uncommon. If using IPM, the parasitoid wasp *Encarsia formosa* should be effective and the entomopathogenic fungus *Verticillium lecanii* can give control if applied when glasshouse humidities are high. The pest is resistant to many pesticides. The insect growth regulators buprofezin and teflubenzuron can give useful control, although whitefly resistance to buprofezin is increasing.

Thrips
Western flower thrips (WFT) (*Frankliniella occidentalis*) is the most common species, but onion thrips (*Thrips tabaci*) can also occur. For description, see under Cucumber, Chapter 9, p. 388. WFT is a major pest, causing direct feeding damage that appears as white flecks on petals and leaves; this can lead to crop rejection. WFT can also transmit tomato spotted wilt virus (TSWV), which can cause a range of symptoms: see p. 405. Biological control methods for WFT are not always reliable on chrysantheum, particularly in the summer months when thrips multiplication is rapid. However, the predatory mite *Amblyseius cucumeris* and the entomopathogenic fungus *Verticillium lecanii* can give useful control. Current research has identified the potential of the predatory bug *Orius laevigatus* and certain predatory mites (*Hypoaspis* spp.) for improving biological control strategies on chrysanthemum, and these may be used commercially in the future. Currently, the majority of chrysanthemum growers rely on regular applications of pesticides to control WFT. WFT is resistant to many pesticides; dichlorvos is currently the most effective and widely used.

Two-spotted spider mite (*Tetranychus urticae*)
For description and damage symptoms, see under Cucumber, Chapter 9, p. 338. It is important to control any infestations before the buds show colour. Where IPM is used, biological control with the predatory mite *Phytoseiulus persimilis* is effective. The pest can be resistant to some of the older acaricides but abamectin, bifenthrin or fenbutatin oxide should be effective. Bifenthrin is incompatible with biological control agents.

Diseases

Blotch (*Septoria chrysanthemella*)
Blotch is an uncommon disease, causing circular, grey-brown leaf spots on older leaves. In severe cases, leaves may turn yellow and fall. To control, increase ventilation and spray at HV with carbendazim (off-label) (SOLA 0009/99) or chlorothalonil.

Brown rust (*Puccinia chrysanthemi*)
This uncommon disease attacks the leaves under humid conditions, causing a yellowish colour on the upper surfaces and, later, rusty circles of spore-bearing pustules on the undersides. Apply oxycarboxin as a fine spray, or chlorothalonil or thiram as HV sprays. These treatments should be applied when the disease first appears and repeated every 10–14 days.

Crown gall (*Agrobacterium tumefaciens*)
Tumours may occur on the roots or above ground, but more commonly large developments of the hardened tissue are found at the base of the stem at soil level. Affected plants may be stunted. The bacterium can survive for long periods in debris. Destroy affected plants, taking cuttings only from healthy mother stock plants. Use sterilized soil and pay attention to hygiene, particularly in the disposal of debris.

Damping-off (*Phytophthora* spp., *Pythium* spp. and *Thanatephorus cucumeris* – anamorph: *Rhizoctonia solani*)
These pathogens of young plants are associated with poor hygiene and failure to sterilize the soil in intensive cropping. Beds must be steam-sterilized after an infected crop. Incorporate etridiazole or propamocarb hydrochloride into compost or drench soil with etridiazole, fosetyl-aluminium or propamocarb hydrochloride to control *Pythium*. These fungicides are also active against *Phytophthora*. If damping-off is caused by *Rhizoctonia*, incorporate quintozene into compost, or into the soil surface layer, or use tolclofos-methyl.

Grey mould (*Botryotinia fuckeliana* – anamorph: *Botrytis cinerea*)
This fungus causes spotting and rotting of buds and flowers ('botrytis flower-rot'), especially if dew is allowed to form on the plants. Losses may also be serious, owing to infection of leaves and stems, especially in stock plants after cuttings have been taken. Botrytis flower-rot may be distinguished from petal blight (p. 404) and ray blight (p. 404) by the masses of grey spores which develop. In humid weather, at bud burst, apply HV sprays of chlorothalonil or iprodione*. If infection has already occurred, remove diseased plants promptly and apply HV sprays, and repeat if humid conditions persist. Stock plants should be treated similarly before cuttings are taken, to reduce infection during propagation. Cuttings may be sprayed with thiram.

Leafy gall (*Rhodococcus fascians*)
Affected plants have a proliferation of short, thickened shoots at the stem base. This infectious disease is easily spread during propagation. Mother stocks can be infected by hands when taking cuttings. In the cropping beds, water splash is probably the main means of spread. Use healthy mother stock plants. Remove and destroy infected plants carefully. Sterilize the soil between crops.

Petal blight (*Itersonilia perplexans*)
This disease is rare on heated crops and is associated with high humidity and with water films that persist on petals overnight. Coalescing, reddish-brown spots usually appear on the tips of the petals and a rot gradually extends inwards to affect the whole bloom. The sporulating fungus appears as a dull, white sheen on the rotted petals. Control is best achieved by ventilating and heating, because flowers are sensitive to chemicals. Alternatively, spray with zineb*, applied as a fine mist at the first sign of flower colour. Chlorothalonil has also given useful results. In humid conditions, repeat applications every 7 days.

Phoma root rot (*Phoma chrysanthemicola*)
The plants are stunted or killed by severe rotting of roots and, ultimately, of the base of the stem. Lower leaves show chlorotic and necrotic areas. Cultivars differ markedly in susceptibility. Soil sterilization with steam will help prevent the disease.

Powdery mildew (*Oidium chrysanthemi*)
The disease is first seen as whitish patches on the upper sides of young leaves and, in severe cases, progresses to cover and disfigure the upper and lower leaf surfaces and even the stems. Apply an appropriate spray (see Table 10.1, p. 379).

Ray blight (*Didymella chrysanthemi*)
Infections occur on flowers, stems or lower leaves. In contrast to petal blight (see above), the inner florets are attacked first; the infection spreads outwards to distort the flower and cause a dark-brown rot. More commonly, stems and lower leaves of cuttings and young plants become stunted with black-brown lesions. Cultivars vary markedly in susceptibility. Remove infected plants and keep the remainder as dry as possible. Apply HV sprays of chlorothalonil or mancozeb every 7 days. Sterilize the soil following an infected crop.

Sclerotinia rot (*Sclerotinia sclerotiorum*)
Light-brown lesions occur on the stem, usually about 10–20 cm above soil level. These lesions often develop rapidly, girdling the stem and causing the plant to wilt. Under humid conditions, a dense, white, fluffy mould develops. Large (0.5–2.0 cm in diameter), irregularly shaped, black sclerotia develop in this mould or, more frequently, in the pith. Remove affected plants carefully to

avoid dropping sclerotia. HV sprays of carbendazim (off-label) (SOLA 0009/99), chlorothalonil or iprodione may give some protection. Hygiene may need to be improved if the disease occurs regularly on a nursery. Sterilize the soil.

Stunt (chrysanthemum stunt viroid)
This virus-like disease is very infectious, being readily spread on hands and knives. It is a notifiable disease when found at a registered premises on chrysanthemum plants intended for planting. The incidence varies according to the amount of stunt present in propagators' stock plants. Symptoms are stunted growth, smaller leaves and flowers and earlier flowering. Plants showing symptoms should be removed carefully. All stock plants should be tested to ensure freedom from stunt.

Tomato spotted wilt virus (TSWV)
This virus disease, which is transmitted principally by western flower thrips (*Frankliniella occidentalis*), can cause severe damage to a crop. Cultivars differ markedly in symptoms: cvs Hurricane and Snowdon often show severe stem necrosis, whereas others show leaf mosaic and/or ring patterns. Virus strain and the time of year also influence symptom expression. Maintaining good control of western flower thrips (see p. 402) is necessary for effective control of TSWV. Remove affected plants.

White rust (*Puccinia horiana*)
This is a widespread and serious disease of chrysanthemum. It is a notifiable disease when found at a registered premises producing chrysanthemum plants intended for planting. Small, pale, slightly raised spots appear on the upper surface of the leaf, with pink/buff pustules turning white on the lower surface 2–4 weeks after infection, and which erupt to release light-brown spores. The rust is favoured by periods of high humidity and the greatest danger of infection is in March/April and September/October; any factor that tends to increase humidity (such as the use of overhead spray lines or tightly-sealed thermal screens/blackouts) should be avoided, particularly during these danger periods. In periods of high risk, or if an outbreak has occurred nearby, apply azoxystrobin (off-label) (SOLA 1536/00), chlorothalonil, mancozeb or oxycarboxin until the threat ceases. Propiconazole (off-label) (SOLA 1211/96) has curative activity and can give good control of established white rust, although resistant strains have recently been reported and these will not be controlled. Chemicals may cause damage, so sprays should not be applied beyond run-off. With a new cultivar or a new pesticide, always check for phytotoxicity on a small batch of plants before extensive use.

Verticillium wilt (*Verticillium albo-atrum* and *V. dahliae*)
This occurs only very rarely if soil sterilization has been effective, but it may be

introduced on infected cuttings. The lower leaves turn yellowish-brown and discoloration of foliage progresses upwards; the plants are stunted and woody; in severe cases, wilting occurs. The onset of symptoms often coincides with flowering. Routine soil sterilization is essential for control. If infection occurs, remove infected plants completely and drench the surrounding soil with carbendazim (off-label) (SOLA 0009/99).

Dianthus (carnation and hybrid pinks)

Pests

Aphids

Glasshouse & potato aphid (*Aulacorthum solani*) and peach/potato aphid (*Myzus persicae*) can both occur. For biological and chemical control, see under *Fuchsia*, p. 412.

Carmine spider mite (*Tetranychus cinnabarinus*)

This serious pest is a reddish-brown mite, which initially causes small, whitish spots on the leaves; as the infestation increases, the leaves become pale, stippled and desiccated. The mites breed on carnations throughout the year. Occasionally, two-spotted spider mite (*T. urticae*) causes similar damage. Biological control on carnation is still under development. HV sprays of abamectin, fenazaquin, fenbutatin oxide or tebufenpyrad should give some control but the pest is difficult to hit under the curled, waxy leaves.

Caterpillars

Caterpillars of various species occur on *Dianthus*, those of carnation tortrix moth (*Cacoecimorpha pronubana*) being the most common. Pheromone traps can be used to monitor activity of adults of this species so that treatments can be timed before damage (leaf webbing and bud damage) occurs. For biological and chemical control, see under *Dendranthema* (chrysanthemum), p. 401.

Grass & cereal mite (*Siteroptes graminum*)

See under *Dianthus*, bud rot, p. 407. Sprays applied against carmine spider mite (see above) may also have some effect.

Western flower thrips (*Frankliniella occidentalis*)

Damage symptoms are white flecks on leaves and petals. Biological control on carnation is still under development. If using pesticides, abamectin, dichlorvos or malathion will give some control. For precautions when using dichlorvos, see under *Fuchsia*, thrips, p. 413.

Diseases

Basal rots (*Alternaria dianthi*, *Fusarium culmorum*, *Phytophthora* spp., *Pythium* spp. and *Thanatephorus cucumeris* – anamorph: *Rhizoctonia solani*)
All of these fungi can cause stem lesions at or just above soil level, and usually within 6 weeks of planting. It is usually necessary to use microscopic examination to identify the pathogen. Affected plants wilt and often die. Sterilize soil between crops or use peat bags or an inert medium. A programme utilizing HV sprays of carbendazim against *Fusarium* and *Thanatephorus*, iprodione against *Alternaria* and *Thanatephorus*, and the broad-spectrum fungicides chlorothalonil and mancozeb*, provides control of the non-oomycete fungi. Treat with etridiazole or propamocarb hydrochloride if *Phytophthora* or *Pythium* occurs.

Bud rot (*Fusarium poae*)
Buds rot before the flowers open. On open blooms, petals may be distorted. The fungus is spread by the mite *Siteroptes graminum*. Remove affected buds. The use of acaricides recommended for control of carmine spider mite (*Tetranychus cinnabarinus*), p. 406, is usually sufficient to control the spread of the disease.

Fusarium wilt (*Fusarium oxysporum* f. sp. *dianthi*)
This is one of the most damaging diseases of carnation and hybrid pinks. The pathogen invades vascular tissues, which become slightly discoloured. All or part of the plant wilts. It is less likely to be a problem if the crop is grown in peat bags.

If the amount of inoculum in soil is kept low by prompt and careful removal of wilted plants and spot treatment of the affected area with a fungicide drench, thorough soil sterilization and scrupulous hygiene between crops will control the disease. Sterilization needs to be effective because carnation can be grown as a 2-year crop. Although substantial losses have occurred in first-year crops, wilt may not begin to spread rapidly until the second year. Metham-sodium, applied before steaming, can give improved control but results have been variable. It is difficult to eradicate the disease once soil inoculum levels have become high; consequently, peat bags on polythene sheets, rockwool or other inert media are increasingly used. Sterilization of the soil beneath the sheet is still advisable because the peat in the bag can become contaminated with the pathogen, but growing in the bag will limit the spread of the disease. Some resistant cultivars of carnation are available.

Soils drenches of carbendazim (off-label) (SOLA 0009/99) may be applied 14 days after planting and then every 1–3 months; early treatment of infection is more likely to be successful. Resistant strains of the pathogen or microbial breakdown of the fungicide may occur with repeated use.

Grey mould (*Botryotinia fuckeliana* – anamorph: *Botrytis cinerea*)
This fungus causes spotting and rotting of buds and flowers; it may be serious if lesions become established on old yellowing leaves, which provide inoculum for

the infection of flowers. Greenhouse ventilation and heating should be regulated to prevent high humidity, which favours the disease. For fungicidal treatments see Table 10.1, p. 379.

Leaf spot or alternaria blight (*Alternaria dianthi*)
This disease is common on plants overwintered with little heat or on cold-stored cuttings. Small, purple spots on leaves and stems enlarge into brown–black sporulating patches with purple edges. Spray at HV with chlorothalonil or iprodione*.

Phialophora wilt (*Phialophora cinerescens*)
This is less common than fusarium wilt. It is a notifiable disease when found on *Dianthus* plants intended for planting. A rapid wilting can occur, often starting on one side of the plant; the foliage turns grey-green and the stem xylem turns brown. For control measures, see under Fusarium wilt (p. 407); soil drenches of carbendazim (off-label) (SOLA 0009/99) are effective against this wilt disease.

Powdery mildew (*Oidium* spp.)
This fungus may cover the leaves and occasionally the calyxes. It is controlled by imazalil or pyrifenox.

Ring spot (*Mycosphaerella dianthi*)
This disease is worst on spray cultivars in winter under humid conditions, when it is difficult to control. Some cultivars are much more susceptible than others. Lesions appear on leaves, stems and calyxes. When they appear on calyxes the affected blooms are unmarketable. The lesions are typically dark-brown with a purple edge and up to 5 mm in diameter. Spores are produced in concentric rings on the spots. The disease is often brought in on cuttings. Avoid conditions of high humidity. HV sprays of difenoconazole (off-label) (SOLAs 1728/97, 1729/97) provide good control. Fortnightly applications are necessary, particularly during the winter months.

Rust (*Uromyces dianthi*)
Rust causes small blisters on the lower leaves and stem, which rupture releasing brown, powdery spores. The disease develops when the foliage remains wet or when dew forms overnight, especially in the autumn and winter; the humidity within the greenhouse should be reduced as much as possible. It is difficult to wet the waxy leaf surface, but a programme based on difenoconazole (off-label) (SOLAs 1728/97, 1729/97), oxycarboxin, chlorothalonil and mancozeb* should give control. Harvest open blooms before applying HV sprays and check for phytotoxicity by spraying a few plants first. Drenches of the systemic fungicide oxycarboxin, applied to plants grown in peat bags, are effective.

Septoria leaf spot (*Septoria dianthi*)
This uncommon disease is usually introduced on cuttings. The symptoms are similar to leaf spot caused by *Alternaria dianthi* (p. 408) but tend to be found on lower leaves. The purple margin is not so obvious and the centre of the spots has minute, black specks rather than the powdery-black appearance of *Alternaria*. Fungicide application is not usually necessary. Control can be achieved by avoiding high humidity or wetting of foliage. Where this is not effective, the fungicides carbendazim (off-label) (SOLA 0009/99) and chlorothalonil should give control.

Stem rot and die-back (*Fusarium culmorum*)
This fungus can occur as a soil saprophyte, both within and outside the greenhouse. It rots the base of the stem and sometimes the branch stems. Woody tissues are not extensively discoloured. The pathogen often invades the stumps left when shoots are removed or when flowers are cut at an internode, causing die-back in the stem; therefore, the stem should be cut at the node. A basal rot of cuttings may occur if they are taken from contaminated plants. Routine sprays of carbendazim (off-label) (SOLA 0009/99) or mancozeb* may be applied every 10–14 days. If established plants are attacked, remove infected shoots.

Euphorbia pulcherrima (poinsettia)

The bracts of poinsettias are very susceptible to injury by pesticides from the time they begin to colour. Both sprays and drenches may cause damage.

Pests

Sciarid flies
Larvae of sciarid flies (especially those of *Bradysia paupera*) can be a major pest, feeding on the roots and causing plants to wilt. Sciarid larvae can also be secondary pests, feeding on diseased root tissue. Preventive control measures are advisable. Biological control with the predatory mite *Hypoaspis* sp. or a drench of entomopathogenic nematodes (*Steinernema feltiae*) are both effective. A drench of teflubenzuron will give curative control, if necessary, and is compatible with IPM. If using a pesticide programme, a drench or compost incorporation with imidacloprid will give persistent control.

Western flower thrips (*Frankliniella occidentalis*)
This pest does not usually breed on poinsettias, but can cause leaf crinkling during feeding if no alternative hosts are available. Poinsettias are often grown following spring bedding plants, so good control of thrips in the preceding crop and cleanup in between crops can reduce the likelihood of infestations

developing. It is not usually necessary to apply specific control measures for thrips on poinsettias.

Whiteflies

Poinsettias are very susceptible to whiteflies. Glasshouse whitefly (*Trialeurodes vaporariorum*) is the most common species but the non-indigenous species tobacco whitefly (*Bemisia tabaci*) can be imported on cuttings – see under Notifiable pests and diseases, tobacco whitefly Chapter 9, p. 329. Biological control of glasshouse whitefly with the parasitoid *Encarsia formosa* is effective. If necessary, fumigation with nicotine will give some control of adult whiteflies, and HV sprays of buprofezin or teflubenzuron will give some control of the nymphs (scales). However, it is difficult to reach the target pest underneath the leaves with sprays; there is also increasing resistance to buprofezin in both whitefly species. If using a pesticide programme, a drench or compost incorporation with imidacloprid will give persistent control.

Diseases

Grey mould (*Botryotinia fuckeliana* – anamorph: *Botrytis cinerea*)

This fungus attacks leaves and bracts under moist conditions, e.g. during propagation or where overhead irrigation is used. If control is not achieved by removing the diseased parts and increasing ventilation, apply a suitable fungicide selected from those listed in Table 10.1, p. 379.

Powdery mildew (*Erysiphe* sp.)

This disease has recently been seen in a few crops, especially of the cv. Sonora. White, mealy patches occur on the bracts, usually where they overlap on nearly mature plants. HV sprays of imazalil have been found to provide good control. Check crop safety before spraying widely.

Pythium root rot (*Pythium* spp.)

This disease can cause serious losses. It is sometimes found in association with damage caused to the stem base by sciarid larvae (see p. 409). Use clean compost and inspect roots regularly. Ensure compost drainage, pH and conductivity are satisfactory. Consider the use of an appropriate fungicide, selected from Table 10.5, p. 390, but check for phytotoxicity by treating just a few plants in the first instance.

Root rots (*Fusarium* spp., *Thanatephorus cucumeris* – anamorph: *Rhizoctonia solani* – and *Thielaviopsis basicola*)

Attention to hygiene will minimize the risk of these diseases occurring. If they do occur, use an appropriate compost drench selected from Table 10.5, p. 390. Tolclofos-methyl, as a compost drench, will control *Rhizoctonia* and a carbendazim drench (off-label) (SOLA 0009/99) will control *Thielaviopsis* (black root rot).

Freesia

Pests

Aphids
Several species can occur, including mottled arum aphid (*Aulacorthum circumflexum*), peach/potato aphid (*Myzus persicae*) and potato aphid (*Macrosiphum euphorbiae*). Direct aphid feeding causes yellow leaf speckling and flower distortion, and *M. euphorbiae* can transmit bean yellow mosaic virus and freesia mosaic virus. Aphids should be controlled before flowering with a HV spray of nicotine, pirimicarb or pymetrozine. Resistant strains of *M. persicae* may not respond to pirimicarb.

Bulb mites (*Rhizoglyphus callae* and *R. robini*)
These tiny, white mites can gain entry to mechanically damaged or diseased tissue and feed on roots, corms and young developing leaves. Damage symptoms are brown streaks in corms and roots, internal tissue breakdown and distorted, ragged or saw-toothed leaves. Store hygiene and removal of damaged or diseased corms in the store will reduce the risk of spread. If necessary, treatment of corms in store with pirimiphos-methyl* will give control (use as on stored grain). For control on the growing crop, a HV spray of pirimiphos-methyl may be used as recommended for control of other pests on protected ornamentals.

Glasshouse whitefly (*Trialeurodes vaporariorum*)
This pest, which is resistant to many pesticides, occurs on freesias only occasionally. HV sprays of the insect growth regulators buprofezin or teflubenzuron can give useful control, although resistance to buprofezin is increasing. Nicotine, used as a spray or fumigant, can give some control of adult whiteflies if temperatures exceed 16°C.

Thrips
Gladiolus thrips (*Thrips simplex*) and western flower thrips (*Frankliniella occidentalis*) can occasionally damage freesias, causing silvery streaks on the leaves and white flecking on the flowers. For control, see under Fuchsia, p. 413.

Diseases

Fusarium yellows (*Fusarium moniliforme* and *F. oxysporum*)
Both pathogens invade freesia roots from the soil, and cause pinkish, superficial

lesions and a reddish-brown rot in the vascular tissues of the corm. Foliage from infected corms becomes yellow and may die. *F. moniliforme* is also seed-borne, and may contaminate healthy seeds during 'chitting' and kill seedlings.

Sterilize seed-sowing and greenhouse soils and use a thiram seed treatment. With crops raised from seed, drench the soil with carbendazim (off-label) (SOLA 0009/99) every 4–6 weeks from the four-leaf stage. Infection of corms may be controlled by dipping in carbendazim (off-label) (SOLA 0009/99) for 15–30 minutes. Alternatively, corms may be soaked in formaldehyde for 30 minutes.

Grey mould (*Botryotinia fuckeliana* – anamorph: *Botrytis cinerea*)
Under humid conditions, this fungus attacks the stems, leaves and flowers of densely grown freesias. Flower spotting may develop after boxing blooms for market and is a common cause of downgrading. Apply a programme of HV sprays (see Table 10.1, p. 379) during autumn and winter to prevent the build-up of the disease.

Fuchsia

Pests

Aphids
Both glasshouse & potato aphid (*Aulacorthum solani*) and peach/potato aphid (*Myzus persicae*) can damage fuchsia plants. If using IPM, these aphids can be controlled by the parasitoid wasps *Aphidius ervi* and *A. colemani*, respectively. If using a pesticide programme, resistant strains of *M. persicae* may not respond to pirimicarb and some other aphicides. A HV spray of pymetrozine, a drench or compost incorporation using imidacloprid or (if temperatures exceed 16°C) a spray or fumigation with nicotine should control resistant strains.

Glasshouse whitefly (*Trialeurodes vaporariorum*)
This can be a serious pest on fuchsia. If using IPM, the parasitoid *Encarsia formosa* gives effective control as long as temperatures are adequate. The pest is resistant to many pesticides. Imidacloprid, used as a drench or incorporated in the compost, should give effective control. Sprays of the insect growth regulators buprofezin and teflubenzuron can also give useful control, although whitefly resistance to buprofezin is increasing. Nicotine, used as a spray or fumigant, can give some control of adult whiteflies if temperatures exceed 16°C.

Large blue flea beetle (*Altica lythri*)
This flea beetle is uncommon but can do severe damage to isolated plants. The pest can invade glasshouses from nearby willow-herbs (*Epilobium* spp.), including rose-bay (*Chamaenerion angustifolium*). The adults are shiny, blue beetles *c.* 5 mm long, and they lay batches of oval, yellow eggs on the leaves. The

yellowish-black larvae feed on the leaves, causing many holes and ragged leaf edges. A spot treatment with a pyrethroid such as cypermethrin or deltamethrin will give effective control. If using IPM, a spray of pyrethrins + resmethrin will be less persistent and, therefore, less harmful to biological control agents.

Thrips
Western flower thrips (*Frankliniella occidentalis*) is a major pest of fuchsia, causing leaf and flower distortion and flecking or silvering of leaves and petals. Onion thrips (*Thrips tabaci*) may also occur. Biological control with the predatory mite *Amblyseius cucumeris* is usually effective, although problems can occur during the summer when thrips multiplication in the flowers is rapid. If using a pesticide programme, dichlorvos (off-label) (SOLA 0625/99) should be effective. Dichlorvos can cause plant damage so should always be tested on a few plants first, if there is no experience of using this product on the crop. Application of dichlorvos should be avoided if the plants are in bud or in flower, as there is a risk of flower abortion or marking or petal drop. A HV spray of abamectin should also give some control.

Two-spotted spider mite (*Tetranychus urticae*)
This is an occasional pest. See under *Dendranthema* (chrysanthemum), p. 402.

Vine weevil (*Otiorhynchus sulcatus*)
See under *Cyclamen*, p. 399.

Diseases

Grey mould (*Botryotinia fuckeliana* – anamorph: *Botrytis cinerea*)
Under very humid conditions, this fungus may attack cuttings, shoots, leaves and buds. It is particularly common at the stem base of cuttings. For control, select a suitable treatment from Table 10.1, p. 379. Avoid high humidity.

Root rots
Fuchsias are susceptible to several root pathogens. *Phytophthora* spp., *Pythium* spp., *Thanatephorus cucumeris* (anamorph: *Rhizoctonia solani*) and *Thielaviopsis basicola* are all commonly found. *Rhizoctonia* can also cause a stem-base decay during rooting of cuttings.

Rust (*Pucciniastrum epilobii*)
Cuttings bought from plant-raisers can introduce the disease on affected leaves. The orange-coloured pustules occur first on the lower leaf surface and can cause defoliation. The disease is favoured by high RH, stagnant air and overhead watering. A routine fungicide programme may be necessary where young plants are grown close together. Use oxycarboxin as HV sprays. Check for phytotoxicity

on a few plants of each cultivar before treating the remainder. Chlorothalonil applied for botrytis control will help prevent rust.

Geranium

See *Pelargonium*

Gerbera

Pests

Aphids
Peach/potato aphid (*Myzus persicae*) and potato aphid (*Macrosiphum euphorbiae*) can occur, and can be controlled biologically with the parasitoid wasps *Aphidius colemani* and *A. ervi*, respectively. For chemical control, see under *Fuchsia*, p. 412.

Caterpillars
See under *Dendranthema* (chrysanthemum), p. 401.

Chrysanthemum leaf miner (*Chromatomyia syngenesiae*)
See under *Dendranthema* (chrysanthemum), p. 401.

Glasshouse whitefly (*Trialeurodes vaporariorum*)
This serious pest is controlled effectively within IPM by the parasitoid wasp *Encarsia formosa*. For chemical control, see under *Fuchsia*, p. 412.

Western flower thrips (*Frankliniella occidentalis*)
Biological control with the predatory mite *Amblyseius cucumeris* is effective and the best option for control.

Diseases

Foot and root rot (*Phytophthora cryptogea*)
Gerberas are very susceptible to this disease and serious losses can occur at any stage of growth, particularly within the first 6 months after planting. The main symptom of the disease is a slow but progressive wilt, which often results in death of the plant. Leaves become red-violet, and then necrotic areas develop. The crown below soil level and the roots rot, sometimes leaving the outer cortex loosely attached to the stele. The disease is soil-borne, but may be introduced on the plants or in contaminated water. Plant healthy plants into sterilized soil or

new compost. Pre-planting incorporation of etridiazole may be useful or compost may be drenched with furalaxyl (see Table 10.5, p. 390). Any plants showing symptoms should be removed.

Powdery mildew (*Erysiphe cichoracearum*)
Typical powdery mildew symptoms are produced on leaves, as well as on flowers and flower stems. The disease can be damaging if left unchecked, leading to the death of leaves and loss of marketable blooms. Bupirimate*, sprayed at the first sign of the disease and thereafter at intervals of 3 weeks, will give satisfactory control.

Hydrangea

Hydrangeas are very susceptible to damage by sprays applied in sunshine or if the plants are dry at their roots.

Pests

Two-spotted spider mite (*Tetranychus urticae*)
Biological control with the predatory mite *Phytoseiulus persimilis* is effective, supplemented by the predatory midge *Feltiella acarisuga* if necessary. A HV spray of fenbutatin oxide, if required, is safe to biological control agents.

Diseases

Grey mould (*Botryotinia fuckeliana* – anamorph: *Botrytis cinerea*)
This disease may occur on dense flower clusters when plants are grown under excessively humid conditions. For control, select suitable fungicides from Table 10.1, p. 379.

Powdery mildew (*Oidium hortensiae*)
Mildew develops on the leaves as a white coating. Select suitable fungicides from Table 10.1, p, 379.

Lisianthus

Pests

Western flower thrips (*Frankliniella occidentalis*)
Biological control with the predatory mite *Amblyseius cucumeris* is effective and the best option.

Diseases

Downy mildew (*Peronospora chlorae*)
This is a major disease of the crop, causing twisting of the flower stem and pale, angular lesions on leaves. Sporulation occurs on the lower leaf surface, especially following periods of prolonged high humidity. Large numbers of oospores are often found within affected tissues. The disease may be introduced on young plants. Control by heat and ventilation to reduce humidity, and by HV sprays of fosetyl-aluminium*, metalaxyl + thiram* or propamocarb hydrochloride*.

Fusarium wilt (*Fusarium oxysporum*)
This seed-borne disease results in poor growth, wilting and plant death. A characteristic orange- or pinkish-coloured sporulation develops at the stem base. Control by using clean or treated seed.

Grey mould (*Botryotinia fuckeliana* – anamorph: *Botrytis cinerea*)
This disease is common and occasionally very damaging. Typically, it causes a stem base rot on which the characteristic grey mould develops. Crop heating, especially where applied from pipes laid on the soil, is an effective preventive treatment. For chemical control, see Table 10.1, p. 379.

Pythium root rot (*Pythium* sp.)
Patches of stunted or wilting plants often develop where the crop is grown frequently. It is sometimes damaging even after soil sterilization. Root tips are shrivelled and brown. Control by improving areas of poor drainage, by soil sterilization and by HV sprays to the stem bases of furalaxyl* or propamocarb hydrochloride.

Matthiola (stock)

Diseases

Downy mildew (*Peronospora parasitica*)
The leaves develop yellow blotches with a greyish, furry coating on the underside; growth of the plant may be checked. Reduce the humidity and keep the leaves as dry as possible during watering. Apply HV sprays of chlorothalonil or metalaxyl + thiram*, or compost drenches of propamocarb hydrochloride.

Root rots (*Pythium* sp. and *Thanatephorus cucumeris* – anamorph: *Rhizoctonia solani*)
These fungi (causing root and stem-base rotting, respectively) can be a problem in unsterilized soil. For control measures, see under *Dendranthema* (chrysanthemum), damping-off p. 403.

Sclerotinia rot (*Sclerotinia sclerotiorum*)
A fluffy, white mycelium forms on the stem, causing the plant to wilt. Large, black sclerotia develop in the mould. For control measures, see under *Dendranthema* (chrysanthemum), p. 404.

Pelargonium

Pests

Aphids
Glasshouse & potato aphid (*Aulacorthum solani*) is the most common species to infest pelargonium. Damage symptoms include yellow blotches to the leaves and leaf distortion. Biological control with the parasitoid wasp *Aphidius ervi* is effective within IPM. If required, a HV spray of pirimicarb will be effective and compatible with IPM.

Glasshouse leafhopper (*Hauptidia maroccana*)
This pest is becoming more of a problem now that IPM is commonly used. For description and damage, see under Tomato, Chapter 9, p. 356. Weed control will reduce sources of infestation. Biological control of the pest is possible with the egg parasitoid *Anagrus atomus* (as on tomato) but research has not been done on ornamentals. A HV spray of buprofezin, as recommended against glasshouse whitefly, will give control of leafhopper nymphs and is compatible with IPM. If using a pesticide programme, a HV spray of cypermethrin or deltamethrin will give control, but neither is compatible with IPM.

Glasshouse whitefly (*Trialeurodes vaporariorum*)
For description, see under Tomato, Chapter 9, p. 356. Regal pelargoniums are particularly susceptible. For biological and chemical control, see under *Fuchsia*, p. 412.

Thrips
Both onion thrips (*Thrips tabaci*) and western flower thrips (WFT) (*Frankliniella occidentalis*) can damage pelargoniums, but WFT is the most common. Ivy-leaf cultivars are particularly susceptible. Damage symptoms include petal flecking, distortion and abortion, and corky patches similar to oedema on the underside of the leaves. Biological control with the predatory mite *Amblyseius cucumeris* is effective. If using a pesticide programme, abamectin, dichlorvos or malathion will give some control. For precautions when using dichlorvos, see under *Fuchsia*, p. 413.

Diseases

Bacterial wilt and leaf spot (*Xanthomonas campestris* pv. *pelargonii*)
This is a serious disease which particularly affects zonal (common) pelargoniums. It produces a black stem rot similar to that caused by *Pythium*, though drier in appearance. Vascular discoloration can occur at stem nodes, and spots or chlorotic sectors may develop on leaves. It is spread by water splash but also on the knives and hands of staff. Discard affected plants. Take cuttings only from disease-free stock plants. There are no chemical control measures.

Grey mould (*Botryotinia fuckeliana* – anamorph: *Botrytis cinerea*)
Leaves, flowers and stems are susceptible to the fungus under humid conditions. 'Soft' plants are particularly vulnerable, especially if debris, such as dead flowers, is allowed to remain on the plant. The lower leaves of recently potted plants are a common infection site. Flower spotting is common on regal pelargoniums. Avoid standing plants too densely. For fungicide control, see Table 10.1, p. 379.

Leaf spot (*Alternaria tenuis*)
Small, water-soaked spots with a central necrotic fleck form on the undersides of the leaves. The lesions increase in diameter to about 3 mm and become necrotic. Some control may be expected from applications of chlorothalonil or iprodione.

Oedema
This unsightly physiological problem is common in the winter months on ivy-leaf cultivars. Corky outgrowths develop on the undersurface of leaves. It is caused by high humidity air around leaves preventing transpiration when the plant is actively imbibing water. To control, increase air-movement around plants.

Root and stem rots
Pelargoniums are susceptible to several soil-borne diseases. *Pythium splendens* causes a black, soft rot at the stem base of cuttings or plants. Fungi such as *Phytophthora* spp., *Thanatephorus cucumeris* (anamorph: *Rhizoctonia solani*) and *Thielaviopsis basicola* are also responsible for the death of plants from root or stem decay. Fresh compost should be used and propagation made from healthy stock. Some control of *Phytophthora* and *Pythium* may be expected from the use of a drench of etridiazole or furalaxyl; propamocarb hydrochloride will help control *Pythium*; tolclofos-methyl will control *Rhizoctonia*; and carbendazim (off-label) (SOLA 0009/99) will give control of *Fusarium* and *Thielaviopsis* rots. Treatments should be applied to a few plants first to check for phytotoxicity.

Rust (*Puccinia pelargonii-zonalis*)
The fungus attacks zonal (common) pelargonium cuttings and plants and can be introduced in to a nursery on fresh stock. Reddish-brown pustules develop in

concentric rings on the undersides of leaves and corresponding yellow areas appear on the upper surface. If the attack is severe, affected leaves turn yellow and die. Remove affected leaves. Apply HV sprays of chlorothalonil or oxycarboxin every 10–14 days. Avoid high humidity and leaf wetness.

Verticillium wilt (*Verticillium albo-atrum* and *V. dahliae*)
This disease particularly affects regal pelargoniums, where it causes stunting, chlorosis of the lower leaves and chlorotic or necrotic leaf sectors. Zonal (common) pelargonium cultivars are affected occasionally, but ivy-leaf cultivars appear to be resistant. Affected plants should be discarded. The disease is most likely to be introduced on infected cuttings.

Primula acaulis, P. malacoides, P. obconica and *P. veris*

Pests

Aphids
Several species, including mottled arum aphid (*Aulacorthum circumflexum*), occur on primulas but the most common species is peach/potato aphid (*Myzus persicae*). Primulas are often grown at cool temperatures unfavourable for biological control (e.g. of *M. persicae* with the parasitoid wasp *Aphidius colemani*). A HV spray of cypermethrin or deltamethrin or pirimicarb will give control if the spray hits the aphids underneath the leaves, although resistant strains of *M. persicae* will not respond to treatment. The new aphid anti-feedant compound pymetrozine has systemic activity and will stop the feeding of all aphid strains of *M. persicae*.

Chrysanthemum leaf miner (*Chromatomyia syngenesiae*)
For description and control, see under *Dendranthema* (chrysanthemum), p. 401. See also under Notifiable pests and diseases, leaf miners, Chapter 9, p. 329.

Glasshouse leafhopper (*Hauptidia maroccana*)
This is becoming a serious pest of primulas. The cool temperatures used for production may be too low for effective use of the egg parasitoid *Anagrus atomus*. For chemical control, see under *Pelargonium*, p. 417.

Two-spotted spider mite (*Tetranychus urticae*)
This is an occasional pest of primulas. For a description, see under Cucumber, Chapter 9, p. 338. Temperatures during production are usually too cool for effective biological control with the predatory mite *Phytoseiulus persimilis*. Owing to the growth habit of the plants, a HV spray with the translaminar acaricide abamectin can be the most effective treatment, as the pest may not be hit underneath the leaves by contact acaricides such as bifenthrin or fenbutatin oxide.

Vine weevil (*Otiorhynchus sulcatus*)
Primulas are particularly susceptible hosts and are often attacked by vine weevil. For description and control, see under *Cyclamen*, p. 399.

Western flower thrips (*Frankliniella occidentalis*)
This pest causes leaf and flower flecking, and can transmit tomato spotted wilt virus which can cause leaf yellowing in the heart of the plants. Virus-infected plants should be rogued to remove the source of infection. If temperatures are adequate, thrips can be controlled biologically with the predatory mite *Amblyseius cucumeris* (which is active at cooler temperatures than some other biological control agents). When using biological control, pesticides used against other pests would then have to be selected for their safety to the predator. For chemical control, abamectin, dichlorvos or malathion may be used. See under *Fuchsia*, thrips, p. 413, for precautions when using dichlorvos.

Diseases

Black root rot (*Thielaviopsis basicola*)
See under *Euphorbia pulcherrima* (poinsettia), root rots, p. 410.

Brown core (*Phytophthora primulae*)
This disease causes roots to develop a 'rat's-tail' appearance. The vascular tissue within roots is characteristically discoloured, becoming purplish-brown. Severely affected plants wilt and die. Remove affected plants. Avoid over-wet compost and avoid standing plants on poorly drained ground. When an outbreak has occurred, surrounding plants can be protected by drenching the roots with furalaxyl.

Grey mould (*Botryotinia fuckeliana* – anamorph: *Botrytis cinerea*)
This disease can cause widespread losses, particularly in *Primula acaulis*. Infection usually originates on chlorotic or necrotic leaves in contact with the compost surface, progressing to cause a stem-base rot. Fallen flower parts also act as foci for infection. Attention to watering and nutrition to reduce occurrence of leaf necrosis will reduce botrytis risk. Ventilate the crop to reduce humidity. For fungicide control measures, see Table 10.1, p. 379.

Leaf spot (*Ramularia agrestis* and *R. primulae*)
Mature spots on infected leaves are up to 3 mm in diameter and have a dark brown centre and a pale yellow margin; a fine, white bloom may develop in the centre of spots. Small, black leaf spots can also occur. Increase ventilation to reduce humidity and avoid prolonged leaf wetness. HV sprays of carbendazim (off-label) (SOLA 0009/99) or zineb* may assist control.

Rhododendron (azalea)

Diseases

Grey mould (*Botryotinia fuckeliana* – anamorph: *Botrytis cinerea*)
Flowers rot in humid conditions, so ample ventilation and prompt removal of all diseased flowers will usually prevent the disease from spreading. However, it is advisable, especially in unheated houses, to apply a programme of suitable fungicides (see Table 10.1, p. 379).

Leaf spot (*Septoria azaleae*)
The fungus causes angular, brown leaf lesions, sometimes with yellow margins, and plants may later become defoliated after infected leaves wither. Because pycnidia in the lesions exude spores when wet, it is important to reduce the humidity in the greenhouse and to dispose of fallen infected leaves. Apply HV sprays of carbendazim (off-label) (SOLA 0009/99) or chlorothalonil.

Rosa (rose)

Systemic compounds may not be effective when applied as soil drenches to old, woody plants; best results will be obtained during active vegetative growth.

Pests

Aphids
The most common species are potato aphid (*Macrosiphum euphorbiae*) and rose aphid (*M. rosae*), both of which are found under the leaves and on buds and stems. If using IPM, the parasitoid wasp *Aphidius ervi* will give control. If necessary, a HV spray of pirimicarb can be used, which is compatible with IPM. If using a pesticide programme, alternative aphicides include cypermethrin, deltamethrin and malathion, but these are incompatible with IPM.

Caterpillars
Several species damage the leaves and buds, some of which web the leaves together. For control of caterpillars, see under *Dendranthema* (chrysanthemum), p. 401.

Glasshouse leafhopper (*Hauptidia maroccana*)
See under *Pelargonium*, p. 417.

Thrips
The most common species is western flower thrips (*Frankliniella occidentalis*). If using IPM, the predatory mite *Amblyseius cucumeris* can be effective. If using a

pesticide programme, abamectin, dichlorvos or malathion should give some control. For precautions when using dichlorvos, see under *Fuchsia*, p. 413.

Two-spotted spider mite (*Tetranychus urticae*)
This can be a major pest under glass. For description, see under Cucumber, Chapter 9, p. 338. Biological control with the predatory mite *Phytoseiulus persimilis* is not always reliable on roses. Research overseas has indicated that the predatory mite *Amblyseius californicus* may be more reliable, and this may be more widely used in the future. A HV spray of fenbutatin oxide is safe in IPM. Other effective acaricides include abamectin and tebufenpyrad, but these have some harmful effects on biological control agents.

Diseases

Black spot (*Marssonina rosae* – anamorph: *Diplocarpon rosae*)
This disease occurs occasionally on roses in greenhouses, when black or purple spots develop on upper surfaces of leaves and leaflets drop prematurely. Spray HV at the first sign of infection with carbendazim (off-label) (SOLA 0009/99), chlorothalonil or dichlofluanid (off-label) (SOLA 0167/95) and repeat at intervals of 10–14 days. Sprays of imazalil may also give some control.

Coniothyrium die-back (*Leptosphaeria coniothyrium* – anamorph: *Coniothyrium fuckelii*)
This disease results in the death of old wood, often by infection through a pruning scar. Affected stems become light-brown or silvery-grey in colour. Black pycnidia (0.5 mm in diameter) of the fungus are often clearly seen on affected tissue. The first obvious symptom of the disease may be the wilting of young lateral shoots on the affected stem. HV sprays of carbendazim (off-label) (SOLA 0009/99), maneb, mancozeb or prochloraz + carbendazim (off-label) (SOLA 2217/99) immediately after pruning should give some reduction in the occurrence of the disease.

Crown gall (*Agrobacterium tumefaciens*)
Galls, which are whitish at first and later brown-black, may be found on roots or stems but more commonly at the base of the stem just below soil level. Large galls completely surrounding the stem may be found, although these do not usually cause an obvious effect on the plant. Occasionally, plants may be stunted and die where the galls become secondarily infected with soft rotting organisms. Remove infected plants.

Downy mildew (*Peronospora sparsa*)
This fungus produces irregularly shaped, purplish spots on leaves and, occasionally, on flower stems. A sparse, whitish-grey, downy fungal growth appears on the undersides of leaves, which typically fall prematurely. Minimize leaf

wetness and reduce the humidity. Spray at 14-day intervals with chlorothalonil, fosetyl-aluminium or metalaxyl + thiram*.

Grey mould (*Botryotinia fuckeliana* – anamorph: *Botrytis cinerea*)
Under excessively humid conditions, flowers and (occasionally) stems may be attacked. Germinating spores may produce spotting of the petals, leading to rotting of the flowers. The fungus can also invade stem wounds, causing die-back of woody tissue. To control the disease, remove infected blooms and decrease the humidity. Apply a suitable treatment selected from Table 10.1, p. 379.

Powdery mildew (*Sphaerotheca pannosa*)
An important disease which disfigures young leaves, stems, buds and petals by distorting, dwarfing and covering them with a typical powdery mould. On some cultivars, dark spots result which may be confused with downy mildew or black spot. To control the disease, spray with a suitable chemical listed in Table 10.1, p. 379, but take precautions to avoid phytotoxicity on sensitive cultivars.

Rust (*Phragmidium mucronatum*)
In summer, rust causes orange-coloured pustules on the leaves and, in autumn, dark-brown ones (often on the same spots). The fungus can perennate as spores on fallen leaves or as mycelium in the wood, from which it emerges as easily overlooked infections of young shoots and buds. Burn all infected debris and apply HV sprays of chlorothalonil, mancozeb or oxycarboxin every 10–14 days. Check cultivar sensitivity before using any pesticide on a large area.

Saintpaulia (African violet)

Do not apply sprays or dips in bright sunshine or under humid, slow-drying conditions, or if the compost is dry. Use tepid water for sprays.

Pests

Western flower thrips (*Frankliniella occidentalis*)
Biological control on *Saintpaulia* is still in development. Dichlorvos is the most effective pesticide. See under *Fuchsia*, thrips, p. 413, for precautions when using dichlorvos.

Diseases

Crown rot (*Phytophthora nicotianae* vars *nicotianae* and *parasitica*)
This disease can cause extensive losses during propagation, and at any stage thereafter. Affected plants have a rather dull, grey appearance, in contrast to

healthy plants. They eventually wilt and collapse. The disease progresses along the leaf in a characteristic wave-shaped front, causing leaf decay. The crown and roots become soft and rotten. Crown rot develops most quickly in wet composts and where the air temperature is above 20°C. Crop hygiene is essential to control this disease and any affected plant must be removed immediately, so that the fungus does not spread in the irrigation system. Sand benches or capillary matting must be disinfected regularly. Empty sand benches can be drenched with fosetyl-aluminium and this fungicide can also be applied safely as a drench at all stages of growth.

Grey mould (*Botryotinia fuckeliana* – anamorph: *Botrytis cinerea*)
The conditions conducive to attack are similar to those described for *Rhododendron* (azalea), p. 421. For chemical control measures, see Table 10.1, p. 379.

Powdery mildew (*Oidium* spp.)
This disease is generally first seen on the calyxes of flowers, but in severe cases the fungus can cover the leaves and flowers in a white, powdery deposit. The disease must be treated as soon as it is first seen, especially in the case of blue-flowered cultivars, where the deposits are most noticeable. Control using suitable fungicides selected from Table 10.1, p. 379.

Zantedeschia aethiopica, Z. elliottiana, Z. rehmannii (arum or calla lily)

Pests

Aphids
Several species, including melon & cotton aphid (*Aphis gossypii*), mottled arum aphid (*Aulacorthum circumflexum*), peach/potato aphid (*Myzus persicae*) and potato aphid (*Macrosiphum euphorbiae*), cause stunting and distortion of the young growth and severe spotting of the flower spathes. Aphids should be controlled before flowering. For chemical control, see under *Dendranthema* (chrysanthemum), p. 401. *A. gossypii* and *M. persicae* can be resistant to certain pesticides.

Thrips
Several species can occur, including glasshouse thrips (*Heliothrips haemorrhoidalis*), onion thrips (*Thrips tabaci*) and western flower thrips (WFT) (*Frankliniella occidentalis*). Damage symptoms include white flecks on the leaves and brown spots and streaks on the flowers. For chemical control of WFT, see under *Fuchsia*, p. 413. The other thrips species are more easily controlled by pesticides, including HV sprays of nicotine or a pyrethroid (such as cypermethrin or deltamethrin) or fumigation with nicotine.

Diseases

Corm rot (*Erwinia carotovora* ssp. *carotovora*)
This is a serious bacterial disease. Plants are attacked at the collar and the 'corms' are rotted, causing yellowing and death of foliage. Badly diseased corms should be destroyed. Others may be saved by washing away all soil when they are dormant, cutting out all diseased tissue and steeping for 2 h in formaldehyde. Plant into sterilized soil without delay. When cleared, disinfect the greenhouse with formaldehyde.

Root rot (*Phytophthora richardiae*)
Infected leaves first become yellowish and streaky, then gradually turn brown and die. Flowers become brown at their tips and may be deformed. Sterilize soil to prevent the spread of this soil-borne disease. Etridiazole or furalaxyl, applied either as a drench to roots or by incorporation into the compost, or fosetyl-aluminium applied as a drench, should give control of this disease.

List of pests cited in the text*

Altica lythri (Coleoptera: Chrysomelidae)	large blue flea beetle
Aphelenchoides ritzemabosi (Tylenchida: Aphelenchoididae)	chrysanthemum nematode
Aphis gossypii (Hemiptera: Aphididae)	melon & cotton aphid
Arion hortensis (Stylommatophora: Arionidae)	garden slug
Aulacorthum circumflexum (Hemiptera: Aphididae)	mottled arum aphid
Aulacorthum solani (Hemiptera: Aphididae)	glasshouse & potato aphid
Bemisia tabaci (Hemiptera: Aleyrodidae)	tobacco whitefly
Brachycaudus helichrysi (Hemiptera: Aphididae)	leaf-curling plum aphid
Bradysia paupera (Diptera: Sciaridae)	a sciarid fly
Cacoecimorpha pronubana (Lepidoptera: Tortricidae)	carnation tortrix moth
Chromatomyia syngenesiae (Diptera: Agromyzidae)	chrysanthemum leaf miner
Clepsis spectrana (Lepidoptera: Tortricidae)	straw-coloured tortrix moth
Deroceras reticulatum (Stylommatophora: Limacidae)	field slug
Ditylenchus dipsaci (Tylenchida: Tylenchidae)	stem nematode
Dysaphis tulipae (Hemiptera: Aphididae)	tulip bulb aphid
Frankliniella occidentalis (Hemiptera: Aphididae)	western flower thrips
Hauptidia maroccana (Hemiptera: Cicadellidae)	glasshouse leafhopper
Heliothrips haemorrhoidalis (Thysanoptera: Thripidae)	glasshouse thrips
Lacanobia oleracea (Lepidoptera: Noctuidae)	tomato moth
Lilioceris lilii (Coleoptera: Chrysomelidae)	lily beetle
Lygus rugulipennis (Hemiptera: Miridae)	tarnished plant bug
Macrosiphoniella sanborni (Hemiptera: Aphididae)	chrysanthemum aphid
Macrosiphum euphorbiae (Hemiptera: Aphididae)	potato aphid
Macrosiphum rosae (Hemiptera: Aphididae)	rose aphid
Myzus persicae (Hemiptera: Aphididae)	peach/potato aphid
Otiorhynchus sulcatus (Coleoptera: Curculionidae)	vine weevil
Phlogophora meticulosa (Lepidoptera: Noctuidae)	angle-shades moth
Phytonemus pallidus (Prostigmata: Tarsonemidae)	cyclamen mite

Polyphagotarsonemus latus (Prostigmata: Tarsonemidae)	broad mite
Rhizoglyphus callae (Astigmata: Acaridae)	a bulb mite
Rhizoglyphus robini (Astigmata: Acaridae)	a bulb mite
Scatella stagnalis (Diptera: Ephydridae)	a shore fly
Scatella tenuicornis (Diptera: Ephydridae)	a shore fly
Siteroptes graminum (Prostigmata: Pygmephoridae)	grass & cereal mite
Steneotarsonemus laticeps (Prostigmata: Tarsonemidae)	bulb scale mite
Tetranychus cinnabarinus (Prostigmata: Tetranychidae)	carmine spider mite
Tetranychus urticae (Prostigmata: Tetranychidae)	two-spotted spider mite
Thrips simplex (Thysanoptera: Thripidae)	gladiolus thrips
Thrips tabaci (Thysanoptera: Thripidae)	onion thrips
Trialeurodes vaporariorum (Hemiptera: Aleyrodidae)	glasshouse whitefly

* The classification in parentheses represents order and family.

List of pathogens/diseases (other than viruses) cited in the text*

Agrobacterium tumefaciens (Gracilicutes: Proteobacteria)†	crown gall
Alternaria alternata (Hyphomycetes)	leaf spot of *Lobelia*
Alternaria dianthi (Hyphomycetes)	alternaria blight or leaf spot of *Dianthus*
Alternaria tenuis (Hyphomycetes)	leaf spot of *Pelargonium*
Botryotinia fuckeliana (Ascomycota)	(common) grey mould
Botrytis cinerea (Hyphomycetes)	– anamorph of *Botrytinia fuckeliana*
Botrytis elliptica (Hyphomycetes)	leaf spot of *Lilium* or lily disease
Botrytis tulipae (Hyphomycetes)	fire of *Tulipa*
Bremia lactucae (Oomycetes)	downy mildew of *Cineraria*
Coleosporium tussilaginis (Teliomycetes)	rust of *Cineraria*
Colletotrichum acutatum (Coelomycetes)	leaf curl of *Anemone*
Coniothyrium fuckelii (Coelomycetes)	– anamorph of *Leptosphaera coniothyrium*
Cylindrocarpon destructans (Hyphomycetes)	brown root rot
Didymella chrysanthemi (Ascomycota)	ray blight of *Dendranthemum*
Diplocarpon rosae (Coelomycetes)	– anamorph of *Marssonina rosae*
Erwinia carotovora ssp. *carotovora* (Gracilicutes: Proteobacteria)†	corm rot of *Cyclamen, Zephyranthes*; soft rot of *Hyacinthus*
Erysiphe aquilegiae var. *ranunculi* (Ascomycota)	powdery mildew of *Anemone*
Erysiphe cichoracearum (Ascomycota)	powdery mildew of *Gerbera*
Fusarium culmorum (Hyphomycetes)	stem rot and basal rot of *Dianthus*
Fusarium moniliforme (Hyphomycetes)	fusarium yellows of *Freesia*
Fusarium oxysporum (Hyphomycetes)	fusarium yellows of *Freesia*, fusarium wilt of *Lisianthus*
Fusarium oxysporum f. sp. *cyclaminis* (Hyphomycetes)	fusarium wilt of *Cyclamen*
Fusarium oxysporum f. sp. *dianthi* (Hyphomycetes)	fusarium wilt of *Dianthus*
Fusarium oxysporum f. sp. *gladioli* (Hyphomycetes)	fusarium yellows of *Gladiolus*
Fusarium oxysporum f. sp. *lilii* (Hyphomycetes)	fusarium wilt of *Lilium*
Fusarium oxysporum f. sp. *narcissi* (Hyphomycetes)	fusarium basal rot of *Narcissus*
Fusarium oxysporum f. sp. *tulipae* (Hyphomycetes)	fusarium bulb rot of *Tulipa*

Fusarium poae (Hyphomycetes)	bud rot of *Dianthus*
Itersonilia perplexans (Basidiomycetes)	petal blight of *Dendranthemum*
Leptosphaeria coniothyrium (Ascomycota)	coniothyrium stem rot and canker of *Rosa*
Marssonina rosae (Ascomycota)	black spot of *Rosa*
Microsphaera begoniae (Ascomycota)	powdery mildew of *Begonia*
Mycosphaerella dianthi (Ascomycota)	ring spot or fairy ring of *Dianthus*
Mycosphaerella macrospora (Ascomycota)	leaf spot of *Iris*
Oidium begoniae (Hyphomycetes)	– anamorph of *Microsphaera begoniae*
Oidium chrysanthemi (Hyphomycetes)	powdery mildew of *Dendranthemum*
Oidium hortensiae (Hyphomycetes)	powdery mildew of *Hydrangea*
Oidium sp. (Hyphomycetes)	powdery mildew of *Dianthus, Saintpaulia*
Penicillium corymbiferum (Hyphomycetes)	bulb rot of *Iris*
Penicillium cyclopium (Hyphomycetes)	bulb rot of *Iris*
Penicillium hirsutum (Hyphomycetes)	bulb rot of *Iris*
Peronospora chlorae (Oomycetes)	downy mildew of *Lisianthus*
Peronospora ficariae (Oomycetes)	downy mildew of *Anemone*
Peronospora parasitica (Oomycetes)	downy mildew of *Matthiola*
Peronospora sparsa (Oomycetes)	downy mildew of *Rosa*
Phialophora cinerescens (Hyphomycetes)	phialophora wilt of *Dianthus*
Phoma chrysanthemicola (Coelomycetes)	phoma root rot of *Dendranthemum*
Phragmidium mucronatum (Teliomycetes)	rust of *Rosa*
Phytophthora cryptogea (Oomycetes)	shanking of *Tulipa*
Phytophthora erythroseptica (Oomycetes)	shanking of *Tulipa*
Phytophthora nicotianae var. *nicotianae* (Oomycetes)	stump rot of *Lilium* and crown rot of *Saintpaulia*
Phytophthora nicotianae var. *parasitica* (Oomycetes)	stump rot of *Lilium* and crown rot of *Saintpaulia*
Phytophthora primulae (Oomycetes)	brown core of *Primula*
Phytophthora richardiae (Oomycetes)	root rot of *Zephyranthes*
Phytophthora spp. (Oomycetes)	basal rot of *Dianthus*
Phytophthora spp. (Oomycetes)	root rot of *Lilium*
Puccinia chrysanthemi (Teliomycetes)	brown rust of *Dendranthemum*
Puccinia horiana (Teliomycetes)	white rust of *Dendranthemum*
Puccinia lagenophorae (Teliomycetes)	rust of *Cineraria*
Puccinia pelargonii-zonalis (Teliomycetes)	rust of *Pelargonium*
Pucciniastrum epilobii (Teliomycetes)	rust of *Fuchsia*
Pythium irregulare (Oomycetes)	root rot of *Iris*
Pythium splendens (Oomycetes)	black stem rot of *Pelargonium*
Pythium spp. (Oomycetes)	root rot of *Dendranthemum, Begonia, Lilium, Tulipa*
Pythium spp. (Oomycetes)	root and stem rot of *Pelargonium*
Ramularia agrestis (Hyphomycetes)	leaf spot of *Primula*
Ramularia primulae (Hyphomycetes)	leaf spot of *Primula*
Rhizoctonia solani (Hyphomycetes)	– anamorph of *Thanatephorus cucumeris*
Rhizoctonia tuliparum (Hyphomycetes	grey bulb rot of *Iris, Tulipa*
Rhodococcus fascians (Firmicutes)†	leafy gall
Sclerotinia sclerotiorum (Ascomycota)	sclerotinia rot of *Dendranthemum, Matthiola*
Septoria azaleae (Coelomycetes)	leaf spot of *Rhododendron* (azalea)

Septoria chrysanthemella (Coelomycetes)	blotch of *Dendranthemum*
Septoria dianthi (Coelomycetes)	septoria leaf spot of *Dianthus*
Sphaerotheca fusca (Ascomycota)	powdery mildew of *Cineraria*
Sphaerotheca pannosa (Ascomycota)	powdery mildew of *Rosa*
Thanatephorus cucumeris (Basidiomycetes)	root rot of *Begonia*; root and stem rot of *Euphorbia* (poinsettia), *Fuchsia*, *Pelargonium*
Thielaviopsis basicola (Hyphomycetes)	black root rot of *Euphorbia* (poinsettia), *Fuchsia*, *Pelargonium*, *Viola*
Tranzschelia pruni-spinosae (Teliomycetes)	plum rust of *Anemone*
Uromyces dianthi (Teliomycetes)	rust of *Dianthus*
Verticillium albo-atrum (Hyphomycetes)	verticillium wilt of *Dendranthemum*, *Pelargonium*
Verticillium dahliae (Hyphomycetes)	verticillium wilt of *Dendranthemum*, *Pelargonium*
Xanthomonas campestris pv. *begoniae* (Gracilicutes: Proteobacteria)†	bacterial blight of begonia
Xanthomonas campestris pv. *campestris* (Gracilicutes: Proteobacteria)†	bacterial leaf spot, leaf and stem rot, wilt of *Cheiranthus*
Xanthomonas campestris pv. *hyacinthi* (Gracilicutes: Proteobacteria)†	yellow disease of *Hyacinthus*
Xanthomonas campestris pv. *pelargonii* (Gracilicutes: Proteobacteria)†	root and stem rot of *Pelargonium*

* For fungi, the classification in parentheses refers to class, although this is not possible within the phylum Ascomycota where classes have yet to be satisfactorily defined (see *Mycological Research*, February 2000). Oomycetes are now classified in Chromista with the brown algae, rather than as true fungi. Some fungi have an asexual (anamorph) and a sexual (teleomorph) state, and the convention is to refer to them by their teleomorph name. However, where anamorph names are still in common use these are listed and cross-referenced to the teleomorph name. Strictly, fungi classified as Coelomycetes and Hyphomycetes should be known as 'hyphomycetous anamorphs' and 'coelomycetous anamorphs' of the relevant teleomorph taxon (e.g. hyphomycetous anamorphic Sclerotiniaceae, for *Botrytis fabae*), respectively. These problems highlight the continual changes in the classification of the fungi.

† Bacteria – the classification in parentheses refers to division and class, or to division only.

Chapter 11
Pests and Diseases of Outdoor Ornamentals, Including Hardy Nursery Stock

A.J. Halstead
Royal Horticultural Society, Wisley, Surrey

J.M. Scrace
ADAS Wolverhampton, West Midlands

Introduction

This chapter concerns the major pests and diseases of the more important plants grown outdoors for their ornamental or amenity value or grown in forest plantations. These include trees, shrubs and herbaceous perennials, referred to by commercial producers as hardy ornamental nursery stock (HONS). Pest and disease control in 'outdoor' ornamentals is complex, owing to the diversity of plant species and cultivars, the variety of cultural systems, and often a production period of several years from propagation to saleable plant. Nursery stock invariably has to be transplanted or moved a number of times during the course of production, and this increases the difficulty in assessing reliably the incidence of pests and diseases, and the effectiveness of pesticide treatment in reducing crop losses. Symptoms of many diseases and some of the pests of HONS are slow to develop following the initial attack, and remedial measures often have to be continued for an extended period before control is evident.

Table 11.1 shows UK production trends in nursery stock over the last decade, and in comparison with the early 1980s. During the 1980s and 1990s the most significant trends have been the great increases in container-grown plants, and in the number of these plants raised under protection (in glasshouses and polytunnels). Between 60% and 70% of holdings producing nursery stock now have at least some container production. It is also interesting to note that 80–90% of the total volume of containerized nursery stock is produced by just 10–20% of the nurseries involved in this type of growing.

The value of exported ornamental plants (including outdoor ornamentals) still trails a long way behind that of imported plants. In both 1995 and 1996 the total value of 'non-edible exports' was less than 10% of that of 'imports'. ADAS estimates that up to 10 million liners, plugs and bare-root plants are now imported just to grow on, at a value of around £3.8 million per annum (and this excludes plants such as forestry and herbaceous species, and rose rootstocks).

Table 11.1 Output of hardy ornamental nursery stock (HONS) in England and Wales*

	Total area (ha)					
	1982	1990	1995	1996	1997	1998
(a) Field-grown						
Roses	1102	925	891	885	837	781
Shrubs	1480	1951	2343	2397	3099	2889
Trees	1338	1506	1782	1675	1350	1258
Herbaceous	274	222	250	209	351	327
Fruit stock	937	778	660	638	743	693
Other HONS	–	1884	1866	1980	1983	1851
Total:	5131	7266	7792	7784	8363	7799
(b) Container grown						
Area (ha)†	353	930	1152	1370	1274	1273
Numbers (millions)	60	158	196	233	216	216
Glass (ha)‡	53 (1983)	176	242	318	375	375

* Data from the MAFF Basic Horticultural Statistics (1998) & Glasshouse Census; MAFF Statistical Review of Container Nursery Stock 1994 (prepared by ADAS).
– Data not available.
† Estimate based on an average of 170 000 containers/ha, as areas figures not available. Excludes nursery stock grown under protection.
‡ Estimated area of nursery stock grown under protection.

Chemical control

There are few crop protection chemicals with recommendations on the product label for use on hardy nursery stock and other outdoor ornamentals. This can be attributed to the high cost of evaluating products on the wide range of plant species and diversity of cultural procedures encountered in the industry. The fungicides listed in Tables 11.4 and 11.5 have label recommendations for use on at least one ornamental crop. In some cases this will refer to a wide-ranging term, such as 'hardy nursery stock' or 'ornamentals', but in others it will refer to a specific ornamental crop, such as rose, chrysanthemum, etc. The treatment of crops not listed on the product label is **entirely at the user's risk** and growers should assess the product on a limited number of plants, on a trial basis, before treating the entire crop.

Under the *Revised Long Term Arrangements for Extension of Use (2000)* pesticides approved for use on any growing crop may be used on commercial holdings and forest nurseries, (a) on ornamental crops where neither the seed nor any part of the plant is to be consumed by humans or animals, and (b) on forest nursery crops prior to final planting out. Again, this use is at the user's risk, and subject to certain conditions. Prominent amongst these is that the pesticide must be used only in the same situation as that on the product label or SOLA, i.e.

outdoor crop or protected crop – approval for use on chrysanthemum, cucumber, lettuce, mushroom and tomato includes 'protected crops' unless otherwise stated. This gives the grower a vast armoury of potential pesticides for use on outdoor ornamentals – one has to look only at the huge number of fungicides and insecticides with approval for use on cereals and other arable crops. However, the vast majority of these have never been evaluated for use on ornamental crops, and they cannot be recommended widely. Where possible, the products mentioned in the tables and text in this chapter have label recommendations for use on at least one ornamental crop, although occasionally mention will be made of a product without such a label recommendation where this is likely to enhance a control programme (or where there is little else available). Mention in the text of use of a pesticide under the provisions of the extension of use arrangements is marked with an asterisk (*).

In addition to the above legislation, there are also specific off-label approvals (SOLAs). These are uses for which approval has been sought by individuals or organizations other than the manufacturers of the pesticide. This usually occurs in cases where there is no product with on-label approval, or which can be used under the extension of use arrangements (outlined above) to control a pest or disease problem on a specific crop or in a specific situation. If a SOLA is granted, then a Notice of Approval is published by MAFF. Anyone intending to use a pesticide under a SOLA must first obtain a copy of the relevant Notice of Approval and comply strictly with the conditions stated within it.

Because of the large number of products available to growers of ornamentals under the extension of use arrangements there are comparatively few SOLAs for ornamentals compared with other horticultural crops, but a few are mentioned within the text and tables.

It should be noted that whilst all of the pesticides mentioned in this chapter can legally be used on outdoor ornamentals, it is illegal to use some of them under protection. Where use under protection is required, reference should be made to the pesticides listed for the control of the same type of pest or disease on an ornamental crop in the Protected Ornamental Crops chapter (Chapter 10). However, there are occasional references in this chapter to pesticides for use under protection where a particular disease is not covered in Chapter 10, e.g. scab of *Malus* and *Pyracantha*.

All pesticides referred to in this chapter are available as professional products. In some cases similar compounds will be available as amateur products (often with a reduced amount of active ingredient). Such products are intended for home and garden use **only**; thus, in the majority of cases, they cannot be used legally on a professional premises.

The number of pesticides available for pest and disease control on plants grown in ornamental amenity areas is also very limited. A few products have approval for use in such areas. If in any doubt as to whether a product is permitted in an amenity area, the user should contact the manufacturer or supplier of the product, or the Pesticides Safety Directorate (PSD) of MAFF.

Pesticide resistance

Wherever possible, repeated use of a single compound should be avoided, to prevent the possibility of the target organism developing resistance. Good anti-resistance strategies always begin with reducing the reliance on chemical control measures, by adopting sound husbandry and cultural control measures (see the following section). Where a chemical control programme is necessary, make full use of formulated mixtures of active ingredients, and apply a sequence of products from different chemical groups.

Pesticide resistance is less well documented in pests and diseases of ornamentals than in those of many other crops. In some cases resistance is suspected but not proven. Documented cases of resistance include that of *Pythium* to furalaxyl and related compounds, and *Botrytis* to carbendazim, iprodione and dichlofluanid (and related compounds); melon & cotton aphid (*Aphis gossypii*) often has resistance to pirimicarb, organophosphate and pyrethroid compounds, and two-spotted spider mite (*Tetranychus urticae*) to organophosphates such as malathion and dimethoate.

Cultural control

Manipulation of cultural factors can often prevent or reduce pest and disease attacks. For example, cultural procedures which contribute to disease problems include the high-density production of container-grown plants and the need for watering systems during the summer months. Overhead watering tends to increase those problems which are spread by water splash and/or humid environments; these include die-back, downy mildew, leaf spots and stem cankers, caused by fungal pathogens (e.g. *Botryotinia fuckeliana* – anamorph: *Botrytis cinerea*, *Glomerella* sp., *Monochaetia* sp., *Peronospora* spp. and *Pestalotiopsis* spp.), bacterial diseases such as fireblight (*Erwinia amylovora*) and leaf blotch of *Syringa* (*Pseudomonas syringae* pv. *syringae*), and leaf nematodes (*Aphelenchoides* spp.). Sub-irrigation systems greatly reduce disease risk. Similarly, poor drainage of the soil, container bed and/or rooting medium increases the spread of soil- and water-borne pathogens causing root, foot and crown rots, e.g. *Phytophthora* and *Pythium*. These pathogens can also be found in untreated sources of water.

There has been a lot of interest recently in 'closed' (i.e. recirculating) irrigation systems, to reduce water use and prevent pesticide and nutrient leaching. Potentially, such systems could lead to a rapid build-up of diseases caused by *Phytophthora*, *Pythium* and some other root rot and wilt pathogens. There has been a great deal of research work (which is still ongoing) into ways of overcoming this problem, and the results of this research are now beginning to be used on nurseries. These include disinfection/sterilization of the water, using techniques such as chlorination, ozonation, UV and heat treatment, and filtration systems such as microfiltration and slow sand filtration.

In recent years there has been increasing use of protective structures to improve plant growth and to avoid weather damage during the winter. However, this 'protection' can also increase the survival and reproduction rate of pests and fungal/bacterial pathogens. Such structures should be well-ventilated whenever weather conditions allow.

Some diseases of outdoor ornamentals are encouraged by wet weather conditions, e.g. scab of *Malus* spp. (*Venturia inaequalis*) and willow anthracnose (*Marssonina salicicola*). The foliage of plants, especially of trees and shrubs, should be examined periodically following a long period of wet weather, particularly in the spring, so that control measures can be implemented at the first signs of trouble. Disease forecasting systems have been, or are being, developed for a range of diseases on fruit, vegetables etc., and some of these have potential for use on ornamental crops, e.g. scab of *Malus*.

Outdoor herbaceous and woody ornamentals which lack vigour owing to poor planting, drought, waterlogging or malnutrition are more susceptible to diseases caused by leaf-spotting and rust fungi than well-grown plants. It is essential, therefore, to plant carefully and to feed, mulch and water as necessary during the growing season to ensure the development of good-quality plants. Also, to help prevent diseases such as downy and powdery mildews and grey mould, avoid overcrowding of plants.

Any cultural or climatic factor that causes wounding or damage increases crop susceptibility to infection by 'wound' pathogens, and pruning under wet conditions tends to increase fungal and bacterial attack. Pruning cuts should be made cleanly, to avoid leaving long stubs or 'snags' (particularly when cutting back the stock on young, grafted trees), as these provide an entry site for *Nectria cinnabarina,* the cause of coral spot – see under *Acer*, p. 452. Wound fungi can also cause serious decay and die-back in established amenity trees.

It is still important to be hygienic on nurseries and in parks and gardens, particularly where perennials are grown, as their permanence favours the build-up of pests and diseases. The grower can do a great deal to prevent such a build-up by observing the following basic principles:

- Control weeds and cut the grass regularly around established trees and shrubs to limit the amount of shelter for pests.
- Clean out hedge bottoms to limit the carryover of pests and diseases from season to season.
- Remove all dead shoots and branches and, where possible, collect and burn fallen diseased leaves.
- Prune as necessary to prevent dense growth and to allow good air circulation. Burn the prunings to kill overwintering eggs and fungal spores.

Pest and disease control during propagation

Crop losses during propagation can be reduced by providing the optimum cultural conditions for germination or rooting of any given species. However, pests and diseases may originate from contamination of seed, cuttings, scions, buds or rootstocks, the water supply, the rooting medium or the aerial environment. To avoid crop losses, not only during germination and rooting but also at later stages of production, it is essential to adopt a hygienic system of propagation. Material must be selected from stock plants that are free from nematodes and other pests, viruses, bacteria and fungal pathogens. These should be controlled on the 'mother' plants, scions, cuttings or seed, using treatments indicated in the text or in Tables 11.2 and 11.3.

Currently, there are few crop protection chemicals with specific recommendations for use during vegetative propagation. A small number of fungicides have recommendations for use either as compost drenches or incorporated into the compost during the rooting of cuttings – see Tables 11.2 and 11.3. However, if possible, many growers tend to avoid this type of fungicide use, as research has shown that it can have an adverse effect on the rooting of certain species or cultivars of ornamental plants.

Propagation under glass exposes plants to common glasshouse pests, including glasshouse whitefly (*Trialeurodes vaporariorum*), two-spotted spider mite (*Tetranychus urticae*) and various species of aphid. In glasshouses and polythene tunnels it is possible to control these and some other pests with biological control agents (e.g. parasitoids or predators); further details are given in Chapter 10. Biological control agents are very susceptible to most pesticides; therefore, if chemicals need to be used for other pests or diseases, ones compatible with a biological control system must be selected.

The humid environment necessary for the rooting of most cuttings also provides ideal conditions for the growth of many pathogens, (e.g. *Botrytis, Cylindrocarpon, Fusarium* and *Thanatephorus* (*Rhizoctonia*)). A programme of fungicide sprays may be applied at this time, often utilizing a number of different compounds such as chlorothalonil, iprodione and prochloraz.

Treatments for sterilizing compost, propagation benches and containers are given in Chapter 10. A number of compounds are available for sterilizing open ground, including chloropicrin, dazomet, metam-sodium and methyl bromide. In most cases, the compounds are applied by specialist contractors.

Micropropagation techniques (applicable to a limited number of hardy nursery stock species) enable rapid multiplication of progeny from a small nucleus of 'mother' plants. It is essential, therefore, that propagules should be freed from viruses and other pathogenic organisms by meristem culture and surface sterilization before inducing proliferation and rooting. However, control of pests and diseases is also important during the 'weaning' or transplanting stage, when the humid environments necessary for plant establishment are conducive to attack by fungi such as *Botryotinia fuckeliana* (anamorph: *Botrytis cinerea*) and *Pythium* spp.

Table 11.2 Fungicides with label recommendations* for the control of damping-off and foot rots (*Pythium* spp.) and root and collar rots (*Phytophthora* spp.)

Compound	Mainly active against**			Ornamental crops on label	Target for application	Application method
	Pythium	*Phytophthora*	Other			
copper ammonium carbonate [Note 1]	✓	✓	–	Seedlings of all crops	Seed, seedlings	(i) Semi-immersion of seed trays immediately after sowing (ii) Drench of seedlings immediately before or after pricking-out
etridiazole [Note 2]	✓	✓	–	Container-grown stock, HONS, tulips	Seeds and seedlings (ornamentals)	(i) Compost incorporation before sowing/planting (ii) Drench before transplanting
					Rooted cuttings and transplants (ornamentals)	(i) Compost incorporation before planting (ii) Compost drench
					Container-grown HONS	(i) Compost incorporation before planting (ii) Compost drench
					Tulips	(i) Soil incorporation before planting (ii) Soil drench after planting
fosetyl-aluminium [Note 3]	–	✓	(downy mildew)	Containerized HONS, glasshouse-grown pot plants	Nursery stock	Drench treatment to rooted cuttings after first potting. Repeat at monthly intervals and at re-potting. A maximum of up to six applications may be required

Contd.

Table 11.2 *Contd.*

Compound	Mainly active against**			Ornamental crops on label	Target for application	Application method
	Pythium	*Phytophthora*	Other			
fosetyl-aluminium Contd.					Glasshouse-grown pot plants:	
					(i) Trays of cuttings	Drench trays at the earliest opportunity
					(ii) Trays of young plants after pricking out	Semi-immersion of tray
					(iii) Established plants	Overhead coarse spray or drench. Repeat after 4–6 weeks if necessary
					Capillary (sand) benches	Treat unstocked benches. Plants can be replaced immediately after treatment
furalaxyl [Note 4]	✓	✓	–	HONS, bedding plants, pot plants	(i) All crops on label, at all growth stages	Compost incorporation
					(ii) Seedlings, cuttings and herbaceous plants.	Compost drench
					(iii) Container-grown HONS	Compost drench

propamocarb hydrochloride† [Note 5]	✓	✓	(downy mildew)	Container-grown ornamentals, container-grown stock, flowers, ornamentals, pot plants, tulips	(i) All crops	Compost incorporation
					(ii) Seeds/seedlings	Drench at seed sowing and/or pricking out, or drench established seedlings
					(iii) Unrooted cuttings	Drench
					(iv) Established plants	Drench
					(v) Tulips	Compost incorporation, dip or drench

* Some of the compounds in the table are available as more than one proprietary product, the label recommendations for which may differ. It is very important, therefore, to check the label of any product intended for use. N.B. Inclusion of a chemical in the table indicates activity against the particular disease cited. It does not imply 'safe' usage on a plant species unless specified on the product label. Test treatments on a few plants before treating the remainder.

** Where a disease appears in parentheses, the compound has recommendations for use against that disease, but on crops other than ornamentals.

† Systemic compound.

Note 1. Also recommended as a foliar spray on chrysanthemum for powdery mildew control (see Table 11.3, p. 438).

Note 2. (i) For best results use as a prophylactic. (ii) Compost incoporation is more effective than drench treatments. However, drench treatments may be made at intervals of 4–8 weeks following compost incorporation. (iii) Do not drench emerged or transplanted seedlings until they are well established. (iv) Wash spray residues off foliage. (v) Do not use on *Escallonia*, *Pyracantha* or *Viola* spp.

Note 3. (i) Although not recommended on the label, there are indications that the fungicide also has some activity against *Pythium*. (ii) The compound is systemic, both upwards and (more unusually) downwards (from a foliar spray). (iii) No specific recommendations for downy mildew control on ornamentals, but can be used at grower's risk under the off-label arrangements as there are recommendations for downy mildew control on other crops (e.g. broad bean, hops, protected lettuce).

Note 4. (i) Protectant activity plus some curative activity if applied before the disease is too advanced. (ii) Do not use on field or border soil. (iii) Fungicide resistance (particularly with *Pythium*) has been confirmed on some nurseries. (iv) Apply protectant compost drenches within 3 days of seedling or planting out. (v) Although not recommended or mentioned on the label, furalaxyl will also have activity against downy mildews. However, as a single compound, rather than a co-formulation, there is an increased chance of fungicide resistance developing. Any applications for downy mildew control should be supported by foliar sprays of another downy mildew fungicide from a different chemical group.

Note 5. (i) Absorbed by the roots and translocated to aerial parts of the plants. (ii) Drench treatment to established plants may be repeated at intervals of 3–6 weeks. (iii) Do not treat young seedlings with overhead drench. (iv) When applied over established seedlings, rinse off foliage with water. (v) Do not use on heathers. (vi) Recommendations for downy mildew control on brassicas; thus, permitted off-label use on ornamentals.

Table 11.3 Fungicides with label recommendations* for the control of grey mould (G), powdery mildews (P), downy mildews (D)**, rusts (R), rose black spot and leaf spots (L), verticillium and fusarium wilts (VF) and *Thanatephorus* (*Rhizoctonia*) (RH)

Compound	Active against***								Frequency	Ornamental crops on label
	G	P	D	R	L	VF	RH	Others		
azoxystrobin† [Note 1]	–	(✓)	–	✓	(✓)	–	–	–		Chrysanthemum (off-label) (SOLA 1536/00)
bupirimate†	–	✓	–	–	–	–	–	–	7–14 days	Begonia, chrysanthemum, rose
bupirimate + triforine†	–	✓	–	✓	✓	–	–	–	10–14 days	Ornamentals, rose
captan [Note 2]	(✓)	–	–	–	✓	–	–	(scab), (*Monilinia*)	7–14 days	Rose
carbendazim† [Note 3]	✓	–	–	–	(✓)	(✓)	–	*Thielaviopsis*, (*Sclerotinia*)	10–14 days (rose) 14 days minimum for SOLA uses	Chrysanthemum, ornamentals, bedding plants (all off-label), pot plants, rose
carbendazim + prochloraz† [Note 4]	(✓)	(✓)	–	–	(✓)	–	–	(*Sclerotinia*)	–	Protected ornamentals (off-label) (SOLA 2217/99)
chlorothalonil [Note 5]	✓	(✓)	(✓)	–	(✓)	–	–	–	7–21 days	Ornamentals
copper ammonium carbonate [Note 6]	–	✓	–	–	(✓)	–	–	(leaf curl)	14 days (chrysanthemum)	Chrysanthemum, seedlings of ornamentals
dichlofluanid	✓	✓	✓	–	✓	–	–	–	10–14 days	Non-edible ornamentals (off-label) (SOLA 0167/93), rose
difenoconazole†	–	–	–	✓	✓	–	–	–	14 days (minimum)	Hybrid pinks, sweet william (off-label) (SOLAs 1728/97, 1729/97)
dodemorph† [Note 7]	–	✓	–	–	–	–	–	–	7–14 days	Rose
fenarimol† [Note 8]	–	✓	–	–	–	–	–	(scab)	10–14 days	Rose

imazalil† [Note 9]	–	✓	–	–	–	–	–	–	7–14 days	Ornamentals, rose
iprodione [Note 10]	✓	–	–	–	–	–	(✓)	(*Sclerotinia*), *Alternaria*	21 days	Pot plants, seeds of flowers and ornamentals
mancozeb [Note 11]	–	–	(✓)	✓	✓	–	–	–	10–14 days	Carnation, chrysanthemum, pelargonium, rose
myclobutanil†	–	✓	–	✓	✓	–	–	(scab), (*Monilinia*)	7–14 days	Rose
oxycarboxin [Note 12]	–	–	–	✓	–	–	–	–	7–14 days	Carnation, chrysanthemum, pelargonium, ornamentals, rose
penconazole [Note 13]	–	✓	–	✓	–	–	–	scab	7–14 days	Ornamental *Malus*, rose
prochloraz [Note 14]	✓	✓	–	–	✓	✓	–	(*Cylindrocarpon*), (*Sclerotinia*), (*Thielaviopsis*)	7–14 days (spray)	Container-grown stock, HONS, ornamentals
propiconazole [Note 15]	–	–	–	✓	–	–	–	–	7 days (minimum)	Chrysanthemum (off-label) (SOLAs 0019/92, 1211/96, 1388/97)
pyrifenox†	–	✓	–	–	(✓)	–	–	scab	7–14 days	HONS, ornamentals
quintozene [Note 16]	✓	–	–	–	–	–	–	*Sclerotinia*	–	Bedding plants, chrysanthemum, dahlia, fuchsia, pelargonium, pot plants
thiram [Note 17]	✓	–	(✓)	✓	–	–	–	(scab)	10–14 days	Carnation, chrysanthemum, ornamentals
tolclofos-methyl [Note 18]	–	–	–	–	–	–	✓	–	–	Ornamentals, seedlings of ornamentals
zineb [Note 19]	✓	–	(✓)	✓	(✓)	–	–	–	10–21 days	Anemone, carnation

Contd.

Table 11.3 *Contd.*

* Many of the compounds in the table are available as a large number of proprietary products, and not all of these products will have label recommendations for use on ornamentals. It is very important, therefore, to check the label of any product intended for use. Unless specifically mentioned in the notes below, all applications are as foliar sprays. N.B. Inclusion of a chemical in the table indicates activity against the particular disease cited. It does not imply 'safe' usage on a plant species, unless specified on the product label. Test treatments on a few plants before treating the remainder.

** See also Table 11.2, p. 435.

*** Where a tick or the name of a disease appears in parentheses, the compound has recommendations for use against that disease but on crops other than ornamentals.

† Systemic compound.

Note 1. For control of white rust on pot chrysanthemums and those grown for cut flowers under protection.

Note 2. Label recommendation is for black spot control.

Note 3. SOLA 0009/99. Allows spray application for the control of foliar diseases. Also, drench treatment for soil- or container-grown crops – specifically for *Thielaviopsis* control but would also have some activity against *Fusarium* and *Verticillium*.

Note 4. Maximum of two applications per crop. Primarily used for control of leaf spots.

Note 5. Protectant activity. Can be redistributed to new growth by water splash.

Note 6. Control of damping-off in seedlings by semi-immersion of seed trays immediately after sowing, or drench immediately before or after pricking out. Also active against bacterial diseases.

Note 7. Do not use on seedling roses.

Note 8. May also have some activity against *Fusarium*.

Note 9. May cause damage if open flowers are sprayed. Also available as a smoke for treatment of protected crops.

Note 10. Seed treatment for seeds of flowers and ornamentals is for control of *Alternaria*.

Note 11. Protectant activity.

Note 12. Avoid spraying when blooms open. Add wetter for first spray if rust is already established.

Note 13. Antisporulant activity.

Note 14. Has a range of recommendations for various ornamentals at various stages of growth. In addition to sprays, there are recommendations for drenches for compost, soil and propagation beds, and a dip for cuttings.

Note 15. SOLAs for use on confirmed outbreaks of white rust.

Note 16. Compost or soil incorporation. Active against the germinating resting structures (sclerotia) of the fungi indicated.

Note 17. Protectant activity. Do not apply to hydrangeas.

Note 18. Protectant activity. Compost incorporation or drench application. Rinse off foliage after drench application. Not recommended for use on heathers.

Note 19. Protectant activity.

Virus and phytoplasma diseases

Virus diseases have been identified in a number of outdoor ornamental species but these have not been listed under hosts in this chapter, because comprehensive descriptions are given in *Virus Diseases of Trees and Shrubs* by J. I. Cooper.

Virus infection is often associated with poor growth and, frequently, depending on environmental conditions, foliage symptoms or leaf and stem distortions are induced. Leaf symptoms include yellow, or pale-green, chlorotic vein-banding, leaf patterns, ring spots, mottles and mosaics.

In the production of ornamental nursery stock, vegetative propagation is probably the most important means of transmitting virus from infected clones to the progeny. However, virus may also be spread by specific vectors, e.g. aphids, nematodes, plants and fungi, or by pollen or seed, or by handling virus-infected plants and tools. Two of the more frequently occurring viruses of hardy nursery stock are the aphid-transmitted cucumber mosaic virus and the nematode- or seed-transmitted arabis mosaic virus.

Other diseases characterized by foliar yellowing (with or without stunting) and ancillary shoot proliferation (witches' broom), or leaf-like petals, can be caused by phytoplasmas. These organisms are often transmitted by leafhoppers.

Sometimes, physiological disorders and damage caused by insects or fungi can be confused with symptoms of virus or phytoplasma infection. For this reason, laboratory diagnosis is always recommended.

Control of virus or phytoplasma diseases depends on:

- propagation from virus-free stock plants;
- insecticidal control of insect vectors to prevent infection;
- routine roguing of plants with poor growth and/or foliar symptoms. Growers who suspect virus infection should submit plants to a diagnostic laboratory for confirmation.

Pests with many hosts

Aphids

These sap-feeding insects can be found on virtually all ornamental plants. Dense colonies may form on the young shoots and undersides of the leaves. The foliage often becomes soiled with the aphids' honeydew and with black, sooty moulds that develop on this sugary excretion. Other signs of attack include discoloured and/or distorted foliage and stunted growth. Some aphids are important vectors of virus diseases. Most infestations, especially on woody plants, begin during the spring as the new growth emerges. Treat at that time or when aphids are first seen, with cypermethrin, deltamethrin, dimethoate, imidacloprid, malathion, nicotine, pirimicarb or pymetrozine. The last two are selective aphidicides that leave most aphid predators and parasitoids unharmed. Check the product label for possible phytotoxic reactions. Aphids on tall deciduous trees, such as birch, lime, oak, sycamore and willow, often cause problems because of the copious quantities of honeydew that fall from the tree. There is little that can be done about this because of the difficulties of spraying tall trees and obtaining adequate cover.

Caterpillars

The main pest species, particularly on deciduous trees and shrubs, is winter moth (*Operophtera brumata*). The pale-green looper caterpillars are active between bud burst and early summer, and may cause extensive holing of the foliage. Tortrix moth caterpillars, mainly those of the highly polyphagous carnation tortrix moth (*Cacoecimorpha pronubana*), bind up the leaves at the shoot tips with silken webbing. The pale-green caterpillars feed out of sight by grazing away the leaf surfaces, and damage is often not noticed until after the moths have emerged. Both types can be controlled by *Bacillus thuringiensis*, cypermethrin, diflubenzuron, nicotine or teflubenzuron. *B. thuringiensis*, diflubenzuron and teflubenzuron are relatively harmless to animals other than caterpillars and so are particularly suitable for use in public areas. Pheromone traps are available for monitoring the flight periods of carnation tortrix moth.

Glasshouse whitefly (*Trialeurodes vaporariorum*)

As with two-spotted spider mite, this sap-feeding pest is mainly a pest of plants in protected situations. For control, see under *Euphorbia pulcherrima* (poinsettia), whiteflies, Chapter 10, p. 410.

Slugs and snails

Various species, e.g. garden slug (*Arion hortensis*), garden snail (*Helix aspersa*) and field slug (*Deroceras reticulatum*), are important pests of ornamentals. Herbaceous plants and annuals are particularly vulnerable, especially in wet seasons. Slugs and snails are controlled by pelleted baits of metaldehyde or methiocarb. The latter is generally the more effective under damp conditions. Alternative controls are aluminium sulfate crystals or, for slugs, biological control with the pathogenic nematode, *Phasmarhabditis hermaphrodita*.

Two-spotted spider mite (*Tetranychus urticae*)

This important pest can occur on many outdoor plants but is mainly a problem on nursery stock that is protected by polythene tunnels or greenhouses. For a description and control in this situation, see under Cucumber, Chapter 9, p. 338, and *Dendranthema* (chrysanthemum), Chapter 10, p. 402, respectively. On outdoor ornamentals, infestation may occur in mid- to late summer. Treat with abamectin, bifenthrin or fenazaquin, or use biological control with *Phytoseiulus persimilis* (a predatory mite).

Vine weevil (*Otiorhynchus sulcatus*)

Vine weevil has become a serious problem in recent years, possibly because of the greater use of peat-based composts. The plump, white, apodous larvae live in the soil where they destroy the roots. Container-grown plants are particularly vulnerable and by the time plants show symptoms of attack (wilting) it is often too late to save them. In nurseries where vine weevil is present treat the potting compost or soil with carbofuran, chlorpyrifos, fonofos or imidacloprid, which should give protection for up to 12 months. The adults are active mainly in late spring and summer, when they make irregular notches in leaf margins; they may also girdle stems. The adults can be controlled by spraying with chlorpyrifos. A pathogenic nematode (*Heterorhabditis megidis*) can be used as a compost drench against vine weevil larvae in late summer.

Diseases with many hosts

Root diseases

Armillaria (honey fungus)

Honey fungus is a major cause of losses of trees and shrubs in amenity plantings. It can also cause problems in fruit trees and bushes and, occasionally, in timber plantations and herbaceous plants. The disease is caused by one of a number of *Armillaria* species. These fungi colonize buried wood, such as tree stumps and wooden posts, and from these nutrient sources can spread through the soil via thin, dark-brown or black, bootlace-like structures known as rhizomorphs (which can be difficult to distinguish from roots). If the tip of a rhizomorph comes into contact with the roots of a susceptible tree or shrub species, it can infect the plant, causing root and stem-base decay. The decline of the plant may occur within a few months, or may take many years, depending on the susceptibility of the plant to *Armillaria* and its size when first infected. Species that are particularly susceptible to honey fungus are listed in Table 11.4.

Because of its method of spread, problems due to honey fungus often show as occasional deaths of plants in a small area over a number of years, or the successive deaths of adjacent plants in a hedge. Affected trees may exude a gummy or watery liquid from the lower stem. However, the key diagnostic feature is the presence of a paper-thin sheet of creamy-white fungal growth (mycelium) beneath the bark of the lower stem or main roots. The affected area will also smell strongly of mushrooms. Occasionally, it is also possible to see either the rhizomorphs or the toadstools (fruiting bodies) of *Armillaria* around dying plants, but neither of these is a reliable indication that honey fungus is the cause of the problem (see p. 444).

Table 11.4 Honey fungus: notably susceptible plant species

Araucaria araucana	monkey puzzle
Betula pendula	silver birch
Betula pubescens	downy birch
Cedrus atlantica	Atlas cedar
Cedrus deodara	deodar cedar
Cedrus libani	Lebanon cedar
Chamaecyparis lawsoniana	Lawson cypress
× *Cupressocyparis leylandii*	Leyland cypress
Cupressus macrocarpa	Monterey cypress
Fragaria spp.	strawberry
Humulus lupulus	hop
Juglans regia	common walnut
Ligustrum ovalifolium	privet
Malus spp.	apple (fruiting and ornamental)
Picea omorika	Serbian spruce
Prunus spp.	cherry, plum (fruiting and ornamental)
Ribes nigrum	black currant
Ribes uva-crispa	gooseberry
Rosa	rose
Rubus agg.	blackberry
Rubus idaeus	raspberry
Salix spp.	willow
Sequoiadendron giganteum	Wellingtonia
Solanum tuberosum	potato
Syringa vulgaris	lilac

For many years it was thought that honey fungus was a single species: *Armillaria mellea*. It is now known, however, that at least five distinct species may be found. The most common of these are *A. mellea, A. ostoyae* and *A. gallica*. The first two are highly pathogenic and can attack healthy trees (*A. ostoyae* being more commonly found on conifers). *A. gallica*, however, usually requires the plant to be weakened or to be declining due to another factor (e.g. drought stress or *Phytophthora* root rot) before it can invade. *A. gallica* usually produces many more rhizomorphs than the other species, and the presence of these structures, therefore, cannot be taken as an indication that honey fungus was the primary cause of death of a plant. Similarly, the toadstools produced by all three species are very alike (brown or honey-brown with whitish flesh and a conspicuous whitish-yellow collar on the stalk just below the cap) and cannot on their own be taken as proof of an attack.

Control of the disease starts with the complete removal of as much of the affected plant as possible. If the problem is with a hedge or with closely spaced plants, it is also prudent to remove the nearest apparently healthy plant on either side, as these may harbour infection. Any large pieces of infected root left behind can act as a nutrient source for rhizomorphs, which will grow out from them. Chemical treatment of the soil is likely to be ineffective if significant root debris remains and unnecessary if thorough excavation has occurred (detached rhizomorphs do not pose a threat). If removal of all large roots is not possible, delay replanting for up to 2 years or plant a resistant species. Plants with a useful level of resistance are listed in Table 11.5.

Table 11.5 Honey fungus: notably resistant plant species

Abies grandis	grand fir
Abies procera	noble fir
Acer negundo†	box elder
Ailanthus altissima	tree of heaven
Berberis	barberry
Buxus sempervirens	box
Castanea sativa	sweet chestnut
Cistus spp.	sun rose
Clematis	clematis
Cotinus coggygria	smoke tree
*Crataegus laevigata**	Midland hawthorn
*Crataegus monogyna**	common hawthorn
Elaeagnus spp.	elaeagnus
Fagus sylvatica	beech
Fallopia baldschuanica	Russian vine
Fargesia (and other genera)	bamboo
*Fraxinus excelsior**	ash
Hedera helix	ivy
*Ilex aquifolium**	holly
Juglans hindsii†	Californian black walnut
Larix decidua	European larch
Larix kaempferi	Japanese larch
Larix × marschlinsii	hybrid larch
Lonicera nitida	shrubby honeysuckle
Lonicera periclymenum	common honeysuckle
Mahonia aquifolium	mahonia
Mahonia japonica	mahonia
Prunus laurocerasus	cherry laurel
Prunus spinosa	blackthorn
Pseudotsuga menziesii	Douglas fir
Quercus ilex	holm oak
Quercus petraea	sessile oak
Quercus robur	English oak
Rhus typhina	stag's horn sumach
Robinia pseudoacacia	false acacia
Sambucus nigra	elder
Tamarix gallica	tamarisk
Taxus baccata	yew
*Tilia × europaea**	common lime

* Serious attacks have occurred on ash, common lime, hawthorn and holly but, considering that they are amongst our most common native trees, these cases are remarkably few and suggest considerable resistance to the disease.

† Box elder and Californian black walnut are probably immune.

It is possible to create a physical barrier to further spread of rhizomorphs by burying sheets of heavy-gauge polythene or PVC. These should extend from just above the soil surface to at least 45 cm below it.

Products containing tar acids have recommendations for honey fungus control on trees and shrubs, but note the above comments regarding the removal of affected material.

Phytophthora

This is a major root pathogen of many nursery stock subjects. Although few hosts are completely resistant to attack, species differ in their tolerance to infection and rate of root rot. Notably susceptible are species of *Chamaecyparis*, *Erica*, *Juniperus* and *Rhododendron*.

Phytophthora cinnamomi is the most damaging species on nursery stock, although other species of *Phytophthora* may also cause severe problems. *Phytophthora* spp. are soil- and water-borne, so badly drained soils and container beds encourage spread of the disease by motile zoospores which infect the 'feeder' roots, causing a rot. *P. cinnamomi* grows most rapidly at temperatures of 20–30°C. *Phytophthora* infection often spreads from roots into the stem base, causing a lesion that reduces or cuts off water uptake. Leaves may wilt, abscise or lose their normal coloration and turn reddish-brown. Foliar symptoms develop most rapidly when the weather is conducive to plant 'stress', e.g. hot, dry, windy or freezing conditions.

Phytophthora spp. can remain viable for many years as 'resting' spores in soil and in the debris of infected roots. A number of control measures may be necessary, including the following:

- Strict nursery hygiene and prompt removal of all diseased plants (and preferably those adjacent to them).
- Use of a clean water supply and rooting medium. If recycled water is used, this may require treatment in some way to reduce the chances of infection. Treatments available include chlorination, ozonation, ultraviolet sterilization and slow sand filtration.
- Selection of propagation material from healthy stock plants.
- Provision of good drainage of container beds or field soil to prevent waterlogging.
- Application of etridiazole, fosetyl-aluminium, furalaxyl or propamocarb hydrochloride as protectant treatments on container-grown plants.
- Disinfection of container beds and propagation areas where there has been a problem. There are now numerous disinfectant products available for use against *Phytophthora* and other root pathogens.
- Avoidance of planting species that are very susceptible to *Phytophthora* into land known to be infested with the pathogen. Soil sterilization can be considered but is often prohibitively expensive, and the chemical may not penetrate to a sufficient depth to prevent a recurrence of the problem.

On a few hosts (e.g. *Ceanothus* and *Ilex*) *Phytophthora* may cause an aerial die-back of branches and stems without apparent root infection. In these cases the disease is best treated using similar cultural and chemical control measures to those for downy mildews (see p. 448, Table 11.2, p. 435, and Table 11.3, p. 438).

Pythium

This fungus is closely related to *Phytophthora*, but is far less damaging to nursery stock. It can cause rotting of cuttings and problems with small liners as a result of root-rotting. Cultural and chemical control measures are very similar to those for *Phytophthora*.

Thanatephorus cucumeris (anamorph: *Rhizoctonia solani*)

This pathogen also has a wide host range. It can produce a number of symptoms, including damping-off of seedlings, stem rot of cuttings, root rot, stem lesions and, occasionally, foliar blight of mature plants. Stem lesions are typically lightish-brown in colour and dry. Foliar blight is a severe problem on heathers, but can also affect azaleas and some other subjects.

In humid conditions, threads (hyphae) of *R. solani* may spread over the surface of the compost or the foliage of affected plants, producing a brown, or silvery, spider-like webbing that can be seen with the unaided eye.

R. solani survives mainly as fragments of hyphae, although it can produce resilient resting structures or sclerotia. It can survive well in soil in the absence of a host plant (i.e. it has good saprophytic ability). It has been found contaminating almost every area of affected nurseries (e.g. capillary matting, compost heaps, Danish trolleys, pots and standing areas).

Cultural measures that help to control the disease include the avoidance of high humidities, overcrowding, overwatering, overfertilizing and deep planting.

Fungicides with recommendations for the control of *Thanatephorus* (*Rhizoctonia*) are listed in Table 11.3, p. 438. Most frequently used are, probably, iprodione and tolclofos-methyl. Many disinfectants claim activity against a range of root pathogens, including *Rhizoctonia*, and can be used to treat standing areas, etc., where an outbreak has occurred.

Thielaviopsis

This fungus causes the disease known as black root rot. It is very damaging to some bedding plants, but is also becoming increasingly common on nursery stock subjects such as *Fuchsia, Ilex*, *Prunus* and *Sambucus*. As its name suggests, the fungus causes a black decay of the roots, the colour coming from the presence of numerous resting spores within the affected tissue. Contamination of standing areas, pots etc. is very common, and the hygiene and disinfection procedures applicable to other root diseases are also relevant for this pathogen.

Chemical control depends on compost drenches of carbendazim (off-label) (SOLA 0009/99).

Foliar diseases

Downy mildew

The symptoms produced by this disease are variable, but often a yellowing or browning of the upper leaf surface is accompanied by the production of a grey, purple or white, downy or felty fungal growth (mycelium and spores) on the undersurface. Leaf distortion may also occur. The amount of sporulation on the undersurface is quite variable, according to the plant species or cultivar attacked, e.g. some of the large-leaved *Hebe* cultivars, such as 'Midsummer Beauty', may have large amounts of sporulation present, whereas with some of the smaller-leaved cultivars the leaves tend to turn brown and abscise without much evidence of fungal growth. On some species, symptoms due to downy mildew may be confused with those caused by leaf nematodes or bacterial pathogens.

Downy mildews are favoured by humid conditions, which aid spore production and infection. The spores (conidia) are short-lived (a few days only) and are readily splash-dispersed or carried on air currents. Some downy mildews on other crops produce a resting spore (oospore) within affected leaves which is capable of surviving in the soil in the absence of the host plant. Knowledge of the function of oospores in downy mildews of ornamental plants is very limited, however.

Much can be done, culturally, to reduce the possibility of infection by manipulating environmental conditions, e.g. lowering RH in glasshouses and polytunnels by venting when possible. Some downy mildews have peak periods for spore production at night, so any overhead watering should be done early in the day to ensure that leaf surfaces are dry at night. The use of capillary sand-beds to irrigate susceptible plants is even more beneficial. Ensure that plants are well spaced. If the market allows, choose more resistant species or cultivars. If an outbreak of downy mildew does occur, dispose of any fallen leaves.

Even if the above precautions are taken, many of the most susceptible subjects, such as *Hebe* and *Rosa*, frequently require fungicide treatment. Products with on-label approval for downy mildew control on ornamentals are listed in Table 11.2, p. 435 and Table 11.3, p. 438. However, on many nurseries downy mildew control programmes are based on HV sprays of products applied under the off-label arrangements. Commonly used fungicides include fosetyl-aluminium* and metalaxyl + thiram*. Additional fungicides used on other crops for downy mildew control, or those used against potato blight, may have activity against downy mildews on ornamentals. All off-label use is, of course, at the user's risk.

Grey mould

Classic symptoms of this disease are browning or die-back of leaves, branches or stems, which may become covered with a grey felt of fungal growth, consisting of fungal threads (hyphae) and large numbers of spores.

The disease is caused by the ubiquitous fungus *Botryotinia fuckeliana* (anamorph: *Botrytis cinerea*), and almost all ornamental plants are susceptible (though to varying degrees). Soft, wilted, ageing or damaged plant parts and fallen flowers or flower parts are most readily attacked by the fungus. Healthy tissue touching infected parts often becomes infected. Cuttings rooted under moist conditions are also vulnerable. Stagnant moist air favours the development of *Botrytis*.

The spores of the fungus are liberated into the air and are carried in air currents. Spore production and infection require humid conditions, and the fungus is active over the temperature range 2–25°C. On decaying plant material, the fungus may produce hard, black resting structures (called sclerotia) which vary considerably in shape but are mostly rather flattened and circular to oval. These vary from 2 to 8 mm in length and can remain dormant for long periods, thus enabling the fungus to survive during adverse conditions. The sclerotia either remain attached to the affected plant or drop to the ground, and eventually germinate to produce more spores.

Because of problems with fungicide resistance, and the limited number of effective fungicides available (particularly for use on plants grown under protection), cultural control measures are very important. Care should be taken to avoid growth checks and to ensure that nutrition is well-balanced. Plants should be handled carefully to avoid or minimize physical damage, and old or moribund leaves or shoots and dead flowers should be removed wherever practicable. Avoid high humidity and still air in protected structures by ventilation (coupled with additional heat if necessary).

Fungicides are frequently used against *Botrytis*, particularly in the winter months on plants grown under protection, and those with label recommendations for use on ornamentals are listed in Table 11.3, p. 438. Resistance to active ingredients from the MBC (e.g. carbendazim) and dicarboximide (e.g. iprodione) groups of fungicides is common. Anti-resistance strategies couple cultural control methods with programmes of fungicides from differing chemical groups. This is more difficult under protection, owing to the limited number of products approved for use in this situation. Far more products are available for use outdoors, although many of these are applied at the grower's risk under the off-label arrangements (e.g. cyprodinil and tebuconazole).

Powdery mildew

Leaves (both surfaces), stems and, occasionally, flowers may be affected. Typical symptoms consist of a mealy, powdery-white or off-white, growth on the affected plant part.

On deciduous species, powdery mildew is often capable of overwintering within the buds or on the wood. Affected buds may produce heavily infected shoots (primaries) in the spring, from which the disease can spread rapidly to adjacent shoots and plants.

On some subjects, particularly in the late summer/autumn, spore cases (cleistothecia) of the sexual state of the mildew may be found on leaves. These are visible as roundish, brown–black structures, 0.5–1.0 mm in diameter, within the fungal growth.

Powdery mildews tend to be active at fairly high temperatures, so are usually a problem in the spring and summer. Although humidity requirements are not as critical as for the downy mildews, high humidity for a few hours in the 24-hour cycle is favourable for spore production and infection. Prolonged wetness of aerial parts of the plant may actually reduce disease development, however. Spores are readily dispersed in dry conditions. Young growth is particularly susceptible, but the disease is also common on older growth. Plants receiving an excess of nitrogen can be affected severely.

There are a number of cultural control measures that may reduce infection. As with all diseases, clean stock material is a good start but, like all foliar diseases, powdery mildew does not respect nursery boundaries. Avoiding long periods of high humidity on protected crops may prevent the development of epidemics, but this is generally not enough to do more than keep the disease at a fairly low level. Ensure adequate watering of plants, and maintain a balanced supply of nutrients.

Fungicides are used frequently on susceptible subjects, and those with a label recommendation for use on ornamentals are listed in Table 11.3, p. 438. Monitor plants closely in the spring for the first signs of infection. Where plants have become infected, dispose of fallen leaves, and on deciduous subjects apply fungicides right up until the natural time for leaf abscission (to try to prevent infection of buds).

Rust

This disease usually occurs as small, discrete eruptions (pustules) on the aerial parts of the plant, especially the leaves. The pustules contain numerous spores which, according to the host and the time of year, may be orange, brown, yellow, black or even white in colour. Usually, the pustules are found on the lower leaf surface, with a corresponding discoloured area or lesion on the upper surface. In a few subjects (e.g. *Pinus* and *Rosa*) the fungus is also capable of producing stem or branch lesions.

Some rusts affect only a single host plant, but others need a second host to complete their life-cycle. The life-cycle can be very complex, involving up to five different types of spore. In some cases more than one spore type can be seen on a single leaf. An example of this occurs on roses in the autumn, where orange pustules (containing the summer spores or uredospores) may be seen alongside black pustules (containing the overwintering spores or teleutospores).

Conditions favouring rust development vary, depending on the species involved, with optimum temperatures differing widely. However, high RH is generally favourable and free (unbound) water on the foliage is often necessary for spore germination and infection. Spores are dispersed from the pustules by water splash and air currents. Rusts on many ornamental plants gradually increase during the growing season, and reach peak levels in late summer and early autumn.

Cultural precautions to reduce the risk of rust include the use, where possible, of more resistant species or cultivars. With many rusts, the time between infection and symptom production (known as the latent period) can vary from 2 to 6 weeks, and the pathogen can often be introduced on bought-in plants and cuttings which initially look healthy. Therefore, be vigilant for the appearance of rust on bought-in stock (and home-raised mother plants). It may be worth roguing severely infected plants, and fallen leaves should certainly be disposed of. Reduce humidity as much as possible in glasshouses and polytunnels.

Fungicides with label recommendations for rust control are listed in Table 11.3, p. 438.

Abies (fir)

Diseases

Needle cast (cause unknown)
Needles develop areas of browning (sometimes affecting the distal half of the leaf, sometimes towards the middle) prior to falling. The problem can seriously disfigure *Abies* spp. grown as Christmas trees. Extensive research (which is still continuing) has not identified a pathogenic cause for the problem.

Phytophthora root rot (*Phytophthora* spp.)
See p. 446 and Table 11.2, p. 438.

Rust (*Pucciniastrum epilobii* and *Melampsorella caryophyllacearum*)
Four rusts have actually been found on *Abies*, but two are of little importance. Of the others, *Pucciniastrum epilobii* has occasionally caused needle cast during the summer months. The alternate hosts of this rust are willow-herbs (*Epilobium* spp.) and rose-bay (*Chamaenerion angustifolium*), and it is reported that it does not spread to *Abies* more than about 50 metres from these hosts. Control of willow-herbs around affected trees, therefore, is the primary control method.

Melampsorella caryophyllacearum infects *Abies* in May and June from its alternate hosts, which are chickweeds (*Stellaria* spp.) and mouse-ears (*Cerastium* spp.). Once established, the fungus can become systemic in the tree, giving rise to tumours and witches' brooms. Whilst it produces spectacular symptoms, the disease is usually of little practical significance in forest stands. In *Abies* grown for Christmas trees, affected trees should be removed, or the witches' brooms pruned out. The alternate hosts should also be controlled, although this may not be as effective as for *P. epilobii*.

Acer (maple, sycamore)

Pests

Aphids (e.g. *Drepanosiphum platanoidis* and *Periphyllus* spp.)
These pale-yellow or black aphids are present from late spring onwards. Honeydew and sooty mould are often associated with infestations. Spray thoroughly shortly after bud burst. See under Aphids, p. 441.

Gall mites (e.g. *Artacris macrorhynchus*, *Eriophyes eriobius* and *E. psilomerus*)
These microscopic mites attack the foliage during the summer. The first-named species causes small, red, pimple-like galls to develop on the upper surface, whereas the others initiate the development of yellowish-pink hairs (erinea) on the undersides. The galls cause little real harm to the plant. There are no recommended controls but removal of the worst-affected leaves may check infestations if done in early summer when galls develop. This is worthwhile only on lightly affected trees.

Horse chestnut scale (*Pulvinaria regalis*)
See under *Aesculus*, p. 453.

Diseases

Coral spot (*Nectria cinnabarina*)
This disease commonly causes die-back of large branches or even whole trees and is recognized by a rash of small, pink, spore-bearing pustules 1–2 mm in diameter which develop on the surface of dead and dying branches. This fungus, unlike *N. galligena* (see under Apple, canker Chapter 8, p. 27) does not induce callus formation. *N. cinnabarina* usually infects via mechanical injury, through frost-damaged tissues, abrasions made by plant 'ties' or dead tissue left as pruning 'snags', and then invades living tissue causing die-back. Spores can be produced by the pustules at any time of the year and a second type of spore is formed, usually in the spring, in dark-red, pinhead-size structures which develop amongst the old pustules. Control by pruning back to healthy wood and by applying a wound paint containing azaconazole + imazalil or octhilinone. Remove and burn all dead wood.

Honey fungus (*Armillaria* spp.)
See p. 443.

Phytophthora root rot (*Phytophthora* spp.)
See p. 446 and Table 11.2, p. 435.

Powdery mildew (*Sawadaea bicornis*)
See p. 449 and Table 11.3, p. 438. The disease can be damaging to nursery stocks as severe attacks often cause defoliation.

Tar spot (*Rhytisma acerinum*)
During the summer large, black spots with a bright-yellow edge develop on the leaves. This disease is more unsightly than harmful, but leaves may fall prematurely and these should be raked up and burnt. If necessary, spray in spring with a copper-containing fungicide. See under *Berberis*, bacterial leaf spot, p. 458, for more details of these fungicides.

Verticillium wilt (*Verticillium dahliae*)
This soil-borne fungus invades roots and causes sudden wilting of leaves, followed by withering and die-back of shoots. Internally, affected shoots develop brown or greenish-brown streaks. Propagate from healthy stock, as there is no reliable eradicant treatment. Infected plants in nurseries should be destroyed and the soil/compost sterilized. Do not plant *Acer* in land infested with *Verticillium*. A test is available from some laboratories, which quantifies the amount of *Verticillium dahliae* within a soil sample. A number of other crops, particularly linseed, potato and strawberry, are hosts to the fungus and may increase inoculum levels within the soil.

Aesculus (horse chestnut)

Pests

Horse chestnut scale (*Pulvinaria regalis*)
Established in the London area since 1964, this sap-feeding pest is now widespread in England. The wide host range includes *Acer* spp., *Cornus kousa,* magnolia, lime, elm, sweet bay and *Skimmia japonica*. It is most easily seen during the summer, when white egg masses may occur in large numbers on the bark. The mature scale is brown, about 4 mm in diameter, and is perched on the edge of an egg mass. Young scales feed on the foliage but move on to the bark during the autumn. Little damage is caused and no honeydew is produced, except sometimes on *Skimmia*. Control on small trees and shrubs by spraying with deltamethrin or malathion in early July, when the more vulnerable newly emerged first-instar nymphs (crawlers), are present. Established trees can tolerate heavy infestations and do not need treatment.

Diseases

Bleeding canker (*Phytophthora* spp.)
This bark disease is caused by *Phytophthora cactorum* or by *P. citricola*. Lesions develop on the stem and ooze a yellow, rusty or brown–black, gummy liquid that

often dries as a brown–black, shiny encrustation. If the affected outer bark is removed, the inner bark is often seen to be stained orange, with a mottled or zoned striation. To control stem lesions, cut out dead and dying bark down to the healthy underlying wood and then remove a strip (at least 50 mm wide) of healthy bark from the periphery of the wound. Use new or sterilized blades. Treat the exposed tissues with a wound paint containing octhilinone. Alternatively, fosetyl-aluminium* can be applied as a foliar spray, or applied directly onto the affected area as a paste.

These *Phytophthora* spp. may also infect and damage roots — see p. 446 and Table 11.2, p. 435.

Coral spot (*Nectria cinnabarina*)
See under *Acer*, p. 452.

Leaf blotch (*Guignardia aesculi*)
In spring, small irregular-shaped discoloured spots appear on the leaves. Later, the spots become brown with yellow margins and coalesce so that the whole leaf withers. Control measures are not usually necessary on established trees, but the disease can cause severe defoliation of young nursery stock. In severe cases, spray at HV with carbendazim* or prochloraz.

Ajuga (bugle)

Diseases

Crown rot (*Phoma exigua* var. *inoxydabilis*)
This fungus causes a firm, brown decay of the crowns, usually leading to plant collapse. Affected plants may also exhibit light-brown leaf spots, and lesions on the stolons. Sprays or drenches of carbendazim* or prochloraz should have some activity against this disease.

Leaf spot (*Ramularia ajugae*)
Leaf spots produced by *Ramularia* are usually also light-brown in colour but they may exhibit a whitish bloom under humid conditions, as a result of the production of large numbers of spores. Control with HV sprays of carbendazim*, chlorothalonil or one of the other fungicides listed in Table 11.3, p. 438, for the control of leaf spots.

Alnus (alder)

Diseases

***Phytophthora* root rot** (*Phytophthora* sp.)
This is a relatively new disease, which has caused considerable concern over the last few years. Common alder (*Alnus glutinosa*) is affected most frequently, although the disease has also been diagnosed on other species. Most cases have involved riverside alders, or trees growing on land that is subject to flooding from rivers. However, the disease has also been found on trees growing well away from any watercourse. Affected trees have thin, yellow foliage and may exhibit tarry or rusty spots on the stems. Eventually, the tree may die. The *Phytophthora* involved is thought to be a new hybrid, related to *P. cambivora*. There are currently few practical control measures – affected trees should be removed.

Althaea (hollyhock)

Diseases

Rust (*Puccinia malvacearum*)
See p. 450. As the rust can overwinter within the stool, it is best to raise new plants at least every other year. Apply HV sprays from the seedling stage onwards, at fortnightly intervals, using one of the fungicides listed in Table 11.3, p. 438, for control of rust.

Alyssum

See under Bedding plants, Chapter 10, p. 380.

Anemone

See Chapter 12, p. 542.

Anemone × *hybrida* (Japanese anemone)

Pests

Leaf nematode (*Aphelenchoides fragariae*)
Leaf nematodes attack the foliage of this and other Japanese anemones in late summer. Dark, black–brown islands develop within the leaves, which are

separated from healthy areas by the larger leaf veins. The plants survive but look unsightly at flowering time. There is no effective control on established plants. Propagate from plants that are free of symptoms and treat rooted propagation material with aldicarb (off-label) (SOLA 1325/95).

Antirrhinum (snapdragon)

Pests

Aphids

Various species, including peach/potato aphid (*Myzus persicae*) and potato aphid (*Macrosiphum euphorbiae*) infest snapdragon. They are typically green, yellow or pinkish in colour and cause stem distortion and stunting. The aphids may be controlled by the sprays listed on p. 441, but do not use malathion.

Diseases

See under Bedding plants, Chapter 10, p. 380.

Aquilegia (columbine)

Pests

Columbine sawfly (*Pristiphora aquilegiae*)

From May to September there are three generations of pale-green caterpillars, which may cause complete defoliation. Control when damage is seen by spraying plants with cypermethrin, malathion or nicotine.

Diseases

Leaf spot (*Marssonina aquilegiae*)

Irregular, pale-brown blotches with a narrow, darker border develop on the leaves. Affected tissues may wither and fall away to give a shot-hole effect. Remove and burn affected leaves. In severe cases, spray HV with carbendazim*, chlorothalonil or prochloraz, or with another fungicide listed for the control of leaf spots in Table 11.3, p. 438.

Powdery mildew (*Erysiphe aquilegiae* var. *aquilegiae*)

See p. 449 and Table 11.3, p. 438.

Aruncus (goat's beard)

Pests

Aruncus sawfly (*Nematus spiraeae*)
The pale-green, gregarious caterpillars may defoliate plants, with two or three generations occurring from May to September. Control as for columbine sawfly, p. 456, when caterpillars are present.

Aster (Michaelmas daisy)

Pests

Cyclamen mite (*Phytonemus pallidus*)
These microscopic mites (on this host sometimes known as 'Michaelmas daisy mite') live and feed inside the buds and in the folds of young leaves. Heavy infestations result in stunted growth, brown scarring along the stems and the production of small, green rosettes instead of flowers. Cuttings propagated under glass can be treated with dicofol + tetradifon*. A wetter should be added. Abamectin will also control this pest. *Aster novi-belgii* is particularly susceptible; *Aster novae-angliae* and *A. amellus* are resistant to this pest.

Diseases

Michaelmas daisy wilt (*Phialophora asteris*)
Some shoots in a clump show browning of the leaves, starting at the base of the stem and spreading upwards. The withered leaves remain hanging on the dead shoots. Destroy badly affected plants and take tip cuttings from healthy plants only. Carbendazim (off-label) (SOLA 0009/99) and prochloraz may have some activity when applied as a drench, but are unlikely to control existing infections.

Powdery mildew (*Erysiphe cichoracearum* var. *cichoracearum*)
See p. 449 and Table 11.3, p. 438.

Azalea

See under *Rhododendron*

Begonia (outdoor types)

See under Bedding plants, Chapter 10, p. 380.

Berberis (barberry)

See also *Mahonia*

Diseases

Bacterial leaf spot (*Pseudomonas syringae* pv. *berberidis*)
Small, round (2–4 mm diameter), purple-black spots develop on the leaves. The spots coalesce and cause yellowing of the foliage and premature leaf fall. Copper fungicides, if applied as protectant sprays, may check disease spread. Copper ammonium carbonate* and copper oxychloride* can be used on ornamentals (both outdoors and under protection). The professional formulation of Bordeaux mixture* can also be used on ornamentals (but outdoors only). Avoid overhead watering.

Honey fungus (*Armillaria* spp.)
See p. 443.

Betula (birch)

Pests

Aphids (various species)
These insects and the sooty mould associated with them are frequently a problem. For control see p. 441.

Diseases

Rust (*Melampsoridium betulinum*)
See p. 450. Young nursery stock or old, weak trees are most susceptible. The disease is unimportant on well-established trees, but protect seedlings by spraying from bud burst to the end of September with one of the fungicides listed for rust control in Table 11.3, p. 438.

Buddleja (butterfly bush)

Pests

Chrysanthemum nematode (*Aphelenchoides ritzemabosi*)
Affected plants sometimes show typical dark necrotic patches on the leaves but more usually on this plant the feeding damage occurs within the buds rather than in expanded leaves. This results in shoot tips drying up and dying; flower spikes are small with many necrotic flowers. Destroy infested plants and start again with clean stock. Propagate from healthy plants and, as an added precaution, treat rooted cuttings with aldicarb (off-label) (SOLA 1325/95).

Diseases

Downy mildew (*Peronospora hariotii*)
See p. 448, Table 11.2, p. 435, and Table 11.3, p. 438. This disease often causes quite angular, brown lesions on the upper leaf surface, which can be confused with those produced by chrysanthemum nematode. Sporulation of the fungus can also be difficult to see, owing to the hairy nature of the leaf undersides.

Buxus (box)

Pests

Box sucker (*Psylla buxi*)
The damage is caused by the pale-green nymphs, which suck sap from the shoot tips in the spring. This results in stunted shoots with the leaves cupped together in a cabbage-like manner. This damage can be tolerated on plants that have already grown to the required height but young plants may need treatment. There are no label recommendations for dealing with this pest but spraying with cypermethrin*, deltamethrin* or dimethoate* against newly emerged nymphs in April, when new growth begins, will prevent damage.

Diseases

Cylindrocladium leaf blight (*Cylindrocladium sp.*)
A recent discovery in the UK, this disease causes browning or purpling of the leaves, coupled with a profuse, whitish, powdery sporulation of the fungus on the leaf undersides under humid conditions. Affected leaves are often shed. To control, dispose of fallen leaves and spray HV with carbendazim* or prochloraz.

Phytophthora root rot (*Phytophthora* spp.)
See p. 446 and Table 11.2, p. 435.

Rust (*Puccinia buxi*)
See p. 450 and Table 11.3, p. 438.

Volutella blight (*Pseudonectria rousseliana* – anamorph: *Volutella buxi*)
The first symptoms of this disease are browning and death of individual leaves or twigs. Later, entire branches may die back – the leaves of such branches often remain attached for some time, although they eventually fall. A blackish internal discoloration is apparent within the affected branches, and the bark sloughs away very easily. The leaves often have pink spore masses of the fungus on their undersides. The fungus is thought to infect in spring via wounds, such as those created during hedge clipping. To control the disease, remove affected branches

and fallen leaves just before growth starts in spring. Spray immediately with prochloraz or with a copper-containing fungicide (for more details on copper fungicides see under *Berberis*, bacterial leaf spot, p. 458). Further fungicide applications in May, June and August may be beneficial.

Calendula (marigold)

See under Bedding plants, Chapter 10, p. 380.

Callistephus (China aster)

Pests

Aphids
Plum leaf-curling aphid (*Brachycaudus helichrysi*) is a frequent pest. The small green aphid infests the shoot tips, causing stunted shoot and leaf growth. Infested leaves often have a purple discoloration. Control as recommended for aphids on p. 441.

Capsids (e.g. *Lygus rugulipennis*)
See under *Dahlia*, p. 473.

Cutworms (e.g. *Agrotis segetum* and *Noctua pronuba*)
These mainly soil-inhabiting caterpillars chew away the outer tissues at the base of the stem. Extensive girdling results in plants wilting and dying. Large larvae are difficult to kill but some control can be achieved by treating the soil with chlorpyrifos or cypermethrin in June/July or when damage is seen.

Diseases

Damping-off, foot and root rot (*Pythium* spp. and *Phytophthora* spp.)
See pp. 446–447 and Table 11.2, p. 435.

Fusarium wilt (*Fusarium oxysporum* f. sp. *callistephi*)
Blackening of the stems occurs above ground level and a pink fungal growth develops on the tissues. Affected plants wilt, usually just before flowering. Destroy diseased plants and grow wilt-resistant cultivars. It may be possible to save adjacent plants by drenching with carbendazim (off-label) (SOLA 0009/99) or prochloraz.

Wirestem, damping-off, foot and root rot (*Thanatephorus cucumeris* – anamorph: *Rhizoctonia solani*)
See p. 447 and Table 11.3, p. 438.

Calluna (heather)

Diseases

Leaf blight, foot and root rot (*Thanatephorus cucumeris* – anamorph: *Rhizoctonia solani*)
See p. 447 and Table 11.3, p. 438. This can be a very damaging pathogen of *Calluna*, causing browning and die-back of the foliage and/or root or stem-base decay.

Needle blight and stem die-back (*Pestalotiopsis* spp.)
The fungus usually infects its host via wounds, or weak or dead tissue, and may be associated with damage from herbicides, weather, or pruning. Infected foliage becomes yellow and then brown and, as the fungus spreads into stems, die-back may kill the plant. The 'pinhead' fruiting bodies (acervuli) produced on infected tissue exude a black mass of spores, which are spread in water splash. Overhead irrigation and dense spacing of crops on container-beds increase disease spread. HV sprays of carbendazim* or prochloraz should be applied as a protectant treatment immediately before or after 'cutting over' the plants. Repeat every 3–4 weeks if wet, humid weather persists. For treatment of cuttings see under *Camellia*, leaf and stem blight, p. 462.

Phytophthora root rot (*Phytophthora* spp.)
See p. 446 and Table 11.2, p. 435. Heather plants are highly susceptible to *Phytophthora*. Infested soil splashed onto leaves and stems can be a means of transferring the pathogen to cutting material.

Camellia

Pests

Cushion scale (*Chloropulvinaria floccifera*)
This insect sucks sap from the larger leaf veins on the undersides of leaves. The pests are flat, yellowish-brown, oval insects up to 3 mm long. Cushion scale lays its eggs in early summer, in a long, white, waxy smear which remains visible long after the eggs have hatched. Large quantities of honeydew are produced, which soil the upper leaf surface and allow the growth of sooty moulds. Control by spraying, when young nymphs are present, with deltamethrin or malathion. Cushion scale usually has one generation per year on outdoor camellias with young nymphs being present from July onwards.

Vine weevil (*Otiorhynchus sulcatus*)
See p. 443.

Diseases

Flower blight (*Ciborinia camelliae*)
Long known in the US, this disease has recently been detected for the first time in the UK. First symptoms are brown specks on the petals (which can be confused with cold damage or grey mould infection). Later, a more advanced brown rot of the flower occurs, particularly at the base. If the calyx is removed, a white ring of fungal growth can be seen around the base of the petals. Infected flowers may remain on the bush or fall to the ground. The fungus produces resting bodies or sclerotia on the petals, which can overwinter in the soil. In the spring the sclerotia germinate to produce spores which re-infect the flowers. Camellia flower blight is a notifiable disease, and its presence should be reported to the Plant Health and Seeds Inspectorate (PHSI) of MAFF.

Gall (*Exobasidium camelliae*)
The flowers and sometimes leaves are converted into large irregular galls. Remove the galls and spray as for Rhododendron, gall, p. 517.

Leaf and stem blight (*Monochaetia karstenii* and *Pestalotiopsis guepini*)
These fungi colonize weak, damaged and dead tissue, causing 'scorched' or necrotic lesions on leaves and stems. The 'pinhead' acervuli develop on the infected tissue and exude a black mass of spores which are spread in water splash. Infected leaves drop prematurely and stem infection causes die-back and often death of young plants. High humidity encourages the disease, and spores carried from 'mother' plants may infect and cause rotting of cuttings during propagation. Remove lesioned leaves and spray HV with carbendazim* or prochloraz. On cuttings propagated under mist, fungal rotting will be reduced if a fungicide drench of prochloraz is applied to the cuttings and rooting medium at insertion and again 2–3 weeks later. If using the compound on a cultivar for the first time, test the application on a small scale first, as some fungicide drenches have been shown to affect the rooting of cuttings.

Leaf blotch, twig blight, canker and die-back (*Glomerella cingulata* f. sp. *camelliae*)
Leaf blotches produced by this fungus are dull-brown, circular to begin with, but then becoming more irregular in shape. Fruiting bodies may occur as tiny black pimples, often in concentric rings within the spots. Infection via leaf scars may lead to twig die-back, whilst stem cankers may occur on other shoots which, if girdled, can cause more extensive die-back. Hybrid camellias are particularly prone to attack. Fallen leaves should be removed, and any blighted twigs or branches pruned out. Spray HV with carbendazim* or prochloraz.

Phytophthora root rot (*Phytophthora* spp.)
See p. 446 and Table 11.2, p. 435.

Campanula (bellflower)

Diseases

Leaf spot (*Ascochyta bohemica*)
Greenish-brown circular spots, up to 13 mm in diameter with a purplish-brown margin, develop on the leaves. These spots may coalesce and the leaves eventually wither. Sometimes the fungus attacks only the stems and petioles, causing the leaves to become distorted and to develop yellow blotches. The disease can be controlled by spraying HV with carbendazim*, chlorothalonil or prochloraz.

Rust (*Coleosporium tussilaginis* f. sp. *campanulae*)
See p. 450 and Table 11.3, p. 438.

Caryopteris

Pests

Capsids (*Lygogoris pabulinus* and *Lygus rugulipennis*)
These sap-feeding pests attack the shoot tips and cause the foliage to be distorted with many small holes. For control see under *Fuchsia*, p. 482.

Castanea (sweet chestnut)

Diseases

Phytophthora root rot or 'ink disease' (*Phytophthora cinnamomi* and *P. cambivora*)
This disease primarily affects the roots, which may produce a blue-black, inky stain (hence the common name). Cankers may form on the large roots and at the base of the trunk. Foliar symptoms are dependent upon the degree of root or stem-base decay. Leaves may be very small, or they may be of normal size but turn yellow and, finally, brown. In some cases, death of the tree can be rapid (within a year of the appearance of foliar symptoms); in others there will be a long, slow decline. Like so many diseases caused by *Phytophthora*, this one is worst on poorly drained soils prone to waterlogging. Such soils should be avoided when planting sweet chestnut. Trees affected by the disease should be removed and destroyed and, if possible, replaced with a more resistant species. Also, see p. 446 and Table 11.2, p. 435.

Catalpa (Indian bean tree)

Diseases

Verticillium wilt (*Verticillium dahliae*)
See under *Acer*, p. 453.

Ceanothus (Californian lilac)

Pests

Scale insects
Various species, e.g. brown scale (*Parthenolecanium corni*) and willow scale (*Chionaspis salicis*), infest *Ceanothus*. The former species has a convex, oval, brown shell, 3–6 mm long when mature; the latter is greyish-white, flatter, 3 mm long and pear shaped. Control as for Cushion scale, p. 461, in early July.

Diseases

Honey fungus (*Armillaria* spp.)
See p. 443.

Phytophthora branch die-back (*Phytophthora citricola*)
Individual branches suffer from a brown die-back, which may occasionally work its way back into the main stem. If a girdling lesion is produced, death of the plant above that point occurs. The roots of affected plants often appear completely healthy. The disease is worst when the foliage remains wet for long periods (e.g. from overhead watering), and spores (sporangia) produced on the surface of affected branches may be splashed on to adjacent plants. Control measures usually applied for downy mildew control are likely to be most effective against this type of *Phytophthora* disease – see p. 448, Table 11.2, p. 435, and Table 11.3, p. 438.

Verticillium wilt (*Verticillium dahliae*)
See under *Acer*, p. 453.

Centaurea (cornflower)

Diseases

Petal blight (*Itersonilia perplexans*)
In wet summers, small, oval, water-soaked spots develop on the outer petals, spread rapidly and spoil blooms. Commence spraying with zineb when the buds first show colour and repeat at weekly intervals.

Powdery mildew (*Oidium* sp.)
See p. 449 and Table 11.3, p. 438.

Rust (*Puccinia cyani*)
See p. 450 and Table 11.3, p. 438.

Cercis (includes Judas tree)

Diseases

Coral spot (*Nectria cinnabarina*)
See under *Acer*, p. 452. This fungus is common on dying plants.

Verticillium wilt (*Verticillium dahliae*)
See under *Acer*, p. 453.

Chaenomeles (Japanese quince)

Pests

Brown scale (*Parthenolecanium corni*)
See under *Ceanothus*, p. 464.

Diseases

Fireblight (*Erwinia amylovora*)
See under Pear, Chapter 8, p. 281.

Chamaecyparis, *Cupressus* and × *Cupressocyparis leylandii* ('false', true and hybrid cypresses)

Pests

Cypress aphid (*Cinara cupressi*)
This comparatively large, greyish-brown aphid, up to nearly 4 mm long, sucks sap from the stems in early to late summer. Sooty mould may develop on the stems and foliage; even light infestations can result in the foliage drying up and turning yellowish-brown from mid-summer onwards. This is particularly seen on the lower parts of hedging conifers. Control is not easy, owing to the difficulty of spraying into compact growth such as occurs on large, clipped hedges. Treatment with suitable insecticides (see p. 441) in early summer may prevent damage.

Diseases

Coryneum canker (*Seiridium cardinale*)
This fungus can cause large cankers on twigs, branches and the main stem, resulting in obvious areas of dead foliage in the crown of affected trees. *Cupressus*

macrocarpa and *C. sempervirens* are very susceptible, and the disease has also been reported quite frequently on × *Cupressocyparis leylandii*. Infection occurs via small areas of damage on the bark. Clusters of small, black fruiting bodies may be seen within the affected areas.

If the damage is not too severe, cut out affected areas and protect with a wound paint based on azaconazole + imazalil or octhilinone.

Fomes root and butt rot (*Heterobasidion annosum*)
See under *Pinus*, p. 506.

Grey mould (*Botryotinia fuckeliana* – anamorph: *Botrytis cinerea*)
See p. 448 and Table 11.3, p. 438. Species with dense, feathery foliage (e.g. *Chamaecyparis pisifera* 'Boulevard') are susceptible to grey mould, particularly following frost damage, when foliar infections may spread into stems and cause die-back.

Kabatina shoot blight (*Kabatina* spp.)
See under *Juniperus*, p. 490.

Needle blight and stem die-back (*Pestalotiopsis* spp.)
See under *Calluna*, p. 461.

Phytophthora root rot (*Phytophthora* spp.)
See p. 446 and Table 11.2, p. 435.

Cheiranthus (wallflower)

Pests

Cabbage root fly (*Delia radicum*)
The white maggots, up to 8 mm long, eat the roots and cause stunted growth and wilting during the summer. For possible control measures, see under Brassica crops, Chapter 7, p. 192, but none of the chemicals listed for vegetable crops carries a recommendation for root fly control on ornamentals.

Flea beetles (*Phyllotreta* spp.)
Seedlings are attacked by tiny, shiny, blue-black beetles which sometimes have a pair of yellow stripes on their elytra. Damaged leaves have holes scalloped in them and heavy attacks may severely check growth during the early stages. Cypermethrin* or deltamethrin*, as used on vegetable brassicas (Chapter 7, p. 198), will protect seedlings if damage is seen, but this usage would be at the grower's risk on ornamental plants.

Diseases

Alternaria leaf spot (*Alternaria* sp.)
The dark-brown leaf spots produced by this pathogen can be controlled by HV sprays of iprodione*. The same compound (with on-label approval) can be used as a seed treatment.

Bacterial leaf spot, leaf rot and stem rot (*Xanthomonas campestris* pv. *campestris*)
This seed- and debris-borne pathogen causes stunting and numerous, tiny, necrotic leaf spots, resulting in leaf distortion and chlorosis. The disease is usually systemic and cannot be controlled by fungicides. Dispose of affected plants and obtain pathogen-tested (i.e. pathogen-free) seed if possible.

Black root rot (*Thielaviopsis basicola* – synanamorph: *Chalara elegans*)
See p. 447 and Table 11.3, p. 438.

Clubroot (*Plasmodiophora brassicae*)
The roots develop irregular swellings and the plants remain stunted. Where the disease is present, its severity can be reduced by raising the pH to 7.0 or above by liming. A test is available from some laboratories to check intended planting sites for the presence of clubroot.

Damping-off, foot and root rot (*Phytophthora* spp. and *Pythium* spp.)
See pp. 446–447 and Table 11.2, p. 435.

Downy mildew (*Peronospora parasitica*)
See p. 448, Table 11.2, p. 435, and Table 11.3, p. 438.

Wirestem, damping-off, foot and root rot (*Thanatephorus cucumeris* – anamorph: *Rhizoctonia solani*)
See p. 447 and Table 11.3, p. 438.

Choisya (Mexican orange blossom)

Pests

Snails
Snails, e.g. garden snail (*Helix aspersa*), and to a lesser extent slugs, climb up the plants at night to feed on the foliage and young stems. The latter are often girdled, resulting in die-back of the shoots. Protect young plants by scattering metaldehyde or methiocarb pellets on the soil.

Two-spotted spider mite (*Tetranychus urticae*)
Heavy infestations of this small, sap feeding pest result in a fine, pale mottling of the upper surface of leaves. For controls see p. 442.

Diseases

Black root rot (*Thielaviopsis basicola* – synanamorph: *Chalara elegans*)
See p. 447 and Table 11.3, p. 438.

Phytophthora root rot (*Phytophthora* spp.)
See p. 446 and Table 11.2, p. 435.

Clarkia (outdoor crops)

See under Bedding plants, Chapter 10, p. 380.

Clematis

Pests

Common earwig (*Forficula auricularia*)
See under *Dendranthema* (chrysanthemum), p. 476.

Diseases

Powdery mildew (*Erysiphe aquilegiae* var. *aquilegiae*)
See p. 449 and Table 11.3, p. 438.

Wilt (*Phoma clematidina*)
This disease is common on large-flowered, hybrid clematis. Small-flowered clematis plants are rarely affected. One or more shoots die back rapidly, often to the base of the plant (although new shoots may sometimes develop). Rotting of the stem base is usually found to be the cause of the problem. The fungus can also cause rotting higher up the stem, as well as root decay and leaf spotting.

Recent HDC-funded research has highlighted a number of cultural measures that will reduce the risk of serious problems from clematis wilt. Good general nursery hygiene is a major factor. Dead plant material should be collected and destroyed on a regular basis. Leaves with spots should be removed when found, and pruning implements should be disinfected between batches of plants.

Moist conditions are favourable for *P. clematidina*, both for disease spread and infection, and the development of stem rots and leaf spots. Therefore, try to avoid lengthy periods of leaf wetness.

HV sprays of carbendazim or prochloraz are frequently used for the control of clematis wilt, although some isolates of *P. clematidina* are resistant to carbendazim. Azoxystrobin* has shown great promise in the HDC-funded work, and can be used outdoors at the grower's risk, although it should be used with caution and evaluated on a small scale first.

Convallaria (lily of the valley)

Pests

Vine weevil (*Otiorhynchus sulcatus*)
See p. 443.

Diseases

Grey mould (*Botryotinia fuckeliana* – anamorph: *Botrytis cinerea*, and *Botrytis paeoniae*)
Brown blotches develop on the leaves, which may rot completely, but no fungal growth may show on the affected tissues. For control use HV sprays of a fungicide listed for grey mould control in Table 11.3, p. 438.

Convolvulus cneorum

Diseases

Phytophthora root rot (*Phytophthora* spp.)
See p. 446 and Table 11.2, p. 435.

Stem rot and foliar blight (*Thanatephorus cucumeris* – anamorph: *Rhizoctonia solani*)
See p. 447 and Table 11.3, p. 438. The fungus causes dry, brown stem lesions, and leaf necrosis that leads to drop. Under humid conditions silvery or brownish mycelial 'webbing' may be visible.

Cordyline (cabbage palm)

Diseases

Phytophthora root rot and crown rot (*Phytophthora* sp.)
See p. 446 and Table 11.2, p. 435. In addition to causing root decay, this pathogen is also sometimes associated with a firm, brown crown rot.

Cornus (dogwood)

Diseases

Bacterial leaf spot and die-back (*Pseudomonas syringae* pv. *syringae*)
Circular or angular, brown spots develop on the leaves. Branch die-back may occur. Avoid overhead watering where the disease is present and treat with a

copper-containing fungicide (see under *Berberis*, bacterial leaf spot, p. 458, for more details of these fungicides). Because this plant is also affected by a number of fungal leaf spots (see below) correct diagnosis is important.

Dogwood anthracnose (*Discula* spp.)
This disease, well known in the US, has only recently caused problems in the UK. It principally affects the flowering dogwoods *Cornus florida* and *C. nuttallii.* The fungus produces a range of symptoms. Leaves may be affected by small spots (brown centres with reddish-purple edges), large blotches (also brown with reddish-purple margins) or complete necrosis (leaf blight). Stem cankers and twig die-back can also occur. Minute fungal fruiting bodies (reddish-brown to black) may just be visible with a hand lens within the affected tissue.

Research work in the US has indicated that chlorothalonil, mancozeb*, propiconazole* and tebuconazole* all have activity against *Discula.* Sprays should be applied at bud burst, with follow-up treatments at least twice thereafter as the leaves are expanding.

As this is a relatively new disease, and the leaf spot symptoms may be confused with those caused by other pathogens (see below), suspected cases of dogwood anthracnose should be confirmed by a diagnostic laboratory.

Fungal leaf spots (*Phyllosticta cornicola* and *Septoria cornicola*)
Both of these fungi cause smallish spots with pale centres and reddish purple margins. Both are readily spread by water-splash. If control is required, prochloraz should have some activity against both pathogens.

Powdery mildew (*Erysiphe tortilis*)
See p. 449 and Table 11.3, p. 438.

Cotinus (smoke tree)

Diseases

Grey mould (*Botryotinia fuckeliana* – anamorph: *Botrytis cinerea*)
Grey mould frequently causes rotting of soft-wood cuttings. Chlorothalonil or iprodione* as sprays, or prochloraz, as a spray or drench, will help prevent decay during rooting of the cuttings.

Powdery mildew (*Oidium* sp.)
See p. 449 and Table 11.3, p. 438.

Verticillium wilt (*Verticillium dahliae*)
See under *Acer*, p. 453.

Cotoneaster

Pests

Brown scale (*Parthenolecanium corni*)
See under *Ceanothus*, p. 464.

Hawthorn webber moth (*Scythropia crataegella*)
Small reddish-brown caterpillars spin a fine but dense silk webbing over the branches. Leaves within the webbed area are grazed, turn brown and die. There is one generation per year, with young larvae active in August/September before overwintering and completing feeding in late April/June. *Crataegus* is also a common host plant. Caterpillars of porphyry knothorn moth (*Numonia suavella*) can also form webs on *Cotoneaster*.

Control by cutting out webbed branches or by treating with cypermethrin, diflubenzuron, or teflubenzuron, or use biological control with *Bacillus thuringiensis*. Avoid spraying insecticides while the shrub is in flower, because of the danger to bees.

Woolly aphid (*Eriosoma lanigerum*)
This pest causes white woolly patches on the stems, which are often swollen at these points. For control, see under Apple, Chapter 8, p. 270, but select a pesticide listed under Aphids, p. 441.

Diseases

Fireblight (*Erwinia amylovora*)
See under Pear, Chapter 8, p. 281.

Honey fungus (*Armillaria* spp.)
See p. 443.

Silver leaf (*Chondrostereum purpureum*)
See under *Prunus*, p. 514.

Crataegus (hawthorn)

Pests

Caterpillars
Caterpillars of various species are associated with hawthorn. For details of hawthorn webber moth (*Scythropia crataegella*), see under *Cotoneaster*, above.

Other caterpillars, e.g. those of brown-tail moth (*Euproctis chrysorrhoea*), common small ermine moth (*Yponomeuta padella*), lackey moth (*Malacosoma neustria*), and vapourer moth (*Orgyia antiqua*) also occur. The first three species all spin silk webs over the infested branches and cause localized defoliation. They are controlled by spraying when webbing or larvae are first noticed, usually in early summer, but in August for brown-tail moth. The chemicals used against hawthorn webber moth (p. 471) are effective. The selective insecticides diflubenzuron, teflubenzuron and *Bacillus thuringiensis* have been used successfully against the brown-tail moth, and would probably also deal with the other species. For brown-tail moth, prune out and destroy the caterpillars' overwintering silken nests.

Diseases

Fireblight (*Erwinia amylovora*)
See under Pear, Chapter 8, p. 281.

Leaf blotch (*Monilinia johnsonii*)
This fungus causes dark-brown to black, irregular blotches on the leaves in the spring. Later, similarly coloured dead leaves hang down limply, and in humid conditions these may become covered with a brown fungal growth. Often, a single leaf is affected in amongst healthy growth. The fungus also affects the fruit and mummifies them. This disease seldom does serious damage and even its cosmetic effect is usually limited, as the affected leaves are quickly masked by the surrounding healthy ones. Control measures, therefore, are unwarranted. It is as well to be aware of the disease, however, as the leaf blackening it causes could be mistaken for symptoms of fireblight.

Leaf spot (*Diplocarpon mespili* – anamorph: *Entomosporium maculatum*)
Numerous, small, brown spots develop on the leaves. Spray with one of the treatments listed for rose black spot and leaf spots in Table 11.3, p. 438.

Powdery mildew (*Podosphaera tridactyla*)
See p. 449 and Table 11.3, p. 438. In the case of newly planted hedges or young soft shoots on trees/bushes, this disease can be very injurious.

Crocus

See Chapter 12, p. 545.

Cupressocyparis

See *Chamaecyparis*, p. 465.

Cupressus

See *Chamaecyparis*, p. 465.

Cytisus (broom)

Pests

Broom gall mite (*Eriophyes genistae*)
This microscopic pest infests the buds and causes them to enlarge into cauliflower-like structures. There are no recommended control measures and badly infested shrubs should be destroyed.

Diseases

Leaf spot (*Pleiochaeta setosa*)
Dark spots develop on the leaves and stems, followed by rapid destruction of the plants in the case of seedlings or cuttings. To control, spray with one of the treatments listed for control of rose black spot and leaf spots in Table 11.3, p. 438.

Dahlia

Pests

Aphids (e.g. *Aphis fabae* and *Brachycaudus helichrysi*)
Black or green aphids infest the shoot tips, flower buds and undersides of the leaves. This can result in stunted growth, poor flowering and transmission of virus diseases. Early treatment is necessary to avoid these problems and several applications of insecticide will be required during the growing season. Suitable chemicals include cypermethrin, deltamethrin, dimethoate, imidacloprid, malathion, nicotine, pirimicarb and pymetrozine.

Capsids (e.g. *Lygocoris pabulinus* and *Lygus rugulipennis*)
Capsids are pale-green, or brown, active insects up to 5 mm long that suck sap from the shoot tips and flower buds. Their toxic saliva causes malformed leaves to develop that have many small holes. The flowers may also be misshapen. Damage can occur from late May onwards. Keep the ground free of weeds that can act as alternative host plants. Control capsids with cypermethrin, deltamethrin or nicotine as soon as their activity is detected.

Caterpillars
Various species, including cabbage moth (*Mamestra brassicae*) and angle-shades moth (*Phlogophora meticulosa*), although usually present in small numbers, may

damage the foliage and flowers. If necessary, treat with cypermethrin, diflubenzuron or teflubenzuron, or use *Bacillus thuringiensis.*

Common earwig (*Forficula auricularia*)
See under *Dendranthemum* (chrysanthemum), p. 476.

Onion thrips (*Thrips tabaci*)
These are thin orange-yellow or black insects up to 2 mm long which live in the flowers and between the folds of new leaves. Their sap-feeding causes flecking of the petals but, more seriously, their feeding at the shoot tips results in misshapen leaves, and scarring and stunting of stems. Damage occurs mainly from late May to July. Control with cypermethrin, deltamethrin, malathion or nicotine.

Two-spotted spider mite (*Tetranychus urticae*)
Large numbers of these small, sap-feeding pests can occur on the undersides of leaves. Affected foliage is at first lightly speckled but it soon becomes yellow or bronzed. For suitable controls see p. 442.

Diseases

Grey mould (*Botryotinia fuckeliana*)
See p. 448. This fungus can be damaging to flowers and buds in moist, warm weather. Remove infected tissue and spray with a compound listed for grey mould in Table 11.3, p. 438.

Sclerotinia rot (*Sclerotinia sclerotiorum*)
Affected plants wilt owing to the presence of the fungus, seen as a dense, white, fungal growth with large, black, resting bodies (sclerotia) within the decaying stems. The fungus may also attack the tubers in store, causing them to rot. Destroy affected plants at the first signs of trouble.

Smut (*Entyloma calendulae* f. sp. *dahliae*)
Circular, brown spots on the lower leaves spread up the plant, and where spots coalesce large areas may be killed and the leaf withers. The disease is likely to be serious only on small bedding cultivars when planted closely. Fungicides containing copper may have some activity (see under *Berberis*, bacterial leaf spot, p. 458, for more details of these fungicides).

Daphne

Diseases

Crown gall (*Agrobacterium tumefaciens*)
Galls develop as irregular or round swellings at intervals along the stems or on

the upper roots. Avoid damaging the plant, because this soil-inhabiting bacterium readily infects wounded tissue; if galls do occur, remove them with a clean cut and burn them.

Leaf spot (*Marssonina daphnes*)
This disease causes spotting on leaves (especially at the base and on the petiole), resulting in appreciable defoliation. Prochloraz and copper-containing fungicides may have activity against this pathogen. Spraying should commence as the leaves open in spring and be repeated once or twice, at fortnightly intervals.

Delphinium

Pests

Delphinium leaf miner (*Phytomyza aconiti*)
The small, white maggots live gregariously within the leaf, causing large, brown, blotch mines. Each mine may contain up to 12 larvae. Damage starts on the lower leaves in May but subsequent generations attack higher up the plant. Monkshood (*Aconitum napellus*) is also attacked. The pupal stage occurs within the mine and is relatively tolerant of sprays. Control the pest by spraying with abamectin, dimethoate or nicotine when signs of damage first appear. With light infestations, removing mined leaves may eliminate the need for spraying.

Diseases

Black blotch (*Pseudomonas syringae* pv. *delphinii*)
This pathogen causes large, black or brownish-black blotches on leaves, stems or even flower buds, but it seems to be very selective as some cultivars are badly attacked whereas others nearby are not affected. It is a difficult disease to check but repeated HV spraying with copper-containing fungicides will protect young plants if applied as soon as the new shoots appear. See under *Berberis*, bacterial leaf spot, p. 458, for more details of copper fungicides. Avoid overhead irrigation, as this will splash the bacteria from plant to plant.

Powdery mildew (*Erysiphe aquilegiae* var. *ranunculi*)
See p. 449 and Table 11.3, p. 438.

Dendranthema (chrysanthemum) (outdoor crops)

Pests

Aphids
Several species occur on chrysanthemum, e.g. melon & cotton aphid (*Aphis gossypii*), peach/potato aphid (*Myzus persicae*) and plum leaf-curling aphid

(*Brachycaudus helichrysi*). These cause stunted growth, poor blooms and problems with honeydew and sooty mould. They can be controlled with cypermethrin, deltamethrin, malathion, nicotine, pirimicarb or pymetrozine. Melon and cotton aphid and peach/potato aphid are often resistant to a wide range of pesticides. A soil drench of imidacloprid or spray of pymetrozine should be used against these aphids. Some cultivars may be damaged by sprays, especially when in bloom; see under *Dendranthema* (chrysanthemum), Chapter 10, p. 401.

Capsids
See under *Dahlia*, p. 473.

Caterpillars (e.g. *Phlogophora meticulosa*)
See under *Dahlia*, p. 473.

Chrysanthemum leaf miner (*Chromatomyia syngenesiae*)
The larvae create sinuous, whitish-brown mines in the foliage. For control use abamectin or nicotine when mines appear.

Chrysanthemum nematode (*Aphelenchoides ritzemabosi*)
Infestations cause wedge-shaped brown or black areas to develop between the leaf veins. Symptoms spread up the plant, especially in wet weather. Control with aldicarb (off-label) (SOLA 1325/95) or use HWT to produce clean cuttings.

Common earwig (*Forficula auricularia*)
Earwigs can damage the foliage but are mainly a problem on the blooms. Cypermethrin or deltamethrin, when used against other pests, will give incidental control of earwigs.

Diseases

Blotch or leaf spot (*Septoria* spp.)
See under *Dendranthema* (chrysanthemum), Chapter 10, p. 403.

Brown rust (*Puccinia chrysanthemi*)
See under *Dendranthema* (chrysanthemum), Chapter 10, p. 403.

Grey mould (*Botryotinia fuckeliana* – anamorph: *Botrytis cinerea*)
See Table 11.3, p. 438, for additional fungicides which may be used on plants outdoors.

Petal blight (*Itersonilia perplexans*)
See under *Centaurea*, p. 464.

Powdery mildew (*Oidium chrysanthemi*)
See p. 449 and Table 11.3, p. 438.

White rust (*Puccinia horiana*)
See under *Dendranthema* (chrysanthemum), Chapter 10, p. 405.

Dianthus (border carnation, pinks)

Pests

Aphids (e.g. *Myzus persicae*)
See under *Dahlia*, p. 473.

Diseases

Fusarium wilt (*Fusarium oxysporum* f. sp. *dianthi*)
See under *Dianthus* (protected crops), Chapter 10, p. 407, for a description, and control with a soil drench of carbendazim (off-label) (SOLA 0009/99) or prochloraz*.

Ring spot (*Mycosphaerella dianthi*)
This fungus causes round, grey spots within which the skin erupts into pustules arranged in roughly concentric rings. Infected leaves wither, stems snap and poor-quality flowers are produced. Difenoconazole (off-label) (SOLAs 1728/97, 1729/97) gives effective control of this disease.

Rust (*Uromyces dianthi*)
See p. 450 and Table 11.3, p. 438. This disease appears as brown spore clusters on both sides of the leaves and stems, usually starting on the lower leaves. To avoid staining, sprays should not be applied to open blooms.

Dianthus barbatus (sweet william)

Diseases

Ring spot (*Mycosphaerella dianthi*)
See under *Dianthus* (border carnation, pinks), above.

Rust (*Puccinia arenariae*)
See p. 450 and Table 11.3, p. 438. In autumn, this disease may spread rapidly and cause severe damage. Do not encourage soft, lush growth as this is attacked easily, and spray at the first sign of the disease.

Digitalis (foxglove)

Pests

Glasshouse leafhopper (*Hauptidia maroccana*)
The adults are pale-yellow with greyish dorsal markings and 3 mm long. Both the adults and the creamy-white nymphs suck sap from the lower leaf surface. They cause a coarse, pale mottling of the upper leaf surface and can be active all year round in sheltered places. Malathion and nicotine are approved for control of leafhoppers on ornamental plants.

Leaf nematode (*Aphelenchoides fragariae*)
This microscopic pest lives within the leaves and causes brown necrotic islands separated from healthy tissues by the larger leaf veins. Symptoms mainly appear on senescing leaves and can spread rapidly in wet conditions where plants are crowded. Destroy badly affected plants and avoid overcrowding. Young plants could be treated with aldicarb (off-label) (SOLA 1325/95).

Diseases

Downy mildew (*Peronospora digitalis*)
See p. 448, Table 11.2, p. 435, and Table 11.3, p. 438. The angular necrotic spots caused by this pathogen can be confused with those caused by leaf nematode. If fungal sporulation cannot be seen on the underside of the spots, laboratory diagnosis would be advisable.

Elaeagnus

Diseases

Coral spot (*Nectria cinnabarina*)
See under *Acer*, p. 452.

Erica (heath)

See *Calluna*, p. 461.

Escallonia

Diseases

Phytophthora root rot (*Phytophthora* spp.)
See p. 446 and Table 11.2, p. 435.

Silver leaf (*Chondrostereum purpureum*)
See under *Prunus*, p. 514.

Eucalyptus (gum tree)

Diseases

Phytophthora root rot (*Phytophthora* spp.)
See p. 446 and Table 11.2, p. 435.

Powdery mildew (*Oidium* sp.)
See p. 449 and Table 11.3, p. 438.

Silver leaf (*Chondrostereum purpureum*)
See under *Prunus*, p. 514.

Euonymus (spindle)

Pests

Black bean aphid (*Aphis fabae*)
Euonymus europaeus and other species of *Euonymus* are one of the primary (overwintering) hosts of this pest. Eggs are laid on the stems in the autumn. The eggs hatch at bud burst and the aphids then feed on the foliage until May. Leaves at the shoot tips become curled as a result of the aphids' sap-feeding activities. Control infestations, as for aphids on p. 441, shortly after bud burst.

Cushion scale (*Chloropulvinaria floccifera*)
See under *Camellia*, p. 461.

Euonymus scale (*Unaspis euonymi*)
This pest attacks the stems and leaves of *Euonymus japonicus*, mainly in coastal regions of southern England. The adults occur beneath whitish-brown, pear-shaped, bark-encrusting scales, each up to 2 mm long. Control by spraying with deltamethrin or malathion, preferably when young scales are present, usually in early July and early September.

Small ermine moths (*Yponomeuta* spp.)
Small, yellowish-grey caterpillars with black dots and black heads live gregariously beneath silken webbing, and cause defoliation in May and June. The most frequent species on spindle is *Yponomeuta cagnatella*. For further details, see under *Crataegus*, caterpillars, p. 471.

Vine weevil (*Otiorhynchus sulcatus*)
The adults often cause severe notching on the foliage of evergreen *Euonymus*. For further details, see under *Rhododendron*, wingless weevils, p. 517.

Diseases

Bacterial leaf spot (*Pseudomonas syringae* pv. *syringae*)
This pathogen causes smallish, brown spots, often circular (but occasionally more angular) and often with a 'greasy', greyish-green margin. If numerous spots are present, leaf distortion can occur and eventually the leaves may fall. Avoid excessive leaf wetness, and spray with a copper-containing fungicide (see under *Berberis*, bacterial leaf spot, p. 458, for more details of these fungicides). To reduce the risk of infection, dispose of fallen leaves.

Leaf spot (*Septoria euonymi*)
This pathogen causes pale, greyish-brown spots on the foliage. Minute fruiting bodies (pycnidia) may be just visible within the affected tissue. Spray with prochloraz or one of the fungicides listed for control of leaf spots in Table 11.3, p. 438.

Powdery mildew (*Oidium euonymi-japonicae*)
See p. 449 and Table 11.3, p. 438.

Fagus (beech)

Pests

Beech aphid (*Phyllaphis fagi*)
Dense colonies of this yellowish-green aphid (which is covered with white, waxy fibres) form on the shoot tips and undersides of the leaves. Peak populations occur in early summer but some aphids persist until late summer. The honeydew they excrete makes the foliage sticky and black with sooty mould. Beech hedges and young trees are more seriously affected than mature trees. Small trees and hedges can be treated when aphids appear in early summer with insecticides listed on p. 441.

Beech scale (*Cryptococcus fagisuga*)
This pest causes a white, powdery substance to appear in the crevices of the trunk and larger branches. Infestations are associated with beech bark disease (see disease section below). Usually, control measures are not required but where specimen trees are heavily infested it may be worth spraying (in September) with deltamethrin against the young nymphs.

Diseases

Beech bark disease (*Nectria coccinea*)
This 'disease' is actually the result of an interaction between a pest and a fungus. Prolonged infestation of beech scale (see p. 480) renders the bark susceptible to attack by *Nectria coccinea*, a weakly pathogenic fungus. In the early stages, small patches of bark are killed, oozing a sticky liquid which dries and turns black. Later, larger areas of bark are affected, which may crack or fall off. Within the affected areas, numerous reddish-brown, pinhead-sized pimples may develop – these are the fruiting structures of the *Nectria* fungus. Other wood-rotting fungi may eventually also colonize the affected tree.

As the disease cannot often gain access to the tree without the damage caused by beech scale (although see following paragraph), control in specimen trees depends on preventing and destroying the pest. In commercial woodlands, affected trees are usually processed before the decay can progress too far.

If the *Nectria* fungus is present but there is no evidence of current or past infestation by beech scale, it is likely that another factor has weakened or damaged the tree, allowing the fungus to colonize it. Drought stress is the most common route (after beech scale infestation) by which *Nectria coccinea* colonizes beech.

Coral spot (*Nectria cinnabarina*)
See under *Acer*, p. 452.

Phytophthora root rot and bleeding canker (*Phytophthora* spp.)
See p. 446 and Table 11.2, p. 435. Also, see under *Aesculus*, p. 453, for symptoms of bleeding canker.

Powdery mildew (*Phyllactinia guttata*)
See p. 449 and Table 11.3, p. 438.

Forsythia

Pests

Bullfinch (*Pyrrhula pyrrhula*)
This pest eats the flower buds during the winter, resulting in shoots that are devoid of flowers except for a few at the shoot tips. No really effective repellents are available. Netting vulnerable shrubs may be possible on a small scale.

Common green capsid (*Lygocoris pabulinus*)
This insect causes distorted foliage with many small holes at the shoot tips, damage occurring from late May onwards. Control is often unnecessary as the flowers are not affected, but capsids could be controlled as described under *Fuchsia*, p. 482.

Diseases

Bacterial blight (*Pseudomonas syringae* pv. *syringae*)
See under *Syringa*, p. 527.

Honey fungus (*Armillaria* spp.)
See p. 443.

Phytophthora root rot (*Phytophthora* spp.)
See p. 446 and Table 11.2, p. 435.

Freesia

See Chapter 10, p. 411.

Fremontodendron

Diseases

Phytophthora collar rot (*Phytophthora citricola*)
See p. 446 and Table 11.2, p. 435. A soft rotting of the tissues occurs at or just above soil level, causing foliar wilt and die-back.

Fuchsia

Pests

Common green capsid (*Lygocoris pabulinus*)
This serious pest of outdoor fuchsias not only causes tattered, distorted leaves at the shoot tips, but also abortion of flower buds. Damage occurs between May and early September. Control with cypermethrin, deltamethrin or nicotine as soon as damage is seen.

Other pests such as aphids and two-spotted spider mite (*Tetranychus urticae*) are described under *Fuchsia* (protected crops), Chapter 10, p. 412.

Diseases

See under *Fuchsia* (protected crops), Chapter 10, p. 413.

Gaillardia (blanket flower)

Diseases

Downy mildew (*Bremia lactucae*)
See p. 448 and Table 11.3, p. 438.

Galanthus (snowdrop)

See Chapter 12, p. 545.

Garrya

Diseases

Leaf spot (*Phyllosticta garryae*)
This disease causes large, grey spots with a purple border, but is likely to be troublesome only on shrubs lacking vigour. To control, remove severely affected leaves and, if necessary, spray HV with prochloraz or one of the compounds listed for leaf spot diseases in Table 11.3, p. 438.

Genista

Pests

Genista aphid (*Aphis genistae*)
Dense colonies of this grey aphid form on the shoots during the summer. Control as for aphids, p. 441.

Gentiana (gentian)

Diseases

Phoma root rot (*Phoma gentianae-sino-ornatae*)
This fungus can cause a severe, black root decay of *Gentiana sino-ornata*. The problem was first recorded in 1989 and the causal fungus (a new species) was identified in 1991. At first the disease caused devastating losses but HDC-funded research revealed that a combination of compost amendment with bark and the application of prochloraz (as a pre-planting dip, followed by compost drenches) gives good control of the disease.

Geranium (herbaceous species)

Diseases

Downy mildew (*Peronospora geranii*)
See p. 448, Table 11.2, p. 435, and Table 11.3, p. 438. In wet seasons this disease may be troublesome on some *Geranium* spp., especially cv. Johnson's Blue. The fungus shows as whitish growth on the underside of the leaves but the most

obvious symptoms are brown blotches on the upper surface which coalesce until the whole leaf shrivels.

Geranium rust (*Uromyces geranii*)
See p. 450 and Table 11.3, p. 438. This disease is found, occasionally, on *Geranium* spp., but it does not occur on pelargoniums, which are attacked by a different species of rust (*Puccinia pelargonii-zonalis*) (see Chapter 10, p. 418).

Geranium (outdoor bedding)

See under *Pelargonium*, p. 417.

Geum (avens)

Pests

Geum sawflies (*Claremontia waldeheimii* and *Monophadnoides rubi*)
Small, pale-green caterpillars, which have bifurcated bristles on their upper surface, cause severe defoliation during early summer. Control is as for Columbine sawfly, p. 456.

Diseases

Downy mildew (*Peronospora gei*)
See p. 448, Table 11.2, p. 435, and Table 11.3, p. 438.

Gladiolus

See Chapter 12, p. 546.

Gleditsia (honeylocust)

Pests

Honeylocust gall midge (*Dasineura gleditchiae*)
This small fly has two or three generations during the summer. It lays eggs on the developing leaves of *G. triacanthos,* causing the leaflets to form pod-like galls, each containing several orange-white larvae. This pest is unlikely to harm tall established trees and should be tolerated. There are no insecticides with label recommendations for gall midges on ornamental plants. The use of dimethoate*

against egg-laying midges in early June and early July should protect small trees from extensive galling.

Godetia

See under Bedding plants, Chapter 10, p. 380.

Hebe (veronica)

Diseases

Downy mildew (*Peronospora grisea*)
See p. 448, Table 11.2, p. 435, and Table 11.3, p. 438. This is one of the most damaging and troublesome diseases in containerized nursery stock production. The symptoms are variable, depending on species and cultivar. Susceptible large-leaved cultivars often show severe leaf distortion, with chlorosis or necrosis of the upper leaf surface and profuse sporulation of the fungus on the under surface. Small-leaved species may have very little sporulation visible, affected leaves simply turning necrotic and dropping.

Cultural control methods, such as wide spacing and the use of sub-irrigation, can help but intensive fungicide programmes (utilizing a range of compounds from different chemical groups) are still required on the most susceptible cultivars. Recent HDC-funded research has shown that azoxystrobin* has good activity against this disease: this compound can be used on *Hebe* at the user's risk.

Leaf spot (*Septoria exotica*)
This shows as white spots with a brown margin. Minute, black spore cases (pycnidia) are sometimes visible. Spray with one of the fungicides recommended for leaf spot diseases in Table 11.3, p. 438.

Phytophthora root rot (*Phytophthora* spp.)
See p. 446 and Table 11.2, p. 435. The foliage of some cultivars becomes chlorotic, and then black, as a consequence of severe root and stem base rotting.

Hedera (ivy)

Diseases

Bacterial leaf spot (*Xanthomonas campestris* pv. *hederae*)
The spots caused by this pathogen differ from those caused by the fungal pathogens described below, in that they are usually dark-brown to black and

'greasy' looking, may be surrounded by a yellow halo and do not contain fruiting bodies. Control is by removal and destruction of affected leaves, avoidance of prolonged leaf wetness and the use of copper-containing fungicides (see under *Berberis*, bacterial leaf spot, p. 458, for more details of these fungicides). The water used for HV spraying of other fungicides will actually spread bacterial leaf spot, so it is important that leaf spots on ivy are correctly identified.

Fungal leaf spot (*Phyllosticta hedericola* and others)
The spots on the leaves are brown but become grey or whitish in the centre, with a purplish-brown margin. They may bear minute, black, fruiting bodies. Control by spraying with carbendazim* or prochloraz, or with another compound listed for control of leaf spot diseases in Table 11.3, p. 438.

Phytophthora root rot (*Phytophthora* spp.)
See p. 446 and Table 11.2, p. 435.

Helianthemum (rock rose)

Diseases

Downy mildew (*Peronospora leptoclada*)
See p. 448, Table 11.2, p. 435, and Table 11.3, p. 438.

Grey mould (*Botryotinia fuckeliana* – anamorph: *Botrytis cinerea*)
See p. 448 and Table 11.3, p. 438.

Helianthus (sunflower)

See under Bedding plants, Chapter 10, p. 380.

Helleborus (hellebore)

Diseases

Leaf spot (*Coniothyrium hellebori*)
The leaves become spotted with round, or elliptical, black blotches, marked by concentric zones which coalesce to form large patches until, eventually, the leaves wither. Smaller spots disfigure the flowers. This disease is worse in wet winters. Remove and burn diseased leaves and flowers. Sprays of carbendazim* or prochloraz, or a copper-containing fungicide may have some activity against this disease.

Hemerocallis (day lily)

Pests

***Hemerocallis* gall midge** (*Contarinia quinquenotata*)
The tiny adult flies are active during May/early July, when they lay eggs on the developing flower buds. Infested buds are abnormally swollen but fail to open as flowers. Several hundred white larvae, each up to 3 mm long, may occur between the petals in a single swollen bud. New to Britain in 1988, this pest is now widespread in gardens in south-eastern England and in East Anglia. Late-flowering cultivars (after mid-July) are not affected. It is difficult to control this pest because the larvae are concealed inside the buds and there is an extended egg-laying period in early summer. Destruction of galled flower buds can reduce future infestations. Treatment with dimethoate during late May and mid June may also give control.

Heuchera (coral flower)

Pests

Vine weevil (*Otiorhynchus sulcatus*)
Heuchera is very susceptible to vine weevil grubs that devour the roots, both on container-grown plants and after planting out. For control see p. 443.

Hosta

Pests

Slugs and snails
See p. 442.

Diseases

Grey mould (*Botryotinia fuckeliana* – anamorph: *Botrytis cinerea*)
See p. 448 and Table 11.3, p. 438. This can cause large brown blotches on the leaves but no fungal growth develops on the spots.

Hyacinthus (hyacinth)

See Chapter 12, p. 548.

Hydrangea

Pests

Common green capsid (*Lygocoris pabulinus*)
This pest attacks the shoot tips during summer, resulting in foliage that is misshapen and tattered with many small holes. Hydrangeas are damaged by some insecticides, such as gamma-HCH and dimethoate. Protection can be given with cypermethrin, deltamethrin or nicotine, but do not spray in bright sunshine.

Two-spotted spider mite (*Tetranychus urticae*)
See p. 442.

Diseases

Honey fungus (*Armillaria* spp.)
See p. 443.

Powdery mildew (*Oidium hortensiae*)
See p. 449 and Table 11.3, p. 438. Powdery mildew causes brown leaf spots that have a white coating.

Hypericum (St. John's wort)

Diseases

Rust (*Melampsora hypericorum*)
See p. 450 and Table 11.3, p. 438. This is one of the subjects on which the colour of the spore pustules changes during the growing season as different spore types are produced. Pustules produced in early/mid-summer are orange, whereas those at the end of the summer are dark brown. Severe infections may result in defoliation. *Hypericum calycinum* is the most susceptible species and some nurseries have ceased growing it because of this disease.

Ilex (holly)

Pests

Cushion scale (*Chloropulvinaria floccifera*)
See under *Camellia*, p. 461.

Holly leaf miner (*Phytomyza ilicis*)
The larvae cause irregular, purplish-yellow blotches in the leaves. Clipped hedges

seem to suffer a higher proportion of mined leaves than bush or tree forms. Heavy infestations can be unsightly but they do not seem to have any obvious effect on the plant's growth. Control is difficult but some reduction in an infestation on small plants may be achieved with dimethoate, plus additional wetting-agent to counteract the waxy nature of the leaf surface. Spray two or three times at 14-day intervals from late May onwards.

Diseases

Phytophthora blight (*Phytophthora ilicis*)
Becoming increasingly common in the UK on *Ilex aquifolium*, this particular species acts as a foliar pathogen. Black spots or blotches develop on the leaves in the autumn, and affected leaves fall. A black die-back of twigs may then develop, which can extend into large branches and, eventually, the main stem – if a girdling lesion develops at this point, the entire plant above the lesion may die. Brown lesions may also develop on the berries.

The fungus is favoured by cool, moist conditions, and is usually most active in the autumn and winter months. In terms of control, the cultural and chemical measures for downy mildews (see p. 448, Table 11.2, p. 435, and Table 11.3, p. 438) are likely to be most effective. Fungicide applications should be made during the time of greatest activity of the fungus, in the autumn and winter.

Iris (bulbous)

See *Iris* Chapter 12, p. 549.

Iris (rhizomatous)

Diseases

Leaf spot (*Mycosphaerella macrospora*)
Frequency of attack varies according to locality, but in south-western counties of England, and elsewhere in wet seasons, the foliage is often marked in early spring by brown, oval-shaped spots which coalesce so that the leaves are badly injured or killed. Control is obtained by spraying with chlorothalonil, prochloraz or zineb*; if necessary, a wetter should be added.

Rhizome rot (*Erwinia carotovora* ssp. *carotovora*)
This bacterial disease causes the rhizome to rot with a yellowish, slimy rot so that the fan of leaves above shows browning of tips, followed by collapse and death. It is sometimes possible to save the plant by cutting away the affected piece of rhizome. Ensure the cut is made into healthy tissue, some distance from the affected area.

Jasminum (jasmine)

Diseases

Grey mould (*Botryotinia fuckeliana* – anamorph: *Botrytis cinerea*)
The fungus commonly causes branch die-back of both *Jasminum officinale* (common white jasmine) and *Jasminum nudiflorum* (winter-flowering jasmine). For more details on symptoms and control measures see p. 448 and Table 11.3, p. 438. A similar, but much rarer, die-back can be caused by the fungus *Sclerotinia sclerotiorum,* but under humid conditions this pathogen produces profuse white mycelium rather than the grey-brown sporulation of *Botrytis*. Both fungi produce resting bodies, or sclerotia; those of *Sclerotinia sclerotiorum* are larger.

Juniperus (juniper)

Pests

American juniper aphid (*Cinara fresai*)
Brownish-grey aphids, 2–4 mm long, suck sap from the younger shoots during the summer, resulting in die-back and heavy coatings of sooty mould. Juniper aphid (*Cinara juniperi*), a smaller species, is also a pest. For control of aphids, see p. 441.

Conifer spinning mite (*Oligonychus ununguis*)
See under *Picea*, p. 504.

Juniper scale (*Carulaspis juniperi*)
This pest produces whitish scales, up to 2 mm long, that encrust the shoots and leaves. Spray against young scales in late June/early July, as for Euonymus scale, p. 479.

Juniper webber moth (*Dichomeris marginella*)
The small, brown caterpillars bind shoots together with silken webbing in May/June. Damaged leaves turn brown, giving the impression of dead patches within the bush. It is difficult to penetrate the webbing with insecticides but those used against hawthorn webber moth on *Cotoneaster* (p. 471) could be tried.

Diseases

Kabatina shoot blight (*Kabatina* sp.)
This disease of juniper can also damage some *Cupressus* species and × *Cupressocyparis leylandii*. Shoot-tip die-back occurs – when examined closely a small constricted area can be found at the base of the die-back, which may

contain minute, black fruiting bodies of the fungus. The disease does not usually progress beyond year-old shoots, although the necrotic tissue can be invaded by other pathogens (such as *Pestalotiopsis* and *Phomopsis*). Carbendazim* and prochloraz may have activity against *Kabatina*, as well as against *Pestalotiopsis* and *Phomopsis*.

Needle blight and stem die-back (*Pestalotiopsis funerea* and *Phomopsis juniperovora*)
These fungi occur separately, or as a complex, and cause similar damage, which is described under needle blight of *Calluna*, p. 461. Yellow or variegated juniper cultivars seem particularly susceptible to damage. The fruiting bodies (acervuli) of *Phomopsis* produce a spore exudate, which appears creamy-white compared with the black mass of spores formed by *Pestalotiopsis*. Apply a HV spray of carbendazim* or prochloraz as a protectant treatment, immediately before or after 'cutting over' young plants. Repeat every 3–4 weeks if wet, humid weather persists. For disease control during propagation see under *Camellia*, leaf and stem blight, p. 462.

Phytophthora root rot (*Phytophthora* spp.)
See p. 446 and Table 11.2, p. 435. This disease can infect most juniper species and cultivars, but in some the root rot spreads very slowly and foliar symptoms may not develop. However, these plants can act as 'symptomless carriers' and spread the fungus to more susceptible species. If *Phytophthora* occurs in only a small proportion of the plants, control measures should be applied to the entire batch.

Laburnum

Pests

Leaf miners
The larva of a fly (*Agromyza demeijerei*) causes an irregular, brown blotch mine along the leaf margin in June/July, with further damage from a second generation in September/October. Another less common species (*Phytomyza cytisi*) causes a linear mine. The larvae of a small moth (*Leucoptera laburnella*) can also tunnel in leaves, causing brown, circular mines in the foliage in June/July and in September. Attacks on small trees can be controlled with dimethoate when signs of attack first appear. Abamectin may give control where leaf-mining flies are the problem.

Diseases

Downy mildew (*Peronospora cytisi*)
See p. 448, Table 11.2, p. 435, and Table 11.3, p. 438.

Honey fungus (*Armillaria* spp.)
See p. 443.

Leaf spot (*Pleiochaeta setosa*)
This causes small, brown spots on the leaves and sometimes die-back of shoots. Control by spraying as for rose black spot, see Table 11.3, p. 438.

Silver leaf (*Chondrostereum purpureum*)
See under *Prunus*, p. 514.

Lamium

Diseases

Downy mildew (*Peronospora lamii*)
See p. 448, Table 11.2, p. 435, and Table 11.3, p. 438. The fungus may show as a white, downy growth on the lower leaf surface but the most obvious symptom is purple spotting on the upper leaf surface.

Larix (larch)

Pests

Larch adelges (*Adelges laricis*)
These small, aphid-like pests are covered with white, waxy fibres. They suck sap from the leaves, which may become covered in sooty moulds. Control is worthwhile only on small, specimen trees; these could be treated with cypermethrin, deltamethrin or pirimicarb against the more vulnerable overwintering nymphs. Spray on a dry, mild day between November and late February.

Diseases

Leaf cast (*Meria laricis*)
In warm, damp weather, this can be a serious disease of young larch trees in nurseries but it is of little significance in dry seasons. The symptoms appear in early May, when the leaflets turn brown from the tip downwards. The needles fall prematurely, soon after becoming completely brown. HV sprays of captan*, sulfur* or zineb* at the end of March and at intervals of 2–3 weeks until the end of July (or earlier in a dry season) have given good control.

Lathyrus (sweet pea)

Pests

Aphids

Aphids, e.g. pea aphid (*Acyrthosiphon pisum*) and peach/potato aphid (*Myzus persicae*), suck sap from the leaves and flowers, and can transmit virus diseases. Control as for *Dahlia*, aphids, p. 473, but do not use malathion and try to avoid spraying open blooms.

Diseases

Damping-off (*Phytophthora* spp. and *Pythium* spp.)
See pp. 446–7 and Table 11.2, p. 435.

Downy mildew (*Peronospora viciae*)
See p. 448, Table 11.2, p. 435, and Table 11.3, p. 438.

Foot and root rot (*Aphanomyces euteiches, Thanatephorus cucumeris* – anamorph: *Rhizoctonia solani*, and *Thielaviopsis basicola* – synanamorph: *Chalara elegans*)
The roots die and often show black patches; discoloration also occurs at the stem bases, which may decay. Raise seedlings in 'sterile' rooting media and ensure that crops are grown in well-drained land.

Fusarium wilt (*Fusarium* spp.)
This causes the leaves to yellow and the plants to wilt. Dark-brown or black marks occur at the stem bases, and there is poor development of root nodules. A red discoloration develops in the inner tissues. Drench treatments of carbendazim* or prochloraz may give some control.

Grey mould (*Botryotinia fuckeliana* – anamorph: *Botrytis cinerea*)
See p. 448 and Table 11.3, p. 438. This can cause spotting and rotting of the flowers in very wet seasons.

Powdery mildew (*Microsphaera trifolii*)
See p. 449 and Table 11.3, p. 438.

White mould (*Ramularia deusta*)
A white, mealy, powdery coating (similar to powdery mildew) develops on both leaf surfaces and on the stems. Diseased tissues become slightly sunken but in severe attacks they become buff-coloured and affected leaves are shed. The disease is most troublesome in wet weather but can be checked by spraying with carbendazim*, chlorothalonil or zineb*.

Laurus (sweet bay)

Pests

Bay sucker (*Trioza alacris*)
The nymphs suck sap from the leaves, causing the leaf margin to thicken and curl over. At first, the damaged parts are pale yellow, but later they dry up and become brown. The nymphs are greyish, with white, waxy fibres radiating from their flattish bodies. There are several generations from May to September. No specific controls have been devised for this pest but cypermethrin* or deltamethrin*, applied (at the user's risk) at the onset of leaf-rolling, may control it.

Scale insects
Bay leaves are frequently attacked by brown soft scale (*Coccus hesperidum*), which coats the foliage with honeydew upon which sooty moulds develop. Brown soft scale has pale yellow-brown, oval shells, up to 3 mm long, found next to the larger veins on the underside of leaves. The main trunk and larger branches may be attacked by horse chestnut scale (*Pulvinaria regalis*) (see under *Aesculus*, p. 453). Control by spraying with deltamethrin, fatty acids or malathion when young nymphs are present. For horse chestnut scale this is in July; however, brown soft scale has several overlapping generations, so all stages in the life-cycle may occur together and more than one treatment may be required.

Diseases

Powdery mildew (*Oidium* sp.)
See p. 449 and Table 11.3, p. 438. Powdery mildew can cause severe distortion of leaves. Pale patches develop on the upper surface, and brown, scaly patches on the underside. A very fine, white growth of the fungus develops on the affected tissues.

Lavandula (lavender)

Pests

Common froghopper (*Philaenus spumarius*)
The nymphs are often numerous in May/June, when they produce and feed within 'cuckoo spit' on the shoots and flower stems. Little real damage is caused and control is often unnecessary. No current products carry label recommendations for control of froghoppers.

Diseases

Grey mould (*Botryotinia fuckeliana* – anamorph: *Botrytis cinerea*)
See p. 448 and Table 11.3, p. 438. The disease is most troublesome on lavender after frost damage.

Honey fungus (*Armillaria* spp.)
See p. 443.

Shab (*Phoma lavandulae*)
Diseased young shoots become chlorotic in May, followed by wilting and death. The fungus produces minute, black, sporing bodies (pycnidia) on the diseased tissues. Destroy infected plants and take cuttings only from those that are healthy. Spray established plants with carbendazim* or prochloraz.

Lavatera

See *Althaea*, p. 455.

Ligustrum (privet)

Pests

Lilac leaf-miner moth (*Caloptilia syringella*)
See under *Syringa*, p. 527. On privet, the pest can be dealt with by clipping the hedge and disposing of the mined leaves.

Privet aphid (*Myzus ligustri*)
This yellowish-green aphid is found on the underside of the leaves from spring to mid-summer. With heavy infestations, the foliage is mottled, rolled lengthwise and drops in mid-summer. For controls see p. 441.

Privet thrips (*Dendrothrips ornatus*)
These thin, sap-feeding insects are pale yellow or brownish-black and up to 2 mm long. They feed on the upper leaf surface during the summer, causing the foliage to develop a dull, silvery discoloration. Control by spraying with cypermethrin, deltamethrin, malathion or nicotine when signs of damage are seen.

Diseases

Honey fungus (*Armillaria* spp.)
See p. 443.

Lilium (lily)

See Chapter 12, p. 550.

Limonium

Diseases

Downy mildew (*Peronospora statices*)
See p. 448, Table 11.2, p. 435, and Table 11.3, p. 438.

Powdery mildew (*Oidium* sp.)
See p. 449 and Table 11.3, p. 438.

Lobelia

See under Bedding plants, Chapter 10, p. 380.

Lonicera (honeysuckle)

Pests

Honeysuckle aphid (*Hyadaphis passerini*)
Heavy infestations of this greyish-black aphid develop on the shoot tips and flowers during the summer. They can cause the flowers to turn brown and the shoots to die back. Control measures are the same as for aphids on *Rosa*, p. 520.

Diseases

Honey fungus (*Armillaria* spp.)
See p. 443.

Powdery mildew (*Oidium* sp.)
See p. 449 and Table 11.3, p. 438.

Lunaria (honesty)

Diseases

White blister (*Albugo candida*)
Blisters or swellings full of white powdery masses of spores develop on leaves and stems. Compounds used for control of downy mildews (see p. 448, Table 11.2,

p. 435, and Table 11.3, p. 438) will also have activity against this disease. Avoid overhead irrigation and high humidity.

Lupinus (lupin)

Pests

Lupin aphid (*Macrosiphum albifrons*)
This pest, first recorded in the UK in 1981, has become widely established on Russell and tree lupins throughout Britain. The aphids are large, whitish-grey insects that form dense colonies on the foliage and flower spikes, often preventing the development of the flowers. The control measures given for aphids on *Dahlia*, p. 473, can be applied.

Diseases

Anthracnose (*Colletotrichum acutatum*)
This pathogen first became a problem on ornamental lupins in the UK in the late 1980s, and anthracnose is now by far the most damaging disease affecting the crop. The fungus causes necrotic lesions on leaf blades, petioles, stems and flower stalks. Under wet conditions a mass of orange-coloured spores is produced on the surface of the lesions. Multiple lesions can result in distorted growth. A symptom which is particularly characteristic of the disease is continual twisting of the petiole to give a 'corkscrew' effect.

The fungus, which can be seed-borne, is favoured by warm temperatures and wet weather. The spores are splash-dispersed, and the disease can spread very rapidly on crops which are overhead irrigated or during warm, rainy weather. The fungus can occur at the very base of the plant, and even on the roots, so the disease can rapidly re-colonize fresh growth if affected plants are cut back.

Cultural control practices include the removal of affected foliage, good air movement around the plants and, if possible, the adoption of low-level irrigation to avoid leaf wetness. Fungicides may need to be applied regularly from the production of new growth in the spring, although at this early stage they should be used with care to avoid plant damage. A programme based around carbendazim*, dichlofluanid* and prochloraz should be used.

Leaf spot (*Pleiochaeta setosa*)
Dark, well-defined spots develop on the leaves, which are soon killed. Spots may also appear on the stems and seed pods; if seedlings are attacked they are usually killed. Destroy all diseased tissue and spray every 14 days with chlorothalonil or prochloraz, or a copper-containing fungicide (see under *Berberis*, bacterial leaf spot, p. 458, for more details of these fungicides).

Powdery mildew (*Microsphaera trifolii*)
See p. 449 and Table 11.3, p. 438.

Magnolia

Pests

Grey squirrel (*Sciurus carolinensis*)
Squirrels bite into the unopened buds, especially of *Magnolia soulangiana,* causing them to drop. Steps should be taken to reduce squirrel numbers.

Horse chestnut scale (*Pulvinaria regalis*)
See under *Aesculus*, p. 453.

Diseases

Bacterial leaf blotch (*Pseudomonas syringae* pv. *syringae*)
Under wet conditions, black blotches develop on the leaves. Overhead irrigation will aggravate this disease. Destroy badly infected plants and apply protectant sprays of a copper-containing fungicide (see under *Berberis*, bacterial leaf spot, p. 458, for more details of these fungicides).

Honey fungus (*Armillaria* spp.)
See p. 443.

Mahonia

Diseases

Bacterial leaf spot (*Pseudomonas syringae* pv. *syringae*)
Purple-black spots and blotches, often with a chlorotic halo, develop on the leaves. If the attack is severe, defoliation may result. Avoid overhead irrigation where the disease is present, and treat with a copper-containing fungicide.

Powdery mildew (*Microsphaera berberidis*)
See p. 449 and Table 11.3, p. 438.

Rusts (*Cumminsiella mirabilissima* and *Puccinia graminis*)
See p. 450 and Table 11.3, p. 438.

Malus (crab apple)

Pests

Aphids (foliar-feeding)
Various species occur on crab apple. For further details, see under Apple, aphids, Chapter 8, p. 260.

Apple leaf miner (*Lyonetia clerkella*)
See under Apple, Chapter 8, p. 262.

Mussel scale (*Lepidosaphes ulmi*)
See under Apple, scale insects, Chapter 8, p. 267.

Winter moth (*Operophtera brumata*)
See under Apple, Chapter 8, p. 269.

Woolly aphid (*Eriosoma lanigerum*)
See under Apple, Chapter 8, p. 270.

Diseases

Blossom wilt (*Monilinia* spp.)
See under Apple, Chapter 8, p. 270.

Canker (*Nectria galligena*)
See under Apple, Chapter 8, p. 271.

Coral spot (*Nectria cinnabarina*)
See under *Acer*, p. 452.

Crown/collar rots (*Phytophthora cactorum* and *P. syringae*)
See under Apple, Chapter 8, p. 272.

Fireblight (*Erwinia amylovora*)
See under Pear, Chapter 8, p. 281.

Honey fungus (*Armillaria* spp.)
See p. 443.

Powdery mildew (*Podosphaera leucotricha*)
See under Apple, Chapter 8, p. 274, and Table 11.3, p. 438.

Scab (*Venturia inaequalis*)
See under Apple, p. 274. Fungicide programmes for the control of this disease on

plants grown under protection should be based around fenarimol*, pyrifenox and thiram.

Specific apple replant disease (*Pythium* spp.)
See under Apple, Chapter 8, p. 276.

Matthiola (stock)

Diseases

Bacterial leaf spot, leaf rot and stem rot (*Xanthomonas campestris* pv. *incanae*)
This pathogen causes almost identical symptoms on *Matthiola* to those caused by *Xanthomonas campestris* pv. *campestris* on *Cheiranthus*. See under *Cheiranthus*, p. 467, for more details of this disease.

Black root rot (*Thielaviopsis basicola* – synanamorph: *Chalara elegans*)
See p. 447 and Table 11.3, p. 438.

Clubroot (*Plasmodiophora brassicae*)
See under *Cheiranthus*, p. 467.

Damping-off, foot and root rot (*Phytophthora* spp. and *Pythium* spp.)
See pp. 446–7 and Table 11.2, p. 435.

Downy mildew (*Peronospora parasitica*)
See p. 448, Table 11.2, p. 435, and Table 11.3, p. 438.

Wirestem, damping-off, foot and root rot (*Thanatephorus cucumeris* – anamorph: *Rhizoctonia solani*)
See p. 447 and Table 11.3, p. 438.

Meconopsis

Diseases

Downy mildew (*Peronospora arborescens*)
See p. 448, Table 11.2, p. 435, and Table 11.3, p. 438.

Myosotis (forget-me-not)

See under Bedding plants, Chapter 10, p. 380.

Narcissus (daffodil)

See Chapter 12, p. 551.

Nemesia

See under Bedding plants, Chapter 10, p. 380.

Nymphaea (water-lily)

Pests

Water-lily aphid (*Rhopalosiphum nymphaeae*)
Dense colonies of brownish-black aphids occur on the lily pads and flower buds. Other waterside plants may also be attacked. All insecticides are toxic to fish and other forms of pond life. In ponds, control has to be confined to forcibly spraying the plants with water to dislodge and drown the aphids.

Water-lily beetle (*Galerucella nymphaeae*)
Brown beetles (6 mm long) and their blackish-yellow larvae form irregular holes (slots) in the lily pads. The flowers are also damaged. Hand-picking the beetles and larvae will reduce infestations in small pools, as will forcible spraying with water. Pesticides are toxic to pond life and should not be used.

Diseases

Crown rot (possibly *Phytophthora* sp.)
This causes a blackening and rotting of the stem; leaves and flowers rapidly collapse and rot. Remove diseased plants. Carbendazim + metalaxyl (off-label) (SOLA 0912/92) is available for the control of this problem.

Leaf spot (*Ramularia nymphaearum*)
Circular, pale brown spots with dark edges develop on the upper leaf surface, and bear small, wart-like pustules of pale, yellowish spores. To control, remove affected leaves.

Paeonia (paeony)

Diseases

Blight (*Botrytis paeoniae*)
Shoot bases become withered and brown, and brown, angular patches develop on leaf tips and flower buds; infected parts are often killed. Cut off wilting stems

below ground level and burn. Spray both herbaceous and tree paeonies with dichlofluanid* every 10–14 days from emergence of foliage until flowering, or use another fungicide listed for control of grey mould in Table 11.3, p. 438.

Honey fungus (*Armillaria* spp.)
See p. 443.

Leaf spot (*Septoria paeoniae*)
Small, light-brown spots with grey centres and purple edges develop on the leaves; in severe attacks, they cause distortion. Similar spots may also occur on the stems. Sprays of carbendazim* or prochloraz may be effective.

Pelargonium

Pests

Capsids
These insects, e.g. common green capsid (*Lygocoris pabulinus*), are pale-green in colour and up to 5 mm long. They suck sap from the shoot tips, causing distorted and tattered foliage. For control see under *Dahlia*, p. 473.

Caterpillars (e.g. *Mamestra brassicae*)
The caterpillars make holes in the foliage and destroy the flowers. For control see under *Dahlia*, p. 473.

Diseases

See Chapter 10, p. 418.

Penstemon

Pests

Leaf nematodes
Two species, chrysanthemum nematode (*Aphelenchoides ritzemabosi*) and leaf nematode (*A. fragariae*), attack *Penstemon*. They live within the foliage, causing purplish-brown discoloured areas, which eventually dry up. Symptoms start on the lowest leaves and gradually work up the stems. For further information, see under *Anemone* × *hybrida*, p. 455.

Petunia

See under Bedding plants, Chapter 10, p. 380.

Philadelphus (mock orange)

Pests

Black bean aphid (*Aphis fabae*)
Dense colonies of black aphids with white waxy patches on their abdomens form on the foliage and young stems in May/June. This can result in die-back of the shoot tips. For controls, see p. 441.

Phlox (perennial and annual)

Pests

Stem nematode (*Ditylenchus dipsaci*)
The nematodes live inside the stem, causing it to be stunted and swollen with a tendency to split at the base. The leaves at the shoot tip are crinkled and often dramatically reduced in width; some will consist of little more than the midrib. Clean stock can be propagated by taking root cuttings or by giving the stool HWT for 1 hour at 44°C. A systemic nematicide, such as aldicarb (off-label) (SOLA 1325/95), may help to check infestations in growing plants. Infested ground should be kept free of phlox and other host plants, e.g. *Aubrietia*, *Dianthus barbatus*, *Gypsophila*, *Helenium*, *Oenothera*, *Primula*, *Solidago* and weeds for 2–3 years.

Diseases

Powdery mildew (*Sphaerotheca fuliginea*)
See p. 449 and Table 11.3, p. 438.

Phlox drummondii (annual)

See under Bedding plants, Chapter 10, p. 380.

Phormium (New Zealand flax)

Pests

Phormium mealybug (*Trionymus diminutus*)
Like the host plant, this mealybug originates from New Zealand and is frost tolerant. Colonies of the flattened, oval, greyish-white, sap-feeding insects, which

are up to 5 mm long, form at the base of leaves or where the leaves are folded upwards. They secrete a white, waxy powder from their bodies. Heavy infestations cause a lack of vigour in the host plant, and small plants are killed. It is very difficult to control, and spraying is not generally recommended as chemicals often fail to reach the pests' hiding places. Avoid propagating from infested plants.

Photinia

Diseases

Powdery mildew (*Podosphaera leucotricha*)
See p. 449 and Table 11.3, p. 438.

Phygelius (phygelia)

Pests

Capsids (e.g. *Lygocoris pabulinus*)
See under *Dahlia* p. 473.

Figwort weevil (*Cionus scrophulariae*)
Both the adult beetles and their slug-like larvae feed on foliage and flowers. The adults are 4–5 mm long and mottled black and white; they also have a distinctive, downwards-pointing snout. The larvae are brownish-yellow and covered with a slimy mucilage. Pupal cases are constructed on the stems and flower stalks and they resemble seed pods. If this pest is a problem in early summer, treatment with chlorpyrifos*, cypermethrin* or deltamethrin* should give control. Avoid spraying plants that are in flower.

Picea (spruce)

Pests

Conifer spinning mite (*Oligonychus ununguis*)
These small, orange-red mites spin a fine silken web between the needles. They are active during the summer months. A fine mottling develops on the foliage, which gradually becomes yellowish-brown and drops prematurely. *Picea albertiana* 'Conica' is particularly susceptible. Other conifers, such as *Cupressus*, *Juniperus*, *Pinus* and *Thuja*, may be attacked. Spray in late May after flushing, with bifenthrin, dicofol or malathion if this pest was a problem in the previous year.

Green spruce aphid (*Elatobium abietinum*)
Dark-green aphids with red eyes suck sap from the foliage during autumn to spring, with the heaviest infestations occurring in mild winters. The old foliage develops a pale, mottled discoloration and drops off in late winter or spring, but the new growth is unaffected. Spray with cypermethrin, deltamethrin, malathion, nicotine or pirimicarb in September, with further treatment during the winter if signs of infestation are seen.

Spruce pineapple-gall adelges (*Adelges abietis*)
This small, sap-feeding insect overwinters on spruce, especially Norway spruce (*Picea abies*), as immature nymphs. When mature, the females lay eggs next to the buds; the next generation of nymphs feeds on the developing buds, causing them to swell into pineapple-like galls. These galls open in late summer to release the adults, the galls then drying up and turning brown. This is unsightly on trees grown for the Christmas market but the galls have little impact on the tree's growth. If necessary, spray the trees on a mild dry day in February with cypermethrin, deltamethrin or pirimicarb to control the overwintering nymphs before eggs are laid on the buds.

Diseases

Bud blight (*Gemmamyces piceae*)
This disease can affect Norway spruce (*Picea abies*), but damage is not usually very severe. The fungus is more damaging on some other species, such as Engelmann spruce (*P. engelmannii*). Buds may either be killed prior to flushing, or the shoots produced may be weakened and curved and bear reduced numbers of needles. By the time the dead buds come to prominence they may be covered in a black, warty 'crust', consisting of the fruiting structures of the causal fungus. Chemical control measures have not been elucidated, but if the disease is known to be a recurring problem consideration should be given to planting a less susceptible species of *Picea* or a different genus.

Fomes root and butt rot (*Heterobasidion annosum*)
See under *Pinus*, p. 506.

Needle rusts (*Chrysomyxa abietis* and *C. rhododendri*)
These rusts may cause a yellow or orange discoloration of needles. *C. rhododendri* produces delicate, white outgrowths on the undersides of affected needles in late summer. Spores from these outgrowths infect rhododendron, the alternate host upon which the fungus overwinters. In contrast, *C. abietis* overwinters on infected needles. This is the more common of the two rusts. Both are most frequently found in Scotland and Northern Ireland, although *C. abietis* can also be found in northern and south-west England. Both rusts are sporadic in

occurrence, and control measures are not recommended for forestry plantations. On ornamental subjects the rust fungicides listed for rust control in Table 11.3, p. 438, should have some effect.

Phytophthora root rot (*Phytophthora* spp.)
See p. 446 and Table 11.2, p. 435.

Pieris

Diseases

Phytophthora root rot (*Phytophthora* spp.)
See p. 446 and Table 11.2, p. 435.

Pinus (pine)

Pests

Pine root aphid (*Stagona pini*)
The aphids occur throughout the year on the roots, amongst masses of bluish-white waxen 'wool'. Severe infestations cause yellowing of the needles and are especially important on container-grown nursery plants, nursery stock and transplants. Treat infested container-grown plants with a drench of chlorpyrifos; severely infested open-bedded plants should be lifted and burnt.

Diseases

Cyclaneusma needle cast and 'radiata yellows' (*Cyclaneusma minus*)
The fungus can affect various species of *Pinus*, but Monterey pine (*P. radiata*) is probably most susceptible. The symptoms on this host are yellowing of needles in the summer (known as 'radiata yellows'). Often, all the needles on a shoot will be affected, apart from a tuft at the end of the branch which remains green. *Cyclaneusma* has also caused severe damage to Scots pine (*P. sylvestris*) in nurseries, in which it causes needle cast. In humid weather, fruiting bodies of the fungus may occur, breaking through the surface of the needle to leave characteristic flaps of epidermal tissue and producing yellow spores. HV sprays of chlorothalonil during the spring and summer months have given good control of *Cyclaneusma*.

Fomes root and butt rot (*Heterobasidion annosum*)
Fomes root and butt rot is the most serious disease of both young and mature conifers in the UK. The fungus may kill pines, particularly on dry, alkaline soils

(pH 6 and above). The fungus commonly rots the roots, and may proceed from the roots into the stem, where it produces an extensive butt rot. Fructifications often appear at the base of trees killed by the disease or on stumps of previously infected ones. The upper surface of the growing fructification is reddish-brown with a white margin, and the undersurface is white and pierced by fine pores. Immature fructifications (in the form of small, white pustules) are commonly found on infected stumps or roots.

The fungus enters the tree by means of airborne spores that infect exposed woody tissue (in forest plantations this is often the cut stump surfaces immediately after thinning or clear-felling). The fungus grows into the stump tissue and down into the root system, and if the roots of healthy trees are in contact with such infected tissues they in turn become infected.

Control is based on the protection of cut stump surfaces, thus preventing infection by airborne spores. Apply stump protectants immediately after all thinning and felling operations. Urea is the standard material used for stump protection and is used as a solution with the addition of a marker dye. Apply the solution liberally to the freshly cut stump surface immediately after felling. The dye is included to assist in complete application and to indicate that the treatment has been done.

In pine plantations, there is an alternative stump treatment. In this case, spore suspensions of a competing fungus, *Peniophora gigantea*, can be applied to the freshly cut surface. This method of biological control is very efficient.

If stump protectants are applied throughout the life of a forestry crop, it can be kept substantially free from the disease and, thus, danger to subsequent crops is much reduced. This is particularly important, as it is very difficult to eradicate the disease once it becomes established within a crop.

If an infected crop is felled and replanted with conifers, rapid infection of the new plants may take place from the stumps of the previously infected crop. Control in crops planted on infested sites can be achieved by the mechanical removal of the stumps of the previous crop, but this is an expensive operation and justified only where the disease is severe. The use of *Peniophora gigantea* for stump treatments in pine areas that are already infested may help to reduce infection in the subsequent crop.

***Lophodermium* needle cast** (*Lophodermium seditiosum*)

The most common 'needle cast' disease of pine; Scots pine (*Pinus sylvestris*) can be affected severely. The disease begins in the late summer or autumn as small yellow spots, which later turn brown, on the needles. In the winter, affected needles may turn a pinkish-brown colour. Defoliation usually occurs in late winter and spring. Fruiting bodies of the fungus are not usually produced until some time after the affected needles have fallen. Sprays of carbendazim*, chlorothalonil or prochloraz, applied in the period July/September, are effective against this disease.

White pine blister rust (*Cronartium ribicola*)
This is a common and damaging disease of five-needled pines. The fungus causes girdling lesions on branches, eventually killing them. Affected shoots are often swollen and distorted; in early summer they produce sac-like outgrowths, containing large numbers of orange-coloured spores. If the infection spreads from a branch into the stem, death of the tree can result.

The alternate hosts of this disease are *Ribes* spp. (mainly black currant). Spores produced on this host in late summer and autumn infect pine needles and progress from them into the shoot.

Fungicidal control of this disease is usually ineffective. Removing affected branches below the lower limit of infection will prevent the fungus colonizing the stem and may save an affected tree.

Pittosporum

Diseases

Leaf spot (*Microsphaeropsis pittospororum*)
This fungus produces small (2–3 mm in diameter) leaf spots, often purple in colour and occasionally with a brown centre. Multiple spotting is often present. *M. pittospororum* is generally regarded as a weak pathogen, requiring some form of damage to the leaf surface in order to colonize the leaf. If it does occur, carbendazim* or prochloraz sprays may be effective.

Phytophthora root rot (*Phytophthora* spp.)
See p. 446 and Table 11.2, p. 435.

Platanus (plane)

Diseases

Anthracnose or leaf scorch (*Apiognomonia errabunda*)
The first symptom is failure of some of the buds to open in spring ('bud blight'), owing to the presence of the overwintering fungus within the buds. The fungus spreads into the twigs, causing die-back ('twig blight'), and the dead tissues show a bright orange-brown discoloration. In the spring, young developing shoots from 1 to 10 cm in length suddenly wilt, turn yellow or brown, die and shed their leaves ('shoot blight'). None of these stages of the disease is, however, as obvious as the slightly later stage that shows on the new or expanded leaves as a brown discoloration, usually along the main veins, and causes the leaves to wither ('leaf blight'). American plane (*Platanus occidentalis*) is very susceptible to anthracnose but oriental plane (*P. orientalis*) is very resistant. The hybrid between the two –

London plane (*P.* × *hispanica*) – has clones that vary in susceptibility, although none is as susceptible as *P. occidentalis*. Affected trees may look unsightly for one or more seasons but recovery is usually rapid and complete. Even severely diseased nursery trees rendered temporarily unsaleable usually recover completely by the end of the growing season. Incidence of the disease is sporadic. Shoot blight develops when temperatures are low immediately following leaf emergence. No chemical or cultural measures can be recommended for control, but on a small scale, it is probably worthwhile collecting and burning fallen leaves in the autumn.

Polemonium (Jacob's ladder)

Diseases

Powdery mildew (*Oidium* sp.)
See p. 449 and Table 11.3, p. 438.

Polygonatum (Solomon's seal)

Pests

Solomon's seal sawfly (*Phymatocera aterrima*)
The adults are black-bodied insects with two pairs of greyish-black wings and they are active during the flowering period. Eggs are laid in the stems, causing purplish scars about 20 mm long. The larvae are whitish-grey with black heads and they grow up to 25 mm long. Infested plants are often completely defoliated. Control as for Columbine sawfly, p. 456, when the larvae are seen. There is only one generation annually.

Populus (poplar)

Pests

Leaf beetles (various species)
See under *Salix* p. 523.

Diseases

Bacterial canker (*Xanthomonas populi*)
This disease causes girdling lesions on twigs and branches, leading to die-back. Large bark cankers may also be produced on the main stem. The bacterium overwinters in bark lesions and infects in spring through wounded tissue

(including natural wounds such as leaf scars). Many poplars with good resistance to bacterial canker have been bred for timber production but fewer resistant species or hybrids are available for amenity use. Affected parts could be pruned out, but the disease is likely to recur at a later date.

Honey fungus (*Armillaria* spp.)
See p. 443.

Leaf spots (*Marssonina* spp.)
Small, irregular, blackish-brown spots occur on the leaves, which fall prematurely. Control may be difficult but dead shoots should be cut out, and smaller trees sprayed at least three times in the spring and at least once in the summer with prochloraz or a copper-containing fungicide (see under *Berberis*, bacterial leaf spot, p. 458, for more details of these fungicides).

Rusts (*Melampsora* spp.)
See p. 450 and Table 11.3, p. 438. Premature defoliation sometimes results from attacks of rust. Poplar clones with good resistance to rust have been bred for use in short-rotation coppice plantations.

Silver leaf (*Chondrostereum purpureum*)
See under *Prunus*, p. 514.

Yellow leaf blister (*Taphrina populina*)
Affected leaves are distorted, and bear large blisters which are bright yellow on the lower surface but remain green on the upper side. Where the disease is troublesome, spray in January or February with a copper-containing fungicide.

Potentilla

Diseases

Powdery mildew (*Sphaerotheca alchemillae*)
See p. 449 and Table 11.3, p. 438.

Primula (outdoor species including polyanthus)

Pests

Bryobia mites (*Bryobia* spp.)
These are small, reddish-black mites with pink legs and they suck sap from the upper leaf surface. This causes a fine, mottled discoloration and, eventually, most

of the green colour is lost. Infestations occur mainly in early summer. Control by spraying with dicofol + tetradifon, dimethoate or malathion.

Caterpillars (e.g. *Noctua pronuba*, and *Phlogophora meticulosa*)
See under *Dahlia*, caterpillars, p. 473.

Two-spotted spider mite (*Tetranychus urticae*)
These are of similar size to bryobia mites but are usually yellowish-green with dark markings and they feed mainly on the undersides of the leaves. The damage symptoms and controls are the same as for bryobia mites. If organophosphate-resistant mites are present, other compounds such as abamectin, bifenthrin or fenazaquin could be used.

Vine weevil (*Otiorhynchus sulcatus*)
Plants often die suddenly, owing to the roots being severed by the larvae. These are plump, white, apodous maggots, with light-brown heads, and are up to 10 mm long. They are most frequently discovered in late autumn to spring, but by then it may be too late to save the plants. For control and preventive measures, see p. 443.

Diseases

Bacterial leaf spot (*Pseudomonas* sp.)
Symptoms consist of dark-brown to black spots, with a yellow margin which can be much broader than the spot itself. Keep the leaves as dry as possible, and apply a copper-containing fungicide (see under *Berberis*, bacterial leaf spot, p. 458, for more details of these fungicides).

Black root rot (*Thielaviopsis basicola* – synanamorph: *Chalara elegans*)
See p. 447 and Table 11.3, p. 438.

Brown core (*Phytophthora primulae*)
This pathogen causes rotting of the roots and crown and produces a brown discoloration in the stele, followed by a collapse of the aerial growth. For control, apply one of the *Phytophthora* treatments listed in Table 11.2, p. 435.

Damping-off, foot and root rot (*Pythium* spp.)
See p. 447 and Table 11.2, p. 435.

Leaf spots (*Phyllosticta primulicola* and *Ramularia* spp.)
Necrotic spots appear on the leaves, making them unsightly and in some cases damaging or killing the foliage. White sporulation is visible in humid conditions on spots caused by *Ramularia*. These diseases are worse on weak plants or in

damp autumns. If required, spray with one of the fungicides listed for the control of leaf spot diseases in Table 11.3, p. 438.

Wirestem, damping-off, root and foot rot (*Thanatephorus cucumeris* – anamorph: *Rhizoctonia solani*) – see p. 447, and Table 11.3, p. 438.

Prunus (ornamental cherry; cherry laurel; Portugal laurel)

Pests

Cherry blackfly (*Myzus cerasi*)
Black aphids infest the shoot tips during spring and early summer, causing severe leaf curling and die-back of the shoots. Cultivars of *Prunus avium* and *P. cerasus* are often affected, but Japanese cherries are usually immune. A winter wash of tar oil can be used against the overwintering eggs during December/January. During the spring, if aphids are seen at the shoot tips, but before severe leaf curling has occurred, small trees can be sprayed with cypermethrin, deltamethrin, malathion or pirimicarb. Some ornamental cherries have been damaged by dimethoate.

Pear slug sawfly (*Caliroa cerasi*)
See under Pear, Chapter 8, p. 279.

Winter moth (*Operophtera brumata*)
See under Apple, Chapter 8, p. 269.

Diseases

Bacterial canker and shot-hole (*Pseudomonas syringae* pv. *mors-prunorum*)
Symptoms differ according to the species of *Prunus* affected. On ornamental cherry, lesions (often with gumming) appear on branches in early spring, and as these cankers develop the shoots die back and some branches tend to become flattened. During summer, brown spots may develop on the infected leaves. Spray with a copper-containing fungicide (see under *Berberis*, bacterial leaf spot, p. 458, for more details of these fungicides) in mid-August and repeat three times at 3- to 4-week intervals to prevent the bacteria infecting bark wounds and leaf scars. Trees should be pruned or cut back between May and the end of August.

On cherry laurel (*Prunus laurocerasus*) and Portugal laurel (*P. lusitanica*) the symptoms are often restricted to chlorotic or necrotic leaf spots, which rapidly drop out to give a 'shot-hole' symptom. Multiple 'shot-holes' are often present, giving the leaves a very ragged appearance. If the symptoms occur on the leaf margin they are easily confused with feeding damage from adult vine weevils, but if the bacterium is the cause the shot-holing is usually also present away from the margins. Copper-containing fungicides can again be used and are often applied throughout the spring, summer and early autumn, particularly if overhead irrigation is used.

Black root rot (*Thielaviopsis basicola* – synanamorph: *Chalara elegans*)
See p. 447 and Table 11.3, p. 438.

Blossom wilt (*Monilinia* spp.)
Following a wet spring, this fungus can cause die-back of young shoots. For a more detailed description of the symptoms and control measures see under Plum and damson, Chapter 8, p. 288.

Downy mildew (*Peronospora sparsa*)
See p. 448, Table 11.2, p. 435, and Table 11.3, p. 438. Sometimes, this disease can be found on cherry laurel (*Prunus laurocerasus*) and Portugal laurel (*P. lusitanica*). It causes irregular necrotic blotches which, on cherry laurel, can be several centimetres in diameter. The edges of the blotches are often a light-green colour, and small amounts of sporulation can sometimes be seen on the underside of the leaf, particularly towards the edge of the lesion. As with many foliar diseases of *Prunus*, affected areas may eventually drop out.

Fungal shot-hole (*Stigmina carpophila, Trochila laurocerasi* and others)
These fungi are found occasionally on leaves of both deciduous and evergreen species. Carbendazim*, prochloraz and other fungicides listed in Table 11.3, p. 438, for the control of leaf spots are likely to be effective. Shot-holing in cherry laurel (*Prunus laurocerasus*) and Portugal laurel (*P. lusitanica*) is much more likely to be due to *Pseudomonas syringae* pv. *mors-prunorum* than to one of these fungi.

Honey fungus (*Armillaria spp.*)
See p. 443.

Peach leaf curl (*Taphrina deformans*)
Young diseased leaves are thick and tinged with red. Older leaves become very puckered and distorted, and the intense red colour later turns white. Diseased leaves drop prematurely and growth is weakened. To control, spray the bare branches before the flower buds burst in January or early February, repeat a fortnight later and spray once again just before leaf fall, with a copper-containing fungicide (see under *Berberis*, bacterial leaf spot, p. 458, for more details of these fungicides).

Powdery mildew (*Podosphaera tridactyla*)
See p. 449 and Table 11.3, p. 438. This disease can cause severe distortion of the leaves of cherry laurel (*Prunus laurocerasus*). Pale blotches may be seen on the upper leaf surface, and brown, scaly patches on the underside. A very fine, white

growth of fungus develops on the affected tissues and, again, the affected areas may eventually drop out.

Silver leaf (*Chondrostereum purpureum*)
This disease can affect a range of ornamental *Prunus* species. The fungus, which is present inside the branches or the main stem, produces toxins that cause the epidermis of the leaf to separate from the cells below. The gap so created becomes filled with air, and produces the silvering effect that gives the disease its name. Brown staining of the heartwood indicates the location of the fungus in the tree. Bracket-shaped fungal fruiting bodies appear on dead branches, stems or roots.

Dead trees should be removed and burnt. Cut-out any diseased branches and prune back to healthy wood in June, July or August, when the trees most readily produce a 'gum barrier' which prevents further penetration of the fungus. All dead wood should be removed and burnt before fruiting bodies of the fungus can form. *C. purpureum* is a wound parasite, and cut branch ends should be protected with a wood paint containing octhilinone.

Specific cherry replant disease
See under Cherry, Chapter 8, p. 285.

Pulmonaria (lungwort)

Diseases

Powdery mildew (*Erysiphe asperifolium*)
See p. 449 and Table 11.3, p. 438.

Pyracantha (firethorn)

Pests

Brown scale (*Parthenolecanium corni*)
See under *Ceanothus*, scale insects, p. 464.

Firethorn leaf-miner moth (*Phyllonorycter leucographella*)
New to Britain in 1989, this tiny moth is now widespread. The larvae feed within the leaves, causing an oval, silvery-brown mine in the centre of the upper leaf surface. Later, the leaf folds upwards so the mine is concealed. Damage, which is most frequently seen in the winter months, is largely cosmetic as even heavy attacks do not affect growth or the ability to produce flowers and berries. If control is required on nursery stock, spray with cypermethrin (off-label) (SOLA 1200/96) or dimethoate when the early stages of mines are detected.

Woolly aphid (*Eriosoma lanigerum*)
See under *Cotoneaster*, p. 471.

Diseases

Fireblight (*Erwinia amylovora*)
See under Pear, Chapter 8, p. 281.

Scab (*Spilocaea pyracanthae*)
This pathogen attacks leaves, shoots, flowers and fruit but it is the 'scabby' lesions on the ornamental fruits that can spoil the beauty of the shrub. For control, spray with a fungicide listed in Table 11.3, p. 438. Where plants are grown under protection, the spray programme should be based around fenarimol*, pyrifenox and thiram. Avoid overhead irrigation. There are considerable differences in cultivar susceptibility to scab.

Pyrethrum

Pests

Chrysanthemum nematode (*Aphelenchoides ritzemabosi*)
See under *Dendranthemum* (chrysanthemum), p. 476.

Diseases

Grey mould (*Botryotinia fuckeliana* – anamorph: *Botrytis cinerea*)
See p. 448 and Table 11.3, p. 438.

Quercus (oak)

Pests

Gall wasps (e.g. *Andricus kollari*, *A. quercuscalicis*, *Biorhiza pallida* and *Neuroterus quercusbaccarum*)
Gall wasps are small insects whose larvae induce the host plant to produce abnormal growths from which the larvae gain nourishment. Galls can be found on roots, leaves, buds, flowers and acorns, with each species of wasp producing galls of a specific type. Many gall wasps have complex life-cycles, with alternate sexual and asexual phases, the larvae of which produce different galls on different parts of the plant or on different specific host plants. None of these galls is harmful to the tree and no control measures have been devised.

Diseases

Oak decline (cause unknown)
This problem, first reported in the 1920s, has undergone a resurgence in the late 1980s and in the 1990s. Typical early symptoms are a thinning of the foliage, with leaves becoming more sparse, smaller and yellow. Symptoms progress into twig and small-branch die-back, followed by death of major limbs. Eventually, the whole tree may succumb.

The cause of the problem is still the subject of research. Environmental stress is thought to be a possible initial cause – drought stress is one possibility, although in Europe (where there is also a problem) wet soils and frost damage have also been implicated. Weakened trees are then thought to be more susceptible to damage from diseases, e.g. powdery mildew and possibly root infection by *Phytophthora*, and pests (e.g. the beetle *Agrilus pannonicus*, which tunnels through the bark). No control measures can be recommended until research provides a greater understanding of the causes of oak decline.

Powdery mildew (*Microsphaera alphitoides*)
See p. 449 and Table 11.3, p. 438.

Rhododendron (including azalea)

Pests

Azalea whitefly (*Pealius azaleae*)
Both adults and the yellowish-green, scale-like nymphs suck sap from the undersides of the leaves of azalea. The upper surface becomes soiled with honeydew and sooty mould. The adults are small, white-winged insects that readily fly up from the plant during the summer months. Control by spraying with cypermethrin in early summer, when adults emerge, or apply imidacloprid as a soil drench.

Caterpillars
Caterpillars are not major pests of rhododendrons but they sometimes damage the foliage. Damage is more likely to be seen in woodland areas where caterpillars fall down from more favoured hosts such as oak. For controls see under *Rosa*, p. 520.

Rhododendron lace-bug (*Stephanitis rhododendri*)
The adults are yellowish to dark brown insects, about 4 mm long. Their wings are ornately veined and carried flat on their backs. Both adults and nymphs suck sap from the undersides of the leaves from May to September. This causes a yellow mottling on the upper surface of the leaves, while the undersides become soiled

with the insects' brown excreta. This pest is a local problem more likely to be seen on plants growing in open, sunny positions. Control by spraying as for rhododendron leafhopper (below) during early summer, if the pest is present.

Rhododendron leafhopper (*Graphocephala fennahi*)
The adults are 6 mm long and turquoise-green with orange-red stripes. They can be disturbed in large numbers from rhododendrons in late summer. Adults and nymphs suck sap from the foliage without causing any obvious symptoms. They are, however, important pests as the females lay overwintering eggs in the flower buds, and this allows infection by the disease bud blast (see below). Where this disease is present, attempts should be made to reduce the leafhopper population by spraying with cypermethrin, malathion or nicotine. Several applications at fortnightly intervals in August–October will be necessary, especially in areas where *Rhododendron ponticum* is established as a naturalized plant.

Wingless weevils (e.g. *Otiorhynchus singularis* and *O. sulcatus*)
The adult beetles are active at night, when they bite out irregular notches from the leaf margins, particularly on low-growing branches. If damage is severe, spray at dusk on a mild evening with chlorpyrifos. Container-grown plants can be damaged by the root-feeding larvae – see p. 443 for protective treatments.

Diseases

Basal and leaf rot of cuttings (*Botryotinia fuckeliana*, *Fusarium* spp., *Glomerella cingulata, Nectria radicicola* – anamorph: *Cylindrocarpon destructans*, *Pestalotiopsis sydowiana* and *Phomopsis ericaceana*)
A range of fungal species has been associated with rotting of rhododendron cuttings during propagation, but *P. sydowiana* and *N. radicicola* (anamorph: *C. destructans*) have occurred most frequently. It is important to control leaf and stem pathogens on stock plants using one of the broad-spectrum fungicides listed in Table 11.3, p. 438, for the control of grey mould, *Fusarium* and leaf spots. On cuttings propagated under polythene film, fungal rotting will be reduced, and the number and quality of rooted cuttings significantly increased, if a fungicide drench of prochloraz is applied to the rooting medium and cuttings at insertion.

Bud blast (*Pycnostysanus azaleae*)
The infected buds turn brown, black or silvery in spring and black bristle-like spore heads protrude from them. The fungus is a wound parasite and it is thought to enter punctures made by rhododendron leafhopper (*Graphocephala fennahi*) as it deposits its eggs in the flower buds. Apart from picking off as many infected buds as possible, spray to control the leafhopper (see above).

Gall or false bloom (*Exobasidium vaccinii* var. *japonicum*)
The young developing leaves or flower buds of azalea turn reddish and swell into

small galls which then become waxy-white owing to the production of large numbers of spores. The complete life-cycle of this fungus is not known. Pick off the galls when they are still green and spray with zineb*.

Honey fungus (*Armillaria* spp.)
See p. 443.

Leaf blight and root rot (*Thanatephorus cucumeris* – anamorph: *Rhizoctonia solani*)
Under humid conditions this pathogen can cause extensive leaf rotting and drop on azalea. HV sprays of iprodione* are most appropriate for this problem; for root-rotting use a compost drench of tolclofos-methyl.

Leaf spot and stem die-back (*Pestalotiopsis sydowiana*)
P. sydowiana usually infects weak or damaged leaves of rhododendron, causing silvery-grey lesions with black pinhead-sized sporing structures which form in concentric rings on the dead tissue. It is seldom serious in well-grown, established bushes but the leaf spots can be disfiguring, and if cutting material is taken from infected plants the fungus causes rotting of cuttings during propagation (see basal rot, p. 517). In young, newly rooted plants the disease may penetrate stems, causing die-back and death of the plant. For control, see under *Camellia* for measures suggested for leaf and stem blight, p. 462.

Petal blight (*Ovulinia azaleae*)
Azalea petals develop small spots that are white on coloured flowers and pale brown on white blooms. Affected tissues take on a water-soaked appearance. The blooms are reduced to wet, slimy masses that wither and remain hanging on the bush. Spray with carbendazim* or zineb* as a routine where this disease is known to be troublesome. Spray as soon as the flower buds show colour and repeat at intervals of 1–2 weeks until just before the flowers open.

Phytophthora root rot (*Phytophthora* spp.)
See p. 446 and Table 11.2, p. 435. Root rot in young plants can cause a rapid drooping of the leaves, stem die-back and death. However, in established field-grown stock, weak growth and chlorotic foliage may be the first indication of root damage, the wilt and die-back symptoms developing as water stress increases.

Powdery mildew (*Microsphaera* spp.)
See p. 449 and Table 11.3, p. 438. Symptoms of powdery mildew on rhododendron are somewhat different from those caused by this disease on most other hosts. Yellow spots or blotches are produced on the upper leaf surface, occasionally with a reddish or purple margin. A very fine, white fungal growth may be seen on the undersurface, but it is often not very obvious. Diseased leaves may drop prematurely.

HDC-funded research has shown that it is important to protect the new growth in spring and, because this tissue is susceptible to scorch, fungicides must be used with care. Penconazole* and pyrifenox have some activity, but epoxiconazole* was shown to be the most effective fungicide in recent trials work. This fungicide has no label recommendation for use on ornamentals and is applied at the grower's risk. There have been reports of phytotoxicity from epoxiconazole on some rhododendron cultivars, and it is vital, therefore, that the product is evaluated on a small number of plants before widespread treatment is undertaken. Indeed, all fungicides should be evaluated for possible phytotoxic effects as there may be differences between cultivars and hybrids. Following the initial application in spring, further sprays are likely to be required during the growing season to maintain control.

Ribes sanguineum (flowering currant)

Diseases

Powdery mildew (*Microsphaera grossulariae*)
See p. 449 and Table 11.3, p. 438.

Robinia (false acacia)

Diseases

Canker and die-back (*Nectria haematococca* var. *brevicona*)
This fungus can colonize wounded or weakened tissues to cause canker and die-back on either the branches or the main stem. Under humid conditions, orange-coloured pustules of the *Fusarium* (anamorph) state of the fungus frequently develop. Prochloraz sprays will have some activity, but it is important to try to identify the initial cause of damage or weakened growth to prevent the problem recurring.

Verticillium wilt (*Verticillium dahliae*)
See under *Acer*, p. 453. Occasionally, *Robinia* is affected by this disease, but not as frequently as hosts such as *Acer* and *Catalpa*.

Rosa (rose)

Pests

Aphids
Colonies of green or pink aphids can be found on the foliage, stems and flower buds from April to September. Common species include rose aphid (*Macro-*

siphum rosae) and rose/grain aphid (*Metopolophium dirhodum*). Treat when necessary with cypermethrin, deltamethrin, dimethoate, imidacloprid, malathion, nicotine, pirimicarb or pymetrozine.

Capsids (e.g. *Lygocoris pabulinus*)
The pale-green insects suck sap from the shoot tips, causing the leaves to develop many small holes. Control infestations by spraying with cypermethrin, deltamethrin, dimethoate or nicotine.

Caterpillars (e.g. *Archips podana*, *Malacosoma neustria*, *Operophtera brumata* and *Orgyia antiqua*)
Many species can attack rose but the most troublesome are winter moth (*Operophtera brumata*) and various tortrix species, which attack the foliage during April/May. Other caterpillars may bore into unopened flower buds. Control caterpillars by spraying with cypermethrin or nicotine. Caterpillars are also controlled by *Bacillus thuringiensis*, diflubenzuron or teflubenzuron, and these are suitable for use in integrated control programmes.

Rose leafhopper (*Edwardsiana rosae*)
The adults are pale yellow insects, up to 3 mm long, which feed by sucking sap from the lower leaf surface. The nymphal stages are creamy-white and feed in the same way. They are active from late April to October and cause a mottled discoloration of the upper leaf surface. In heavy infestations, particularly on wall-trained climbers, much of the green colour may be lost from the foliage. To control, spray with malathion or nicotine.

Rose leaf-rolling sawfly (*Blennocampa phyllocolpa*)
Damage occurs in May/June. Leaflets become tightly rolled downwards along their length, in response to chemicals injected by the female when eggs are laid. The eggs later give rise to small, pale-green caterpillars that eat the rolled leaflets. There is one generation per year. This is a very troublesome pest in gardens. Control is difficult, since the damage is caused by the adults rather than the larvae, and adults can emerge from late April to June. The adults can be controlled with most contact insecticides, such as cypermethrin and deltamethrin, but there are no approved label recommendations for controlling this pest.

Rose slug sawfly (*Endelomyia aethiops*)
Two generations of larvae occur in May/June and July/August. The larvae are whitish-green, with light brown heads, and up to 15 mm long. They graze away the upper or lower leaf surface, damaged areas later drying up and turning white or brown. Control is relatively easy with nicotine, rotenone, or other contact insectides effective against caterpillars.

Other foliar-feeding sawflies (e.g. *Allantus cinctus, Arge ochropus, Arge paganus, Cladius difformis* and *C. pectinicornis*)
The larvae of these species all feed exposed on the leaf surface and are controlled as for rose slug sawfly (see p. 520).

Rose thrips (*Thrips fuscipennis*)
These are thin yellow or dark-brown insects, up to 2 mm long, which suck sap from the leaves and petals. Heavy infestations cause brown markings on the petals and cause the foliage to become silvery. Control is as for rose leafhopper (p. 520), but severe attacks by thrips on outdoor roses are not common.

Two-spotted spider mite (*Tetranychus urticae*)
Heavy infestations of this tiny sap-feeding pest result in the leaves becoming bronzed and falling prematurely. Suitable acaricides include abamectin, bifenthrin or fenpropathrin. Fenazaquin should not be used on roses. Strains of the mite resistant to organophosphate compounds may occur. Biological control with *Phytoseiulus persimilis* (a predatory mite) can be effective on outdoor roses as part of an integrated control programme.

Diseases

Black mould (*Chalaropsis thielavioides*)
This fungus can cause extensive failures following budding. It grows prolifically beneath the bud shield, preventing union of the bud with the rootstock. The fungal growth is white at first, but later turns black as large numbers of spores are produced. The source of the *Chalaropsis* may be the soil, or contaminated rootstocks or bud sticks. Spraying the rootstocks and plants which are to be the source of the bud wood with carbendazim* or prochloraz prior to the budding operation may reduce the risk of bud failures due to this pathogen. Where possible, avoid storing budwood at high humidity for any length of time because if it is contaminated with *Chalaropsis*, such conditions will encourage the fungus to multiply.

Black spot (*Diplocarpon rosae*)
This causes circular black spots on the leaves, and in some cultivars the fungus can be found in scabby lesions on the stems. The fungus overwinters in the bud scales, and on stems and old leaves that remain on bushes, but can also survive for a few months in fallen leaves. Nevertheless, winter treatments against the fungus do not give good control. Start spraying immediately after pruning and repeat at fortnightly intervals throughout the season, using a fungicide listed in Table 11.3, p. 438. As the disease is worse on weak bushes, feeding, mulching and watering should be carried out as necessary. Applications of a foliar feed during the growing season are also beneficial.

Crown gall (*Agrobacterium tumefaciens*)
See under *Daphne*, p. 474.

Downy mildew (*Peronospora sparsa*)
See p. 448, Table 11.2, p. 435, and Table 11.3, p. 438. Downy mildew is a very damaging disease on roses grown as nursery stock (particularly under protection), and can result in rapid and spectacular defoliation of affected plants. The symptoms on the upper leaf surface usually consist of angular purple blotches. The sporulation of the fungus on the undersurface can be very difficult to see, and it is possible to confuse the disease with the early stages of black spot. Control begins with the manipulation of environmental conditions so that they are less favourable to the pathogen. Fungicides which have performed well in recent HDC-funded disease control trials include cymoxanil + mancozeb + oxadixyl* and fosetyl-aluminium* (both of which can be used as HV sprays on outdoor roses at the user's risk under the extension of use arrangements).

Honey fungus (*Armillaria* spp.)
See p. 443.

Powdery mildew (*Sphaerotheca pannosa*)
See p. 449 and Table 11.3, p. 438.

Rust (*Phragmidium mucronatum* and *P. tuberculatum*)
See p. 450 and Table 11.3, p. 438. This disease can affect stems as well as leaves. The rootstock *R. dumetorum* 'Laxa' is very susceptible to rust infection and unless this disease is controlled, the quality of the budded plants may be reduced substantially.

Verticillium wilt (*Verticillium dahliae*)
This pathogen has been isolated from wilting roses. Therefore, propagate only from healthy scions and root stocks and dig up and burn infected bushes. Do not replant in infected ground. For further details, see under *Acer*, p. 453. Affected roses do not often exhibit the vascular staining found in many other species affected by this disease.

Salix (willow)

Pests

Aphids
Several aphid species attack willow, causing the foliage to be soiled with honeydew and sooty mould. Large willow aphid (*Tuberolachnus salignus*) is a large, dark-brown aphid, which forms dense colonies on the lower stems in early

autumn. Wasps (*Vespula* spp.) are often attracted to the aphid colonies by the large quantities of honeydew that are excreted. For control of aphids see under *Rosa*, aphids, p. 519.

Leaf beetles (e.g. *Lochmaea caprea, Phyllodecta* spp. and *Plagiodera versicolora*)
The adults range in size from 3 mm to 6 mm long and are greyish-brown, metallic blue or greenish-black in colour. The yellowish-black larvae are up to 6 mm long and feed in groups on willow leaves; poplar is also a frequent host plant. Adults and larvae graze the lower leaf surface, causing the remaining damaged tissues to dry up. There are two generations during May to September. Control is worthwhile only if young trees are being damaged by the early summer generation, when chlorpyrifos, cypermethrin or deltamethrin could be used.

Sawflies – foliar-feeding (e.g. *Nematus pavidus* and *N. salicis*)
Several generations of gregarious caterpillars occur during the summer months and defoliation on willows may be severe. Control by spraying as for Columbine sawfly, p. 456, when larvae are present.

Sawflies – gall-forming (*Pontania* spp.)
Various species of gall-forming sawflies attack willows and cause the development of swollen yellow-green or red structures in the leaf blade. The galls sometimes resemble bean seeds and each one contains a single larva. There are two generations in May/June and August/September. No real damage is caused and control measures are not required.

Diseases

Anthracnose (*Marssonina salicicola*)
This fungal infection appears as small, brown spots on leaves and small, blackish cankers on shoots of various *Salix* spp., but is most troublesome on weeping willow (*S.* 'Chrysocoma') and in severe attacks the tree may lose its weeping habit. The disease is worse in wet seasons. The leaves fall prematurely and should be raked up and burnt. On large trees control is often impossible, and feeding to encourage vigour and resistance may be tried. The fungus overwinters in cankers but, because they are so small and numerous, it is not feasible to remove cankered shoots. Some control may be achieved by spraying with prochloraz or a copper-containing compound (see under *Berberis*, bacterial leaf spot, p. 458, for more details of these fungicides). One application should be made as the leaves unfold and one more in summer.

Black canker (*Glomerella miyabeana*)
This pathogen is found on *Salix alba vitellina, S. alba* spp. 'Cardinalis' and *S. americana,* amongst others. Symptoms can easily be confused with those of

willow scab (see below). Leaf spots occur in late spring/early summer. Affected leaves blacken and die, often remaining attached to the tree. The fungus can spread into the stem to produce flattened, elongated cankers, on which two types of fruiting body may be produced, one in the spring and one in autumn. Both appear as minute 'pimples' on the affected wood.

Control measures may be necessary on nursery trees, particularly where overhead watering is used. Captan*, carbendazim*, prochloraz, zineb* or a copper-containing fungicide can be applied; continued applications may be necessary throughout the growing season.

Honey fungus (*Armillaria* spp.)
See p. 443.

Rust (*Melampsora* spp.)
See p. 450 and Table 11.3, p. 438. Resistant clones are available of the *Salix* spp. commonly grown in short-rotation coppice, where planting of mixtures of clones is practised.

Scab (*Venturia saliciperda* – anamorph: *Pollaccia saliciperda*)
This disease is mainly a problem on *Salix alba vitellina*, contorted willow (*S. matsudana* 'Tortuosa') and crack willow (*S. fragilis*). Irregularly shaped spots occur on the leaves in spring, which enlarge until the leaves blacken and die. Affected leaves may fall or remain attached. Black lesions are produced on twigs, and if they become girdling the shoot may be killed. Shoot tips can blacken and curve over. In wet conditions, dark-brown fungal growth may be produced on the leaves, particularly near the veins.

The fungus overwinters on twig lesions. It produces spores on these in the spring, which infect the young leaves. The further development of the disease is favoured by cool, wet weather. Control may be necessary on nursery trees – spray with one or more of the fungicides listed for scab control in Table 11.3, p. 438.

Watermark disease (*Erwinia salicis*)
This is a major disease of cricket-bat willow (*Salix alba* 'Coerulea'). It also affects other *S. alba* varieties and some other species. The bacterium causes leaf death (leaves turn reddish-brown but remain on the tree) and internal staining of the affected branches. It is thought that bacteria on leaf surfaces of infected or healthy trees, or from infected branches (in the form of sticky bacterial 'ooze'), can infect via wounds (including natural wounds, such as leaf scars).

The disease may damage ornamental trees, but affected cricket-bat willows have brittle wood and, therefore, are useless for the production of cricket bats. Affected trees should be removed and burnt, and cutting implements sterilized. Legislation is in place, enabling compulsory felling of affected trees in some counties.

Salvia (sage)

Diseases

Downy mildew (*Peronospora lamii*)
See p. 448, Table 11.2, p. 435, and Table 11.3, p. 438.

Phytophthora root rot (*Phytophthora* spp.)
See p. 446 and Table 11.2, p. 435.

Powdery mildew (*Erysiphe cichoracearum*)
See p. 449 and Table 11.3, p. 438.

Wirestem and root rot (*Thanatephorus cucumeris* – anamorph: *Rhizoctonia solani*)
See p. 447 and Table 11.3, p. 438.

Sambucus (elder)

Pests

Elder aphid (*Aphis sambuci*)
This black aphid forms dense colonies on the young stems and foliage in May/June, often causing die-back. For control see p. 441.

Two-spotted spider mite (*Tetranychus urticae*)
Heavy infestations cause a pale mottling of the foliage and early leaf fall in late summer. For control see p. 442.

Diseases

Black root rot (*Thielaviopsis basicola* – synanamorph: *Chalara elegans*)
See p. 447 and Table 11.3, p. 438.

Phytophthora root rot (*Phytophthora* spp.)
See p. 446 and Table 11.2, p. 435.

Senecio

Diseases

Phytophthora root rot (*Phytophthora* spp.)
See p. 446 and Table 11.2, p. 435.

Solidago (golden rod)

Diseases

Powdery mildew (*Erysiphe cichoracearum*)
See p. 449 and Table 11.3, p. 438.

Sorbus (rowan, whitebeam)

Pests

Pear leaf blister mite (*Eriophyes pyri*)
This microscopic pest (on *Sorbus*, often known as 'sorbus blister mite' and sometimes considered to be the subspecies *E. pyri sorbi*) lives within the leaves during the summer months. Symptoms first show in late May as pale green blotches on the leaves but these gradually turn brown. No specific controls have been devised for this pest on *Sorbus*.

Diseases

Fireblight (*Erwinia amylovora*)
See under Pear, Chapter 8, p. 281.

Phytophthora root and collar rot (*Phytophthora* spp.)
See p. 446 and Table 11.2, p. 435.

Silver leaf (*Chondrostereum purpureum*)
See under *Prunus*, p. 514.

Spiraea

Diseases

Bacterial leaf spot (*Pseudomonas syringae* pv. *syringae*)
This disease produces circular or angular, brown spots, often with a purple margin. Branch die-back may also occur. Avoid overhead irrigation where the disease occurs and spray with a copper-containing fungicide (see under *Berberis*, bacterial leaf spot, p. 458, for further details of these fungicides).

Powdery mildew (*Sphaerotheca* sp.)
See p. 449 and Table 11.3, p. 438.

Stachys

Diseases

Powdery mildew (*Erysiphe galeopsidis*)
See p. 449 and Table 11.3, p. 438.

Stranvaesia

Diseases

Fireblight (*Erwinia amylovora*)
See under Pear, Chapter 8, p. 281.

Syringa (lilac)

Pests

Lilac leaf-miner moth (*Caloptilia syringella*)
There are two generations, with damage occurring in late May/June and in August/September. The white larvae (known as 'lilac leaf miners') start feeding gregariously as leaf miners, causing a brown blotch mine. As they grow larger they vacate the mine and roll up the leaf from the tip, using silk to hold it in place. Small infestations can be dealt with by picking off affected leaves; otherwise, spray with cypermethrin (off-label) (SOLA 1200/96) or dimethoate when mining commences.

Privet thrips (*Dendrothrips ornatus*)
See under *Ligustrum*, (privet) p. 495.

Diseases

Blight (*Pseudomonas syringae* pv. *syringae*)
Small, angular, brown spots develop first on leaves, then on young flowers and, finally, on the foliage; shoots blacken and wither away. Overhead irrigation and dense spacing of container plants increase disease spread. For control, cut back the affected portions to a healthy bud and spray with a copper-containing fungicide (see under *Berberis*, bacterial leaf spot, p. 458, for more details of these fungicides).

Honey fungus (*Armillaria* spp.)
See p. 443.

Phytophthora root rot (*Phytophthora* spp.)
See p. 446 and Table 11.2, p. 435.

Tagetes (African/French marigold)

See under Bedding plants, Chapter 10, p. 380.

Taxus (yew)

Pests

Vine weevil (*Otiorhynchus sulcatus*)
Young plants are very susceptible to larval damage, especially those growing in containers. For control see p. 443.

Diseases

Phytophthora root rot (*Phytophthora* spp.)
See p. 446 and Table 11.2, p. 435. Early symptoms consist of purpling and bronzing of the foliage. The disease can cause serious problems in formal planting schemes, such as hedges and mazes. It is vital to ensure that the soil structure and drainage are good prior to planting yews, so that rapid establishment occurs.

Thuja (western red cedar)

Pests

Conifer spinning mite (*Oligonychus ununguis*)
See under *Picea*, (spruce) p. 504.

Cypress aphid (*Cinara cupressi*)
See under *Chamaecyparis* etc., p. 465.

Diseases

Fomes root and butt rot (*Heterobasidion annosum*)
See under *Pinus*, (pine) p. 506.

Needle blight (*Didymascella thujina*)
This disease can cause serious damage to nursery trees. Individual needles or

fronds turn brown and, at a later stage, dark-brown spots appear on the upper surface of the needles. In a heavy attack most of the needles may become infected on the lower fronds and considerable die-back occurs.

These symptoms should not be confused with winter bronzing, which is a natural occurrence during severe winters, when the whole plant turns reddish-brown but recovers its natural colour in spring. Control can be achieved by HV sprays of prochloraz applied at 3-weekly intervals during the growing season.

***Phytophthora* root rot** (*Phytophthora* spp.)
See p. 446, and Table 11.2, p. 435.

Tilia (lime)

Pests

Lime leaf aphid (*Eucallipterus tiliae*)
This aphid infests the undersides of leaves that become soiled with honeydew and sooty mould. There is no effective control for tall trees but small specimens could be treated as for aphids on rose – see under *Rosa*, aphids, p. 519.

Lime nail-gall mite (*Eriophyes tiliae*)
The galls are red, upright structures, up to 15 mm long, which form on the upper surface of leaves during summer. The mites live inside these hollow structures. The subspecies *E. tiliae lateannulatus* is associated with small-leafed lime (*Tilia cordata*) and produces much smaller galls. No harm is caused by nail-gall mites, and attempts at control are not worthwhile.

Diseases

Bleeding canker (*Phytophthora* spp.)
See p. 446, under *Aesculus*, p. 453, and Table 11.2, p. 435.

Coral spot (*Nectria cinnabarina*)
See under *Acer*, p. 452.

Verticillium wilt (*Verticillium dahliae*)
See under *Acer*, p. 453.

Tulipa (tulip)

See Chapter 12, p. 555.

Ulmus (elm)

Diseases

Dutch elm disease (*Ophiostoma novi-ulmi*)
This is the most well-known tree disease of the last 20–30 years. It is transmitted by elm bark beetles (*Scolytus* spp.) or by natural root-grafting between trees. There has been a resurgence of the disease in recent years, as many of the suckers arising from previously affected trees are now of a size where they, too, can be affected. The symptoms are all too familiar: rapid wilting and death of trees in spring or summer or die-back of individual branches, often with the ends of shoots hooked over into 'shepherd's crooks'. Prominent vascular staining is often present within affected branches.

There are currently specific off-label approvals for the use of metam-sodium (SOLA 0592/93) (to sever the root grafts which can transmit the disease from tree to tree) and thiabendazole (SOLA 0990/95) (as an injection to try to arrest the infection in trees with early symptoms, or to protect nearby trees). However, if symptoms have been present for some time, the tree should be removed as soon as possible. The movement of trees is subject to EU legislation, with certain areas having 'Protected Zone' status. There is no effective insecticide treatment for the bark beetles.

The closely related fungus *Ophiostoma ulmi* causes a similar, but much milder, disease from which trees often recover.

Vaccinium

Diseases

Phytophthora root and crown rot (*Phytophthora* spp.)
See p. 446 and Table 11.2, p. 435.

Verbascum (mullein)

Diseases

Powdery mildew (*Erysiphe verbasci*)
See p. 449 and Table 11.3, p. 438.

Veronica

See under *Hebe*, p. 485.

Viburnum

Pests

Aphids (e.g. *Aphis fabae* and *A. viburni*)
Both species cause leaf curling in the spring and early summer. *Viburnum carlesii* is particularly susceptible. For control of aphids see under *Rosa*, p. 519.

Viburnum beetle (*Pyrrhalta viburni*)
The larvae form holes in the foliage in late May/June with further damage being caused by adults in July/August. Overwintering eggs are laid in the shoots, causing canker-like pits in the bark. The adults are greyish-brown and *c.* 7 mm long; the larvae are yellowish, with extensive black markings. *Viburnum opulus* and *V. lantana* are the main hosts, but *V. tinus* is also attacked. Contact insecticides such as chlorpyrifos*, cypermethrin* or deltamethrin* should give control.

Viburnum whitefly (*Aleurotuba jelinekii*)
The small, white-winged adults are present throughout the summer on *Viburnum tinus*. Other types of *Viburnum* are not affected. The overwintering stage is a black scale-like pupa, which has a white, waxy fringe. All stages in the life-cycle occur on the undersides of the leaves. This is not a very damaging pest but heavy infestations may soil the foliage with honeydew and sooty mould. Control is as for Rhododendron, Azalea whitefly, p. 516.

Diseases

Honey fungus (*Armillaria* spp.)
See p. 443.

Leaf spot and die-back (*Phoma exigua* var. *viburni*)
This disease most commonly affects *Viburnum tinus*. Pale or purplish spots develop on the leaves. Similar-coloured lesions may develop on stems and branches and, if girdling occurs, the tips of the shoots die back. Affected cuttings turn black from the base upwards and die. Remove affected parts. Spray with carbendazim* or prochloraz.

Vinca (periwinkle)

Diseases

Blackleg (*Phoma exigua* var. *inoxydabilis*)
This fungus attacks *Vinca minor*, causing black spots on leaves and extensive black lesions on the trailing stems. Fruiting bodies develop within the affected

tissue, but as these are very small and also black in colour they can be difficult to see. As the spores are splash-dispersed the disease is worst in wet weather or where overhead irrigation is used. Many of the fungicides recommended in Table 11.3, p. 438, for leaf spot control will have some activity against this disease. In recent HDC-funded research work, flusilazole + carbendazim* and a mixture of thiophanate-methyl* and mancozeb* gave the best control. These products can be used outdoors on *Vinca* at the user's risk under the extension of uses arrangements.

Rust (*Puccinia vincae*)
See p. 450 and Table 11.3, p. 438. Numerous pustules, producing brown masses of spores, develop on the undersides of leaves on shoots which take on an erect habit and do not flower. Although the fungus is said to be systemic, the disease does not occur every year on infected plants.

Viola odorata (violet)

Pests

Aphids (e.g. *Myzus ornatus*)
See under *Dahlia*, p. 473.

Two-spotted spider mite (*Tetranychus urticae*)
See p. 442.

Violet leaf midge (*Dasineura affinis*)
Damage is caused by the larvae infesting the leaves. Their presence causes the leaves to become greatly thickened and prevents them from unrolling. The larvae are whitish-orange maggots up to 2 mm long. Symptoms are most readily seen in the winter, since galled leaves remain on the plant whereas normal foliage dies down. Control is difficult as there are often three or four generations per year and the larvae are well protected by the galled leaf. On a small scale, hand removal of affected leaves can be effective; otherwise dimethoate* could be tried when galling commences.

Diseases

Black root rot (*Thielaviopsis basicola* – synanamorph: *Chalara elegans*)
See p. 447 and Table 11.3, p. 438.

Viola tricolor (pansy, viola)

See under Bedding plants, Chapter 10, p. 380, and under Aphids, p. 441.

Weigela

Pests

Chrysanthemum nematode (*Aphelenchoides ritzemabosi*)
The microscopic nematodes live within the foliage, causing brown necrotic areas separated by the larger leaf veins from uninfested parts of the leaf. This is a troublesome pest where plants are grown in close proximity, such as in nursery beds. Avoid overhead watering which provides ideal conditions for the pest to spread and do not propagate from infested stock. Rooted cuttings can be treated with aldicarb (off-label) (SOLA 1325/95).

Yucca

Diseases

Leaf spot (*Coniothyrium concentricum*)
Large, brown, well-defined leaf spots with grey centres are caused by this disease, and minute, pinhead-sized sporing structures often develop in concentric rings on the infected tissue. To control, remove severely infested leaves and spray with carbendazim* or prochloraz to protect young developing leaves.

Zinnia

See under Bedding plants, Chapter 10, p. 380.

List of pests cited in the text*

Acyrthosiphon pisum (Hemiptera: Aphididae)	pea aphid
Adelges abietis (Hemiptera: Adelgidae)	spruce pineapple-gall adelges
Adelges laricis (Hemiptera: Adelgidae)	larch adelges
Agrilus pannonicus (Coleoptera: Buprestidae)	a jewel beetle
Agromyza demeijerei (Diptera: Agromyzidae)	laburnum leaf-miner fly
Agrotis segetum (Lepidoptera: Noctuidae)	turnip moth
Aleurotuba jelinekii (Hemiptera: Aleyrodidae)	viburnum whitefly
Allantus cinctus (Hymenoptera: Tenthredinidae)	banded rose sawfly
Andricus kollari (Hymenoptera: Cynipidae)	marble gall wasp
Andricus quercuscalicis (Hymenoptera: Cynipidae)	acorn cup gall cynipid
Aphelenchoides fragariae (Tylenchida: Aphelenchoididae)	leaf nematode
Aphelenchoides ritzemabosi (Tylenchida: Aphelenchoididae)	chrysanthemum nematode
Aphis fabae (Hemiptera: Aphididae)	black bean aphid
Aphis genistae (Hemiptera: Aphididae)	genista aphid
Aphis gossypii (Hemiptera: Aphididae)	melon & cotton aphid

Aphis sambuci (Hemiptera: Aphididae)	elder blackfly
Aphis viburni (Hemiptera: Aphididae)	viburnum aphid
Archips podana (Lepidoptera: Tortricidae)	fruit tree tortrix moth
Arge ochropus (Hymenoptera: Argidae)	large rose sawfly
Arge paganus (Hymenoptera: Argidae)	variable rose sawfly
Arion hortensis (Stylommatophora: Arionidae)	garden slug
Artacris macrorhynchus (Prostigmata: Eriophyidae)	maple bead-gall mite
Biorhiza pallida (Hymenoptera: Cynipidae)	oak-apple gall wasp
Blennocampa phyllocolpa (Hymenoptera: Tenthredinidae)	leaf-rolling rose sawfly
Brachycaudus helichrysi (Hemiptera: Aphididae)	plum leaf-curling aphid
Bryobia spp. (Prostigmata: Tetranychidae)	bryobia mites
Cacoecimorpha pronubana (Lepidoptera: Tortricidae)	carnation tortrix moth
Caliroa cerasi (Hymenoptera: Tenthredinidae)	pear slug sawfly
Caloptilia syringella (Lepidoptera: Gracillaridae)	lilac leaf-miner moth
Carulaspis juniperi (Hemiptera: Diaspididae)	juniper scale
Chionaspis salicis (Hemiptera: Diaspididae)	willow scale
Chloropulvinaria floccifera (Hemiptera: Coccidae)	cushion scale
Chromatomyia syngenesiae (Diptera: Agromyzidae)	chrysanthemum leaf miner
Cinara cupressi (Hemiptera: Lachnidae)	cypress aphid
Cinara fresai (Hemiptera: Lachnidae)	American juniper aphid
Cinara junipari (Hemiptera: Lachnidae)	juniper aphid
Cionus scrophulariae (Coleoptera: Curculionidae)	figwort weevil
Cladius difformis (Hymenoptera: Tenthredinidae)	an antler sawfly
Cladius pectinicornis (Hymenoptera: Tenthredinidae)	an antler sawfly
Claremontia waldeheimii (Hymenoptera: Tenthredinidae)	a geum sawfly
Coccus hesperidum (Hemiptera: Coccidae)	brown soft scale
Contarinia quinquenotata (Diptera: Cecidomyiidae)	hemerocallis gall midge
Cryptococcus fagisuga (Hemiptera: Eriococcidae)	beech scale
Dasineura affinis (Diptera: Cecidomyiidae)	violet leaf midge
Dasineura gleditchiae (Diptera: Cecidomyiidae)	honeylocust gall midge
Delia radicum (Diptera: Anthomyiidae)	cabbage root fly
Dendrothrips ornatus (Thysanoptera: Thripidae)	privet thrips
Deroceras reticulatum (Stylommatophora: Limacidae)	field slug
Dichomeris marginella (Lepidoptera: Gelechiidae)	juniper webber moth
Ditylenchus dipsaci (Tylenchida: Tylenchidae)	stem nematode
Drepanosiphum platanoidis (Hemiptera: Callaphididae)	sycamore aphid
Edwardsiana rosae (Hemiptera: Cicadellidae)	rose leafhopper
Elatobium abietinum (Hemiptera: Aphididae)	green spruce aphid
Endelomyia aethiops (Hymenoptera: Tenthredinidae)	rose slug sawfly
Eriophyes eriobius (Prostigmata: Eriophyidae)	maple felt gall mite
Eriophyes genistae (Prostigmata: Eriophyidae)	broom gall mite
Eriophyes psilomerus (Prostigmata: Eriophyidae)	sycamore felt gall mite
Eriophyes pyri (Prostigmata: Eriophyidae)	pear leaf blister mite
Eriophyes tiliae (Prostigmata: Eriophyidae)	lime nail-gall mite
Eriosoma lanigerum (Hemiptera: Pemphigidae)	woolly aphid
Eucallipterus tiliae (Hemiptera: Callaphididae)	lime leaf aphid
Euproctis chrysorrhoea (Lepidoptera: Lymantriidae)	brown-tail moth
Forficula auricularia (Dermaptera: Forficulidae)	common earwig
Galerucella nymphaeae (Coleoptera: Chrysomelidae)	water-lily beetle
Graphocephala fennahi (Hemiptera: Cicadellidae)	rhododendron leafhopper
Hauptidia maroccana (Hemiptera: Cicadellidae)	glasshouse leafhopper
Helix aspersa (Stylommatophora: Helicidae)	garden snail

Hyadaphis passerini (Hemiptera: Aphididae)	honeysuckle aphid
Lepidosaphes ulmi (Hemiptera: Diaspididae)	mussel scale
Leucoptera laburnella (Lepidoptera: Lyonetiidae)	laburnum leaf-miner moth
Lochmaea caprea (Coleoptera: Chrysomelidae)	a willow leaf beetle
Lygocoris pabulinus (Hemiptera: Miridae)	common green capsid
Lygus rugulipennis (Hemiptera: Miridae)	tarnished plant bug
Lyonetia clerkella (Lepidoptera: Lyonetiidae)	*larva* = apple leaf miner
Macrosiphum albifrons (Hemiptera: Aphididae)	lupin aphid
Macrosiphum euphorbiae (Hemiptera: Aphididae)	potato aphid
Macrosiphum rosae (Hemiptera: Aphididae)	rose aphid
Malacosoma neustria (Lepidoptera: Lasiocampidae)	lackey moth
Mamestra brassicae (Lepidoptera: Noctuidae)	cabbage moth
Metopolophium dirhodum (Hemiptera: Aphididae)	rose/grain aphid
Monophadnoides rubi (Hymenoptera: Tenthredinidae)	a geum sawfly
Myzus cerasi (Hemiptera: Aphididae)	cherry blackfly
Myzus ligustri (Hemiptera: Aphididae)	privet aphid
Myzus ornatus (Hemiptera: Aphididae)	violet aphid
Myzus persicae (Hemiptera: Aphididae)	peach/potato aphid
Nematus pavidus (Hymenoptera: Tenthredinidae)	lesser willow sawfly
Nematus salicis (Hymenoptera: Tenthredinidae)	willow sawfly
Nematus spiraeae (Hymenoptera: Tenthredinidae)	aruncus sawfly
Neuroterus quercusbaccarum (Hymenoptera: Cynipidae)	oak leaf spangle-gall cynipid
Noctua pronuba (Lepidoptera: Noctuidae)	large yellow underwing moth
Numonia suavella (Lepidoptera: Pyralidae)	porphyry knothorn moth
Oligonychus ununguis (Prostigmata: Tetranychidae)	conifer spinning mite
Operophtera brumata (Lepidoptera: Geometridae)	winter moth
Orgyia antiqua (Lepidoptera: Lymantriidae)	vapourer moth
Otiorhynchus singularis (Coleoptera: Curculionidae)	clay-coloured weevil
Otiorhynchus sulcatus (Coleoptera: Curculionidae)	vine weevil
Parthenolecanium corni (Hemiptera: Coccidae)	brown scale
Pealius azaleae (Hemiptera: Aleyrodidae)	azalea whitefly
Periphyllus spp. (Hemiptera: Chaitophoridae)	maple aphids
Philaenus spumarius (Hemiptera: Cercopidae)	common froghopper
Phlogophora meticulosa (Lepidoptera: Noctuidae)	angle-shades moth
Phyllaphis fagi (Hemiptera: Callaphididae)	beech aphid
Phyllodecta spp. (Coleoptera: Chrysomelidae)	willow leaf beetles
Phyllonorycter leucographella (Lepidoptera: Gracillaridae)	pyracantha leaf-miner moth
Phyllotreta spp. (Coleoptera: Chrysomelidae)	flea beetles
Phymatocera aterrima (Hymenoptera: Tenthredinidae)	Solomon's seal sawfly
Phytomyza aconiti (Diptera: Agromyzidae)	*larva* = delphinium leaf miner
Phytomyza cytisi (Diptera: Agromyzidae)	*larva* = laburnum leaf miner
Phytomyza ilicis (Diptera: Agromyzidae)	*larva* = holly leaf miner
Phytonemus pallidus (Prostigmata: Tarsonemidae)	cyclamen mite
Plagiodera versicolora (Coleoptera: Chrysomelidae)	a willow leaf beetle
Pontania spp. (Hymenoptera: Tenthredinidae)	bean gall sawflies
Pristiphora aquilegiae (Hymenoptera: Tenthredinidae)	columbine sawfly
Psylla buxi (Hemiptera: Psyllidae)	box sucker
Pulvinaria regalis (Hemiptera: Coccidae)	horse chestnut scale
Pyrrhalta viburni (Coleoptera: Chrysomelidae)	viburnum beetle
Pyrrhula pyrrhula (Passeriformes: Fringillidae)	bullfinch
Rhopalosiphum nymphaeae (Hemiptera: Aphididae)	water-lily aphid
Sciurus carolinensis (Rodentia: Scuiridae)	grey squirrel

Scolytus spp. (Coleoptera: Scolytidae)	bark beetles
Scythropia crataegella (Lepidoptera: Yponomeutidae)	hawthorn webber moth
Stagona pini (Hemiptera: Pemphigidae)	pine root aphid
Stephanitis rhododendri (Hemiptera: Tingidae)	rhododendron lace-bug
Tetranychus urticae (Prostigmata: Tetranychidae)	two-spotted spider mite
Thrips fuscipennis (Thysanoptera: Thripidae)	rose thrips
Thrips tabaci (Thysanoptera: Thripidae)	onion thrips
Trialeurodes vaporariorum (Hemiptera: Aleyrodidae)	glasshouse whitefly
Trionymus diminutus (Hemiptera: Pseudococcidae)	phormium mealybug
Trioza alacris (Hemiptera: Triozidae)	bay sucker
Tuberolachnus salignus (Hemiptera: Lachnidae)	large willow aphid
Unaspis euonymi (Hemiptera: Diaspididae)	euonymus scale
Yponomeuta cagnatella (Lepidoptera: Yponomeutidae)	a small ermine moth
Yponomeuta padella (Lepidoptera: Yponomeutidae)	common small ermine moth
Yponomeuta spp. (Lepidoptera: Yponomeutidae)	small ermine moths

* The classification in parentheses represents order and family.

List of pathogens/diseases (other than viruses) cited in the text*

Agrobacterium tumefaciens (Gracilicutes: Proteobacteria)†	crown gall
Albugo candida (Oomycetes)	white blister of *Lunaria*
Alternaria sp. (Hyphomycetes)	leaf spot of *Cheiranthus*
Aphanomyces euteiches (Oomycetes)	foot and root rot of *Lathyrus*
Apiognomonia errabunda (Ascomycota)	anthracnose or leaf scorch of *Platanus*
Armillaria gallica (Basidiomycetes)	honey fungus
Armillaria mellea (Basidiomycetes)	honey fungus
Armillaria ostoyae (Basidiomycetes)	honey fungus
Armillaria spp. (Basidiomycetes)	honey fungus
Ascochyta bohemica (Coelomycetes)	leaf spot of *Campanula*
Botryotinia fuckeliana (Ascomycota)	(common) grey mould
Botrytis cinerea (Hyphomycetes)	– anamorph of *Botrytinia fuckeliana*
Botrytis paeoniae (Hyphomycetes)	grey mould of *Convallaria* and *Paeonia*
Bremia lactucae (Oomycetes)	downy mildew of *Gaillardia*
Chalara elegans (Hyphomycetes)	– synanamorph of *Thielaviopsis basicola*
Chalaropsis thielavioides (Hyphomycetes)	black mould of *Rosa*
Chondrostereum purpureum (Basidiomycetes)	silver leaf
Chrysomyxa abietis (Teliomycetes)	needle rust of *Picea*
Chrysomyxa rhododendri (Teliomycetes)	needle rust of *Picea*
Ciborinia camelliae (Ascomycota)	flower blight of *Camellia*
Coleosporium tussilaginis f. sp. *campanulae* (Teliomycetes)	rust of *Campanula*
Colletotrichum acutatum (Coelomycetes)	anthracnose of *Lupinus*
Coniothyrium concentricum (Coelomycetes)	leaf spot of *Yucca*
Coniothyrium hellebori (Coelomycetes)	leaf spot of *Helleborus*
Cronartium ribicola (Teliomycetes)	blister rust of *Pinus*
Cumminsiella mirabilissima (Teliomycetes)	rust of *Mahonia*
Cyclaneusma minus (Ascomycota)	needle cast of *Pinus*
Cylindrocarpon destructans (Hyphomycetes)	– anamorph of *Nectria radicicola*

Cylindrocladium sp. (Hyphomycetes)	leaf blight of *Buxus*
Didymascella thujina (Ascomycota)	needle blight of *Thuja*
Diplocarpon mespili (Ascomycota)	leaf spot of *Crataegus*
Diplocarpon rosae (Ascomycota)	black spot of *Rosa*
Discula spp. (Coelomycetes)	anthracnose of *Cornus*
Entomosporium maculatum (Coelomycetes)	– anamorph of *Diplocarpon mespili*
Entyloma calendulae f. sp. *dahliae* (Ustomycetes)	smut of *Dahlia*
Erwinia amylovora (Gracilicutes: Proteobacteria)†	fireblight
Erwinia carotovora ssp. *carotovora* (Gracilicutes: Proteobateria)†	rhizome rot of *Iris*
Erwinia salicis (Gracilicutes: Proteobacteria)†	watermark disease of *Salix*
Erysiphe aquilegiae var. *aquilegiae* (Ascomycota)	powdery mildew of *Aquilegia, Clematis*
Erysiphe aquilegiae var. *ranunculi* (Ascomycota)	powdery mildew of *Delphinium*
Erysiphe asperifolium (Ascomycota)	powdery mildew of *Pulmonaria*
Erysiphe cichoracearum (Ascomycota)	powdery mildew of *Salvia, Solidago*
Erysiphe cichoracearum var. *cichoracearum* (Ascomycota)	powdery mildew of *Aster* (Michaelmas daisy)
Erysiphe galeopsidis (Ascomycota)	powdery mildew of *Stachys*
Erysiphe tortilis (Ascomycota)	powdery mildew of *Cornus*
Erysiphe verbasci (Ascomycota)	powdery mildew of *Verbascum*
Exobasidium camelliae (Basidiomycetes)	gall of *Camellia*
Exobasidium vaccinii var. *japonicum* (Basidiomycetes)	gall or false bloom of *Rhododendron* (azalea)
Fusarium oxysporum f. sp. *callistephi* (Hyphomycetes)	fusarium wilt of *Callistephus*
Fusarium oxysporum f. sp. *dianthi* (Hyphomycetes)	fusarium wilt of *Dianthus*
Fusarium spp. (Hyphomycetes)	fusarium wilt of *Lathyrus*
Gemmamyces piceae (Ascomycota)	bud blight of *Picea*
Glomerella cingulata (Ascomycota)	rotting of cuttings of *Rhododendron*
Glomerella cingulata f. sp. *camelliae* (Ascomycota)	leaf blotch, twig blight, canker and die-back of *Camellia*
Glomerella miyabeana (Ascomycota)	black canker of *Salix*
Guignardia aesculi (Ascomycota)	leaf blotch of *Aesculus*
Heterobasidion annosum (Basidiomycetes)	root and butt rot of Cupressaceae, *Pinus, Thuja*
Itersonilia perplexans (Hyphomycetes)	petal blight of *Centaurea, Dendranthemum*
Kabatina spp. (Coelomycetes)	shoot blight of *Cupressus*, × *Cupressocyparis, Juniperus*
Lophodermium seditiosum (Ascomycota)	needle cast of *Pinus*
Marssonina aquilegiae (Coelomycetes)	leaf spot of *Aquilegia*
Marssonina daphnes (Coelomycetes)	leaf spot of *Daphne*
Marssonina salicicola (Coelomycetes)	anthracnose of *Salix*
Marssonina spp. (Coelomycetes)	leaf spot of *Populus*
Melampsora hypericorum (Teliomycetes)	rust of *Hypericum*
Melampsora spp. (Teliomycetes)	rust of *Populus, Salix*
Melampsorella caryophyllacearum (Teliomycetes)	rust of *Abies*
Melampsoridium betulinum (Teliomycetes)	rust of *Betula*
Meria laricis (Hyphomycetes)	needle cast of *Larix*
Microsphaera alphitoides (Ascomycota)	powdery mildew of *Quercus*
Microsphaera berberidis (Ascomycota)	powdery mildew of *Mahonia*
Microsphaera grossulariae (Ascomycota)	powdery mildew of *Ribes*

Microsphaera spp. (Ascomycota)	powdery mildew of *Rhododendron*
Microsphaera trifolii (Ascomycota)	powdery mildew of *Lathyrus, Lupinus*
Microsphaeropsis pittospororum (Coelomycetes)	leaf spot of *Pittosporum*
Monilinia johnsonii (Ascomycota)	leaf blotch of *Crataegus*
Monilinia spp. (Ascomycota)	blossom wilt of *Malus, Prunus*
Monochaetia karstenii (Coelomycetes)	leaf and stem blight of *Camellia*
Mycosphaerella dianthi (Ascomycota)	ring spot of *Dianthus*
Mycosphaerella macrospora (Ascomycota)	leaf spot of *Iris*
Nectria cinnabarina (Ascomycota)	coral spot
Nectria coccinea (Ascomycota)	beech bark disease
Nectria galligena (Ascomycota)	canker of *Malus*
Nectria haematococca var. *brevicona* (Ascomycota)	canker and die-back of *Robinia*
Nectria radicicola (Ascomycota)	basal rot of *Rhododendron* cuttings
Oidium chrysanthemi (Hyphomycetes)	powdery mildew of *Dendranthema*
Oidium euonymi-japonicae (Hyphomycetes)	powdery mildew of *Euonymus*
Oidium hortensiae (Hyphomycetes)	powdery mildew of *Hydrangea*
Oidium sp. (Hyphomycetes)	powdery mildew of *Centaurea, Cotinus, Eucalyptus, Laurus, Limonium, Lonicera, Polemonium*
Ophiostoma novi-ulmi (Ascomycota)	Dutch elm disease
Ophiostoma ulmi (Ascomycota)	Dutch elm disease
Ovulinia azaleae (Ascomycota)	petal blight of *Rhododendron* (azalea)
Peronospora arborescens (Oomycetes)	downy mildew of *Meconopsis*
Peronospora cytisi (Oomycetes)	downy mildew of *Laburnum*
Peronospora digitalis (Oomycetes)	downy mildew of *Digitalis*
Peronospora gei (Oomycetes)	downy mildew of *Geum*
Peronospora geranii (Oomycetes)	downy mildew of *Geranium*
Peronospora grisea (Oomycetes)	downy mildew of *Hebe*
Peronospora hariotii (Oomycetes)	downy mildew of *Buddleja*
Peronospora lamii (Oomycetes)	downy mildew of *Lamium, Salvia*
Peronospora leptoclada (Oomycetes)	downy mildew of *Helianthemum*
Peronospora parasitica (Oomycetes)	downy mildew of *Cheiranthus, Matthiola*
Peronospora sparsa (Oomycetes)	downy mildew of *Prunus, Rosa*
Peronospora statices (Oomycetes)	downy mildew of *Limonium*
Peronospora viciae (Oomycetes)	downy mildew of *Lathyrus*
Pestalotiopsis funerea (Coelomycetes)	needle blight, stem die-back of *Calluna, Chamaecyparis,* × *Cupressocyparis, Cupressus, Erica, Juniperus*
Pestalotiopsis guepini (Coelomycetes)	leaf and stem blight of *Camellia*
Pestalotiopsis sydowiana (Coelomycetes)	leaf spot of *Rhododendron*
Pestalotiopsis spp. (Coelomycetes)	needle blight, stem die-back of conifers and heathers
Phialophora asteris (Hyphomycetes)	wilt of *Aster* (Michaelmas daisy)
Phoma clematidina (Coelomycetes)	wilt of *Clematis*
Phoma exigua var. *inoxydabilis* (Coelomycetes)	blackleg of *Vinca*; crown rot of *Ajuga*
Phoma exigua var. *viburni* (Coelomycetes)	leaf spot and die-back of *Viburnum*
Phoma gentianae-sino-ornatae (Coelomycetes)	root rot of *Gentiana*
Phoma lavandulae (Coelomycetes)	shab of *Lavandula*
Phomopsis ericaceana (Coelomycetes)	rotting of cuttings of *Rhododendron*

Phomopsis juniperovora (Coelomycetes)	needle blight, stem die-back of *Juniperus*
Phomopsis spp. (Coelomycetes)	rotting of cuttings of *Rhododendron* and other evergreen species
Phragmidium mucronatum (Teliomycetes)	rust of *Rosa*
Phragmidium tuberculatum (Teliomycetes)	rust of *Rosa*
Phyllactinia guttata (Ascomycota)	powdery mildew of *Fagus*
Phyllosticta cornicola (Coelomycetes)	leaf spot of *Cornus*
Phyllosticta garryae (Coelomycetes)	leaf spot of *Garrya*
Phyllosticta hedericola (Coelomycetes)	leaf spot of *Hedera*
Phyllosticta primulicola (Coelomycetes)	leaf spot of *Primula*
Phytophthora cactorum (Oomycetes)	bleeding canker of *Aesculus*; crown rot of *Malus*
Phytophthora cambivora (Oomycetes)	'ink' disease of *Castanea*
Phytophthora cinnamomi (Oomycetes)	'ink' disease of *Castanea*
Phytophthora citricola (Oomycetes)	bleeding canker of *Aesculus*; branch die-back of *Ceanothus*; collar rot of *Fremontodendron*
Phytophthora ilicis (Oomycetes)	blight of *Ilex*
Phytophthora primulae (Oomycetes)	brown core of *Primula*
Phytophthora spp. (Oomycetes)	foot and root rot of *Abies, Acer, Alnus*, azalea, *Buxus, Calluna, Camellia, Choisya*, conifers, *Convolvulus, Erica, Escallonia, Eucalyptus, Forsythia, Hebe, Hedera, Hibiscus, Picea, Pieris, Salvia, Sambucus, Senecio, Sorbus, Syringa, Taxus, Thuja, Vaccinium*; bleeding canker of *Fagus, Tilia*; damping-off, foot and root rot of *Cheiranthus, Lathyrus, Matthiola* and other seedlings
Phytophthora syringae (Oomycetes)	collar rot of *Malus*
Plasmodiophora brassicae (Plasmodiophoromycetes)	clubroot of *Cheiranthus, Matthiola*
Pleiochaeta setosa (Hyphomycetes)	leaf spot of *Cytisus, Laburnum, Lupinus*
Podosphaera leucotricha (Ascomycota)	powdery mildew of *Malus, Photinia*
Podosphaera tridactyla (Ascomycota)	powdery mildew of *Crataegus, Prunus*
Pollaccia saliciperda (Hyphomycetes)	– anamorph of *Venturia saliciperda*
Pseudomonas sp. (Gracilicutes: Proteobacteria)†	bacterial leaf spot of *Primula*
Pseudomonas syringae pv. *berberidis* (Gracilicutes: Proteobacteria)†	bacterial leaf spot of *Berberis*
Pseudomonas syringae pv. *delphinii* (Gracilicutes: Proteobacteria)†	black blotch of *Delphinium*
Pseudomonas syringae pv. *mors-prunorum* (Gracilicutes: Proteobacteria)†	bacterial canker of *Prunus*
Pseudomonas syringae pv. *syringae* (Gracilicutes: Proteobacteria)†	bacterial blight of *Cornus, Euonymus, Forsythia, Magnolia, Mahonia, Philadelphus, Spiraea, Syringa*
Pseudonectria rousseliana (Ascomycota)	leaf blight and die-back of *Buxus*
Puccinia arenariae (Teliomycetes)	rust of *Dianthus barbatus*
Puccinia buxi (Teliomycetes)	rust of *Buxus*

Puccinia chrysanthemi (Teliomycetes)	brown rust of *Dendranthema*
Puccinia cyani (Teliomycetes)	rust of *Centaurea*
Puccinia graminis (Teliomycetes)	rust of *Mahonia*
Puccinia horiana (Teliomycetes)	white rust of *Dendranthema*
Puccinia malvacearum (Teliomycetes)	rust of *Althaea, Lavatera*
Puccinia vincae (Teliomycetes)	rust of *Vinca*
Pucciniastrum epilobii (Teliomycetes)	rust of *Abies*
Pycnostysanus azaleae (Hyphomycetes)	bud blast of *Rhododendron*
Pythium spp. (Oomycetes)	damping-off, foot and root rot of *Cheiranthus, Geranium, Lathyrus, Matthiola, Primula* and other seedlings; specific replant disease of *Malus, Prunus*
Ramularia ajugae (Hyphomycetes)	leaf spot of *Ajuga*
Ramularia deusta (Hyphomycetes)	white mould of *Lathyrus*
Ramularia nymphaearum (Hyphomycetes)	leaf spot of *Nymphaea*
Ramularia spp. (Hyphomycetes)	leaf spot of *Primula*
Rhizoctonia solani (Hyphomycetes)	– anamorph of *Thanatephorus cucumeris*
Rhytisma acerinum (Ascomycota)	tar spot of *Acer*
Sawadaea bicornis (Ascomycota)	powdery mildew of *Acer*
Sclerotinia sclerotiorum (Ascomycota)	sclerotinia rot of *Dahlia*
Seiridium cardinale (Coelomycetes)	canker of *Cupressus*, × *Cupressocyparis*
Septoria cornicola (Coelomycetes)	leaf spot of *Cornus*
Septoria euonymi (Coelomycetes)	leaf spot of *Euonymus*
Septoria exotica (Coelomycetes)	leaf spot of *Hebe*
Septoria paeoniae (Coelomycetes)	leaf spot of *Paeonia*
Septoria spp. (Coelomycetes)	blotch or leaf spot of *Dendranthema*
Sphaerotheca alchemillae (Ascomycota)	powdery mildew of *Potentilla*
Sphaerotheca fuliginea (Ascomycota)	powdery mildew of *Phlox*
Sphaerotheca pannosa (Ascomycota)	powdery mildew of *Rosa*
Sphaerotheca sp. (Ascomycota)	powdery mildew of *Spiraea*
Spilocaea pyracanthae (Hyphomycetes)	scab of *Pyracantha*
Stigmina carpophila (Hyphomycetes)	shot-hole of *Prunus*
Taphrina deformans (Ascomycota)	peach leaf curl
Taphrina populina (Ascomycota)	yellow leaf blister of *Populus*
Thanatephorus cucumeris (Basidiomycetes)	web blight, leaf blight, foot and root rot of *Calluna*, azalea, *Convolvulus, Erica*; wirestem, damping-off, foot and root rot of *Cheiranthus, Geranium, Lathyrus, Matthiola, Primula, Salvia* and seedlings
Thielaviopsis basicola (Hyphomycetes)	black root rot of *Cheiranthus, Choisya, Geranium, Lathyrus, Matthiola, Primula, Prunus, Sambucus, Viola odorata*
Trochila laurocerasi (Ascomycota)	shot-hole of *Prunus*
Uromyces dianthi (Teliomycetes)	rust of *Dianthus*
Uromyces geranii (Teliomycetes)	rust of *Geranium*
Venturia inaequalis (Ascomycota)	scab of *Malus*
Venturia saliciperda (Ascomycota)	scab of *Salix*

Verticillium dahliae (Hyphomycetes)	wilt of *Acer, Catalpa, Ceanothus, Cercis, Cotinus, Rhus, Robinia, Rosa, Tilia*
Volutella buxi (Hyphomycetes)	– anamorph of *Pseudonectria rousseliana*
Xanthomonas campestris pv. *campestris* (Gracilicutes: Proteobacteria)†	bacterial leaf spot, leaf and stem rot, wilt of *Cheiranthus*
Xanthomonas campestris pv. *hederae* (Gracilicutes: Proteobacteria)†	bacterial leaf spot of *Hedera, Primula*
Xanthomonas campestris pv. *incanae* (Gracilicutes: Proteobacteria)†	bacterial spot, wilt of *Matthiola*
Xanthomonas populi (Gracilicutes: Proteobacteria)†	bacterial canker of *Populus*

* For fungi, the classification in parentheses refers to class, although this is not possible within the phylum Ascomycota where classes have yet to be satisfactorily defined (see *Mycological Research*, February 2000). Oomycetes are now classified in Chromista with the brown algae, rather than as true fungi. Plasmodiophoromycetes are now classified as Protozoa rather than as true fungi. Some fungi have an asexual (anamorph) and a sexual (teleomorph) state, and the convention is to refer to them by their teleomorph name. However, where anamorph (or synanamorph) names are still in common use these are listed and cross-referenced to the teleomorph name. Strictly, fungi classified as Coelomycetes and Hyphomycetes should be known as 'hyphomycetous anamorphs' and 'coelomycetous anamorphs' of the relevant teleomorph taxon (e.g. hyphomycetous anamorphic Sclerotiniaceae, for *Botrytis fabae*), respectively. These problems highlight the continual changes in the classification of the fungi.

† Bacteria – the classification in parentheses refers to division and class.

Chapter 12
Pests and Diseases of Outdoor Bulbs and Corms

M.J. Lole
ADAS Wolverhampton, West Midlands

J.B. Briggs
ADAS, Guildford, Surrey

Introduction

The UK bulb industry produces dry bulbs for sale, together with cut flowers, from both the field and the glasshouse. About 4100 ha of bulbs were planted in 1998, of which some 3800 ha were down to *Narcissus*, by far our most important bulb crop. Approximately 90 ha of gladioli are grown, together with smaller areas of anemone, tulip, iris and lily. There is some UK interest in the production of the smaller types of bulb, such as crocus, muscari and snowdrop; however, most such stocks are imported for flower production and the remainder are for garden and amenity use. The export trade in narcissus bulbs has resulted in the need to grow stocks substantially free from pests and diseases, in order to meet the strict phytosanitary requirements of importing countries.

The number of pesticides actually approved for use on bulbs and corms is now very small. However, as these are ornamental crops many products can be used under the provisions of the *Revised Long Term Arrangements for Extension of Use (2000)*. Specifically, these arrangements allow the use of pesticides approved for use on any growing crop to be used on commercial holdings on ornamental crops, including bulbs, subject to certain restrictions. Mention in the text of use of a pesticide under the provisions of these Arrangements is marked with an asterisk (*).

Anemone

Pests

Aphids

Mottled arum aphid (*Aulacorthum circumflexum*), peach/potato aphid (*Myzus persicae*) and certain other species of aphid cause direct damage by feeding on the flower bracts and petals; aphids can also transmit virus diseases. As soon as aphids are seen, spray with nicotine. Strains of *M. persicae* resistant to carbamate, organochlorine, organophosphorus and/or pyrethroid insecticides will be controlled by nicotine.

Caterpillars
The caterpillars (larvae) of several species of moth, including angle-shades moth (*Phlogophora meticulosa*), eat the flowers and foliage. If necessary, spray with a pyrethroid, e.g. deltamethrin or lambda-cyhalothrin*, ideally during the early stages of larval development to minimize the extent of damage.

Cutworms
These are the caterpillars of certain noctuid moths, e.g. turnip moth (*Agrotis segetum*) and garden dart moth (*Euxoa nigricans*), which (apart from the earliest growth stages) typically feed at or just below ground level; they are most troublesome in July and August, especially in hot, dry summers. Cutworms do most damage to seedlings but older plants are also attacked. Cutworms are more readily killed when they are small and still feeding above ground level on the foliage or flower buds. Heavy rainfall or irrigation when caterpillars are at the vulnerable foliar-feeding stage can dislodge and/or drown them. Otherwise, it may be necessary to use an insecticide, such as deltamethrin or lambda-cyhalothrin*.

Slugs
Slugs occasionally damage the leaves, stems and flowers. To obtain control, apply pellet baits of methiocarb or metaldehyde when the slugs are active on the soil surface.

Symphylids
Open-ground symphylids (*Symphylella* spp.) and, occasionally, the glasshouse symphylid (*Scutigerella immaculata*), will feed on the roots, corms and underground stems, resulting in patchy growth. Where damage occurs regularly, spray the soil with gamma-HCH* and lightly incorporate before drilling the seed or planting the corms.

Diseases

Downy mildew (*Plasmopara* spp. and *Peronospora* spp.)
These fungi cause the leaves to turn a dull greyish-yellow. Often, infected leaves curl downwards, which makes plants look stunted (not to be confused with symptoms of leaf curl – see p. 544). Large numbers of spores are produced on the undersides of leaves, and these are readily dispersed in wet and windy weather. Resting spores are produced which infest the soil and remain viable for several years.

Foliar sprays are usually uneconomic and a long crop rotation is the usual method of control. If this is not possible, soil sterilants (such as dazomet or metam-sodium) will kill many of the resting spores.

Grey mould (*Botryotinia fuckeliana* – anamorph: *Botrytis cinerea*)
This is essentially a disease of poorly grown plants, or is a result of weather damage; it is common, for example, after severe gales in the south-west of England. The fungus establishes within the crown of plants and eventually kills them. Spores form prolifically on the dead tissue and are subsequently splashed by rain onto nearby plants.

Several fungicides are effective; dithiocarbamates and iprodione are likely to give good control.

Leaf curl (*Colletotrichum* spp.)
This disease was reported in Australia in 1979 and appeared shortly afterwards in Britain. All stages of growth of the anemone plant are affected. A leaf-curling symptom appears, reminiscent of hormone damage. Young plants show a characteristic down-cupping and curling of the leaves, which often fail to open fully. The petioles of older plants become twisted, causing the leaves to hang down; also, new shoots may show tip necrosis and die-back. On mature plants, distinct brown lesions occur on both the petioles and the flower stalks, and orange spore masses may be seen. As yet there is no satisfactory method of control.

Leaf spot and die-back (*Mycocentrospora acerina*)
This disease is most commonly seen in early spring after prolonged periods of wet weather, and dead plants often occur in circular or oval patches. Seedlings develop grey-black spots on the petioles and leaflets. In time the spots coalesce, resulting in extensive tissue necrosis and, eventually, plant death. Occasional plants show rotting of the tap root. As yet, there is no satisfactory control, but MBC fungicides may have some beneficial effect.

Plum rust (*Tranzschelia discolor*)
This disease sometimes affects anemones in the south-west of England. The plants are distinguished by their thickened and more upright appearance; they are also sterile and so do not flower. The undersides of the leaves display aecia and cluster-cups, the latter producing the spores that infect plum (*Prunus domestica* ssp. *domestica*) and bullace (*P. domestica* ssp. *institia*) trees later in the year. Spores from leaves of infected plum and bullace trees in the vicinity of anemones may in turn cause infection to anemones. Vulnerable anemone plants may be protected by spraying in the winter months with oxycarboxin*.

Powdery mildew (*Erysiphe ranunculi*)
This is a common infection when day temperatures are high and there are heavy

night-time dews. Leaves appear whitish and, occasionally, small black bodies (cleistothecia) are formed in the white superficial fungal mycelium. These are probably of lesser importance than the masses of dry, powdery spores that are blown from infected leaves on to nearby plants.

Occurrence of this disease is not sufficiently frequent for control measures to be tested but fungicides, e.g. bupirimate, fenarimol and imazalil, used to control other powdery mildews should be effective.

Crocus

Pests

Gladiolus thrips (*Thrips simplex*)
This pest may feed on crocus corms in store, producing a yellowish-brown discoloration beneath the skin; attacked corms make poor growth. For control, see under *Gladiolus*, p. 546.

Tulip bulb aphid (*Dysaphis tulipae*)
See under *Tulipa* (tulip), p. 555.

Diseases

There is a growing commercial interest in the smaller decorative bulbs, of which crocus is the most popular. Many of the diseases to affect such bulbs, e.g. grey mould, dry rot and fusarium rot, have similar development patterns and control methods to corresponding diseases of the bigger bulbs: tulip fire (p. 557), dry rot of *Gladiolus* (p. 547) and basal rot of *Narcissus* (p. 553), respectively.

Galanthus (snowdrop)

The former practice of removing rich stands of snowdrop bulbs from private woodlands for immediate sale has now ceased, as has the importation of bulbs harvested from the wild. However, a large-scale industry supplying replacements has yet to arise.

Pests

Stem nematode (*Ditylenchus dipsaci*)
Infested plants develop leaf 'spickels', are stunted and distorted, and when cut across the bulbs show brown rings of damage. There are no recommendations for chemical or hot water treatment of snowdrop bulbs, so destruction of infested stocks is advised.

Tulip bulb aphid (*Dysaphis tulipae*)
See under *Tulipa* (tulip), p. 555.

Diseases

Grey mould (*Botrytis galanthina*)
This disease has been known for over a century. It is first seen when infected shoots appear above ground, turning grey as thousands of fungal spores are produced. The fungal rot will extend to the bulb but spores are spread by water splash on to other plants. The life-cycle of the fungus may be assumed to be similar to that of *B. tulipae*. Snowdrops are frequently sold 'green', soon after flowering, and disease control is then difficult. However, dormant bulbs may be treated in a dip of carbendazim for 30 minutes, and this should give reasonable control. The growing crop should be protected with dithiocarbamate sprays.

Gladiolus

Pests

Aphids
Several species, e.g. peach/potato aphid (*Myzus persicae*), potato aphid (*Macrosiphum euphorbiae*) and shallot aphid (*Myzus ascalonicus*), all infest the foliage; they also transmit virus diseases. Control is as for *Anemone*, aphids, p. 542.

Tulip bulb aphid (*Dysaphis tulipae*) attacks the dry corms in store. Control is as for *Tulipa* – tulip bulb aphid, p. 555.

Caterpillars
Foliage-feeding caterpillars of cabbage moth (*Mamestra brassicae*) and angle-shades moth (*Phlogophora meticulosa*) are troublesome occasionally. Control is as for *Anemone*, caterpillars, p. 543.

Gladiolus thrips (*Thrips simplex*)
The dark-brown adults and yellow nymphs of this very small, elongate insect suck the sap of the stems and leaves, thereby producing pale-yellow or silvery streaks. They will also feed on flowers, producing small, white flecks on the petals, thereby spoiling their value. The thrips overwinter in store on the corms, feeding under the scales, where they will continue to breed at temperatures above 10°C. Infested corms have rough, greyish-brown patches on the surface.

Control infestations in the field by spraying with nicotine as soon as thrips are found. To prevent a build-up of the pest in store treat regularly with pirimiphos-methyl; then reduce the store temperature to about 10°C to limit further breeding.

Slugs

These sometimes graze the emerging foliage during wet periods. For control, see under *Anemone*, p. 543.

Diseases

Botrytis rot (*Sclerotinia draytoni*)

Small sunken lesions, with a cavity beneath the brown rot, may be seen in infested corms. If corms are stored in damp conditions these lesions will enlarge and subsequently develop into a brown, spongy mass. Black sclerotia (resting bodies) may also form. Corm infection can lead to a spotting of the flowers. Shoots from infected bulbs may become infected and, once above ground, the fungus forms spores that are spread to nearby plants, principally by water splash. Secondary leaf spotting then develops, which will eventually maintain the disease until the flowers become infected. The disease is perpetuated by short crop rotations and by infection passing from the parent to the daughter corms.

Control involves several practices. Corms should be lifted and dried rapidly – this induces a wound reaction and effectively prevents lesions from enlarging. Corm dips are effective, but see under Fusarium yellows below. A cold dip for 30 minutes should be adequate.

Leaf infection may be prevented or controlled by routine cyclic spraying of carbendazim, iprodione or dithiocarbamate. Captan has been used successfully in America.

Dry rot (*Stromatinia gladioli*)

This is the most common and serious disease of gladioli in this country. Symptoms of the disease differ according to the size of the corm and time of attack. Very small plants (derived from cormels) are killed soon after emergence, in a manner reminiscent of 'damping-off'. With large corms, shoots emerge apparently healthy but at any time during growth infected ones turn yellow, then brown and finally die in the space of 1 or 2 weeks. Such shoots are easily pulled away from the corm and, at the base, many thousands of minute, black sclerotia may be seen. These sclerotia infect soil to such an extent that it becomes 'sick' and remains unsuitable for growing gladioli for many years. The fungus also causes a dry rot of corms in store.

There is no effective treatment of the growing plants. However, corm infection may be controlled by treating corms with hot water containing formaldehyde for 30 minutes at 53°C. Soil sterilization can also be effective.

This fungus also attacks *Acidanthera*, *Crocus*, *Freesia*, *Galanthus*, *Lapeyrousia*, *Montbretia* and *Narcissus*.

Fusarium yellows (*Fusarium oxysporum* f. sp. *gladioli*)

In many countries, this is the most serious disease of gladioli. However, it is less severe in Britain, probably because of the lower soil temperatures. Nevertheless, it remains as a potential threat. The first symptom is a yellowing of the leaf tip, followed by die-back, which is similar to dry rot. However, there are some fundamental differences between the two diseases. With 'yellows' the stem base will appear clean and is free from sclerotia. The roots of plants with 'yellows' are usually brown or absent, unlike those with dry rot which look healthy. Fusarium yellows disease usually coincides with hot weather and high soil temperatures. Corms frequently rot away in the ground. Infected corms in store often become mummified.

The disease is controlled by HWT, with both formaldehyde and thiabendazole* added.

Storage rot (*Penicillium gladioli*)
This is a storage disease and is associated with damaged corms or with badly stored corms. Affected corms usually show only one, but occasionally two or three, relatively large (mainly lateral) lesions. The lesions are reddish-brown and affected corms are soft and wet to the touch. Care in avoiding damage at lifting, and in subsequent handling, is important because the fungus attacks through wounds. Treatment with thiabendazole, primarily to control other diseases, will also control this rot.

Hyacinthus (hyacinth)

Pests

Large narcissus fly (*Merodon equestris*)
Attacks occasionally develop on hyacinth bulbs; for a description and control measures, see under *Narcissus* (daffodil), p. 552.

Stem nematode (*Ditylenchus dipsaci*)
Symptoms of damage are similar to those seen on *Narcissus* (p. 553), except that definite leaf 'spickels' do not develop. HWT of infested bulbs for 4 hours at 45°C, which is based on the Dutch recommendation, is the only effective control measure currently available.

Diseases

Yellow disease (*Xanthomonas hyacinthi*)
Although almost all hyacinths used in the forced flower trade are imported, there is no special reason why bulbs should not be produced out of doors in the UK. Several pathogens attack hyacinths but the limiting condition is yellow disease. Control of this bacterial infection is obtained largely by good cultural methods, especially during storage.

Iris

Pests

Aphids

Species infesting the foliage include peach/potato aphid (*Myzus persicae*) and potato aphid (*Macrosiphum euphorbiae*). In addition, tulip bulb aphid (*Dysaphis tulipae*) is found occasionally on bulbs in store. For control, see under *Anemone*, p. 542, and *Tulipa* (tulip), tulip bulb aphid, p. 555.

Potato tuber nematode (*Ditylenchus destructor*)

This nematode, which closely resembles stem nematode (*D. dipsaci*) in size and shape, attacks bulbous iris but produces damaging symptoms in the bulb only. Attacked bulbs show dark, reddish-brown streaks on the scales, extending upwards from the base plate. The emerging foliage generally lacks vigour and bulb yields are depressed. HWT for 3 hours at 44.4°C, preceded by warm storage and a pre-soak, is the only effective control measure available.

Slugs

These attack the bulbs, or graze the foliage and occasionally feed on the flowers. For control, see under *Anemone*, p. 543.

Thrips

Gladiolus thrips (*Thrips simplex*) and iris thrips (*Frankliniella iridis*) feed on the foliage, causing a distinctive silvering. They also attack flowers, producing white flecks on the petals. For control, see under *Gladiolus*, p. 546.

Diseases

Blue mould (*Penicillium corymbiferum*)

This fungus is found on almost all bulbous irises and it attacks the bulbs through small wounds. Even with great care at lifting and during grading some mechanical damage is done. If these bulbs are subsequently stored in high humidities, rotting will start. This may extend throughout the bulb, which becomes hard and chalky. Sometimes, however, other organisms turn the bulb into a wet, rotten mass. Losses can be considerable, especially where aggressive strains are present.

Careful handling is the most important measure to avoid infection.

Fusarium basal rot (*Fusarium oxysporum* f. sp. *gladioli*)

This disease is seldom found in the UK, probably because of the cool climate, but it does represent a serious threat. The symptoms resemble narcissus basal rot, and infected plants are stunted with yellowing leaves. Infected bulbs produce either no roots or brown, rotting ones, and bulbs, when split, will show degrees of browning that extend from the root plate.

Control is obtained by dipping bulbs in a warm-water suspension (20–30°C) of carbendazim (off-label) (SOLA 0009/99) or thiabendazole* as soon after lifting as possible.

Grey bulb rot (*Rhizoctonia tuliparum*)
This is a potentially destructive disease of iris and tulip. See under *Tulipa* (tulip), p. 556, for details and control measures.

Ink disease (*Drechslera iridis*)
This disease takes its name from the inky markings it causes on *Iris reticulata*. It also attacks Dutch iris (hybrids of the bulbous iris, *Iris xiphium*), causing a severe leaf disease that reduces both flower quality and bulb yield. The spots themselves are typically black but affected leaves also turn yellow and die prematurely.

Regular spraying with dithiocarbamate fungicides* gives good control.

Leaf spot (*Mycosphaerella macrospora*)
This is the most common and most serious leaf disease, which is distinguished from ink disease by the brown, rather than black, spotting of the leaves. These spots will coalesce and cause premature leaf death. Flower quality is affected by spotting, and premature leaf senescence will affect flower quality and yield in the subsequent season.

The fungus attacks and enters the leaves several weeks before any spotting is seen. It is important, therefore, to begin spraying soon after emergence. Dithiocarbamate* fungicides are effective; good results have also been obtained with a flowable formulation of chlorothalonil*.

Lilium (lily)

Pests

Aphids
Tulip bulb aphid (*Dysaphis tulipae*) can be troublesome on bulbs in store and, together with other species, e.g. peach/potato aphid (*Myzus persicae*) and shallot aphid (*M. ascalonicus*), will infest the foliage and transmit virus diseases. Melon & cotton aphid (*Aphis gossypii*) is a major problem under protection. For control, see under *Anemone*, p. 542, and *Tulipa* (tulip), tulip bulb aphid, p. 555.

Lily beetle (*Lilioceris lilii*)
This bright-red beetle, which has black legs and antennae, is found locally in England. The larvae are orange in colour but frequently covered with a black, slimy accumulation of faeces. Both the adults and larvae feed on the leaves,

flowers and seed pods, causing considerable damage. To control infestations, spray with a contact insecticide such as cypermethrin or nicotine.

Lily thrips (*Liothrips vaneeckei*)
Adults and nymphs of this thrips feed on the bulbs in store; infestations result in light-brown patches on the scales, which are often invaded by fungal rots. Control is as for thrips on *Gladiolus*, p. 546.

Slugs
Lily bulbs are very susceptible to damage caused by species such as field slug (*Deroceras reticulatum*) and garden slug (*Arion hortensis*). For control, see under *Anemone*, p. 543.

Diseases

See under Bulbs (forced) – *Lilium* (lily), Chapter 10, p. 394.

Narcissus (daffodil)

Narcissus is the most important bulb crop grown in the UK. The UK is also the world's largest producer, growing about half of the global area. Some of the crop is forced for flower production but most is grown to produce bulbs. Approximately 40% of the dry bulb crop is sold in the home and export markets. Bulb production and forcing are predominant in eastern England; outdoor flower production is most important in the south-west.

Pests

Aphids
Direct damage caused by aphids feeding on foliage or flowers is rare, but many species can act as vectors of virus diseases. This is of major importance in virus-tested stock, which should be protected against the aphid vectors by being grown in aphid-proofed houses or, on a field scale, by isolation from other stocks and by regular treatment with insecticides, such as the pyrethroid deltamethrin, that have anti-feeding properties.

Bulb mites (*Rhizoglyphus callae* and *R. robini*)
These shiny, pearly-white, globular mites infest bulbs already damaged by pests, diseases etc. They extend the original damage and prevent recovery from minor cuts and bruising. The mites are easily controlled by routine HWT.

Bulb scale mite (*Steneotarsonemus laticeps*)
These very small, slow-moving, translucent-brown mites feed between the bulb scales. Infested bulbs show brown streaks extending down to the base plate as

long, narrow scars. In the field, infested bulbs rarely show symptoms of mite damage but such bulbs gradually lose vigour and yields may be reduced.

Routine HWT for controlling stem nematode (see p. 553) will eradicate the mites in bulb stocks. Alternatively, bulbs may be fumigated with methyl bromide, but this is a task for a contractor. Bulbs which show symptoms of attack during forcing can be treated by drenching them with endosulfan.

Large narcissus fly (*Merodon equestris*)

The yellowish-white maggot tunnels into the centre of the bulb, destroying much of the tissue. The maggot, usually one per bulb, enters through the base plate after eggs hatch in May and June, but damage is not obvious until September or October. The fully grown maggot is about 18 mm long and leaves the bulb in March to pupate. Infested bulbs are softer than usual (especially around the neck) and, if replanted, will probably not flower but will instead throw up a clump of abnormally narrow leaves ('grass') or may not grow at all.

The maggots are easily killed by routine HWT as for stem nematode (see p. 553). No completely effective replacements for aldrin, which was withdrawn in 1989, have become available for crops grown for 2 years, and control currently depends upon a combination of cultural measures (e.g. early harvest, rotation), HWT and dipping in a solution containing chlorpyrifos.

Migratory nematodes (*Longidorus* spp., *Paratrichodorus* spp., *Trichodorus* spp. and *Xiphinema* spp)

These nematodes are not important from the point of view of direct damage to narcissus, but they are all potential vectors of virus diseases. Examples are: arabis mosaic virus (AMV) (vectors = *Xiphinema* spp.); raspberry ringspot virus (RRV) (vectors = *Longidorus* spp.); strawberry latent ringspot virus (SLRV) (vectors = *Xiphinema* spp.); tobacco rattle virus (TRV) (vectors = *Trichodorus* spp. and *Paratrichodorus* spp.); and tomato black ring virus (TBRV) (vectors = *Longidorus* spp.). There is no satisfactory control measure available, so careful site selection is important. The nematode vectors tend to be found in sandy soils.

Root-lesion nematodes (*Pratylenchus* spp.)

These very small nematodes feed on the roots, making small, brown or black lesions which allow entry of fungal rots. The attacked roots eventually decay and the bulbs make poor growth, damage usually occurring in patches. Attacks are confined mainly to bulbs grown in Cornwall and in the Isles of Scilly. Control measures must be directed against the nematode in the soil. The fumigant nematicide 1,3-dichloropropene or the soil sterilant metam-sodium are recommended where root-lesion nematodes are a problem.

Small narcissus flies (*Eumerus strigatus* and *E. tuberculatus*)

The greyish-white maggots (larvae) of these pests frequently infest bulbs that have been damaged mechanically or attacked by diseases and other pests. They are secondary pests, merely extending damage already done. They are sometimes mistaken for larvae of large narcissus fly (see p. 552), but are smaller (up to 9 mm long) and, usually, there are several in each infested bulb. Control measures taken against the primary pests and diseases will reduce small narcissus fly infestations.

Stem nematode (*Ditylenchus dipsaci*)
This is the most important pest of *Narcissus*, producing damage symptoms on the foliage, flower stems and in the bulb. The leaves of attacked plants are usually pale, stunted and distorted, with characteristic pale-yellow localised swellings ('spickels'). Infested bulbs often flower late, or they may rot in the ground if the infestation is severe. Attacked bulbs are soft and, when cut across, show rings of brown tissue where the nematodes have been feeding in the scales.

All narcissus stocks should be inspected closely during the growing season and infested plants and bulbs removed and destroyed. After lifting, discard all soft or suspect bulbs and hot-water-treat the stock for 3 hours at 44.4°C with formadehyde plus wetter.

Whether or not stem nematode was found during the growing season, it is advisable to hot-water-treat all narcissus bulbs every 2 years as a routine in the period between lifting and replanting.

Where a 1st-year outdoor flower crop is required, HWT damage can be minimized by warm-storing the bulbs for 7 days at 30°C and then soaking them in cold water for at least 3 hours before hot-water-treating for 3 hours at 46.7°C.

On nematode-infested land the interval between bulb crops should be at least 5 years after groundkeepers have been removed. No other susceptible host plant (e.g. bluebell, onion, snowdrop and strawberry) should be grown in the interval. It may be considered worthwhile sterilizing small areas of infested land with 1,3-dichloropropene or metam-sodium.

Swift moths
The caterpillars of garden swift moth (*Hepialus lupulinus*) and ghost swift moth (*H. humuli*) can sometimes damage narcissus bulbs by feeding on the outer scales. The normal diet of the caterpillars is the roots of grasses, especially couch grass (*Elytrigia repens*), so problems usually occur only where control of grass weeds is poor. There are no effective insecticides that might control an established infestation. The best advice, therefore, is to keep sites as weed free as possible.

Diseases

Basal rot (*Fusarium oxysporum* f. sp. *narcissi*)
This is the most serious and limiting disease of *Narcissus*. The fungus is active all the year round but most infection takes place from early summer until autumn when soil temperatures are at their highest. The fungus penetrates the bulbs, either through the roots or directly into the root plate. Infected bulbs, when cut

longitudinally, have a chocolate-brown appearance and are soft to the touch. In store, the fungus may produce masses of pinkish spores on the root plate and, in handling, these spores will easily be dislodged to infect other bulbs. In the field, infection is less easy to see but leaf tip yellowing or early senescence are the most common symptoms. Although cvs Carlton and Golden Harvest are most susceptible, the fungus is often found on other cultivars. Control is difficult and rests upon a series of good husbandry measures (because the fungus is most active between 25 and 35°C, rapid drying after lifting followed by cool storage is required, coupled with a long rotation, minimal handling-damage and good storage), supported by formaldehyde with or without additional thiabendazole or prochloraz in the HWT tank. Cold formaldehyde dips may also be used at lifting, providing the stock is free from stem nematode.

Fire (*Sclerotinia polyblastis*)
This is less common than in previous years because flowers are picked before they open. The fungus survives on trash from the previous season and in the soil. In spring, fruiting bodies discharge spores which cause a spotting of the delicate floral parts. Tazetta cultivars are particularly susceptible.

Control is achieved by spraying with a fungicide but many formulations leave unsightly deposits. Two cycles of spraying with a dithiocarbamate* at flowering should be sufficient to control the disease.

Leaf scorch (*Stagonospora curtisii*)
This disease is common in the south-west of England, and is most usually seen as a necrosis of the leaf tip. The life-cycle is very similar to that of the fungus causing smoulder (see below), but the leaf lesions are usually lighter in colour. Although the lesions occur anywhere on leaves, and suggest secondary spread from the leaf tip, the infection more usually arises as the shoot passes through the neck of the bulb. The fungus survives in the dry, papery scales on the outside of the bulbs and does not usually cause widespread rotting.

Control is achieved by spraying zineb* in an oil-based tank-mix. The oil formulation gives great tenacity, which is needed in the south-western climate, but good results have also been achieved with a flowable formulation of chlorothalonil*. HWT itself lessens the number of primary infectors, which are also affected by storage temperatures.

Smoulder (*Sclerotinia narcissicola*)
This disease causes a leaf and flower spot which, in some years, reaches considerable proportions. The fungus survives from one season to another in the form of small, black resting bodies in soil, as well as in and on the outer scales of bulbs. These begin to grow in the autumn and may infect the shoots that emerge on 'primaries'. Spores develop on these, and are then distributed by wind and water splash to nearby plants to cause secondary spotting. Sometimes, shoots of infected bulbs escape infection but the fungus develops in the mother bulb and

infects the developing daughter bulbs. It is unusual for the disease to appear in the first year of a crop but it is frequent in the second year, especially where the ridges were re-formed in autumn (because infected trash is buried near the emerging shoots).

A good measure of control is obtained from HWT; however, in biennial cropping, spraying with dithiocarbamates* may be necessary.

White mould (*Ramularia vallisumbrosae*)
This leaf disease is largely confined to the south-west. It usually appears early in the year as small, sunken spots or streaks on leaves. These streaks and spots are usually grey-green or yellowish. In moist weather they enlarge, and coalesce to affect large areas of leaf. A white powdery mass of spores may develop on the underside of infected leaves. Also, small pycnidia and black sclerotia form in the leaf. The fungus survives in leaf debris.

Many cultivars are attacked but, usually, it is the late-flowering ones (such as cvs Cheerfulness and Double White) that are most susceptible. The disease usually assumes serious proportions only in bulbs that are kept planted for several years. It is rare in the first season and not at all common in the second. A spray programme based on a tank mix of mancozeb* with benomyl* should be effective.

Tulipa (tulip)

Pests

Stem nematode (*Ditylenchus dipsaci*)
The strain or race of stem nematode that damages tulips is particularly virulent and has a wide host range, including narcissus.

Infested tulips show fine cracks and purplish streaks on the stems below the flower head. When an attack is severe, the stem is distorted and cracked, the flower head is bent over and the outer petals often fail to colour. Infested bulbs have glossy, greyish or brownish patches on the side above the base plate but, when cut across, the bulbs seldom show distinct rings of damage (cf. typical symptoms in *Narcissus*, p. 553).

As light infestations are difficult to detect in dormant bulbs, tulip stocks should be examined at flowering time for signs of stem nematode attack. Infested plants and their bulbs should be removed and destroyed. It may be worth while sterilizing infested soils before replanting with other bulbs. HWT of tulip bulbs, as practised for narcissus, is not practicable as tulips can be severely damaged by the process. Thorough rogueing, strict attention to hygiene and close examination of newly acquired stocks will help to limit stem nematode attacks.

Tulip bulb aphid (*Dysaphis tulipae*)
This is a common pest of tulip and other bulbs in store, where it feeds in dense

colonies beneath the outer scales. When infested bulbs are planted, the aphids feed on and damage the emerging shoots, distorting and checking growth. Regular fumigation of bulb stores with smokes of nicotine will prevent a build-up of these aphids.

Other aphid species, including black bean aphid (*Aphis fabae*) and peach/potato aphid (*Myzus persicae*), occasionally infest tulips growing in the field. They rarely cause direct damage but they will transmit virus diseases; for control see under *Anemone*, aphids, p. 542.

Diseases

Fusarium bulb rot or sour (*Fusarium oxysporum* f. sp. *tulipae*)
This disease became important principally as a rot of Darwin hybrid tulips. Extensive work in Holland has shown that there are two serious effects of the disease. First, bulbs rot; secondly, some of the bulbs produce ethylene which, in store, causes a flower bud blast which is not seen until flowering time. Although dipping in fungicides such as thiabendazole* should give some control, attention to aspects of crop husbandry are very important. Bulbs are less susceptible when lifted before the tunic is membranous. Also, the bulbs should be handled carefully to lessen physical damage, and planting should be delayed until soil temperatures have fallen below 15°C.

Grey bulb rot (*Rhizoctonia tuliparum*)
This is a serious and persistent soil-borne disease of tulip and iris. The first sign of damage is often the appearance in the spring of bare patches within the crop. When bulbs that have failed to appear are dug up, they are found to have rotted from the tip downwards, often leaving vigorous roots intact. Rotted tissues remain fairly firm, turn grey and may be packed with fungal mycelium. Sclerotia (resting bodies) are produced up to 7 mm diameter. These are at first pale but they later darken with age.

No spore-bearing structures have been identified in the fungus, which therefore persists from year to year in the mycelial or sclerotial states. Transfer of the organism is assumed to occur in lightly diseased bulbs or in soil attached to other, non-host, plant species. Sclerotia survive in soil for 3–5 years.

Control is difficult. Where the disease has been confirmed, removal and replacement of the topsoil should be considered. A pre-planting application of quintozene may also prove beneficial.

Pythium root rot (*Pythium* sp.)
This is principally a disease of forced tulips. Both in boxes and border soil, areas of bulbs of stunted growth suggest *Pythium* attack, especially if such bulbs have virtually no root yet the bulb itself is seemingly sound. Sometimes, apparently healthy roots will have red lesions that develop into a complete rot in which only

the outer cylinder of tissue is left. Prevention is the only means of control. Boxing and border soil should not be allowed to become too wet. A pre-planting drench of etridiazole, or etridiazole incorporated into the boxing soil, is usually very effective.

Tulip fire (*Botrytis tulipae*)
Tulip fire has become far less significant in the past 20 years. 'Primaries' (infected shoots) appear above ground and spores from these cause secondary leaf spots which, in wet conditions, develop into aggressive lesions. Small leaf spots do not enlarge and are unimportant but aggressive ones lead to such losses of leaf that flower quality is poor and the subsequent yield of bulbs is much reduced. Although the fungus lives on bulbs, some shoots escape infection, with the result that daughter bulbs develop normally. However, as they enlarge and as the mother bulb scales rot, the fungus moves on to the daughter bulb. Small, black resting bodies form on the tunic and on the base of the flower stalk, perpetuating the disease. A few bulbs will have lesions on the fleshy scales underneath the tunic.

Control is achieved largely by dipping the bulbs during storage for 30 minutes in carbendazim and iprodione* (but not if a benzimidazole-tolerant *Fusarium*-causing sour is present in the stock). Good rogueing for primaries, and several foliar sprays of dithiocarbamate fungicides or iprodione*, will prevent leaf infection. It is important not to use benzimidazole formulations as foliar sprays, because tolerance by *Botrytis* would remove the excellent control gained in the annual dip.

Yellow pock (*Corynebacterium flaccumfasciens* pv. *oortii*)
This bacterial disease occurs from time to time in imported tulip bulbs. Yellow spots occur on the bulb in store and severely infected bulbs die without producing a shoot. In the growing plant, leaf splitting (which looks similar to frost damage) is seen. However, if the tissues are examined and the stem split carefully, yellow spots may be noticed in the conducting vessels. Eventually, these plants wither and die.

List of pests cited in the text*

Agrotis segetum (Lepidoptera: Noctuidae)	turnip moth
Aphis fabae (Hemiptera: Aphididae)	black bean aphid
Aphis gossypii (Hemiptera: Aphididae)	melon & cotton aphid
Arion hortensis (Stylommatophora: Arionidae)	garden slug
Aulacorthum circumflexum (Hemiptera: Aphididae)	mottled arum aphid
Deroceras reticulatum (Stylommatophora: Limacidae)	field slug
Ditylenchus destructor (Tylenchida: Tylenchidae)	potato tuber nematode
Ditylenchus dipsaci (Tylenchida: Tylenchidae)	stem nematode
Dysaphis tulipae (Hemiptera: Aphididae)	tulip bulb aphid
Eumerus strigatus (Diptera: Syrphidae)	a small narcissus fly

Eumerus tuberculatus (Diptera: Syrphidae)	a small narcissus fly
Euxoa nigricans (Lepidoptera: Noctuidae)	garden dart moth
Frankliniella iridis (Thysanoptera: Thripidae)	iris thrips
Hepialus humuli (Lepidoptera: Hepialidae)	ghost swift moth
Hepialus lupulinus (Lepidoptera: Hepialidae)	garden swift moth
Lilioceris lilii (Coleoptera: Chrysomelidae)	lily beetle
Liothrips vaneeckei (Thysanoptera: Phlaeothripidae)	lily thrips
Longidorus spp. (Dorylaimida: Longidoridae)	needle nematodes
Macrosiphum euphorbiae (Hemiptera: Aphididae)	potato aphid
Mamestra brassicae (Lepidoptera: Noctuidae)	cabbage moth
Merodon equestris (Diptera: Syrphidae)	large narcissus fly
Myzus ascalonicus (Hemiptera: Aphididae)	shallot aphid
Myzus persicae (Hemiptera: Aphididae)	peach/potato aphid
Paratrichidorus spp (Dorylaimida: Trichodoridae)	stubby-root nematodes
Phlogophora meticulosa (Lepidoptera: Noctuidae)	angle-shades moth
Pratylenchus spp. (Tylenchida: Pratylenchidae)	root-lesion nematodes
Rhizoglyphus callae (Astigmata: Acaridae)	a bulb mite
Rhizoglyphus robini (Astigmata: Acaridae)	a bulb mite
Scutigerella immaculata (Symphyla: Scutigerellidae)†	*glasshouse symphylid*
Steneotarsonemus laticeps (Prostigmata: Tarsonemidae)	bulb scale mite
Symphylella spp. (Symphyla: Symphylidae)†	*open-ground symphylids*
Thrips simplex (Thysanoptera: Thripidae)	gladiolus thrips
Trichodorus spp (Dorylaimida: Trichodoridae)	stubby-root nematodes
Xiphinema spp (Dorylaimida: Longidoridae)	dagger nematodes

* The classification in parentheses represents order and family, except (†) where order is replaced by class.

List of pathogens/diseases (other than viruses) cited in the text*

Botryotinia fuckeliana (Ascomycota)	(common) grey mould
Botrytis cinerea (Hyphomycetes)	– anamorph of *Botrytinia fuckeliana*
Botrytis galanthina (Hyphomycetes)	grey mould
Botrytis tulipae (Hyphomycetes)	tulip fire
Colletotrichum spp. (Coelomycetes)	leaf curl
Corynebacterium flaccumfasciens pv. *oortii* (Firmicutes)†	yellow pock
Drechslera iridis (Hyphomycetes)	ink disease
Erysiphe ranunculi (Ascomycota)	powdery mildew
Fusarium oxysporum f. sp. *gladioli* (Hyphomycetes)	fusarium basal rot, fusarium yellows
Fusarium oxysporum f. sp. *narcissi* (Hyphomycetes)	basal rot of *Narcissus*
Fusarium oxysporum f. sp. *tulipae* (Hyphomycetes)	fusarium bulb rot or sour
Mycocentrospora acerina (Hyphomycetes)	leaf spot and die-back
Mycosphaerella macrospora (Ascomycota)	leaf spot
Penicillium corymbiferum (Hyphomycetes)	blue mould
Penicillium gladioli (Hypomycetes)	storage rot of *Gladiolus*
Peronospora spp. (Oomycetes)	downy mildew
Plasmopara spp. (Oomycetes)	downy mildew
Pythium sp. (Oomycetes)	pythium root rot
Ramularia vallisumbrosae (Hyphomycetes)	white mould
Rhizoctonia tuliparum (Hyphomycetes)	grey bulb rot
Sclerotinia draytoni (Ascomycota)	botrytis rot
Sclerotinia narcissicola (Ascomycota)	smoulder

Sclerotinia polyblastis (Ascomycota)	fire of *Narcissus*
Stagonospora curtisii (Coelomycetes)	leaf scorch
Stromatinia gladioli (Ascomycota)	dry rot of *Gladiolus*
Transzchelia discolor (Teliomycetes)	plum rust
Xanthomonas hyacinthi (Gracilicutes: Proteobacteria)†	yellow disease

* For fungi, the classification in parentheses refers to class, although this is not possible within the phylum Ascomycota where classes have yet to be satisfactorily defined (see *Mycological Research*, February 2000). Oomycetes are now classified in Chromista with the brown algae, rather than as true fungi. Some fungi have an asexual (anamorph) and a sexual (teleomorph) state, and the convention is to refer to them by their teleomorph name. However, where anamorph names are still in common use, these are listed and cross-referenced to the teleomorph name. Strictly, fungi classified as Coelomycetes and Hyphomycetes should be known as 'hyphomycetous anamorphs' and 'coelomycetous anamorphs' of the relevant teleomorph taxon (e.g. hyphomycetous anamorphic Sclerotiniaceae, for *Botrytis fabae*), respectively. These problems highlight the continual changes in the classification of the fungi.

† Bacteria – the classification in parentheses refers to division and class.

Selected Bibliography and Further Reading

General publications

Alford, D. V. (1999). *A Textbook of Agricultural Entomology*. Blackwell Science: Oxford.
Anon. (1985). *Plant Physiological Disorders*. MAFF Reference Book 223. HMSO: London.
Anon. (1999). *Using Pesticides*, 2nd edition. British Crop Protection Council: Farnham.
Buczacki, S. T. & Harris, K. M. (1981). *Collins Guide to the Pests, Diseases and Disorders of Garden Plants*. Collins: London.
Carter, D. J. (1984). *Pest Lepidoptera of Europe*. Dr W Junk: Dordrecht.
Copping, L. G. (ed.) (1998). *The BioPesticides Manual*. BCPC: Farnham.
Edwards, C. A. & Heath, G. W. (1964). *The Principles of Agricultural Entomology*. Chapman & Hall: London.
Gratwick, M. (ed.) (1992). *Crop Pests in the UK. Collected edition of MAFF leaflets*. Chapman & Hall: London.
Heaney, S., Slawson, D., Hollomon, D. W., Smith, M., Russell, P. E. & Parry, D. (eds) (1994). *Fungicide Resistance*. BCPC Monograph No. 60. BCPC: Farnham.
Hope, F. (1980). *Recognition and Control of Pests and Diseases of Farm Crops*. Blandford Press: Poole.
Ingram, D. & Robertson, N. (1999). *Plant Disease – A Natural History*. Harper Collins: London.
Jones, D. G. (ed.) (1998). *The Epidemiology of Plant Diseases*. Kluwer: Dordrecht.
Jones, F. G. W. & Jones, M.G. (1984). *Pests of Field Crops*, 3rd edition. Edward Arnold: London.
Jones, G. D. (1987). *Plant Pathology. Principles and Practice*. Open University Press: Milton Keynes.
Marshall, G. (ed.) (1996). *Diagnosis in Crop Production*. BCPC Symposium Proceedings No. 65. BCPC: Farnham.
Moreton, B. D. (1969). *Beneficial Insects and Mites*. MAFF Bulletin 20, 6th edition. MAFF: London.
Seymour, P. (1989). *Invertebrates of Economic Importance in Britain. Common and Scientific Names*, 4th edition. HMSO: London.
Southcombe, T. (ed.) (1999). *Boom and Fruit Sprayers Handbook*. British Crop Protection Council: Farnham.
Southcombe, T. (ed.) (1999). *Hand-held and Amenity Sprayers Handbook*. British Crop Protection Council: Farnham.
Southey, J. F. (ed.) (1978). *Plant Nematology*, 3rd edition. HMSO: London.
Whitehead, A. G. (ed.) (1998). *Plant Nematode Control*. CAB International: Wallingford.
Whitehead, R. (ed.) (2000) (new edition each year). *The UK Pesticide Guide 2000*. CAB International: Wallingford.

Chapter 1 Pest and disease management

Beaumont, A. (1947). The dependence on the weather of the dates of potato blight epidemics. *Transactions of the British Mycological Society* **31**, 45–53.
Cavalieri, L. F. & Kocak, H. (1994). Chaos in biological control systems. *Journal of Theoretical Biology* **169**, 179–87.

Crute, I. R., Holub, E. B. & Burdon, J. J. (eds) (1997). *The Gene-for-Gene Relationship in Plant–Parasite Interactions.* CAB International: Wallingford.

Dent, D. (1991). *Insect Pest Management.* CAB International: Oxford.

Dowley, L. J., Bannon, E., Cooke, L. R., Keane, T. & O'Sullivan, E. (eds) (1995). *Phytophthora infestans 150.* EAPR: Dublin.

Frahm, J., Volk, T. & Johnen, A. (1996). Development of the PRO_PLANT decision-support system for plant protection in cereals, sugarbeet and rape. *EPPO Bulletin* **26**, 609–22.

Greig-Smith, P., Frampton, G. & Hardy, T. (1992). *Pesticides, Cereal Farming and the Environment. The Boxworth Project.* HMSO: London.

Jones, D. G. (ed.) (1998). *The Epidemiology of Plant Diseases.* Kluwer Academic Publishers: Dordrecht.

Kreuss, A. & Tscharntke, T. (1994). Habitat fragmentation, species loss and biological control. *Science* **264**, 1581–84.

Kromp, B, (1999). Carabid beetles in sustainable agriculture: a review on pest control efficacy, cultivation impacts and enhancement. *Agriculture, Ecosystems and Environment* **74**, 187–228.

Large, E. C. (1952). The interpretation of progress curves for potato blight and other plant diseases. *Plant Pathology* **1**, 109–17.

Lucas, J. A., Bowyer, P. & Anderson, H. M. (eds) (1999). *Septoria on Cereals: a Study of Pathosystems.* CAB International: Wallingford.

Matthews, G. A. (1992). *Pesticide Application Methods*, 2nd edition. Longman: Harlow.

McKinlay, R. G. & Atkinson, D. (1995). *Integrated Crop Protection: Towards Sustainability?* BCPC Symposium Proceedings No. 63. BCPC: Farnham.

Morgan, D. & Walters, K. F. A. (1999). Developing innovative computerised technologies to aid rational pest management. *Aspects of Applied Biology* **55**, 67–71.

Morgan, D., Walters, K. F. A., Oakley, J. N. & Lane, A. (1998). An internet-based decision support system for the rational management of oilseed rape invertebrate pests. *Proceedings Brighton Crop Protection Conference – Pests & Diseases 1998*, **1**, 259–64.

Oakley, J. N., Walters, K. F. A., Ellis, S. A., Green, D. B., Watling, M. & Young, J. E. B. (1996). Development of selective aphicide treatments for integrated control of summer aphids in winter wheat. *Annals of Applied Biology* **128**, 423–36.

Ridgway, R. L., Inscoe, M. & Arn, H. (eds) (1990). *Insect Pheromones and Other Behaviour-modifying Chemicals.* BCPC Monograph No. 51. BCPC: Farnham.

Smith, L. P. (1956). Potato blight forecasting by 90 per cent humidity criteria. *Plant Pathology* **5**, 83–87.

Speight, M. R., Hunter, M. D. & Watt, A. D. (1999). *Ecology of Insects: Concepts and Applications.* Blackwell Science: Oxford.

van der Plank, J. E. (1966). *Plant Diseases: Epidemics and Control.* Academic Press: London.

Zadoks, J. C. (1981). EPIPRE: a disease and pest management system for winter wheat developed in the Netherlands. *EPPO Bulletin* **11**, 365–69.

Zadoks, J. C. & Schein, R. D. (1979). *Epidemiology and Plant Disease Management.* Oxford University Press: Oxford.

Chapter 2 Pests and diseases of cereals

A'Brook, J. & Evans, H. M. (1983). Epidemiology of barley yellow dwarf virus. *Report, Welsh Plant Breeding Station, 1982,* 170–72.

Bateman, G. L. (1977). Effects of organomercury treatment of wheat seeds on *Fusarium* seedling disease in inoculated soil. *Annals of Applied Biology* **85**, 195–201.

Brokenshire, T. & Cooke, B. M. (1978). The effect of inoculation with *Selenophoma donacis* at different growth stages on spring barley cultivars. *Annals of Applied Biology* **89**, 211–17.

Chamswarng, C. & Cook, R. J. (1985). Identification and comparative pathogenicity of *Pythium* species from wheat roots and wheat-field soils in the Pacific Northwest. *Phytopathology* **75**, 821–27.

Coskun, H., Bateman, G. L. & Holloman, D. W. (1985). Population structure of MBC-resistant eyespot in sixteen wheat and barley crops in 1984. *ISPP Chemical Control Newsletter No.* 6, pp. 14–15.

Derron, J. O. & Goy, G. (1990). La mouche jaune des chaumes (*Chlorops pumilionis* Bjerk.): biologie, nuisibilité, moyens de lutte. *Revue Suisse d'agriculture* **22**, 101–5.

Empson, D. W. & Gair, R. (1982). *Cereal Pests.* MAFF Reference Book 186, 2nd edition. HMSO: London.

Fisher, N. & Griffin, M. J. (1984). Benzimadazole (MBC) resistance in *Septoria tritici. ISPP Chemical Control Newsletter No. 5*, pp. 8–9.

Gair, R., Jenkins, J. E. E., Lester, E. & Bassett, P. (1987). *Cereal Pests & Diseases*, 4th edition. Farming Press: Ipswich.

Glen, D. M., Spaull, A. M., Mowat, D. J., Green, D. B. & Jackson, A. W. (1993). Crop monitoring to assess the risk of slug damage to winter wheat in the UK. *Annals of Applied Biology* **122**, 161–72.

Green, D. B. (1996). Managed applications of molluscicides for slug control in winter wheat. *Proceedings Brighton Crop Protection Conference – Pests & Diseases 1996*, **1**, 197–202.

Griffiths, D. J. & Peregrin, W. T. H. (1960). Control of halo blight in oats. *Plant Pathology* **9**, 10–14.

Hassan, Z. M., Kramer, C. L. & Eversmeyer, M. G. (1986). Summer and winter survival of *Puccinia recondita* and infection of soil borne uredospores. *Transactions of the British Mycological Society* **86**, 365–72.

Jones, D. G. & Clifford, B. C. (1978). *Cereal Diseases, their Pathology and Control.* BASF: Hadleigh.

Jordan, V. W. L. & Alien, E. C. (1984). Barley net blotch: influence of straw disposal and cultivation methods on inoculum potential, and on incidence and severity of autumn disease. *Plant Pathology* **33**, 547–59.

Malik, M. M. S. & Battes, C. C. V. (1960). The infection of barley by loose smut (*Ustilago nuda* (Jens.) Rostr.). *Transactions of the British Mycological Society* **43**, 117–25.

Malone, J. P. & Lorimer, R. (1975). The incidence of pathogenic fungi in Northern Ireland barley and oat seed samples. *Plant Pathology* **24**, 140–43.

Mann, J. A., Harrington, R., Morgan, D., Walters, K. F. A., Barker, I., Tones, S. J. & Foster, G. N. (1996). Towards decision support for control of barley yellow dwarf vectors. *Proceedings British Crop Protection Council Conference – Pests & Diseases* **1**, 179–84.

Mantle, P. G. & Shaw, S. (1977). A case study of the aetiology of ergot disease of cereals and grasses. *Plant Pathology* **26**, 121–7.

Martinez, M. & Pillon, O. (1993). La Zabre des céréales: un ravageur encore mal connu. *Phytoma – La Défense des Végétaux* **446**, 26–34.

Masri, S. S. & Ellingboe, A. H. (1966). Primary infection of wheat and barley by *Erysiphe graminis. Phytopathology* **56**, 389–95.

Oakley, J. N., Cumbleton, P. C., Corbett, S. J., Saunders, P., Green, D. I., Young, J. E. B. & Rogers, R. (1998). Prediction of wheat blossom midge activity and risk of damage. *Crop Protection* **17**, 145–49.

Oakley, J. N., Walters, K. F. A., Ellis, S. A., Green, D. B., Watling, M. & Young, J. E.B. (1996). Development of selective aphicide treatments for integrated control of summer aphids in winter wheat. *Annals of Applied Biology* **128**, 423–36.

Proeseler, G., Kegler, H. & Schwahn, P. (1986). Weitere Hinweise zum Gerstengelbmosaik-Virus. *Nachrichtenblatt für den Pflanzenschutz in der DDR* **40**, 25–27.

Scott, P. R., Benedikz, P. W., Jones, H. G. & Ford, M. (1985). Some effects of canopy structure and microclimate on infection of tall and short wheats by *Septoria nodorum. Plant Pathology* **34**, 578–93.

Sheridan, J. E. (1971). The incidence and control of mercury resistant strains of *Pyrenophora avenae* in British and New Zealand seed oats. *New Zealand Journal of Agricultural Research* **14**, 469–80.

Simkin, M. B. & Wheeler, B. E. J. (1974). The development of *Puccinia hordei* on barley cv. Zephyr. *Annals of Applied Biology* **78**, 225–35.

Wolfe, M. S., Minchin, P. N. & Slater, S. E. (1984). Powdery mildew of barley. *Annual Report of the Plant Breeding Institute, 1983*, pp. 91–95.

Yeates, J. S. & Parker, C. A. (1985). Rate of natural senescence of seminal root cortical cells of wheat, barley, and oats, with reference to invasion by *Gaeumannomyces graminis*. *Transactions of the British Mycological Society* **86**, 683–85.

Young, J. E. B. & Cochrane, J. (1993). Changes in wheat bulb fly (*Delia coarctata*) populations in East Anglia in relation to crop rotations, climatic data and damage forecasting. *Annals of Applied Biology* **123**, 485–98.

Young, J. E. B., Talbot, G. A., Kilpatrick, J. B. & Saunders, P. J. (1994). The influence of soil cultivations in set-aside on oviposition of wheat bulb fly. *Aspects of Applied Biology* **40**, 233–36.

Chapter 3 Pests and diseases of oilseeds, brassica seed crops and field beans

Alford, D. V. (1979). Observations on the cabbage stem flea beetle, *Psylliodes chrysocephala*, on winter oil-seed rape in Cambridgeshire. *Annals of Applied Biology* **93**, 117–23.

Alford, D. V., Walters, K. F. A., Williams, I. H. & Murchie, A. K. (1996). A commercially viable low-cost strategy for the management of seed weevil populations on winter oilseed rape in the UK. *Proceedings Brighton Crop Protection Conference – Pests & Diseases 1996*, **2**, 609–14.

Biddle, A. J., Smart, L. E., Blight, M. M. & Lane, A. (1996). A monitoring system for pea and bean weevil (*Sitona lineatus*). *Proceedings Brighton Crop Protection Conference – Pests & Diseases 1996*, **1**, 173–78.

Davies, J. M. Ll., Gladders, P., Young, C., Dyer, C., Hiron, L., Locke, T., Lockley, D., Ottway, C., Smith, J., Thorpe, G. & Watling, M. (1999). Petal culturing to forecast sclerotinia stem rot in winter oilseed rape: 1993–1998. *Aspects of Applied Biology* **56**, 129–34.

Ellis, M. B. (1968). *Descriptions of Pathogenic Fungi and Bacteria, No. 162, Alternaria brassicae*. Commonwealth Mycological Institute: Kew.

Ellis, M. B. & Waller, J. M. (1974). *Descriptions of Pathogenic Fungi and Bacteria, No. 432, Botrytis fabae*. Commonwealth Mycological Institute: Kew.

Ellis, S. A., Raw, K., Oakley, J. N., Parker, W. E. & Lane, A. (1995). Development of an action threshold for cabbage aphid in oilseed rape. *Proceedings 9th International Rapeseed Congress, Cambridge, UK*, pp. 1037–39.

Evans, E. J. & Gladders, P. (1981). Diseases of winter oilseed rape and their control, east and south-east England, 1977–81. *Proceedings British Crop Protection Conference – Pests and Diseases 1981* **1**, 505–12.

Evans, K. A. & Scarisbrick, D. H. (1994). Integrated insect pest management in oilseed rape crops in Europe. *Crop Protection* **13**, 400–12.

Fitt, B. D. L., Gladders, P., Turner, J. A., Sutherland, K. G., Welham, S. J. & Davies, J. M. Ll. (1997). Prospects for developing a forecasting scheme to optimise use of fungicides for disease control on winter oilseed rape in the UK. *Aspects of Applied Biology* **48**, *Optimising Pesticide Applications*, pp. 135–42.

Garthwaite, D. G. & Thomas, M. R. (1999). *Pesticide Usage Survey Report 159, Arable Crops 1998*. MAFF: London.

Gladders, P., Ellerton, D. R. & Bowerman, P. (1991). Optimising the control of chocolate spot. *Aspects of Applied Biology* **27**, *Production and Protection of Legumes*, pp. 105–10.

Gladders, P., Jones, A. M., Lockley, K. D., Young, C. S., Turley, D., Fitt, B. D. L. & Towns, H. (1999). Occurrence and importance of diseases of winter linseed. *Aspects of Applied Biology* **56**, 177–82.

Gladders, P. & Symonds, B. V. (1995). Occurrence of canker (*Leptosphaeria maculans*) in eastern England 1977–1993. *IOBC/WPRS Bulletin* **19**, 1–11.

Glen, D. M., Jones, H. & Fieldsend, J. K. (1990). Damage to oilseed rape seedlings by the field slug, *Derocarus reticulatum*, in relation to glucosinolate concentration. *Annals of Applied Biology* **117**, 197–207.

Hebblethwaite, P. D. (ed.) (1983). *The Faba Bean* (*Vicia faba* L.). *A Basis for Improvement.* Butterworth: London.

Hughes, J. M. & Evans, K. A. (1995). Factors determining the distribution of brassica pod midge (*Dasineura brassicae*). *Proceedings 9th International Rapeseed Congress, Cambridge, UK,* pp. 1028–30.

Jay, C. N. & Smith, H. G. (1995). The effect of beet western yellows virus on the growth and yield of oilseed rape. *Proceedings 9th International Rapeseed Congress, Cambridge, UK,* pp. 664–66.

Jellis, G. J., Davies, J. M. L. & Scott, E. S. (1984). Sclerotinia on oilseed rape: implications for crop rotation. *Proceedings British Crop Protection Conference – Pests & Diseases 1984*, **2**, 709–16.

John, M. E. & Holliday, J. M. (1984). Distribution and chemical control of *Psylliodes chrysocephala* and *Ceutorhynchus picitarsis* in winter oilseed rape. *Aspects of Applied Biology* **6**, 281–92.

Lane, A. & Walters, K. F. A. (1994). Pest control requirements of oilseed rape under reformed CAP. *Aspects of Applied Biology* **40**, *Arable Farming under CAP Reform,* pp. 171–81.

Lane, A. & Walters, K. F. A. (1995). Prospects for a decision support system for pest of oilseed rape in the UK. *Proceedings of the 9th International Rapeseed Congress, Cambridge, UK,* pp. 1019–21.

Mercer, P. C., Hardwick, N. V., Fitt, B. D. L. & Sweet, J. B. (1994). Diseases of linseed in the UK. *Plant Varieties and Seeds* **7**, 135–50.

Oakley, J. N., Corbett, S. J., Parker, W. E. & Young, J. E. B. (1996). Assessment of risk and control of flax flea beetles. *Proceedings Brighton Crop Protection Conference – Pests & Diseases 1996*, **1**, 191–96.

Parker, W. E. & Biddle, A. J. (1998). Assessing the damage caused by black bean aphid (*Aphis fabae*) on spring beans. *Proceedings Brighton Crop Protection Conference – Pests & Diseases 1998*, **3**, 1077–82.

Paul, V. H. & Rawlinson, C. J. (1992). *Diseases and Pests of Rape.* 1. Auflage. Th. Mann: Gelsenkirchen-Buer.

Perryman, S. A. M., Fitt, B. D. L. & Gladders, P. (1999). Effect of diseases on the yield of winter linseed. *Aspects of Applied Biology* **56**, 211–18.

Punithalingam, E. & Holliday, P. (1972). *Descriptions of Pathogenic Fungi and Bacteria, No. 331, Leptosphaeria maculans.* Commonwealth Mycological Institute: Kew.

Punithalingam, E. & Holliday, P. (1975). *Descriptions of Pathogenic Fungi and Bacteria, No. 461, Ascochyta fabae.* Commonwealth Mycological Institute: Kew.

Rawlinson, C. J., Muthyalu, G. & Turner, R. H. (1978). Effect of herbicides on epicuticular wax of winter oilseed rape (*Brassica napus*) and infection by *Pyrenopeziza brassicae*. *Transactions of the British Mycological Society* **71**, 441–51.

Souter, S. D., Castells-Brooke, N. I. D., Antoniw, J. F., Welham, S. J., Fitt, B. D. L., Evans, N., Gladders, P., Turner, J. A. & Sutherland, K. (1999). Forecasting light leaf spot (*Pyrenopeziza brassicae*) of winter oilseed rape (*Brassica napus*) on the Internet. *Aspects of Applied Biology* **55**, 37–42.

Sylvester-Bradley, R. & Makepeace, R. J. (1984). A code for stages of development in oilseed rape (*Brassica napus* L.). *Aspects of Applied Biology* **6**, 399–419.

Wafford, J. D., Gladders, P. & McPherson, G. M. (1986). The incidence and severity of

Brussels sprout diseases and the influence of oilseed rape. *Aspects of Applied Biology* **12**, 1–12.

Walters, K. F. A. & Lane, A. (1991). Incidence and severity of insects damaging linseed in England and Wales 1988–1989. *Aspects of Applied Biology* **28**, *Production and Protection of Linseed,* pp. 121–28.

Walters, K. F. A. & Lane, A. (1992). Incidence of pollen beetles in winter rape and evaluation of thresholds for control. *Proceedings of the Brighton Crop Protection Conference – Pests and Diseases 1992*, **2**, 545–50.

Walters, K. F. A. & Lane, A. (1994). The development of an improved management strategy for cabbage seed weevil on oilseed rape. In: S. R. Leather, N. J. Mills, A. D. Watt & K. F. A. Walters (eds) *Individuals, Populations and Patterns in Ecology*, pp. 187–97. Intercept Press: Andover.

Way, M. J., Cammell, M. E., Taylor, L. R. & Woiwod, I. P. (1981). The use of egg counts and suction trap samples to forecast the infestation of spring-sown field beans, *Vicia faba,* by the black bean aphid, *Aphis fabae*. *Annals of Applied Biology* **98**, 21–34.

Welham, S. J., Fitt, B. D. L., Turner, J. A., Gladders, P. & Sutherland, K. (1999). Relationships between regional weather and incidence of light leafspot (*Pyrenopeziza brassicae*) on winter oilseed rape in England and Wales. *Aspects of Applied Biology* **56**, 51–59.

West, J. S., Biddulph, J. E., Fitt, B. D. L. & Gladders, P. (1999). Epidemiology of *Leptosphaeria maculans* in relation to forecasting stem canker severity on winter oilseed rape in the UK. *Annals of Applied Biology* **135**, 535–46.

Winfield, A. L. (1992). Management of oilseed rape pests in Europe. *Agricultural Zoological Review* **5**, 51–95.

Chapter 4 Pests and diseases of forage and amenity grass and fodder crops

Bowen, R. & Plumb, R. T. (1979). The occurrence and effects of red clover necrotic mosaic virus in red clover (*Trifolium pratense*). *Annals of Applied Biology* **91**, 227–36.

Brooks, D. D. (1965). Wild and cultivated grasses as carriers of the take-all fungus. *Annals of Applied Biology* **55**, 307–16.

Catherall, P. L. (1979). Virus diseases of cereals and grasses and their control through plant breeding. *Report of the Welsh Plant Breeding Station for 1978*, pp. 205–26.

Clements, R. O. & Cook, R. (1996). Pest damage to established grass in the UK. *Agricultural Zoology Reviews* **7**, 157–79.

Cook, R. & Yeates, G. W. (1993). Nematode pests of grassland and forage crops. In: K. Evans, D. L. Trudgill & J. M. Webster (eds), *Plant Parasitic Nematodes in Temperate Agriculture*, pp. 305–50. CAB International: Wallingford.

French, N., Nichols, D. B. R. & Wright, A. J. (1990). Yield response of improved upland pasture to the control of leatherjackets under increasing rates of nitrogen. *Grass and Forage Science* **45**, 99–102.

Glen, D. M., Cuerden, R. & Butler, R. C. (1991). Impact of the field slug *Deroceras reticulatum* on establishment of ryegrass and white clover in mixed swards. *Annals of Applied Biology* **119**, 155–62.

Gray, E. G. & Copeman, G. J. F. (1975). The role of snow moulds in winter damage to grassland in northern Scotland. *Annals of Applied Biology* **81**, 235–39.

Holmes, S. J. I. (1983). The susceptibility of agricultural grasses to pre-emergence damage caused by *Fusarium culmorum* and its control by fungicide and seed treatment. *Grass and Forage Science* **38**, 209–14.

Holmes, S. J. I. (1985). Barley yellow dwarf virus in ryegrass and its detection by ELISA. *Plant Pathology* **34**, 214–20.

Kendall, D. A., George, S. & Smith, B. (1996). Occurrence of barley yellow dwarf viruses

in some common grasses (Gramineae) in south west England. *Plant Pathology* **45**, 29–37.
Lam, A. (1984). *Drechslera siccans* from ryegrass fields in England and Wales. *Transactions of the British Mycological Society* **83**, 305–11.
Lam, A. (1985). *Drechslera andersenii* sp. nov. and other *Drechslera* spp. on ryegrass in England and Wales. *Transactions of the British Mycological Society* **85**, 595–602.
Lewis, G. C., Lavender, R. H. & Martyn, T. M. (1996). The effect of propiconazole on foliar fungal diseases, herbage yield and quality of perennial ryegrass. *Crop Protection* **15**, 91–95.
Lewis, G. C. & Thomas, B. J. (1991). Incidence and severity of pest and disease damage to white clover foliage at 16 sites in England and Wales. *Annals of Applied Biology* **118**, 1–8.
Mühle, E. (1971). *Krankheiten und Schädlinge der Futtergräser*. 1. Auflage. S. Hirzel: Leipzig.
Murray, P. J. & Clements, R. O. (1995). Distribution and abundance of three species of *Sitona* (Coleoptera, Curculionidae) in grassland in England. *Annals of Applied Biology* **127**, 229–37.
Nye, I. W. B. (1959). The distribution of shoot-fly larvae (Diptera, Acalypterae) within pasture grasses and cereals in England. *Bulletin of Entomological Research* **50**, 53–62.
Potter, L. R. (1993). The effects of white clover mosaic virus on vegetative growth and yield of clones of S.100 white clover. *Plant Pathology* **42**, 797–805.
Priestley, R. H., Thomas, J. E. & Sweet, J. B. (1988). *Diseases of Grasses and Herbage Legumes*. National Institute of Agricultural Botany: Cambridge.
Smith, J. D., Jackson, N. & Woolhouse, A. R. (1989). *Fungal Diseases of Amenity Turf Grasses*. E. & F. N. Spon: London.
Thomas, J. E. (1991). Diseases of established grassland. *Strategies for weed, disease and pest control in grassland: Proceedings of the British Grassland Society Conference, Gloucester*, pp. 3.1–3.12.
Winfield, A. L. (1961). Observations on the biology and control of the cabbage stem weevil *Ceutorhynchus quadridens* (Panz) on Trowse mustard (*Brassica juncea*). *Bulletin of Entomological Research* **52**, 589–600.
Wright, C. E. (1967). Blind seed disease of ryegrass. *Euphytica* **16**, 122–30.
York, C. A. (1998). *Turfgrass Diseases and Associated Disorders*. Sports Turf Research Institute: Bingley.

Chapter 5 Pests and diseases of potatoes

Anon. (1989). *Potato Pests*. MAFF Reference Book 187. HMSO: London.
Anon. (1997). *NFU-retailer ICM Protocol for Fresh Market Potatoes*. National Farmers Union: London.
Beaumont, A. (1947). The dependence on the weather of the dates of outbreaks of potato blight epidemics. *Transactions of the British Mycological Society* **31**, 44–53.
Bowden, J., Cochrane, J., Emmett, B. J., Minall, T. E. & Sherlock, P. L. (1983). A survey of cutworm attacks in England and Wales, and a descriptive population model for *Agrotis segetum* (Lepidoptera : Noctuidae). *Annals of Applied Biology* **102**, 29–47.
Bradshaw, N. J. & Vaughan, T. B. (1996). The effect of phenylamide fungicides on the control of potato late blight (*Phytophthora infestans*) in England and Wales from 1978 to 1992. *Plant Pathology* **45**, 249–69.
Broadbent, L. (1953). Aphids and virus diseases in potato crops. *Biological Reviews* **28**, 350–80.
Burton, W. G. (1989). *The Potato*, 3rd edition. Longman: Harlow.
Cock, L. J. (1990). Potato blight. In: J. S. Gunn (ed.) *Crop Protection Handbook – Potatoes*, pp. 35–54. BCPC: Farnham.

Collier, G. F., Wurr, D. C. E. & Huntington, V. C. (1978). The effect of nutrition on the incidence of internal rust spot in the potato. *Journal of Agricultural Science, Cambridge* **91**, 241–43.

Collier, G. F., Wurr, D. C. E. & Huntington, V. C. (1980). The susceptibility of potato varieties to internal rust spot. *Journal of Agricultural Science, Cambridge* **94**, 407–10.

Davies, H. V. (1998). Physiological mechanisms associated with the development of internal necrotic disorders of potato. *American Journal of Potato Research* **75**, 37–44.

Foster, S. P., Denholm, I., Harding, Z., Moores, J. D. & Devonshire, A. L. (1998). Intensification of insecticide resistance in UK field populations of the peach–potato aphid, *Myzus persicae* in 1996. *Bulletin of Entomological Research* **88**, 127–30.

Gibson, R. W. (1974). The induction of top-roll symptoms on potato plants by the aphid *Macrosiphum euphorbiae*. *Annals of Applied Biology* **76**, 19–26.

Harris, P. M. (ed.) (1991). *The Potato Crop: the Scientific Basis for Improvement*. Chapman & Hall: London.

Haydock, P. P. J. & Evans, K. (1998). Management of potato cyst nematodes in the United Kingdom: an integrated approach? *Outlook on Agriculture* **27**, 259–66.

Ingram, D. S. & Williams, P. H. (1991). *Advances in Plant Pathology. Vol. 7, Phytophthora infestans, the Cause of Late Blight of Potato*. Academic Press: London.

Iritani, W. M., Weller, L. D. & Knowles, N. R. (1984). Factors influencing incidence of internal brown spot in Russet Burbank potatoes. *American Potato Journal* **61**, 335–43.

Large, E. C. (1959). Potato blight forecasting in England and Wales. *Proceedings IVth International Congress of Crop Protection 1957, Hamburg, Germany*, pp. 215–20.

McKinlay, R. G. & Franklin, M. F. (1983). Potato aphid and leaf roll virus control: insecticide granules and sprays. *Proceedings of the 10th International Congress of Plant Protection 1983*, **3**, 1204–18.

Moores, G. D., Devine, G. J. & Devonshire, A. L. (1994). Insecticide resistance due to insensitive acetylcholinesterase in *Myzus persicae* and *Myzus nicotianae*. *Proceedings of the Brighton Crop Protection Conference – Pests and Diseases 1994*, **1**, 413–18.

Parker, W. E. (1996). The development of baiting techniques to detect wireworms (*Agriotes* spp., Coleoptera: Elateridae) in the field, and the relationship between bait-trap catches and wireworm damage to potato. *Crop Protection* **15**, 521–27.

Parker, W. E. (1998). Forecasting the timing and size of aphid populations (*Myzus persicae* and *Macrosiphum euphorbiae*) on potato. *Aspects of Applied Biology* **52**, *Protection and Production of Sugar Beet and Potatoes*, pp. 31–8.

Port, C. M. & Port, G. M. (1986). The biology and behaviour of slugs in relation to crop damage and control. *Agricultural Zoology Reviews* **1**, 255–99.

Radcliffe, E. B. (1982). Insect pests of potato. *Annual Review of Entomology* **27**, 173–204.

Schepers, H. T. A. M. & Bouma, E. (1997). Proceedings of the workshop on the European network for the development of an integrated control strategy of potato late blight. *PAV Special Report No. 1*, January 1997.

Schepers, H. T. A. M. & Bouma, E. (1998). Proceedings of the workshop on the European network for the development of an integrated control strategy of potato late blight. *PAV Special Report No. 3*, January 1998.

Schepers, H. T. A. M. & Bouma, E. (1999). Proceedings of the workshop on the European network for the development of an integrated control strategy of potato late blight. *PAV Special Report No. 6*, January 1999.

Schepers, H. T. A. M. & Bouma, E. (2000). Proceedings of the workshop on the European network for the development of an integrated control strategy of potato late blight. *PAV Special Report No. XX*, January 2000.

Slawson, D. D. (1997). *Fungicide Resistance in Potato Pathogens – Potato Blight* (*Phytophthora infestans*). British Potato Council/The Fungicide Resistance Action Group–UK (FRAG–UK).

Smith, L. P. (1956). Potato blight forecasting by 90% humidity criteria. *Plant Pathology* **5**, 83–7.

Trudgill, D. L. (1986). Yield losses caused by potato cyst nematodes: a review of the current position in Britain and prospects for improvement. *Annals of Applied Biology* **108**, 181–98.

Walker, D. F. (1998). Potatoes in the next millennium. *Aspects of Applied Biology* **52**, *Protection and Production of Sugar Beet and Potatoes*, pp. 7–10.

Chapter 6 Pests and diseases of sugar beet

Anon. (1982). *Pests, Diseases and Disorders of Sugar Beet.* BM Press: Sartrouville. (Distributed by IACR-Broom's Barn, Higham, Suffolk, UK.)

Asher, M. J. C. & Williams, G. E. (1991). Forecasting the national incidence of sugar beet powdery mildew from weather data in Britain. *Plant Pathology* **40**, 100–107.

Asher, M. J. C. (1998). Progress towards the control of rhizomania in the UK. *Aspects of Applied Biology* **52**, 415–22.

Asher, M. J. C. (1999). Foliar disease control in 1999. *British Sugar Beet Review* **67**(2), 20–21.

Asher, M. J. C. & Dewar, A. M. (1999). Rhizomania and other pests and diseases in 1998. *British Sugar Beet Review* **67**(1), 13–17.

Baker, A. & Dunning, R. A. (1975). Association of populations of Onchyiurid Collembola with damage to sugar beet seedlings. *Plant Pathology* **24**, 150–54.

Benada, J., Sedivy, J. & Spacek, J. (1987). *Atlas of Diseases and Pests in Beet.* Elsevier: Amsterdam.

Cochrane, J. & Thornhill, W. A. (1987). Variation in annual and regional damage to sugar beet by pygmy beetle (*Atomaria linearis*). *Annals of Applied Biology* **110**, 231–38.

Cooke, D. A. (1984). The relationship between numbers of *Heterodera schachtii* and sugar beet yields on a mineral soil, 1978–81. *Annals of Applied Biology* **104**, 121–29.

Cooke, D. A. (1987). Beet cyst nematode (*Heterodera schachtii)* and its control on sugar beet. *Agricultural Zoology Reviews* **2**, 135–83.

Cooke, D. A. (1989). Damage to sugar beet crops by ectoparasitic nematodes, and its control by soil-applied granular pesticides. *Crop Protection* **8**, 63–70.

Cooke, D. A. (1991). The effect of beet cyst nematode, *Heterodera schachtii*, on the yield of sugar beet in organic soils. *Annals of Applied Biology* **118**, 153–60.

Cooke, D. A. & Scott, R. K. (eds) (1993). *The Sugar Beet Crop.* Chapman & Hall: London.

Cotton, J., Cooke, D. A., Darlington, P. & Hancock, M. (1992). Surveys of beet cyst nematode (*Heterodera schachtii)* in England. *Annals of Applied Biology* **120**, 95–103.

Dewar, A. M. (1994). The virus yellows warning scheme – an integrated pest management system for beet in the UK. In : S. R. Leather, N. J. Mills, A. D. Watt & K. F. A. Walters (eds), *Individuals, Populations and Patterns in Ecology,* pp. 173–85. Intercept Press: Andover.

Dewar, A. M. & Cooke, D. A. (1986). Recent developments in control of nematode and arthropod pests of sugar beet. *Aspects of Applied Biology* **13**, 89–99.

Dewar, A. M. & Haylock, L. A. (1995). The long hot summer spawns a red menace. *British Sugar Beet Review* **63**(4), 20–22.

Dewar, A. M., Haylock, L. A. & Ecclestone, P. M. J. (1996). Strategies for controlling aphids and virus yellows in sugar beet. *Proceedings Brighton Crop Protection Conference – Pests & Diseases 1996* **1**, 185–90.

Dewar, A. M., Haylock, L. A., Campbell, J., Harling, Z., Foster, S. P. & Devonshire, A. L. (1998). Control in sugar beet of *Myzus persicae* with different insecticide-resistance mechanisms. *Aspects of Applied Biology* **52**, 407–14.

Dunning, R. A. (1957). Mirid damage to seedling beet. *Plant Pathology* **6**, 19–20.

Dunning, R. A. (1961). Mangold fly incidence, economic importance and control. *Plant Pathology* **10**, 1–9.

Dunning, R. A. & Winder, G. H. (1965). Sugar beet seedling populations and protection from wireworm injury. *Proceedings 3rd British Insecticide and Fungicide Conference, Brighton, 1965*, pp. 88–89.

Francis, C. (1998). Sugar beet in the next millennium. *Aspects of Applied Biology* **52**, 1–5.

Franklin, M. T. (1965). A root-knot nematode, *Meloidogyne naasi*, on field crops in England and Wales. *Nematologica* **11**, 79–86.

Garthwaite, D. G. & Thomas, M. R. (1999). *Pesticide Usage Survey Report 159: Arable Crops 1998*. MAFF: London.

Green, R. E. (1977). Mouse damage. *British Sugar Beet Review* **45**(1), 30.

Harrington, R., Dewar, A. M. & George, B. (1989). Forecasting the incidence of virus yellows in sugar beet in England. *Annals of Applied Biology* **114**, 449–69.

Jaggard, K. W., Limb, M. & Proctor, G. H. (eds) (1995). *Sugar Beet: A Grower's Guide*, 5th edition. SBREC/MAFF: London.

Jones, F. G. W. & Dunning, R. A. (1972). *Sugar Beet Pests*. MAFF Bulletin 162, 3rd edition. HMSO: London.

Maughan, G. L., Cooke, D. A. & Gnanasakthy, A. (1984). The effects of soil-applied granular pesticides on the establishment and yield of sugar beet in commercial fields. *Crop Protection* **4**, 446–57.

Payne, P. A., Asher, M. J. C. & Kershaw, C. D. (1994). The incidence of *Pythium* spp. and *Aphanomyces cochlioides* associated with sugar beet growing soils of Britain. *Plant Pathology* **43**, 300–8.

Smith, H. G. (1989). Distribution and infectivity of yellowing viruses in field-grown sugar beet plants. *Annals of Applied Biology* **114**, 481–87.

Smith, H. G. & Hallsworth, P. B. (1990). The effects of yellowing viruses on yield of sugar beet in field trials 1985–1987. *Annals of Applied Biology* **116**, 503–11.

Smith, H. G., Hallsworth, P. B. & Stevens, M. (1998). Aphid infectivity and virus yellows forecasting. *British Sugar Beet Review* **65**(1), 20–22.

Thornhill, W. A. & Edwards, C. W. (1985). The effects of pesticides and crop rotation on the soil-inhabiting fauna of sugar beet fields. Part 1: The crop and macro-invertebrates. *Crop Protection* **4**, 409–23.

Way, M. J., Cammell, M. E., Taylor, L. R. & Woiwod, I. P. (1981). The use of egg counts and suction trap samples to forecast the infestation of spring-sown field beans, *Vicia faba*, by the black bean aphid, *Aphis fabae*. *Annals of Applied Biology* **98**, 21–34.

Werker, R. (1998). The future in virus yellows forecasting. *British Sugar Beet Review* **66**(1), 36–8.

Werker, A. R., Dewar, A. M. & Harrington, R. (1998). Modelling the incidence of virus yellows in sugar beet in relation to numbers of migrating *Myzus persicae*. *Aspects of Applied Biology* **52**, 115–20.

Whitehead, A. G. & Hooper, D. J. (1970). Needle nematodes (*Longidorus* spp.) and stubby-root nematodes (*Trichodorus* spp.) harmful to sugar beet and other field crops in England. *Annals of Applied Biology* **65**, 339–50.

Whitney, E. D. & Duffus, J. E. (eds) (1986). *Compendium of Beet Diseases and Insects*. APS Press: St Paul.

Chapter 7 Pests and diseases of field vegetables

Babadoost, M., Derie, M. L. & Gabrielson, R. L. (1996). Efficacy of sodium hypochlorite treatments for control of *Xanthomonas campestris* pv. *campestris* in brassica seeds. *Seed Science and Technology* **24**, 7–15.

Bertolini, P. & Tian, S. P. (1997). Effect of temperature of production of *Botrytis allii* conidia on their pathogenicity to harvested white onion bulbs. *Plant Pathology* **46**, 432–38.

Blake, F. (1999). *Organic Farming and Growing*, 3rd edition. WBC Book Manufacturers: Bridgend.

Channon, A. G. (1963). Studies on parsnip canker. I. The causes of the disease. *Annals of Applied Biology* **51**, 1–15.

Channon, A. G. (1964). Studies on parsnip canker. III. The effect of sowing date and spacing on canker development. *Annals of Applied Biology* **54**, 63–70.

Cheah, L. H., Page, B. B. C. & Shepherd, R. (1997). Chitosan coating for inhibition of *Sclerotinia* rot of carrots. *New Zealand Journal of Crop and Horticultural Science* **25**, 89–92.

Clarkson, J. & Kennedy, R. (1997). Quantifying the effect of reduced doses of propiconazole (Tilt) and initial disease incidence on leek rust development. *Plant Pathology* **46**, 952–63.

Claxton, J. R., Arnold, D. L., Blakesley, D. & Clarkson, J. M. (1995). The effects of temperature on zoospores of the crook root fungus *Spongospora subterranea* f. sp. *nasturtii*. *Plant Pathology* **44**, 765–71.

Coaker, T. H. & Finch, S. (1971). Cabbage root fly, *Erioischia brassicae*. *Report of the National Vegetable Research Station for 1970*, pp. 23–42.

Corbiere, R., Molinero, V., Lefebvre, A. & Spire, D. (1995). Detection du mildiou du pois (*Peronospora viciae*) par methode immunoenzymatique (ELISA) dans les lots de semences. *EPPO conference on new methods of diagnosis in plant protection, Wageningen, the Netherlands, 25–28 January 1995, Bulletin–OEPP* **25**, 47–56.

Crowton, O. W. B. & Kennedy, R. (1999). Effects of humidity and wetness duration on the germination and infection of *Erysiphe cruciferarum*. *Proceedings 1st International Powdery Mildew Conference, 29 August–2 September 1999*, p. 60.

Emmett, B. J. (1980). Key for the identification of lepidopterous larvae infesting brassica crops. *Plant Pathology* **29**, 122–23.

Entwistle, A. R. (1992). Controlling *Allium* white rot (*Sclerotium cepivorum*) without chemicals. *Phytoparasitica* **20**, 121S–125S.

Fry, W. E. & Fohner, G. R. (1985). Construction of predictive models: I. Forecasting disease development. In: C. A. Gilligan (ed.), *Advances in Plant Pathology*, Vol. 3. *Mathematical Modelling of Crop Disease*. Academic Press: London.

Gaag, D. J. van der & Frinking, H. D. (1997). Factors affecting germination of oospores of *Peronospora viciae* f. sp. *pisi in vitro*. *European Journal of Plant Pathology* **103**, 573–80.

Halmer, P. (1988). Technical and commercial aspects of seed pelleting and film-coating. In: T. J. Martin (ed.) *Applications to Seeds and Soil*. BCPC Monograph No. 39, pp. 191–204. BCPC: Thornton Heath.

Harris, K. M. & Scott, P. R. (eds) (1989). Crop protection information – an international perspective. *Proceedings of the International Crop Protection Information Workshop*. CAB International: Wallingford.

Hill, D. S. (1987). *Agricultural Insect Pests of Temperate Regions and Their Control*. Cambridge University Press: Cambridge.

Humpherson-Jones, F. M. (1991). The development of weather related forecasts for vegetable crops in the UK. Problems and prospects. *Bulletin OEPP/EPPO Bulletin* **21**, 425–29.

Humpherson-Jones, F. M. (1993). Effect of surfactants and fungicides on clubroot (*Plasmodiophora brassicae*). *Annals of Applied Biology* **122**, 457–65.

Jesperson, G. D. & Sutton, J. C. (1987). Evaluation of a forecaster for downy mildew of onion (*Allium cepa L.*). *Crop Protection* **6**, 95–103.

de Jong, P. D., Daamen, R. A. & Rabbinge, R. (1995). The reduction of chemical control of leek rust, a simulation study. *European Journal of Plant Pathology* **101**, 687–93.

Kennedy, R., Wakeham, A. J. & Cullington, J. E. (1999). Production and immunodetection of ascospores of *Mycosphaerella brassicicola*: ringspot of vegetable crucifers. *Plant Pathology* **48**, 297–307.

Kirk, W. D. (1992). *Insects on Cabbage and Oilseed Rape*. Naturalists' Handbooks 18. Richmond Publishing: Slough.

Knight, J. D. (ed.) (1999). Information technology for crop protection. *Aspects of Applied Biology* **55**, 88 pp.

Kohl, J., Molhoek, W. M. L., van der Plas, C. H. & Fokkema, N. J. (1995). Suppression of sporulation of *Botrytis* spp. as a valid biocontrol strategy. *European Journal of Plant Pathology* **101**, 251–59.

Lewis, T. (ed.) (1997). *Thrips as Crop Pests*. CAB International: Wallingford.

Madeira, A. D., Fryett, K. P., Rossall, S. & Clarke, J. A. (1993). Interaction between *Ascochyta fabae* and *Botrytis fabae*. *Mycological Research* **97**, 1217–22.

Mathieu, D. & Kushalappa, A. C. (1993). Effects of temperature and leaf wetness duration on the infection of celery by *Septoria apiicola*. *Phytopathology* **83**, 1036–40.

Maude, R. B. (1970). The control of *Septoria* on celery seed. *Annals of Applied Biology* **65**, 249–54.

Maude, R. B. & Presly, A. H. (1977). Neck rot (*Botrytis allii*) of onions. II. Neck rot in stored onion bulbs and control of the disease. *Annals of Applied Biology* **86**, 181–88.

McGaughey, W. H. (1985). Insect resistance to the biological insecticide *Bacillus thuringiensis*. *Science* **229**, 193–95.

McKinlay, R. G. (ed.) (1992). *Vegetable Crop Pests*. Macmillan Press: London.

Paulus, A. O., Nelson, J., Ganey, J. & Snyder, M. (1977). Systemic fungicides for control of phycomycetes on vegetable crops. *Proceedings 9th British Insecticide and Fungicide Conference* **3**, 929–35.

Petrie, G. A. (1988). Races of *Albugo candida* (white rust and staghead) on cultivated Cruciferae in Saskatchewan. *Canadian Journal of Plant Pathology* **10**, 142–50.

Scherm, H. & Bruggen, A. H. C. van (1994). Weather variables associated with infection of lettuce by downy mildew (*Bremia lactucae*) in coastal California. *Phytopathology* **84**, 860–65.

Schoneveld, J. A. (1994). Effect of irrigation on the prevention of scab in carrots. *Acta Horticulturae* **354**, 135–44.

Smilde, W.D., van Nes, M. & Frinking, H. D. (1996). Effects of temperature on *Phytophthora porri in vitro*, *in planta*, and in soil. *European Journal of Plant Pathology* **102**, 687–95.

Stegmark, R. (1994). Downy mildew on peas (*Peronospora viciae* f. sp. *pisi*). *Agronomie* **14**, 641–47.

Stewart, A. & Franicevic, S.C. (1994). Infected seed as a source of inoculum for *Botrytis* infection of onion bulbs in store. *Australasian Plant Pathology* **23**, 36–40.

Sutton, J. C., James, T. D. W. & Rowell, P. M. (1986). BOTCAST: a forecasting system to time the initial fungicide spray for managing *Botrytis* leaf blight of onions. *Agriculture, Ecosystems and Environment* **18**, 123–43.

Taylor, J. D. (1972). Field studies on halo-blight on beans (*Pseudomonas phaseolicola*) and its control by foliar sprays. *Annals of Applied Biology* **70**, 191–97.

Taylor, J. D. & Dudley, C. L. (1977). Seed treatment for the control of halo-blight on beans (*Pseudomonas phaseolicola*). *Annals of Applied Biology* **85**, 223–32.

Walsh, J. A. (1992). The epidemiology and control of watercress yellow spot virus. *Recent Advances in Vegetable Virus Research. 7th Conference ISHS Vegetable Virus Working Group*, Athens, Greece, 12–16 July 1992, 2 pp.

Webster, M. A. & Dixon, G. R. (1991). Boron, pH and inoculum concentration influencing colonisation by *Plasmodiophora brassicae*. *Mycological Research* **95**, 74–79.

Xue, L., Charest, P. M. & Jabaji-Hare, S. H. (1998). Systemic induction of peroxidases, 1, 3-beta-glucanases, chitinases, and resistance in bean plants by binucleate *Rhizoctonia* species. *Phytopathology* **88**, 359–65.

Chapter 8 Pests and diseases of fruit and hops

Alford, D. V. (1984). *A Colour Atlas of Fruit Pests. Their Recognition, Biology and Control.* Wolfe Scientific: London.

Anon. (1999) (New edition each year). *Assured Produce Scheme Protocols: Generic Crop Protocol; Crop Specific Protocols for Bush Fruit (202), Strawberries (212), Top Fruit (214), Stone Fruit (215).* Checkmate International (Assured Produce Registrar): Oxford.

Berrie, A. M., Xu, X.-M., Harris, D. C., Roberts, A. L., Evans, K., Barbara, D. J. & Gessler, C. (eds) (1997). Proceedings of 4th Workshop on Integrated Control of Pome Fruit Diseases. *IOBC/WPRS Bulletin* **20**(9), 276 pp.

Blommers, L. H. M. (1994). Integrated pest management in European apple and pear orchards. *Annual Review of Entomology* **39**, 213–41.

Briolini, G., Nguyen, T. X. & Verzone, D. (eds) (1994). Proceedings International Colloquium on Integrated Control in Pear. *IOBC/WPRS Bulletin* **17**(2), 158 pp.

Butt, D. J. (ed.) (1994). Proceedings 3rd Workshop on Integrated Control of Pome Fruit Diseases. *Norwegian Journal of Agricultural Sciences.* Supplement No. 17. Agricultural University of Norway.

Byrde, R. J. W. & Willets, H. J. (1977). *The Brown Rot Fungi of Fruit: their Biology and Control.* Pergamon Press: Oxford.

Croft, B. A. (1990). *Arthropod Biological Control Agents and Pesticides.* John Wiley & Sons: New York.

Cross, J. V., Solomon, M. G., Blommers, L., Campbell, C. A. M., Easterbrook, M. A., Jay, C. N., Jolly, R., Jenser, G., Kuhlmann, U., Lilley, R., Olivella, E., Toepfer, S. & Vidal, S. (1999). Biocontrol of pests of apple and pears in Northern and Central Europe: 2 Parasitoids. *Biocontrol Science and Technology* **9**, 277–314.

Ellis, M. A., Converse, R. H., Williams, R. N. & Williamson, B. (eds) (1991). *Compendium of Raspberry and Blackberry Diseases and Insects.* APS Press: Minnesota.

Frankenhuyzen, A. van (1992). *Schadelijke en Nuttige Insekten en Mijten in Fruitgewassen,* 2nd edition. Nederlandse Fruittelers Organisatie: Wageningen.

Gordon, S. C., Woodford, A. T. & Birch, A. N. E. (1997). Arthropod pests of *Rubus* in Europe: pest status, current and future control strategies. *Journal of Horticultural Science* **72**, 831–62.

Helle, W. & Sabelis, M. W. (eds) (1985). *Spider Mites: their Biology, Natural Enemies and Control.* (World Crop Pests Volume 1A). Elsevier: Amsterdam.

Helle, W. & Sabelis, M. W. (eds) (1985). *Spider Mites: their Biology, Natural Enemies and Control.* (World Crop Pests Volume 1B). Elsevier: Amsterdam.

Jones, A. L. & Aldwinckle, H. S. (1990). *Compendium of Apple and Pear Diseases.* APS Press: Minnesota.

Lindquist, E. E., Sabelis, M. W. & Bruin, J. (eds) (1996). *Eriophyoid Mites: their Biology, Natural Enemies and Control.* (World Crop Pests Volume 6). Elsevier: Amsterdam.

Locke, T., Berrie, A. M., Cooke, L. & Edwards, J. (1999). *Fungicide Resistance in Apple and Pear Pathogens.* Fungicide Resistance Action Group – UK. (Leaflet distributed by the Apple and Pear Research Council), 8 pp.

Maas, J. L. (1998). *Compendium of Strawberry Diseases,* 2nd edition. APS Press: Minnesota.

MacHardy, W. E. (1996). *Apple Scab: Biology, Epidemiology, and Management.* APS Press: Minnesota.

Massee, A. M. (1954). *The Pests of Fruit and Hops,* 3rd edition. Crosby Lockwood & Son: London.

Milaire, H. G., Baggiolini, M., Gruys, P. & Steiner, H. (eds) (1974). Les organismes auxiliaires en verger de pommiers. *OILB Brochure No. 3,* 242 pp.

Minks, A. K. & Harrewijn, P. (eds) (1987). *Aphids: their Biology, Natural Enemies and Control.* (World Crop Pests Volume 2A). Elsevier: Amsterdam.

Minks, A. K. & Harrewijn, P. (eds) (1987). *Aphids: their Biology, Natural Enemies and Control.* (World Crop Pests Volume 2B). Elsevier: Amsterdam.
Minks, A. K. & Harrewijn, P. (eds) (1988). *Aphids: their Biology, Natural Enemies and Control.* (World Crop Pests Volume 2C). Elsevier: Amsterdam.
Neve, R. A. (1991). *Hops.* Chapman & Hall: London.
Pearson, R. C. & Goheen, A. C. (1988). *Compendium of Grape Diseases.* APS Press: Minnesota.
Polesny, F., Müller, W. & Olszak, R. W. (eds) (1996). Proceedings of an International Conference on Integrated Fruit Production. *IOBC/WPRS Bulletin* **19** (4), 442 pp.
Smith, I. M., Dunez, J., Phillips, D. H., Lelliott, R. A. & Archer, S. A. (eds) (1988). *European Handbook of Plant Diseases.* Blackwell Science: Oxford.
Snowdon, A. L. (1990). *Post-harvest Diseases and Disorders of Fruits and Vegetables.* Wolfe Scientific: London.
Solomon, M. G. (1987). Fruit and hops. In: A. J. Burn, T. H. Coaker & P. C. Jepson (eds). *Integrated Pest Management*, pp. 329–60. Academic Press: London.
Solomon, M. G., Cross, J. V., Fitzgerald, J. D., Campbell, C. A. M., Jolly, R. L., Olszak, R. W., Niemczyk, E. & Vogt, H. (2000). Biocontrol of pests of apples and pears in northern and central Europe – 3. Predators. *Biocontrol Science & Technology* **10**, 101–38.
Spencer, D. M. (1978). *The Powdery Mildews.* Academic Press: London.
Steiner, H. (1985). *Nützlinge im Garten.* Ulmer: Stuttgart.
Sterk, G. (1991). *De Gîntegreerde Bestrijding in de Fruitteelt.* Opzpelomgsstation van Gorsem: St. Truiden.
Van der Geest, L. P. S. & Evenhuis, H. H. (eds) (1991). *Tortricid Pests: their Biology, Natural Enemies and Control.* (World Crop Pests Volume 5). Elsevier: Amsterdam.
Wormald, H. (1946). *Diseases of Fruits and Hops*, 2nd edition. Crosby Lockwood & Son: London.

Chapter 9 Pests and diseases of protected vegetables and mushrooms

Albajes, R., Gullino, M. L., van Lenteren, J. C. & Elad, Y. (eds) (1999). *Integrated Pest and Disease Management in Greenhouse Crops.* Kluwer Academic Publishers, London.
Bennison, J. A. & Corless, S. P. (1993). Biological control of aphids on cucumbers: further development of open rearing units or 'banker plants' to aid establishment of aphid natural enemies. *IOBC/WPRS Bulletin* **16**, 5–8.
Blancard, D. (1994). *A Colour Atlas of Tomato Diseases.* Manson Publishing: London.
Blancard, D., Lecoq, H. & Pitrat, M. (1994). *A Colour Atlas of Cucurbit Diseases.* Manson Publishing: London.
Elad, Y. (1997). Effect of filtration of solar light on the production of conidia by field isolates of *Botrytis cinerea* and on several diseases of greenhouse grown vegetables. *Crop Protection* **16**, 635–42.
Elad, Y. & Shtienberg, D. (1995). *Botrytis cinerea* in greenhouse vegetables: chemical, cultural, physiological and biological controls and their integration. *Integrated Pest Management Reviews* **1**, 15–29.
Fletcher, J. T. (1992). Disease resistance in protected crops and mushrooms. *Euphytica* **63**, 33–49.
Fletcher, J. T., White, P. F. & Gaze, R. H. (1989). *Mushrooms – Pest and Disease Control.* Intercept: Andover.
Hussey, N. W., Read, W. H. & Hesling, J. J. (1969). *The Pests of Protected Cultivation.* Arnold: London.
Jarvis, W. R. (1992). *Managing Diseases in Greenhouse Crops.* The American Phytopathological Society: St Paul.

Jarvis, W. J., Shipp, J. L. & Gardiner, R. B. (1993). Transmission of *Pythium aphanidermatum* in greenhouse cucumber by the fungus gnat *Bradysia impatiens* (Diptera: Sciaridae). *Annals of Applied Biology* **122**, 23–29.

Maude, R. B. (1996). *Seed Borne Diseases and their Control.* CAB International: Wallingford.

Sampson, C. & Walker, P. (1998). Improved control of *Liriomyza bryoniae* using an action threshold for the release of *Diglyphus isaea* in protected tomato crops. *Mededelingen van de Faculteit Landbouwwetenchappen Rijksuniversiteit Gent* **63**, 415–22.

Stanghellini, M. E. & Rasmussen, S. L. (1994). Hydroponics – a solution for zoosporic fungi. *Plant Disease* **78**, 1129–38.

Wardlow, L. R. & O'Neill, T. M. (1992). Management strategies for controlling pests and diseases in glasshouse crops. *Pesticide Science* **36**, 341–47.

Chapter 10 Pests and diseases of protected flowering ornamentals

Albajes, R., Gullino, M. L., van Lenteren, J. C. & Elad, Y. (eds) (1999). *Integrated Pest and Disease Management in Greenhouse Crops.* Kluwer Academic Publishers: London.

Alford, D. V. (1991). *A Colour Atlas of Pests of Ornamental Trees, Shrubs and Flowers.* Wolfe Scientific: London.

Alford, D. V. & Backhaus, G. F. (1996). Proceedings of the Second International Workshop on Vine Weevil (*Otiorhynchus sulcatus* Fabr.) (Coleoptera: Curculionidae). *Mitteilungen aus der Biologischen Bundesanstalt für Land- und Forstwirtschaft* **316**, 1–122.

Gratwick, M. & Southey, J. F. (eds) (1986). *Hot-water Treatment of Plant Material.* MAFF Reference Book 201, 3rd edition. HMSO: London.

Hausbeck, M. K., Pennypacker, S. P. & Stevenson, R. E. (1996). The effect of plastic mulch and forced heated air on *Botrytis cinerea* on geranium stock in a research greenhouse. *Plant Disease* **80**, 170–73.

Hussey, N. W., Read, W. H. & Hesling, J. J. (1969). *The Pests of Protected Cultivation.* Arnold: London.

Jacobsen, R. J. (1993). Integrated pest management in spring bedding plants: a successful package for commercial crops. *IOBC/WPRS Bulletin* **16** (8), 105–12.

Köhl, J., Gerlagh, M., DeHaas, B. H. & Krijger, M. C. (1998). Biological control of *Botrytis cinerea* on cyclamen with *Ulocladium atrium* and *Gliocladium roseum* under commercial growing conditions. *Phytopathology* **88**, 568–75.

Moorhouse, E. (1992). A review of the biology and control of vine weevil, *Otiorhynchus sulcatus. Annals of Applied Biology* **121**, 431–54.

Sirjusingh, C. & Sutton, J. C. (1996). Effects of wetness duration and temperature on infection of geranium by botrytis. *Plant Disease* **80**, 160–65.

Wardlow, L. R., Davies, P. J. & Brough, W. (1993). Integrated pest management techniques in protected ornamental plants. *IOBC/WPRS Bulletin* **16** (8), 149–57.

Wohanka W., Luedtke, H., Ahlers, H. & Luebke, M. (1999). Optimisation of slow filtration as a means for disinfecting nutrient solutions. *Acta Horticulturae* **481**, 539–44.

Chapter 11 Pests and diseases of hardy ornamentals

Alford, D. V. (1991). *A Colour Atlas of Pests of Ornamental Trees, Shrubs and Flowers.* Wolfe Scientific: London.

Ann, D. (1997). Fungicidal control of *Phoma* on *Vinca minor. HDC Project News* **45**, 14.

Bevan, D. (1987). *Forest Insects.* HMSO: London.

Buczacki, S. T. & Harris K. (1998). *Collins Photoguide to the Pests, Diseases and Disorders of Garden Plants.* Harper Collins: London.

Carter, C. & Winter, T. (1998). *Christmas Tree Pests*. Forestry Commission Field Book 17. HMSO: London.

Cooper, J. I. (1979). *Virus Diseases of Trees and Shrubs*. Institute of Terrestrial Ecology: Cambridge.

Coyier, D. L. & Roane, M. K. (eds) (1986). *Compendium of Rhododendron and Azalea Diseases*. APS Press: Minnesota.

Darlington, A. (1968). *Plant Galls in Colour*. Blandford Press: London.

Gratwick, M. & Southey, J. F. (eds) (1986). *Hot-water Treatment of Plant Material*. MAFF Reference Book 201, 3rd edition. HMSO: London.

Greenwood, P. & Halstead, A. (1997). *The RHS Pests and Diseases*. Dorling Kindersley: London.

Hansen, E. M. & Lewis, K. J. (eds) (1997). *Compendium of Conifer Diseases*. APS Press: Minnesota.

Horst, R. K. (1983). *Compendium of Rose Diseases*. APS Press: Minnesota.

Kenyon, D. & Dixon, G. (1997). New fungicides for rhododendron. *HDC Project News* **43**, 6.

Litterick, A. M. & Holmes, S. J. (1994). Integrated control of root diseases on ornamental ericaceous plants. *Proceedings Brighton Crop Protection Conference – Pests and Diseases 1994* **2**, 807–10.

Litterick, A. M. & McQuilken, M. P. (1997). Chemical control of binucleate *Rhizoctonia* during propagation of *Calluna vulgaris* in Scotland. *Crop Protection* **16**, 173–78.

O'Neill, T. M. (1999). Hebe downy mildew – refining fungicide treatment. *HDC News* **55**, 24–25.

Peace, T. R. (1962). *Pathology of Trees and Shrubs with Special Reference to Britain*. Clarendon Press: Oxford.

Pettitt, T. (1998). Slow sand filtration for HNS. *HDC News* **53**, 22–23.

Phillips, D. H. & Burdekin, D. A. (1982). *Diseases of Forest and Ornamental Trees*. Macmillan: London.

Pirone, P. P. (1978). *Diseases and Pests of Ornamental Plants*, 5th edition. John Wiley & Sons: New York.

Reed, P. J., Dickens, J. S. W. & O'Neill, T. M. (1996). Occurrence of anthracnose *(Colletotrichum acutatum)* on ornamental lupin in the United Kingdon. *Plant Pathology* **45**, 245–48.

Roberts, S. (1997). Bacterial diseases of hardy nursery stock. *HDC Project News* **47**, 4–6.

Sinclair, W. A., Lyon, W. H. & Johnson, W. T. (1987). *Diseases of Trees and Shrubs*. Cornell University Press: New York.

Strouts, R. G. & Winter, T. G. (1994). *Diagnosis of Ill-health in Trees*. HMSO: London.

Van de Graaf, P. *et al.* (1999). Clematis wilt. *HDC News* **56**, 24–25.

Chapter 12 Pests and diseases of outdoor bulbs and corms

Alford, D. V. (1991). *A Colour Atlas of Pests of Ornamental Trees, Shrubs and Flowers*. Wolfe Scientific: London.

Bergman, B. H. H. (1965). Field infection of tulip bulbs by *Fusarium oxysporum*. *Netherlands Journal of Plant Pathology* **71**, 129–35.

Bergman, B. H. H. & Noordmeyer-Luyk, C. E. (1973). Influence of soil temperature on field infection of tulip bulbs by *Fusarium oxysporum*. *Netherlands Journal of Plant Pathology* **79**, 221–28.

Gratwick, M. & Southey, J. F. (eds) (1986). *Hot-water Treatment of Plant Material*. MAFF Reference Book 201, 3rd edition. HMSO: London.

Hodson, W. E. H. (1928). The bionomics of the bulb mite *Rhizoglyphus echinops* Fumouze and Robin. *Bulletin of Entomological Research* **19**, 187–200.

Hodson, W. E. H. (1934). The bionomics of the bulb scale mite *Tarsonemus approximatus* Banks var. *narcissi*. Ewing. *Bulletin of Entomological Research* **25**, 177–85.

Lane, A. (1984). *Bulb Pests*. MAFF Reference Book 51, 7th edition. HMSO: London.

Linfield, C. A. (1986). A comparison of the effects of temperature on the growth of *Fusarium oxysporum* f. sp. *narcissi* in solid and liquid media. *Journal of Phytopathology* **166**, 278–81.

Moore, A. (1960). Tulip eelworm. *Agriculture, London* **66**, 452–58.

O'Neill, T. M. & Mansfield, J. W. (1982). The causes of smoulder and the infection of narcissus by species of *Botrytis*. *Plant Pathology* **31**, 65–79.

O'Neill, T. M., Mansfield, J. W. & Lyon, G. D. (1982). Aspects of narcissus smoulder epidemiology. *Plant Pathology* **31**, 101–19.

Price, D. (1970). The seasonal carry-over of *Botrytis tulipae* (Lib.) Lind.: the cause of tulip fire. *Annals of Applied Biology* **65**, 49–58.

Price, D. (1970). Tulip fire caused by *Botrytis tulipae* (Lib.) Lind.: the leaf spotting phase. *Journal of Horticultural Science* **45**, 233–38.

Price, D. (1977). Some pathological aspects of narcissus basal rot, caused by *Fusarium oxysporum* f. sp. *narcissi*. *Annals of Applied Biology* **86**, 11–17.

Price, D. & Briggs, J. B. (1974). The control of *Botrytis tulipae* (Lib.) Lind.: the cause of tulip fire, by fungicide dipping. *Experimental Horticulture* **26**, 36–9.

Price, D. & Briggs, J. B. (1976). The timing of hot-water treatment in controlling *Fusarium oxysporum* basal rot of narcissus. *Plant Pathology* **25**, 197–200.

Southey, J. F. (1957). Observations on races of *Ditylenchus dipsaci* infesting bulbs. *Journal of Helminthology* **31**, 39–46.

Winfield, A. L. (1970). Factors affecting the control by hot-water treatment of stem nematode *Ditylenchus dipsaci* (Kuhn) Filipjev in narcissus bulbs. *Journal of Horticultural Science* **45**, 447–56.

Winfield, A. L. (1972). Observations on the control of the stem nematode *Ditylenchus dipsaci* of tulips with hot-water or thionazin treatment. *Journal of Horticultural Science* **47**, 357–64.

Woodville, H. C. (1960). Further experiments on the control of bulb fly in narcissus. *Plant Pathology* **9**, 68–70.

Glossary

Abiotic Non-biological.
Acaricide A pesticide active against mites.
Acervulus (*pl.* **acervuli**) A layer of closely packed fungal hyphae giving rise to short, conidia-bearing conidiophores.
Adventitious Arising from an abnormal position, such as roots arising from a stem or leaf.
Aecial Pertaining to an aecium.
Aecidiospore – *see* Aeciospore.
Aecidium (*pl.* **aecidia**) – see Aecium (*pl.* aecia).
Aeciospore One of several spores formed in a chain-like series within an aecium.
Aecium (*pl.* **aecia**) One of the spore-bearing structures of a rust fungus.
Aestivation Summer diapause.
Alternate host Either of two dissimilar host plants of a pest or pathogen that requires both to complete its full life-cycle.
Anamorph The asexual (non-sexual, imperfect) reproductive state of a fungus.
Anthesis Flowering.
Antibiosis The production (by a living organism) of a substance which diffuses into its surroundings and there is toxic to individuals of another species.
Aphicide An insecticide active against aphids.
Apodous Without legs, legless.
Apothecium (*pl.* **apothecia**) A saucer- or cup-shaped ascocarp, bearing asci on the open surface.
Ascocarp A general term for the fruiting body of an ascomycetous fungus.
Ascospore A sexual spore produced within an ascocarp.
Ascosporic Pertaining to ascospores.
Ascus (*pl.* **asci**) A sac-like structure in which the sexual spores of ascomycetous fungi are formed.
Awn The stiff, bristle-like appendage often present on the flowering glumes (q.v.) of cereals and grasses.
Axil The angle (of a leaf) between its upper side and the stem from which it arises; most buds arise within the axil.
Axillary Pertaining to the axil.
Bacilliform Rod-shaped, as a bacterium.
Ballistospore A spore ejected forcefully from its point of origin.
Basidiospore A fungal spore formed on a basidium (q.v.).
Basidium (*pl.* **basidia**) A cell of a basidiomycete fungus in which nuclear fusion and meiosis occur, and on which basidiospores (usually four) are formed.
Binucleate With two nuclei.
Biotic Biological.
Calyx The outermost part of a flower, typically consisting of green leaf-like sepals that, in the bud stage, enclose and protect the other floral parts.
Casing In mushroom production, the layer of nutritionally inert material overlaid on the compost to enable mushroom sporophores to be produced. Currently, this is a mixture of peat and lime. In the UK deep-dug black peat and sugar-beet lime predominate.
Caterpillar A typically elongated insect larva with an obvious head and distinct legs, as found in the majority of butterflies, moths and sawflies.
Cauda The tail-like projection at the posterior end of the abdomen of an aphid.

Certified stock Cultivated plant material guaranteed to be of a defined quality, especially free of virus.
Chlamydospore A thick-walled fungal spore capable of surviving conditions that are unfavourable to growth of the fungus as a whole, formed by thickening the wall of a hyphal cell.
Chlorosis A disease of a plant, characterized by the yellow (chlorotic) condition of parts that are normally green.
Chlorotic – *see* Chlorosis.
Chog In mushroom production, that part of the mushroom stipe left after the mushroom has been cut for marketing. This can be either left attached on the bed or discarded after the mushroom has been plucked and cut. In the disease context this is most likely to be on the bed.
Cleistothecium (*pl.* **cleistothecia**) An ascocarp in which the asci are completely enclosed by the outer wall.
Coleoptile The protective sheath surrounding the plumule (q.v.) in cereal and grass seedlings.
Conidiophore A specialized conidia-producing fungal hypha.
Conidium (*pl.* **conidia**) The asexual spore of certain fungi, cut off externally at the apex of specialized hyphae called conidiophores.
Corolla The petals of a flower joined at the base, often into a tube.
Cortex Parenchyma tissue surrounding the vascular system in stems and roots and bounded on the outside by the epidermis.
Cross-pollination The transfer of pollen from the anther of one plant to the stigma of another.
Cultivar A variety of cultivated plant, produced by breeding or selection.
Cuticle The more or less waterproof layer, composed mainly of cutin covering the aerial parts of a plant.
Cutin A fatty, hydrophilic (water-repellent) polymer, forming the main component of the plant cuticle.
Diapause A period of suspended development or growth, as in hibernation.
Dikaryon A fungal cell containing two complementary haploid nuclei, usually of opposite mating type.
Dikaryotic Pertaining to a dikaryon (q.v.).
Ectoparasite A parasite that develops on the outside of the body of its host.
Ectoparasitoid A parasitoid that develops on the outside of the body of its host.
Elytron (*pl.* **elytra**) The hardened, horny forewing of a beetle and certain other insects; sometimes, as in a beetle, called the 'wing case' as it protects the hindwing.
Endoparasite A parasite that develops within the body of its host.
Endoparasitoid A parasitoid that develops within the body of its host.
Entomopathogenic A pathogen that attacks an insect.
Epidemiology The science of epidemics.
Epidermis The outer layer of cells of a plant or animal.
Epigaeic Ground dwelling.
Erinium (*pl.* **erinia**) A plant leaf gall containing a proliferation of often enlarged hairs.
Flagellum (*pl.* **flagella**) A long, whip-like structure that serves to propel a motile cell, such as a zoospore.
Forma specialis (*pl.* **formae speciales**) A physiological form (race) of a fungal species with a specific host range that is different from those of other 'formae speciales' of the same species.
Frass The solid excreta produced by an insect, especially a caterpillar.
Genotype The genetic make-up of an organism.
Glume One of a pair of shoot-enclosing bracts at the base of a cereal or grass spikelet.
Green bridge A temporal vegetative link between the harvesting of one crop and the

germination of a following crop, allowing the survival of, for example, viruliferous aphids in cereals.

Groundkeeper A potato tuber remaining in the ground after harvest, and capable of producing a plant in a following crop.

Haploid A cell or organism with just one set of chromosomes.

Haulm The stems of a potato plant.

Hibernaculum The structure (e.g. a silken cocoon) in which an animal hibernates.

Hibernation Dormancy during the winter, i.e. winter diapause.

Honeydew A sugary substance excreted through the anus of certain sap-feeding insects, such as aphids.

Hypha (*pl.* **hyphae**) The filament of a fungal body, composed of one or more cylindrical cells, and increasing in length by growing at the tip and giving rise to new hyphae by lateral branching.

Hyphal Pertaining to a hypha.

Hypocotyl That part of the stem of a seedling below the cotyledons.

Imperfect stage – *see* Anamorph.

Inoculum The infective material that will produce disease.

Instar The larval or nymphal stage in the development of, for example, an insect or mite, between two moults.

Internode That part of a plant shoot between two nodes (q.v.).

Interveinal The area of a leaf between the major veins.

Lamina The blade of a leaf.

Larva The immature growth stage of an insect, which is very different in appearance from the adult into which it metamorphoses during a pupal stage; insect larvae, unlike nymphs (q.v.) lack compound eyes. Also, the first (typically six-legged) juvenile stage in the development of a mite between the egg and first eight-legged nymphal stage.

Latent/incubation period The time between infection and symptom appearance (latent period) or sporulation by a fungal pathogen (incubation period). It is often cited in this way if the first symptom appearance is sporulation.

Lenticel A small, often elliptical, pore that develops in the woody stems of plants when the epidermis is replaced by cork; the lenticel is packed with loosely arranged cells, allowing gaseous exchange between the stem interior and the atmosphere.

Lesion Site of damage or injury caused by a pathogen.

Life-cycle The succession of stages through which an organism develops. In an insect this is often egg–larva–pupa–adult; in a fungus this is often spore–mycelium–fruiting body.

Maggot The worm-like, but stubby, larva of certain insects, e.g. certain Diptera (flies), with no appendages and without an obvious head; the anterior end of the body is often noticeably tapered.

Meristem The area of a plant in which active cell division occurs and from which new tissue derives.

Microorganism A minute organism that is invisible or virtually invisible to the unaided (naked) eye.

Microsclerotium (*pl.* **microsclerotia**) A very small sclerotium (q.v.).

Mould A general term for the visible mycelial or spore mass of a microfungus.

Mycelial Pertaining to mycelium (q.v.).

Mycelium A collective term for the mass of hyphae forming the vegetative part of the body of a fungus.

Necrotic Dead, usually dark-coloured, plant tissue.

Nematicide A pesticide active against nematodes.

Neonate larva The stage of a larva between egg hatch and the commencement of feeding.

Node The part of a plant stem from whence one or more leaves arise.

Nymph The immature growth stage of an insect, usually similar in general appearance to the adult and often sharing similar feeding habits; unlike insect larvae (q.v.) nymphs typically possess compound eyes. Also, the immature (typically eight-legged) growth stage of a mite.

Obligate parasite A parasite that is incapable of feeding other than parasitically.

Oestrogen A substance producing changes in the female sexual organs.

Oestrogenic Pertaining to oestrogen.

Oogonium (*pl.* **oogonia**) The female sex organ of certain fungi, containing one or more oospheres.

Oosphere A large, spherical, naked, sedentary, female egg, formed within an oogonium.

Oospore A thick-walled fungal resting spore produced from a fertilized oosphere.

Oviposition The act of egg laying.

Paedogenesis Parthenogenetic reproduction in which the adult stage is omitted.

Paedogenetic larva One that gives rise to further individuals by the process of paedogenesis (q.v.).

Parasite An organism that lives within or upon another (the host), obtaining food from it but giving nothing in return.

Parasitoid Typically, an insect parasite in which only the larval stage is parasitic and the adult free-living – a parasitoid (unlike a parasite) usually kills its host.

Parenchyma Thin-walled cellular tissue, permeated by air spaces, typically found in the cortex and pith.

Parthenogenesis Reproduction that does not involve fertilization.

Parthenogenetic Pertailing to parthenogenesis.

Pathogen A parasite (e.g. bacterium, fungus or virus) that causes disease.

Pathogenicity The infective capacity of a pathogen.

Pathotype A population of a nematode that can attack only some of the range of host plants capable of being attacked by the whole species.

Pathovar A form of a plant pathogen (usually a bacterium) with a different host range from that of other morphologically identical pathovars of the same pathogen.

Perennate To survive (usually in a dormant state) between epidemics.

Perennation The survival of fungi (usually in a dormant state) between epidemics.

Pericarp The wall of a plant ovary after it has matured into a fruit.

Periderm The corky cambium and its products, e.g. cork.

Perimedullary zone The inner phloem and storage parenchyma, between the vascular ring and the pith.

Perithecium (*pl.* **perithecia**) An enclosed ascocarp with a neck and pore at the top, through which the ascospores are released.

Petiole Leaf stalk.

pH A quantitative expression for acidity, ranging from 0 to 14; pH 7 is neutral, < 7 is acid and > 7 is alkaline.

Phloem The conducting tissue within a plant within which sugars are transported.

Photosynthesis In green plants, the synthesis of organic compounds from water and carbon dioxide, using the energy of sunlight absorbed by chlorophyll.

Phytoalexin A toxic chemical, produced metabolically as a response to infection by a potential pathogen, and involved in resistance to that and other pathogens.

Phytophagous Plant-feeding.

Phytoplasma The living matter of a plant cell.

Phytotoxin A poisonous chemical, toxic to plants.

Pith The central storage parenchyma (q.v.).

Pollination The transfer of pollen from the anther of a plant to the stigma (cf. Cross-pollination).

Pollinizer A plant whose pollen is used to pollinate another by cross-pollination; sometimes called a 'pollinator', but the latter term is better reserved for a pollen-transferring organism such as a bee (i.e. an organism that effects pollination).

Polyphagous Feeding on many different kinds of food.

Predator An animal that feeds by preying on other animals.

Primary host As in certain aphids, the typically woody hosts upon which winter eggs are laid (cf. Secondary host).

Pseudothecium (*pl.* **pseudothecia**) The sexual fruiting body of an ascomycetous fungus.

Pupa (*pl.* **pupae**) The pre-adult stage of an insect in which metamorphosis occurs. Also, the pre-adult 'resting' stage of certain mites.

Puparium (*pl.* **puparia**) In higher Diptera (true flies), the barrel-shaped structure, formed from the cast-off final-instar larval skin, within which the pupa is formed.

Pycnidiospore An asexual spore produced from a pycnidium.

Pycnidium (*pl.* **pycnidia**) A minute, hollow, asexual fruiting body lined with conidia-bearing conidiophores.

Pycniospore – *see* Spermatium (*pl.* spermatia).

Pycnium (*pl.* **pycnia**) One of the spore-bearing structures of a rust fungus.

Race A population of a pest or pathogen that can attack only some of the range of host plants capable of being attacked by the whole species.

Repellent A chemical used to repel pests from plants, usually by their unpleasant smell or taste.

Resistant (pest) A pest able to withstand exposure to certain pesticides that would normally be expected to kill it.

Resistant (plant) A plant able to withstand attack by a pest or pathogen.

Rhizomorph A root-like structure composed of fungal hyphae.

Riparian Of, on, a riverbank.

Roguing The removal and destruction of diseased or pest-infested plants.

Rotation The season-by-season alternation of different crops on an area of land.

Saprophyte An organism that feeds on dead or decaying organic matter.

Saprophytic Feeding on dead or decaying organic matter.

Sclerotial Pertaining to a sclerotium.

Sclerotium (*pl.* **sclerotia**) A compact, hard, tissue-like mass of fungal hyphae, capable of remaining dormant for long periods during adverse conditions before germinating to produce a fungal growth.

Secondary host As in certain aphids, the typically herbaceous summer host upon which breeding occurs parthenogenetically and eggs are not laid (cf. Primary host).

Semiochemical A chemical used in interspecific or intraspecific communication.

Sepal One of several leaf-like structures forming the calyx of a flower.

Siphunculus One of a pair of pores or tube-like structures on the abdomen of an aphid, through which alarm pheromones and other defensive compounds are discharged.

Sooty mould A black, soot-like fungal growth that often develops on the leaves and other parts of plants infested by aphids or other honeydew-excreting insects.

Sorus A spore mass erupting through, or replacing, host tissue.

Spathe The bract enclosing the inflorescence of some monocotyledonous plants, e.g. arum lily.

Spermatium (*pl.* **spermatia**) A haploid spore produced by a rust fungus.

Spiracle A breathing pore in an insect – an opening of the tracheal system.

Sporangiophore A fungal hypha bearing one or more sporangia, and sometimes morphologically distinct from a vegetative hypha.

Sporangium (*pl.* **sporangia**) A plant organ within which spores are produced.

Spore As in bacteria and fungi, a single-celled or multi-celled reproductive body that becomes detached from the parent and gives rise directly or indirectly to a new individual.

Sporidium (*pl.* **sporidia**) The haploid sexual spore (basidiospore) developing from a basidium.

Sporing Spore-producing.

Sporophore A general term for the spore-producing and spore-bearing structure of a fungus.

Sporulate To produce spores.

Sporulation The act of spore production.
Steckling A young sugar beet plant.
Stele The vascular cylinder in plant stems or roots, consisting of (for example) the phloem and xylem.
Stipe The stalk of a mushroom.
Stolon A horizontally growing plant stem that roots at the nodes.
Suberin A complex material found in the walls of plant cork cells, rendering them impervious to water.
Suberization The deposition of suberin (q.v.).
Synanamorph One of two or more morphologically distinct anamorphs of the same fungus.
Systemic Of a pathogen: capable of spreading throughout the host. Of a pesticide: capable of movement within the plant tissue; e.g. from the roots to the aerial parts, or from one leaf to another (cf. Translaminar).
Teleutospore – *see* Teliospore.
Teliospore A thick-walled resting spore of a rust fungus.
Tobamovirus A virus in the tobacco mosaic virus group.
Translaminar Movement of a pesticide within a leaf from the site of deposition but no further (cf. Systemic).
Tuber The swollen end of an underground stem, bearing buds in axils of scale-like rudimentary leaves (= stem tuber, e.g. potato), or a swollen root (= root tuber, e.g. dahlia).
Tuberization The formation of tubers.
Unsuberized Lacking suberin (q.v.).
Uredinial stage The stage at which uredosori are produced.
Urediniospore A non-sexual dispersal spore produced by a rust fungus.
Uredinium (*pl.* uredinia) A sorus (q.v.) producing urediniospores.
Urediospore – *see* Urediniospore.
Uredosorus (*pl.* uredosori) – *see* Uredinium (*pl.* uredinia).
Uredospore – *see* Urediniospore.
Ustilospore A spore of a smut fungus from which the basidiospores are formed.
Varietal Pertaining to a variety; here, in relation to plants, pertaining to a cultivar (q.v.).
Vascular Pertaining to the vascular system (q.v.).
Vascular system The continuous conducting tissue within a plant, consisting of, for example, the phloem and xylem.
Vector An organism, such as an aphid or a nematode, capable of transmitting a disease.
Viroid A virus-like organism.
Viruliferous A vector that is carrying virus.
Virus A non-cellular microorganism (consisting of only DNA or RNA, with a protein coat) capable of replication but only when in association with a more complex organism.
Volunteer A crop plant arising from seed remaining in the ground after harvest, and typically appearing in a following crop.
Xylem The lignified, water-conducting and supporting tissue in plants, and the major component of wood.
Zoosporangium (*pl.* zoosporangia) A sporangium that produces zoospores.
Zoospore A naked, motile spore (swarm spore) produced within a zoosporangium and bearing one or more flagella.

Pest Index

Disease, Pathogen and Disorder Index

General Index

Lightning Source UK Ltd.
Milton Keynes UK
01 June 2010

154978UK00001B/8/P